The
Electronics Assembly
Handbook

IFS

THE ELECTRONICS ASSEMBLY HANDBOOK

Edited by
Frank Riley
and The Staff of
Electronic Packaging and Production

IFS Publications

Springer-Verlag
Berlin   Heidelberg   New York
London   Paris   Tokyo

Frank Riley
The Bodine Corporation
PO Box 3245
317 Mountain Grove Street
Bridgeport, CT 06605
USA

Electronic Packaging & Production
Cahners Publishing Company,
A Division of Reed Publishing (USA) Inc.
1350 E. Touhy Avenue
PO Box 5080
Des Plaines, IL 60017-5080
USA

British Library Cataloguing in Publication Data

The Electronics Assembly Handbook

1. Electronic apparatus and appliances—
Assembly
I. Riley, Frank   II. Electronic Packaging & Production
621.3815'1   TK7870

ISBN 0-948507-41-1 IFS Publications
ISBN 3-540-19449-5 Springer-Verlag Berlin
ISBN 0-387-19449-5 Springer-Verlag New York

© 1988 **IFS Ltd,** 35-39 High Street, Kempston,
          Bedford MK42 7BT, UK
and **Springer-Verlag** Berlin   Heidelberg   New York
          London   Paris   Tokyo

Phototypeset by Wagstaffs Typeshuttle, Henlow, Bedfordshire
Printed and bound by Short Run Press Ltd, Exeter   Printed in England

Acknowledgements

The majority of the articles comprising this work originally appeared in *Electronic Packaging and Production*, a monthly publication of Cahners Publishing Company, USA. IFS Publications would like to express its thanks to Cahners Publishing for allowing these articles to be used and for its assistance in the preparation of the book. *Electronic Packaging and Production* is published monthly.

In addition, the following publishers/organisations granted permission for some of the other papers to be reprinted in the book, and IFS Publications would like to express its appreciation of their cooperation.

Assembly Engineering
Hitchcock Publishing Company
25W550 Geneva Road
Wheaton, IL 60188
USA

Assembly Automation
IFS Publications
35/39 High Street
Kempston
Bedford MK42 7BT
UK

AT & T Technologies Inc.
Engineering Research Center
Princeton, NJ 08540
USA

Machine and Tool Blue Book
Hitchcock Publishing Company
25W550 Geneva Road
Wheaton, IL 60188
USA

Printed Circuit Assembly
PMS Industries
1790 Hembree Road
Alpharetta, GA 30201
USA

Society of Manufacturing Engineers
One SME Drive
PO Box 930
Dearborn, MI 48121
USA

IFS Conferences
35/39 High Street
Kempston
Bedford MK42 7BT
UK

The General Electric Company PLC.,
1 Stanhope Gate
London W1A 1EH
UK

Preface

When offered the challenge of collating and editing a series of papers on electronics assembly, I accepted the task with mixed feelings. As one who has spent his life in the field of mechanized assembly I have viewed the developments of mechanized electronics assembly and testing with professional interest and intellectual curiosity. Having spent several years in the Signal Corps at the end of the Second World War this observation of the changing world of electronics manufacturing has spanned the transition from vacuum tube to transistor, from relay to microprocessor and from cable to satellite. This long observation tends to confirm the old cliché that the more things change, the more they remain the same.

This book is intended to assist those given the responsibility of assembly, joining and testing the amazing array of electronic components into functional devices of the highest quality at the lowest cost.

The emphasis on quality cannot be understated. From the most inexpensive consumer item to the most complex military or space unit, quality is imperative to survival in the manufacturing arena of a world economy.

The emphasis on quality is matched with the expectation that electronic goods will be manufactured with ever-increasing capability at even lower costs.

This triology of lower cost, increased quality (or reliability) with ever-increasing functioning capability, is almost unique to the electronics industry. While much of this has been and will be achieved through breakthroughs in product design and new or improved materials, these must be packaged and assembled in usable forms in the most efficient and inexpensive ways.

This collection of articles and papers is specifically aimed at this required efficiency in the assembly of electronic components and the verification of their functioning at each instrumental step in the manufacturing process.

In assembling these papers, several assumptions have been made. First and most importantly, this book is intended to serve those producing or manufacturing engineers who are involved in the assembly and testing of electronic components rather than in the design or fabrication of these components. To do this the emphasis has been given to those papers reflecting the latest commercially available equipment rather than that of academic research. Great emphasis is placed on the implementation and integration of available tools for efficient assembly in a manner consistent with the human resources, volume of production and good product design that will ensure profitable manufacturing.

The book contains the opinions of many practitioners of electronics assembly. Much of this may be subjective or biased toward specific solutions. It is hoped, however, that exposure to a broad spectrum of such opinions is beneficial to the reader.

It is impossible to utilize a collection of such articles without commercial references. These are given without endorsement but as indicative of the range of commercially available tools for successful electronics assembly.

We have drawn on a wide number of papers and sources. It would be unfair, however, not to note the special importance of those articles which originally were prepared for *Electronic Packaging & Production* magazine. They form the foundation for the book.

In attempting to cover all areas of concern to those involved with electronics assembly, it soon became apparent that the book would become unwieldy. It was possible to devote only a short space to the questions of 'rework' (which has been mentioned briefly in the chapter on Soldering) and to 'burn in' which is included in the chapter on Testing. Each of these topics could stand a full treatment not possible here.

Finally, no attempt has been made to cover thick-film production on laser trimming. There is an underlying assumption that the technology of inserting axial lead devices or DIP units in bandoliers or magazine tubes is a mature technology requiring little coverage in this text. Our emphasis is on the emerging world of practical technology.

Frank Riley

CONTENTS

CHAPTER 3 – COMPONENT INSERTION

Robotic Assembly

Robotic Application

CHAPTER 4 – SURFACE-MOUNT TECHNOLOGY

The Demands of SMT

CHAPTER 5 – HYBRID CIRCUITS

CHAPTER 6 - SOLDERING AND CLEANING

Solder Quality Definition

Wave Soldering

Adhesives

The Role of Solder Paste

Cleaning Soldered Assemblies

Repairing Soldered Assemblies

CHAPTER 7 – TEST AND INSPECTION

Verification of Incoming Components and Circuit Boards

Testing Completed Assemblies

Automating the Test Function

CHAPTER 8 – ENVIRONMENTAL CONCERNS IN ELECTRONICS ASSEMBLY

Clean Rooms

Personal Discipline

Protection Against Electrostatic Damage

CHAPTER 9 – THE ROLE OF CAE/CAD/CAM IN ELECTRONICS MANUFACTURING

CHAPTER 10 – AUTOMATING ELECTRONICS MANUFACTURING

Physical Integration

Case Histories in Electronic Assembly Automation

CHAPTER 1

THE CHANGING WORLD OF ELECTRONICS ASSEMBLY

Electronics manufacturing is the fastest growing segment of the manufacturing world. Electromechanical and mechanical devices are being replaced or displaced by electronic devices at an unbelievable rate.

The capability of electronic circuits grows steadily while the unit sales price plummets. Many manufacturers of consumer electronics have moved to third world countries to reduce assembly labor costs.

Miniaturization and increasing demands on reliability for automation, telecommunication, space and defense needs are forcing new looks at manual assembly of electronic devices. Capital equipment is often better suited than manual labor to achieve the needs of this changing market-place.

Electronic production and manufacturing engineers will be needed to specify, install and manage this equipment. Many of these engineers will have to come from outside the electronic industry. Others who have grown up within established electronic concerns are finding new requirements facing them at an ever-increasing rate.

Certain types of automated equipment have been used for many years. These include sequential axial lead insertion machinery for high-volume production, X-Y positioning tables for small lot production of axial and radial lead devices and leaded components available in magazine tubes.

NC drilling machines are routinely used to prepare printed circuit boards for component induction. This book is not intended to cover these established areas of electronics assembly.

The 1980s have seen developments that have radically altered the very nature of electronics assembly. The rapid growth of surface-mount technology, the availability of robotic parts transfer devices, are typical of the changes occurring. The manufacturing engineer must also address management demands for shorter lead times, reduced inventory levels and high yield levels.

Lastly, all equipment decisions must be consistent with possible total integration at some later date.

SYSTEMS, VISION AND COMPONENTS IN ELECTRONICS ASSEMBLY

Electronics manufacturing in the past decade has both experienced and driven change. On printed circuit board (PCB) lines, competition, accelerated new product introduction and product redesigns, increasing board densities, stiffer quality requirements, and price/cost pressures are forcing manufacturing managers to rethink their entire manufacturing strategy. The new strategy is a systems approach to electronics assembly.

In PCB assembly, a systems approach means many things. In design, it means using a computer and design station to find rapidly the best configuration for a single board. In production, it means being able to download CAD data for an automatic driller/router to make the bare board. In assembly, it means being able to call up a new program and start up a whole new board assembly sequence in seconds. In test, it means high-speed testing of a variety of parameters on an in-line basis. In quality, it means being able to sense, track, analyze and stop defects quickly.

The future of electronic assembly lies in the ability to identify defects consistently and automatically at speeds beyond the capabilities of human operators. By catching mistakes after each step, defective boards are identified and can be repaired prior to subsequent value-added manufacturing steps.

Selecting the right system

Manual, semi-automated, and automated component insertion/placement are mature technologies. The real decision, however, lies in selecting the best approach for an application. After making sure that the product design has been optimized for automated assembly, the following factors should be considered:

- Product volume – Unusually high or low volumes tend to automatically exclude most assembly options.
- Board sizes – Reasonably standard (all the same size) boards can be handled by automated equipment easily; a variety of board sizes generally will dramatically increase equipment costs. How uniform and repeatable are overall dimensions, mounting holes and pads, and handling locating features?
- Product mix – Are a number of different components required for each different board type? Will a single operator/machine/work cell be required to handle all possible component configurations?
- Component mix – Are all components through-hole (leaded), surface mount, or a hybrid mix? Are separate solder paste/adhesive application stations needed, and can they operate at or near line speeds? Will components be mounted on one or both sides of the board?
- Component orientation – Should components be purchased in loose form, on reels, in pre-oriented styrofoam or loosely mounted trays, magazines/tubes? How important is accurate component presentation to the assembly method selected? Is there a need for any pre-insertion/placement processing, such as sorting, orienting, lead forming and/or cutting, taping/sequencing, and electrical verification to determine that values, tolerances and polarity are correct?
- Component clearances – Depending on

First published in *Assembly Engineering*, May 1985.
© 1985 Hitchcock Publishing Co.

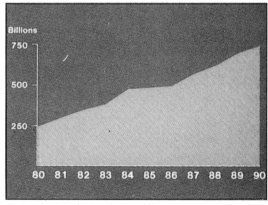

Total electronic production (Courtesy of Universal Instruments Corp.)

how the component is to be held (by the leads, body edges, or on top by a vacuum pickup), adequate clearances are needed to place and release the component. As board densities increase, the problem is compounded.

- PCB handling – Will operators, conveyors or machines transport boards between operations? Is part tracking required? Is buffer storage required to maintain flow? Is static protection needed?
- PCB inspection – What testing is required, and can it be performed adequately by operators or are vision systems and automatic test equipment necessary?

Automation

Automation, or semi-automation, offers tremendous potential to increase throughput.

It also offers a means of bringing consistency and accuracy to the assembly process.

As the US electronics industry moves towards higher board densities, greater product mix, and improved quality levels, automation is the only answer. Surface mount assembly, for example, is not practical without a high degree of automation. And once any amount of automation is introduced, it becomes almost habit forming; the only way to push any of these areas to higher levels is through increased automation.

For years, the drive was to push component insertion machines faster and faster to increase productivity. Today, however, the drive for speed has given way to an even greater need for accuracy. In many cases, users are willing to trade some of today's machine speeds for increased placement accuracy.

Robotic assembly

In 1984, according to Intelledex, approximately 800–1000 robots were purchased for use in the electronics industry. In 1982, probably no more than 100 were purchased by the same industry. The Robots 8 conference gave evidence of the electronics indus-

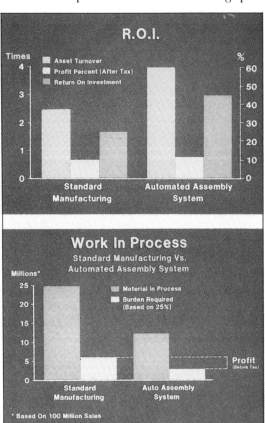

Depending on product volume, mix, and other factors, automated assembly can accelerate ROI and decrease WIP

Multiwire's approach to high-density PCBs can eliminate the step of board wiring, and improve impedance control compared with board etching

try's growing use of robotics in the past two years; 50% of the attendees were interested in small parts handling and assembly. Two years prior, interest was negligible.

Estimates for 1990 range from 3600 to 4400 units per year; this would represent an installed base of 13,000–14,000 robots in the electronics industry. The reasons behind this move include: an expanding economy; increasing cost pressures, especially in the disk drive and personal computer markets; increased competition in the telecommunications industry; increased foreign competition; the need to increase product reliability; and shorter product life cycles and rapid product growth curves.

Probably the greatest single reason behind the popularity of robots for electronics assembly, is their rapidly growing capability. They are becoming faster and can integrate vision, force sensing, sophisticated control and communications, easier programming, and the automotive industry proven economics.

Common applications for robotics in electronics assembly today include: disk media handling, machine loading/unloading; disk stack assembly; solder masking; test equipment station load/unload; board assembly, especially odd-form components (connectors, switches, relays, telephone jacks, capacitors, resistor networks, etc.). In the future, developing applications will include: a wide variety of assembly tasks in telecommunications and personal computers; PBX, CBX and mainframe industries; military electronics; semiconductors; and component and connector assembly.

Near the end of this decade, approximately 60% of all robotic electronic assembly operations will still be in the area of board assembly. As a natural progression, however, robots will be used in increasing numbers to perform final product assembly.

Manufacturers with low-to-medium board volumes (500,000 boards/year or less) or with high board mix (100+ board types) will be major users of robots because automated insertion equipment will still be too costly, and generally handle only one type of component. At the other end of the spectrum, robots offer manual assembly operations manufacturing control, repeatability, and reduced work-in-process inventories.

In electronics assembly, the era of faster cycle times has given way to the need for greater placement accuracy. While high-volume surface mount device users will turn to dedicated placement equipment, others will be able to use robotic SMD placement because it can respond quickly to the changes still taking place in the industry.

Future robots

Robots need to be communications compatible with all other assembly equipment at all levels of control hierarchy. A MAP-type system is critical to the long-term growth of both the robotics and electronics industries. Other needs include faster speeds (without sacrificing accuracy or repeatability; vision, grippers, and force sensors to find parts in 'flexible' parts presentation devices; and, sophisticated cell control to ease integration with the other equipment needed in a workcell for electronics assembly.

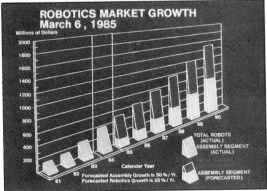

Robotics market growth (Courtesy of Intelledex)

		Low 100,000	Medium	High 500,000	Volume of PCBs/Yr
Low		Manual	Semiautomatic Dedicated	Dedicated	
Medium	100	Manual Semiautomatic Robotic	Semiautomatic Robotic Dedicated	Dedicated	
High	500	Manual Semiautomatic Robotic	Semiautomatic Robotic		
Number of Different PCBs/Yr.				Assembly Techniques — CEERIS	

Assembly techniques based on mix and volume

Engineers, therefore, are faced with three distinct responsibilities: selection of the appropriate level of automation, ensuring that it operates with enormous attention to detail and the discipline that it implies; and lastly to ensure that all of these installations are capable at some later date of being integrated into some cohesive unified system.

MANAGING AUTOMATED ELECTRONICS ASSEMBLY

Automatic assembly machines keyed dramatic productivity improvements in the 1970s. The 1980s, however, are characterized by new marketing and manufacturing constraints, as well as new technological and management solutions.

In electronics particularly, foreign competition for populating printed circuit boards (PCBs) is fierce; high quality assemblies are routinely achieved at very low cost. Domestically, high interest rates in the early 1980s provided the incentive for reducing the volume of component inventory and the value of work-in-process (WIP). Ever-shortening product lifetimes also required that final products be manufactured for immediate shipment. Otherwise, rapid obsolescence could result in large write-offs.

These financial constraints and market challenges favor the introduction of just-in-time (JIT) or pull management organizations as well as flexible manufacturing systems. In addition, the technologies required for fully automated flexible assembly and integrated management are becoming available. These solutions are now being implemented at world-class domestic assembly operations, and will become increasingly common in the 1990s. The US answer to foreign challenges is to employ high-technology solutions and advanced manufacturing management concepts. These are domains where US capabilities are best demonstrated.

Understanding the system

An assembly system implies that automatic assembly equipment is integrated through an automated materials handling system connecting all assembly cells, and a computerized monitoring system at the supervisory level. Implementation of these modules additionally requires that: boards be designed for manufacturability; computer integrated manufacturing (CIM) become a major long-term commitment; automated board identification be implemented through barcode or similar tracking procedures; and in-line quality control and the highest possible yields be achieved.

The concept of an assembly environment that builds hundreds of thousands – even millions – of complex assemblies per year with hundreds of different boards in lot sizes as low as 10–100 PCBs is very difficult for industrialists used to high-volume, low-mix straight-line production.

In the USA, the machining and metal-working industries have been among the first to recognize the need for flexibility to reduce setup costs caused by small lot size and high-mix production requirements. Just-in-time materials delivery, 'pull' manufacturing scheduling, and computerized warehouses with guided vehicles have also initiated a changed consciousness within the production management community. The chief management difference between metalworking and electronics is often related to production volume. High mix in conjunction with high volume creates totally different problems for electronics industries – especially if the profit margin for each unit is relatively small.

When dozens, or even hundreds, of different assemblies must be scheduled through a cost-effective, automated shop, the data and feedback loop requirements become too great for people to manage effectively without

First published in *Assembly Engineering*, May 1986.
Author: Charles-Henri Mangin, CEERIS International Inc.

computer assistance. Ultimately, the computer system must drive the shop in order for the schedules to be met, let alone met profitably. The electronics assembly shop of 1990 will employ not only automatic inserters (auto-inserters) and placement equipment, but will use robots for kitting parts for these machines, and other robots for assembly (especially odd-form assembly), automatic storage and retrieval systems (AS/RS) for materials handling and storage, robotic test loaders and high technology in-line inspection/quality control equipment.

The glue that holds all of this high-powered hardware together is the emerging local area network (LAN) software. A computer system with two-way communication between all the assembly and materials handling elements on the shop floor plus management interface, i.e. scheduling, CAD/CAM, reporting, etc., is essential to optimize the dynamics of change such that flexibility in production scheduling is maintained. If it is not maintained, the scheduling/throughput snarls will defeat the purpose of the shop and threaten profitability.

The integration of the flow of information and its associated data banks is a major management issue. Existing structures may oppose its implementation. However, it is the responsibility of management to commit itself to a more productive organization.

Local area networks

Computer-aided manufacturing (CAM) is an often misunderstood, underutilized technology. In electronics, the application of computer-aided design (CAD) is limited in its CAM application to generating drill and artwork tapes for numerical control (NC) programmable equipment. Only recently have dedicated efforts been made to utilize pattern programs for insertion machines and bills of materials/reference designator files for manufacturing use. With the advent of computer integrated manufacturing (CIM), increasing pressure exists to utilize and integrate these three concepts.

The assembly operation of tomorrow demands that CAD/CAM/CIM become routine

in the operation of a high-volume/high-mix electronics assembly environment. It suggests an environment in which all work cells, AGVs, AS/RS, and people are in an intimate information loop. This requires a local area network that will only be optimized if, considering the mix of computer-controlled equipment types, the system is integrated and interfaced so that the various elements (CPU, barcode devices, machine controllers, etc.) can talk to one another.

Assuming that the interface technology can connect the variety of intelligent equipment, the best, most reliable topography must be set up to ensure the most efficient, least interfering operation. A broadband LAN utilizing coaxial cable or fiber optics in a bus topology provides decentralized cotrols, and can operate despite failure of one of the elements. This system is also easily expandable and can handle digital, voice and video information. Message collision detection also should be included to prevent timely data loss from coincidental transmission. The broadband capability permits frequency assignments to help avoid message bashing in the loop. The bus-type loop is also the lowest cost in terms of hardware, installation and maintenance.

Supervisory level controls

In-house developed production management systems do exist. The problems to date have been mainly cost and lack of availability for most potential users.

Universal Instruments and several other vendors are now announcing supervisory level packages for simultaneous control and monitoring of the assembly modules and materials handling equipment. A typical system architecture is described in the accompanying diagram showing supervisory level architecture, based on the system from Universal Instruments.

The functional dynamics of this supervisory or shop-floor control system may be briefly described as a closed-loop, multi-directional information and command executive. The central process host (CPH) coordinates incoming, outgoing, and fixed data to execute a production schedule. By organizing

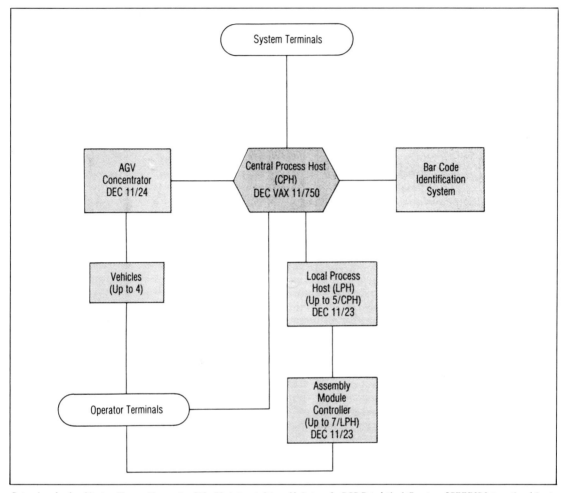

Supervisory level architecture (Source: 'Economics of Flexible Automated Assembly Systems for PCB Populating'. Courtesy of CEERIS International Inc.)

the variables of parts availability, revision changes, priorities, equipment availability and WIP, a continuous allocation of transporter (AGV) instructions, kitting instructions, machine program down-loading commands and upstream kit planning may be performed.

The local process host (LPH) receives machine program instructions from the CPH for a given batch, then takes the pattern programs from its library and downloads the right programs to the available machines in the respective assembly modules as required for each module. The program loading done by the module operator as part of setup is performed by keying the correct command

for program input to the particular machine. The pattern program is then downloaded from the LPH to the module controller and assembly begins.

The same basic logic exists for materials handling except that instead of pattern program downloading, AGVs are given routing instructions to a specific destination. The AGV operating system will use a local process host arrangement for routing, but the AGVs will be part of the barcode identification system as well. Each AGV will be reading barcodes at module locations to inform the CPH as to its location. These locations include load and unload stations at machines, stockroom, quality control areas, etc. The AGV will read the barcode at each machine in the module until it finds the one it has been told to visit, or the routing program may be defined by route/destination decisions in the guidance system software.

Operator terminals will exist wherever data input is required. Schedule updates, materials and MRP input, engineering instruction and program changes, and management reporting stations are the major terminal positions.

High-quality assembly

A pull-type manufacturing environment is highly sensitive to any process disruption because of the lack of buffer inventory of products. Rework is not deemed acceptable because it builds up WIP and increases the lead time.

Chances of having a good board at first-pass test are extremely slim when one considers the cumulative rate of defects of incoming PCBs and components, assembly machines, solder-cleaning and manual operations. Statistics on populated board yields are often misleading since they usually include a fair amount of visual inspection and touch-up before in-circuit test: the traditional first ATE level test.

Average yield statistics are not always a satisfactory indicator since the rate and source of defects vary greatly depending upon the type of board (single vs multilayer), the nature of the board (digital, analog or hybrid), the real estate (circuit density), the component mix (digital ICs, transistor/diodes and passives) and the number of solder joints. Since the smooth operation of a pull assembly line requires the highest possible yields (above 85% at in-circuit), a precise understanding of the origin of defects is a prerequisite to improving assembly. Integrated systems require a high-quality process, not an expensive test system. The achievable and required yields at each process step are summarized in the accompanying table. These yields are achieved today in world-class operations through good assembly management prac-

tice. They also result in drastic reductions of the costs of inspection, test, and rework.

The technologies are available for implementing automated and integrated systems. The real issues are management-related. Integration goes against most current structures and only good management practice ensures the successful deployment of these advanced assembly systems.

This text is intended to expose production engineers to new emerging developments, trends, manufacturing materials and equipment they will use in the years to come. These views may even be contradictory as one technique struggles to dominate another.

In serving this highly competitive industry, that of electronics assembly, there are no national barriers. While the economics of one country may dictate a focus on specific areas of electronics, the technology is a world technology. Equipment selection must therefore be consistent with world competition if one is to survive.

One last caution. Pioneers in any technology suffer rapid adolescence. Prototypes are not noted for durability. Since post-installation applications support is notoriously weak in electronic equipment, engineers can ill afford to have chronic equipment failure as they try to bring this equipment on stream.

Process Step	Yield %
bare boards released to assembly	99.90
components released to assembly	99.95
autoinsertion	99.80
computerized vision inspection prior to soldering	99.00
soldering	99.90
first-pass yield at ATE:	
• analog boards	above 90
• digital boards	above 85

Required and achievable process yields (Source: 'Test, Rework and Inspection Management'. Courtesy of CEERIS International Inc.

CHAPTER 2
CIRCUIT BOARDS

In attempting to keep this book to a manageable size, certain decisions were made concerning the contents. As mentioned in the Preface, no attempt was made to discuss thick-film construction, alumina substance nor flexible circuits. Neither was there intended to be a discussion of component fabrication. Printed circuit boards, however, do not stay in tight little compartments. For many reasons an electronics assembly engineer will want to know about their manufacture, the chemicals involved (and their environmental hazards) and the necessary cleaning procedures. While cleaning is described in depth in the chapter on soldering some mention must be made here of the cleaning prior to component insertion.

It should be mentioned that no effort is made to describe hole drilling. Drilling follows along the lines of mechanical drilling, usually using CNC machinery. It goes without saying that component insertion on automatic machinery is totally dependent on hole location accuracy for success.

PCB PRODUCTION EQUIPMENT

Although present technological pressure is toward high density fine-line PCBs and high-volume automatic component-insertion methods, much of today's activity involves processing and assembling simple print-and-etch boards with the accent on manual assembly. Five major production operations encompass the fabrication of print-and-etch board assemblies: screen printing, etching, drilling, component insertion and soldering.

Estimates of annual US sales of screen-printing machines vary from $5 million to over $15 million, with growth expanding at a predicted yearly rate ranging from 8% to 12%. Machine prices range from $25,000 to over $150,000, reflecting availability of equipment in varied degrees of automation, from semiautomatic to fully automatic.

Typical of many screen printers, this Cugher system features automatic transport, registration and printing, requiring only that the operator preload one panel while the other is being printed

Screen printers

The latest advances in screen printers include increased automation, greater emphasis on operator convenience, modular machine design and microprocessor programmable control. There have also been improvements and refinements in registration systems as well as contact and peel-off devices.

Besides automatic printing, the most recent automation improvement has been the addition of loading and unloading mechanisms. Although screen printing remains primarily an operator-dependent process, automatic load and unload is said to increase the machine's cycle time up to 20% over manual handling. Modern printers can apply resist at rates exceeding 1000 impressions per hour.

Today's screen printers are also claimed to have greater structural integrity and torsional rigidity together with improved electronic sensors and readout devices for better repeatability of setup and screen/panel positioning.

Future improvements in screen-printing machines, per se, appear limited. Aside from computerized controls and a tie-in to a host computer, significant advances are not anticipated. Suppliers agree that screen printers have reached the stage of being basically highly accurate holding fixtures for the workpiece and screen. As such, any additional improvements in the process must come from advances in screen making and resist technology.

As a process, however, screen printing is expected to survive even through some drastic changes in the PCB production process. Suppliers predict that the present cumbersome operations of etching and plating may eventually be eliminated and replaced by the direct screening of conductive polymers.

Screen printing is still often referred to as an 'art' because of the wide variation in success, especially when attempting to screen finer lines. Machine suppliers report having witnessed production processes where 0.005in. lines spaced at 0.006in. were being consistently maintained over an 18 × 24in. panel, whereas others were having difficulty imaging 0.012in.

Printing fine lines and ensuring a successful screen-printing operation involves three major factors in addition to an accurae screen-printing machine. First, the screen itself determines the final image and therefore controls the degree of resolution. Screen making is critical to the overall process.

First published in *Electronic Packaging & Production*, November 1984, under the title 'PCB demands pace production equipment market expansion'.
Author: Howard W. Markstein, Western Editor.

Second, the resist ink must have the proper viscosity and rheology in order to reproduce and maintain a fine, sharp line on the PC panel. Third, high resolution screen printing must be performed in a controlled environment similar to the thick-film hybrid circuit process.

In general, successful screen-printing operations take place in a controlled environment and have quality support services and engineering talent available to identify problems and enforce standards. All too frequently the capabilities of screen printing are underestimated by management. Screen printing is far from a cheap and dirty process where unskilled personnel are involved.

Etching equipment

Annual US sales of stand-alone PCB etching systems are estimated to be between $10 million and $16 million. This figure represents the market for conveyorized production etching systems and does not include other wet processing equipment such as photoresist developers and stripping equipment (the total wet processing market approaches $35 million). The annual growth rate for etching systems is estimated to be 10%.

Most PC etchers are within a price range of $20,000–$50,000, but some sophisticated systems can cost more than $250,000. Prices vary in accordance with conveyor width, speed, provision for heating and drying, number of process chambers, sump sizes, number of pumps, type of process (alkaline or acid),

Vertical wet processing systems convey panels and inner layers vertically through all stages. The system shown, from Circuit Services, has an etch chamber, flood rinse, spray rinse and dryer

degree of automatic controls, and modules for pollution control, water conservation and etchant regeneration.

Recent developments and improvements in etching systems include spiral nozzles as an alternative to oscillating manifolds, magnetic-drive pumps, outboard sumps, conveyors designed to process flexible circuits, regeneration of cupric chloride and alkaline etchants, etchant copper removal, vertical processing design, and the use of programmable controllers to improve process control.

Today's systems feature a high degree of automatioin. Computerized monitoring of the etchant for pH, specific gravity, etc., has resulted in a uniformity of processing and minimum rework or scrap by maintaining the etchant's chemical balance. These automatic etchant replenishing systems are now available for all three of the major etchants: peroxide/sulfuric, cupric chloride and alkaline/ammoniacal. Another feature is automatic panel sensing for controlling the spray sequence to conserve etchant and rinse water. Future improvements in etching systems will tend toward increased automation and control. Etchers will be configured to interface with a customer's automatic process control monitoring system and host computer.

Drilling machines

The annual dollar-volume market for drilling and routing systems was estimated to be between $80 million and $100 million for 1983. Machine prices vary according to system size, number of spindles and degree of automation. Modern drilling/routing systems are fully computerized, capable of communications networking and can be tied in with CAD/CAM.

Most drilling machines presently offered are designed for the high-precision drilling of high-density multilayer boards. While simple print-and-etch boards do not require precision drilling, for multilayers it is one of the most critical processing operations. Hence, modern drillers are large, stable systems equipped with precision indexing mechanisms and high-speed precision spindles. Drilling parameters such as spindle speed and

Modern drilling systems, such as this one from Excellon Automation, feature computer control, high-speed air-bearing spindles and a granite support base for stability. Outfitting with high-torque spindles also allows rooting

been expended in designing drill bits to accommodate the trend towards smaller diameter holes. Various geometries and material combinations are being evaluated to balance such factors as impact stability, torsional strength and bit wear (life).

Tool management is seen as a major factor in providing a fully automated system. While machine loading and unloading is a relatively simple problem involving materials handling, controlling a large-drill-bit inventory is highly complex. Modern drillers feature automatic drill-bit changing but require an operator to replenish the drill-bit holder frequently. Future machines are expected to incorporate tool management systems capable of automatically supplying the system with various drills as needed, and doing so over an extended time period without manual intervention.

feed are controlled to provide smear-free, accurate holes.

With all their precision, however, drilling machines are basically tool holders, and the quality of output is primarily determined by the quality of the drill bit. Much research has

MODERN PLATERS AND ETCHERS

Simply defined, plating and etching are two PCB processing steps that involve wet chemistry for the addition and subtraction of metal. These seemingly opposing functions represent major operations in PCB fabrication. As such, users are searching for more sophisticated, cost-effective and automated equipment to keep pace with today's advances in PCB design and high-volume production requirements. The most demanding feature is that of providing a high degree of process control, both by the use of automatic process monitoring systems and through the overall reliability of machine design.

The various types of PCB plating systems in use today, from the most simple to the more automated, include hand-serviced tank lines, tank lines employing a manually operated hoist, conveyorized pattern platers, automated tab of finger platers, and tank lines with a fully automated and programmable sidearm or overhead hoist. A number of these systems can be configured for rotary, U-type or straight-line work flow. The systems are also adaptable to meet the variety of plating requirements for today's PCBs; namely, copper, tin, tin-lead, nickel, gold, palladium and palladium-nickel.

Some of the more significant advances in PCB plating systems over the past few years are:

- Improved hoist control by the incorporation of user-friendly microcomputer controllers. A number of hoists on the same line can be controlled as to immersion time and sequence.
- Higher current density acid copper cells for faster plating speed.
- Fully enclosed machines for the smaller systems.
- The acceptance of continuous flow, high-speed tab platers with improved drag-out control and plating thickness uniformity.
- Pulse plating and rapid impingement systems for high-speed deposition.
- The development of plating solutions that accommodate higher current densities and are compatible with high-speed plating requirements.

First published in *Electronic Packaging & Production*, February 1985, under the title 'Modern platers and etchers offer improved process control'. Author: Howard W. Markstein, Western Editor.
© 1985 Reed Publishing (USA) Inc.

Modern PCB plating lines, such as shown here from Robbins & Craig, feature high-speed plating and a programmable automatic hoist, this one of sidearm design. Rapid impingement of the plating solution reduces the plating time and offers increased uniformity or deposit

Process control is achieved in modern plating systems by accurate control of all variables such as chemical balance of the plating bath, bath temperature, applied current and panel immersion time. Automatic

An overhead hoist is shown transferring a rack of PC panels in a copper plating line. The plating system is supplied by Process Automation and features a chain-driven hydraulic hoist with anti-sway mechanism

programmable hoist controllers ensure a predetermined immersion cycle, while automatic bath controllers (chemical robots) monitor and replenish the bath as needed. These systems effectively maintain pH, brightener level, copper content, and reducing and chelating agents. Plating bath temperature is also automatically controlled as is current by the use of automatic rectifier control systems. Modern plating systems, therefore, employ sophisticated techniques in an effort to maximize process control.

System throughput

The throughput of PCB plating systems has been maximized by improved process control and high-speed plating methods achieved by higher current densities, and by agitation or rapid impingement (velocity) plating techniques. Throughput is also determined by the number of processing tanks or process steps, and by the required thickness of plating.

According to Technic, a leading plating system supplier, the rate of production is determined not so much by how fast PCBs can be conveyed through a plating system, but by how much current density can be effectively applied without degrading the quality of the plated deposit. Current density directly influences how quickly a certain thickness of metal can be plated. However, the amount of current density that can be successfully applied is dependent upon the interrelation of other factors.

Technic says the geometry of the surfaces to be plated, the area that is to be plated, the mechanical limitations of the plating system, the concentration of metal in the plating bath, temperature, pH, and the agitation level, all affect the maximum effective current density, and therefore the total throughput.

Tab platers

Sel-Rex, a supplier of plating chemicals and plating machines, says that the acceptance of high-speed tab platers has been a significant factor in the production capability of the PCB industry. Its systems are now fully automated with automatic load and unload, automatic plating solution balance control and feature high-speed deposition. Modern in-line tab platers can sequentially strip solder, scrub, nickel plate, rinse, surface activate, gold strike and gold plate PCB edge connector tabs at speeds up to 20 linear board feet per minute.

Technic says automatic tab platers were first introduced in 1979, and have now become the most advanced automatic plating systems available. These systems feature a mechanical arrangement that conveys PCBs vertically through a series of enclosed process chambers. The machine also has a means of

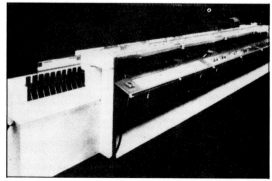

Automatic tab platers have become an accepted method for processing PCB edge connector tabs. This machine from Sel-Rex provides constant thickness distribution. High current densities allow shorter plating cell lengths and compact machine design

circulating plating solutions from reservoirs into process chambers in a continuous cycle. Another requirement is that of continuous electrical contact, usually via metal brushes, that supply high current to the tabs. Tab platers also use special, high-speed plating solutions, most of which are proprietary formulations.

Choosing a plating system

The major factors a buyer should consider when purchasing a plating system, whether tab or batch hoist and tank systems, are as follows:

- Return on investment (ROI) – plating quality and yield, not lowest price system.
- System quality – ease of operation and minimum downtime.
- Flexibility – ease of changeover for different board sizes, types, and plating requirements.
- Capacity or throughput – the system must meet present and anticipated production requirements.
- Conservation – rate of consumption per quantity and quality of finished product, such as gold solution, water, power, etc.
- Floor space restrictions.
- Programming simplicity.

Etching systems

Etching is one of the more harsh operations in PCB processing, and its efficiency is judged by the speed and uniformity of dissolving copper from a clad laminate. Today's etching machines are designed to minimize undercutting and etch fine-line circuit patterns on multicircuit panels at high production rates. Over the past years, the process has been improved, refined and automated such that the user now has a wide choice of etching machines, etchant chemistry, modularity options, automation features and process controls.

Equipment suppliers emphasize that the numerous advances in etchant system design are all aimed at providing increased process control, especially in a high-production environment, while also incorporating features that ensure machine reliability, minimum

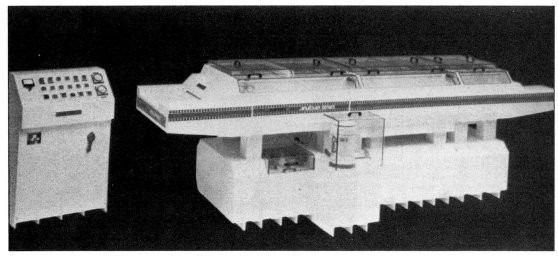

Modern etchers place emphasis on process control, and their sleek design, such as this one from Transworld Circuit Products, belies the harsh chemical environment within

down-time and ease of maintenance. The list of technology advances and machine features includes: spiral nozzles or oscillating manifolds for more uniform etchant dispersion; magnetic-drive pumps; automated chemistry controls including pH and specific gravity; rinse water purification and recirculation; use of stainless steel, PVC, polypropylene, Teflon, Kynar and other materials for improved machine construction; and the use of programmable controllers to monitor and control the important process parameters in large production systems.

Al Rowe, vice president of Transworld Circuit Products says, that the major automated features and process control improvements in PCB etchers are feed-and-bleed systems and optical sensing. The former

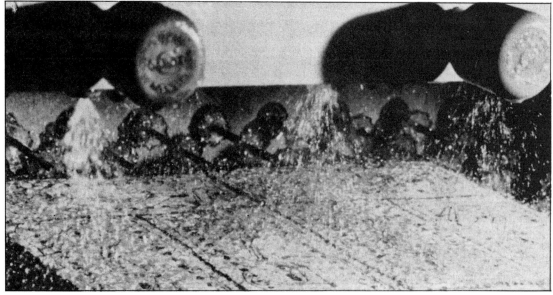

PCB production involves numerous wet processing steps including cleaning, developing, stripping, plating and etching. Chemcut's alkaline etcher shown here has a spray oscillation system for uniform coverage

involves specific gravity sensors for alkaline etchants and regenerative systems for cupric chloride. These features eliminate draining, filling and reheating, while also providing a more cost-effective chemistry.

"Specific gravity sensors coupled with pH monitors have fairly well removed the need for constant bath sampling and manual bath replenishing for optimum balance," says Rowe.

Optical sensing has provided today's etching systems with a degree of 'intelligence' in determining the presence of work on the conveyor and to inform operators of a conveyor jam, malfunction, or if a process station has shut down. Rowe says optical probes with delay on and off relays can also be used for maximum rinse water conservation. The rinse cycle is only activated when panels are present on the conveyor.

The most common etching systems available consist of a horizontal conveyor transporting the PC panels through the etch and rinse chambers. These systems also include reel-to-reel processing of thin laminate and flexible circuitry. However, vertical etching systems are also available, as are non-conveyorized batch systems and rotary bottom etchers. The latter has the PCB panel mounted upside down and it continually rotates with an eliptical motion as the spray impinges upon the bottom surface. This concept is said to provide greater fine-line definition.

Western Technology Associates has developed a system similar in concept to the bottom etcher, but of vertical orientation of the panel. Called the VRP (vertical, rotary, planetary), the concept is being expanded into a new machine capable of high production via continuous transport.

Howard A. Rosenberg, technical director at Western Technology Associates, says that both accuracy and precision in etching can only be achieved by exposing the surface to be etched to a totally controlled and uniform flow of etchant for a specific period of time.

Rosenberg contends that today's horizontal and vertical etchers do not etch to full potential because the PC panels are conveyed stationary.

"Simply exposing the workpiece to a pressurized etchant spray or immersion will invariably produce uneven etch due to conditions we commonly know as puddling, tear dropping, nozzle patterning, and tiger striping," says Rosenberg.

At Circuit Services, marketing manager Robert J. LaBombard says that vertical etchers provide more uniform processing of the panel side to side and top to bottom. This results in minimum undercutting and eliminates over-etching. He advises that vertical etchers, because of the panels' orientation, results in less carryover of etchant into the rinse waters.

LaBombard claims that vertical processing has also proven to be advantageous for photoresist developing, thereby reinforcing the opinion of the overall improved processing features of vertical systems. This is especially important since the quality of the etched pattern is dependent upon the quality of the developed resist pattern.

"For fine-line etching," says LaBombard, "the photoresist development step is critical. Tests by Circuit Services have shown that resist sidewalls are more vertical, with fewer troublesome 'feet' extending from the sidewalls when developed vertically compared to horizontally."

Choosing an etcher

The major factors a buyer should consider when purchasing an etcher are as follows:

- *Throughput.* Production volume, or throughput, is determined by chamber size or length of etching area, type of etchant used, and the thickness of copper to be etched. These factors establish the required dwell time of the panel within the etch chamber.

- *Conveyor width.* The width of conveyor needed is determined by the panel size.

- *Process control.* Precise control of the operating parameters is essential and includes: temperature, conveyor speed, etchant flow rate and chemical balance.

- *Uniformity of etch.* The equipment must be capable of etching the pattern in a uniform manner across the width and length of the panel. The finer the lines and spaces, the more important this consideration becomes.

- *Chemistry compatability.* Construction materials used for the etcher must be compatible with the selected chemistry.

- *Construction.* The equipment should be of heavy duty, leak-proof construction having proper chemical isolation, and anti-pollution rinsing.

- *Operator safety.* Fluid level interlocks and over-temperature sensing devices must be provided.

- *Maintenance.* The need for servicing and downtime should be minimal, but ease of servicing must be considered.

CHEMICALS IN PCB MANUFACTURING

They come in several forms including powders, liquids and pastes. They come in many hues of pastel shades as well as clear. They seal, coat, stick, selectively attack, bond, plate and clean. They are electronic chemicals; without them, printed circuit boards could not be fabricated and assembled.

In a July 1984 study prepared for the prestigious consulting firm of Arthur D. Little Inc., Arthur P. Lagace, president of Lagace & Associates, grouped electronic chemicals into the following categories: substrates, image-producing materials, wet chemicals, plating chemicals, gases and packaging materials. To these, cleaning, joining and bonding chemicals and materials could be added. The size and growth of these product markets are summarized in the table on the facing page.

The following first six product category descriptions are taken from the Arthur D. Little/Lagace study.

Substrates

Most PCB substrates are made of thermoset polymers reinforced with fibrous materials to provide strength and stiffness. The polymers used include epoxy, polyester, phenolic, polyimides, and melamine. The principal reinforcing material is fiberglass cloth. PCB substrates maybe faced with copper foil on one side or on both sides; or may be multilayer

laminates, the fastest-growing type. PCB substrates may be rigid or flexible. Flexible substrates are gaining wider use because they allow greater freedom of design and ease of assembly. These flexible substrates are made of copper laminated to thin films of polyimide.

Certain physical properties of PCB substrates, particularly the temperature properties of the polymers, limit the usefulness of current products. In addition, processors are looking for lower-cost laminates and improved processing economics as well as waste reduction and control.

General Electric's Plastics Group has recently introduced into the PCB marketplace an amorphous thermoplastic resin, Ultem, which makes an excellent substrate material for injection-molded boards. The Ultem polyethermide resin is similar to other polyimide materials widely used for flexible circuits and for high-performance/high-temperature rigid circuit boards. The material has the thermal stability to withstand circuit board processing, yet can be injection molded at 350–400°C. It is characterized by excellent electrical properties including high surface and volume resistivity and high dielectric strength.

Combining Ultem resin substrates with its own designs and plating technology, Circuit-Wise Inc. produces injection-molded, single-sided circuit boards, double-sided planar circuit boards and multilayer boards. All exterior board dimensions and through holes

First published in *Electronic Packaging & Production*, August 1985, under the title 'Chemicals play vital roles in PCB manufacturing'.
Author: S. Leonard Spitz, East Coast Editor.
© 1985 Reed Publishing (USA) Inc.

Product category	1983	1988	Growth, %/yr
Substrates	540	860	10
Image-producing	80	130	10
Wet chemicals	100	140	7
Plating chemicals	120	170	7
Gases	20	40	15
Packaging materials	20	30	8
Other	20	30	8
Total	**900**	**1,400**	9

US PCB chemicals market by product category, 1983 – 1988 (millions of 1983 dollars). (Source: Arthur D. Little Inc.)

are molded in a single step. The multilayer boards feature layers with molded bosses that fit into molded holes in adjoining layers, eliminating the possibility of through-hole misalignment between internal and external layers.

Multilayer interconnection packages, molded of Ultem thermoplastic resin, allow for 100% electrical testing of each layer prior to lamination

Injection-molded using polyetherimide resin, these double-sided planar circuit boards feature molded-in through-holes and recessed traces on both sides

Image-producing materials

The principal categories of image-producing materials are photoresists and screen-printable resist inks. The function of a resist is to adhere to a surface and protect the underlying material from chemical reaction taking place in adjacent uncovered areas.

Photoresists are light-sensitive polymers which are applied as liquids or as thin adhesive dry films to a copper-faced PCB laminate. After the resist is applied to the substrate it is exposed through appropriate artwork to UV light, which forms an image of photochemically altered material in the resist. Development of the image consists of removing soluble areas of the coating to expose portions of the surface for chemical treatment.

A single-sided, injection-molded PCB from Circuit-Wise Inc. has three-dimensional features on its unplated side and can hold numerous electronic components

Photoresists are either positive or negative working. Negative developing photoresists, in which the unexposed areas remain soluble and are removed, are based on cinnamates and methyl methacrylates copolymerized with other monomers from the acrylate class. The polymeric composition is made photosensitive with a quinone sensitizer which produces free radicals that initiate crosslinking upon exposure to UV light. After imaging, the uncrosslinked polymer is dissolved with solvents or aqueous alkaline systems and the desired pattern remains on the substrate, ready to resist the action of the etch solution.

In positive resist systems, the areas exposed to light become soluble. Positive resists are phenolic polymers blended with diazonapthoquinones. On exposure to light, the diazo rearranges to a carboxylic acid that can be dissolved by alkalis. The oleophilic unphotolyzed diazo in the unexposed zones is not dissolved. The consumption of positive imaging photoresists is smaller, but is increasing, especially for the manufacture of fineline circuitry used in high-speed military and computer applications.

Positive resists have superior resolution and can be used with projected imaging systems. Disadvantages of positive resists include longer exposure, high cost and critical development conditions. Negative imaging photoresists have better acid resistance but do not have the resolution capability of positive resists.

One specific negative photoresist system comes from W.R. Grace & Co. and is known as Accutrace. This primary imaging resist system consists of five basic steps: chemical preclean; resist application; registration and imaging; aqueous development; and UV post-exposure.

The chemical preclean utilizes an alkaline section to remove oils or other organic contaminants, followed by an acid copper cleaner to remove oxidation. Pumped directly from a disposable pail in which the polymer is shipped, the resist is applied by a rubber coating roll. For inner-layer applications, resist thickness can be varied from 0.0008 to 0.0212in. For plated-through hole applications, the resist thickness can be varied from 0.0012 to 0.0018in.

The exposure mode is off-contact because the resist is liquid, and occurs after mechanical registration between artwork and panel. Exposure time, using a 5kW mercury-xenon lamp, ranges between 5 and 11 seconds. This translates to a minimum production rate of 180 sides/hour including registration time. A highly collimated light source is capable of accurately imaging 0.003in. lines and spaces with straight resist sidewalls and less than 0.001in. shift in the positioning of any element on the artwork with respect to the imaged position on the panel.

A spray developer containing 0.5% sodium carbonate solution produces the image in 20 seconds at 24°C. The spraying action of the developer solution removes and flushes away the uncured liquid resist. After developing, the panel is rinsed with water and dried at 49°C.

The last step in the imaging process calls for a brief UV post-exposure to harden the resist and allow for panel stacking.

Wet chemicals

Wet chemicals are used in a number of the process steps in the manufacture of finished PCBs, including etching of the copper, solder plate removal, and several cleaning operations. These chemicals tend to be standard industrial etchants and cleaners. However, some suppliers provide special blends and proprietary mixtures for electronic applications. Some of the wet chemicals used in PCB production are listed in the table below.

With particular reference to etching, today's hydrogen peroxide/sulfuric acid etch was first used as a replacement for chromic-acid-based pickling solutions in the basic copper and brass industries. Conversion to peroxide/sulfuric pickle eliminated problems such as handling and disposing of spent pickle and treating large amounts of chrome in the wastewater treatment systems. The replacement of persulfate-type etchants for preplate cleaning and etching represents the first application of this technology for PCBs.

"The key to the use of hydrogen peroxide

Copper etchant materials	Solder plate removal	Solvents
Ammonium bicarbonate Ammonium chloride Ammonium hydroxide Ammonium phosphate Chromium sulfate Cupric chloride	Fluoboric acid	Fluorocarbons Methylene chloride Perchlorethylene Trichlorethylene Ferric chloride Hydrogen peroxide Persulfates Sodium perchlorate Sulfuric acid

Selected wet chemicals used in PCB production (Source: Arthur D. Little Inc.)

as an etchant," explains Mark Koch of Electrochemicals, "is the ability to stabilize hydrogen peroxide in the presence of copper. The stabilizer must carefully control the catalytic decomposition of peroxide without accumulating in the etching bath. To be effective, the stabilizer of choice needs to be consumed at the same rate as the peroxide to prevent a bath imbalance."

Electrochemicals' Perma-Etch replenisher is a proprietary hydrogen peroxide formulation containing the necessary amounts of stabilizer, catalyst and inhibitor to maintain an efficient and stable etching process. The catalyst or etch promoter provides an etch rate of 0.00101in. per minute. The etching system is compatible with most resists, results in minimal undercutting, and in the case of tin-lead solder plate exhibits excellent reflow properties.

Plating chemicals

A wide variety of plating chemical compounds are used in the manufacture of PCBs. Plating

Plating step	Purpose
Electroless copper	Through-hole plating
Copper plating	Through-hole plating
Solder (tin-lead) plating	To provide a metal etch resist and a solderable substrate for subsequent soldering of components
Nickel electroplating	To produce an undercoat for precious and non-precious metals
Tin-nickel electroplating	To provide a metal etch resist
Tin electroplating	To provide a metal etch resist, solderability and electrical contact
Gold electroplating	To provide corrosion-resistant contacts and high conductivity
Rhodium electroplating	To provide corrosion-resistant contacts, high conductivity and low noise

Key plating steps and their functions (Source: Arthur D. Little Inc.)

steps are used to provide through-hole conductivity in double-sided and multilayer boards, to prepare boards for soldering, and to provide good corrosion-resistant gold contacts on the final product. Some key plating steps and their functions are summarized in the table below.

Total expenditures for plating chemicals are heavily influenced by the cost of the precious metals used in the formulations, particularly the gold, palladium and rhodium. Technical work is being directed to the development of non-precious systems and alloys containing little or no precious metal. Other developments include an increased use of electroless copper in place of plating steps to provide heavier deposition layers and the substitution of less expensive systems for palladium catalyst electroless deposition systems.

Actively pushing exotic metal plating technology, Engelhard Corp. has been in the forefront of developing precious-metal electroplating salts and solutions, dating back four decades. Its electroplating products and processes include custom-designed proprietary gold, silver, platinum, palladium, rhodium, and ruthenium baths as well as anodes and precious-metal salts.

When asked in 1984 about Engelhard's involvement in electronic chemistry, Cyrus H. Holley, president of the Specialty Chemicals Division, had this to say. "We are bullish about the electronics industry and we are developing a significant niche in it. Our concentration currently is in three technologies: thin-film, thick-film, and electroplating.

"The most important product area is highly sophisticated electronic circuitry, where miniaturization and speed are important. To increase the speed of switching, the industry is turning to microwave frequencies. We are in the unique position of combining technologies to build these kinds of circuits."

Gases

The major application of gases in PCB processing is hole desmearing. As part of PCB production, holes are drilled through the substrate. These holes are then plated to provide a conductive path from one face of

the board to the opposite face. Because the heat of drilling can cause the polymer to soften, the inside of the drilled holes may become smooth, making metal adhesion difficult. Plasma processing roughens the polymer and promotes metal adhesion.

Packaging materials

Packaging chemicals for PCBs consist mostly of protective coatings applied in the final stages of board assembly. These coatings are applied to ensure that the substrate maintains good electrical properties, and in some cases good mechanical properties, in hostile environments.

Silicones, varnishes, epoxies, acrylics, polyurethanes, and polyesters are all used to encapsulate PCBs. The resulting tight envelope conforms to the irregular configuration of a loaded PCB and shields the interconnected assembly against humidity, salt spray, dust, corrosion, acid, and solvent attack, as well as against unwanted vibration. The conformally coated package thus provides significantly increased reliability and improved performance over an assembled board without such protection.

The coatings can be custom blended depending on user needs and can be sprayed onto vertically oriented boards without running. Formulations in varying viscosities are available for alternative methods of application such as dipping and flooding. The coats are either transparent or opaque and (particularly in the case of the former) add a cosmetic dimension to an otherwise plain board.

Conap, Dow Corning, Loctite and Miller-Stephenson are firms which actively research and supply conformal coatings to PCB manufacturers. "Reliability is king," declares Robert E. Batson, manager of Loctite's Electronics Division, when describing the conformal coating system his company recently brought out. He neither diminishes nor puts down cost savings, just arranges the pecking order according to his own perception: reliability, quality, productivity, with cost savings a distant fourth and performance a given.

Batson contends that his company has taken a different route to conformal coating in three respects. First, the product, conformal coating Shadowcure 361, invented in 1982 by Loctite chemist Larry Nativi, is a UV-cured urethane acrylate which sets up in 5–30 seconds. Second, because 361 combines UV and anaerobic technologies, the material cures even in areas behind opaque objects shaded to the UV source. Third, Loctite has taken the system approach in which PCB manufacturers can be supplied with not only the conformal coating but also the capital equipment with which to apply it.

CLEANING PCBs FOR HIGHER QUALITY

"Cleaning technology will become more and more important as we attempt to crowd more functions in a smaller space on PCBs," says Paul O'Hern, product manager-surface treatment, Chemcut Corp.

Evangelically, O'Hern continues, "Most emphasis seems to be aimed at cleaning during and after the final assembly process. The industry has not spent much time or effort in examining cleaning processes and techniques during the actual manufacture of the board itself. It is becoming more important to control the cleanliness of the board at each step of the manufacturing process, from the time the raw materials arrive until the final board is shipped."

Some of these PCB manufacturing steps for which cleaning is vital include: cleaning laminated copper prior to electroless copper plate; cleaning electroless copper prior to or after imaging; cleaning electroplated copper before solder or tab plating; cleaning solder before reflowing; removing residues after solder stripping; and cleaning boards after reflow.

Board assembly-related cleaning processes are the focus of this section; not to de-emphasize the more generic subject of cleaning at all appropriate steps in manufacturing, but rather to home in on one important operation as an example for all other operations to emulate.

Benefits of cleaning

Cleaning is to overall PCB manufacturing what soldering is to reliable interconnections.

First published in *Electronic Packaging & Production*, September 1985.
Author: S. Leonard Spitz, East Coast Editor.

Cleaning prevents electrical faults, the most obvious of which is current leakage which reduces insulation resistance. Ionic contamination, absorbed organic materials, and particulates such as metallic chips number among the prime contributors to current leakage. Non-conductive contaminants such as dirt films, rosin and adhesive residues can interfere in the making and breaking of contacts.

Cleaning assures adequate coating adhesion. Entrapped conductive or corrosive materials under organic coatings such as photoresists, solder masks and conformal coatings pose a hazard to good bond strength. While interfacial contamination in electroplating does not necessarily prevent coating adhesion, severe solderability problems can result.

Cleaning eliminates corrosion hazards. The corrosion process can totally consume small conductors or cause embrittlement of larger parts. In addition, the corrosion products themselves are usually electrically conductive in the presence of moisture and may cause shorts.

Cleaning enhances appearance while easing both visual and automatic inspection operations. Defects such as heat damage, delamination, and blistering stand out clearly against clean surfaces. And to many, a sparkling-clean surface connotes good workmanship.

Design for cleaning

The layouts of PCBs should provide adequate spacing around and under components to permit penetration of the cleaning solution and its subsequent removal by drainage and blow-off methods. If not carefully considered, cleaning surface-mount assemblies (SMAs) can be complicated by the PCB design and the component layout. One prominent cleaning system manufacturer, Electrovert, recommends consideration of the following six guidelines for ease of flux or contamination removal when designing a PCB:

- Plated through holes should not be placed beneath the component bodies to prevent flux from travelling to the topside of the PCB during wave soldering. Fan-out patterns are preferred.
- Substrate cutouts should be avoided or kept to a minimum.
- The ratio of board thickness to width should be sufficient to preclude the use of stiffeners in wave soldering. Flux becomes entrapped between stiffener and substrate, leaving residue on the substrate after cleaning.
- If surface area permits, allow for maximum conductor spacing.
- Use solder masks that maintain good adhesion after several reflow solder processes.
- In component layout design, consider lead extension direction and component staggering. These parameters affect cleaner flow uniformity and turbulence underneath and over the assembly while it passes through a conveyorized cleaning system.

Solvent vs aqueous cleaning

Solvent and aqueous cleaning represent the basic two methods of choice, although a combination of the two is possible.

Solvent cleaning involves any number of solvents including chlorocarbons, fluorocarbons, alcohols, and aromatics. It is classified into two categories: cold cleaning and vapor degreasing.

Cold cleaning generally refers to small manual operations and applications in which materials are sensitive to hot solvents.

Vapor degreasing uses chloro or fluorocarbons either alone or in mixtures with each other or with other solvents. Vapor degreasing is more effective than cold cleaning.

Aqueous cleaning is divided into two major

PCBs conveyed through boiling immersion section of an in-line cleaner (Courtesy of Unique Industries Inc.)

categories: water only and water with additives. Water only refers to tap, softened or deionized water, and is used with water soluble (organic acid) fluxes.

The cleaning capabilities of water can be enhanced by the addition of saponifiers. A saponifier is an alkaline material which reacts with rosin and oil to form a washable soap. Small amounts of saponifiers remove non-polar dirt when water-soluble fluxes are used; larger amounts are needed to remove rosin-type fluxes.

Lower capital equipment costs and lower energy demand are usually concomitants of solvent cleaning. The solvents have excellent wetting and solvency action, and are generally recyclable. Solvent systems may be selected for removing particular contaminants, are specifically designed for applications associated with SA flux, and offer short drying cycles.

In general, aqueous cleaning chemicals are biodegradable and need only to have their pH adjusted before disposal. Other desirable characteristics of aqueous cleaning include: low foaming; low toxicity; and low volatility. In addition, aqueous cleaning is non-flammable, removes ionic soils and rosin flux, is not detrimental to platings and substrate material, rinses freely, and is inexpensive.

Solvent cleaning equipment must be well ventilated to meet OSHA safety requirements. Solvents in the water contribute to total toxic organic levels which must meet federal pollution laws. Solvents are frequently hazardous, flammable and expensive.

Certain plastics are not compatible with chlorinated solvents. Polycarbonate and styrenic materials used in device packaging can be affected by these solvents.

Water promotes oxidation of metals, thus when used in electronic circuits it must be completely removed. Water is also a vehicle for ion transfer under inadequate cleaning conditions. Saponifiers are corrosive and once used must be removed.

Aqueous systems require high energy levels to heat wash solutions and dry boards after cleaning. In high-volume operations, drying time can be an inhibiting factor.

The solvency of aqueous solutions is demonstrably less effective than solvents for cleaning surface-mount devices.

When PCBs were first manufactured in quantity, the flux used was rosin, and deflux-

ing was accomplished with solvents. In the late 1960s, environmental concerns, desire for faster soldering speeds, emergence of water-soluble fluxes, and the development of saponifiers for rosin-type flux caused a shift in cleaning equipment preference to aqueous systems. During the 1970s, aqueous cleaning found favor with all segments of electronic assembly manufacturers.

As the 1970s drew to a close, new concerns about aqueous cleaning were voiced. These included rising energy costs, lack of military approval, and insufficient scientific evidence effects from solvents.

The 1980s ushered in new fluxes, such as the SA (synthetic activation) type as well as new solvent blends. Advanced, efficient cleaning equipment emerged along with surface-mount components, which caused the pendulum to swing back towards solvent systems.

Every solvent, including water, organic

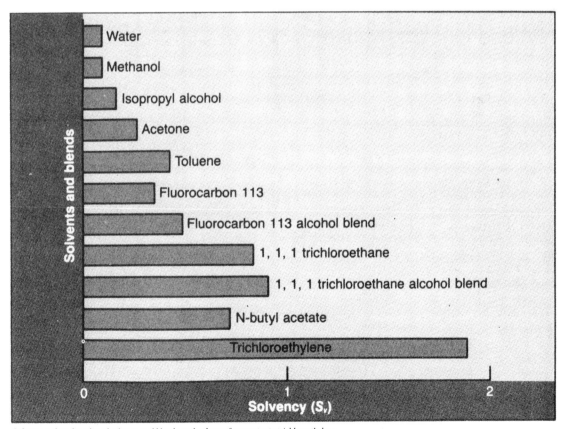

Solvency values for selected solvents and blends rank solvents from water to trichloroethylene

compounds and polymers' is characterized by a solubility parameter based upon a dispersion (non-polar), polar and hydrogen bonding component. The solubility of any substance in a solvent depends upon how closely the solubility parameter of the solvent matches the same parameter in the material to be dissolved (the solute). If the dispersion (δd), polar (δp), and hydrogen (δh) bonding factors are known, it is possible to predict which solvent or solvent blend will be most effective.

In work done at Dow Chemical USA and reported in a paper by W.L. Archer, S.M. Dallessandro and T.D. Cabelka presented at NEPCON West 1985, solvency values (S_v) for various solvents can be determined from the equation:

$$S_v = \frac{1}{\delta_a \text{ rosin} - \delta_a \text{ solvent}}$$

where δ_a is a function of δp and δh. The calculated solvency values for a number of solvents and blends are shown in the histogram. These S_v values can be used to select the best solvent or blend for removal of a particular soil, e.g. solder flux.

The China Lake Naval Weapons Center studied flux removal techniques and concluded that not one cleaning choice is necessarily best. They further showed that cleaning in both solvent and aqueous systems improved results dramatically.

The issue of solvent vs aqueous cleaning has historically provoked a mini-controversy over their respective merits. Looking through the bias set according to their product association, our experts firmly planted their respective feet on both sides of the issue.

Art Gillman, president of Unique Industries Inc., offered some neutral ground: "Today the choice of solvent vs aqueous is made primarily on the basis of flux choice and component types to be cleaned."

Occupying the same neutral territory, Chemcuts' Paul O'Hern has this to say: "When done properly, there is very little difference on final board cleanliness between an aqueous cleaning process and a solvent cleaning process. Aqueous cleaning processes generally require a little more thought and preparation in the planning stage to implement successfully."

From John Hibbert at Westek comes this observation: "Water is an excellent solvent for the water soluble or dissoluble contaminants (polar). When detergents and other chemicals are added to water, it even removes the non-water soluble (non-polar) contaminants. Water has various grades from tap through deionized which, so far, are superior to conventional solvents in removing ionic contaminants."

Also coming down on the side of aqueous systems, Jeanne Kitazaki of Clemelex said, "Aqueous cleaning can combine several cleaning steps into one by mixing appropriate ingredients. Solvents are limited to specific applications for a particular soil."

On the other hand, George Henzel, vice president of Cobehn Inc., argues that "solvent cleaning allows the producer to select the solvent best designed to remove the specific contaminants present in his operation. Therefore, the producer does not have to rely on just one medium acting on a wide variety of contaminants, but can match the solvent to the specific problem contaminant."

John Lott of Dow Chemical agrees that solvent cleaning is the preferred method. He explains that "aqueous-based products may not wet many substrate materials, and their solvency has not been demonstrated as effective in cleaning surface-mount devices."

Solvent cleaning equipment

The vapor degreaser is the most common solvent cleaner, with both batch and in-line machines available. The size and type of equipment depends upon the size and shape of the work, the method of work racking and handling, the hourly production rate, the amount and type of soil normally present, the degree of cleanliness required, and the utility services available.

The heat requirement constitutes a major consideration in the design of a vapor degreaser. The designer develops a heat balance which takes into account the heat required to bring the work load up to degreasing temperature, the heat required for internal fluxing of solvent, the heat losses

Family of batch-type vapor degreasers features low-profile, internal hand-spray system and low liquid-level control

by radiation and convection, and the sensible heat removal required for control of wash sump temperature.

Water cleaning equipment

Kitchen and cafeteria dishwashers serve as role models for batch and in-line aqueous systems, respectively. Metallic and polypropylene cabinets are used with the plastic construction that is proving to be more energy efficient. Design variations between equipment are mainly in the spray configuration, the volume of liquid pumped, and the location of air knives.

Two exceptionally modern in-line aqueous cleaning contenders are Electrovert's Century 2000 C and Westek's Formula Series. As an example of state-of-the-art features available

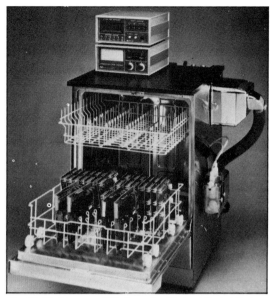

The 630μP batch aqueous PCB cleaner from Electronic Controls Design processes 500 10x10in. board assemblies per eight-hour shift

in a stainless steel, sleekly elegant package, the Century 2000 C system offers microprocessor control for process steps such as automatic start/stop, conveyor speed, water and dryer temperature, cleaning cycle, detergent injection, and liquid-level control.

Directly related to their cost of generation, the purity and quality of water is a factor for some consideration in the selection of aqueous systems. With proper processing, soft water is sufficient to assure conformance with stringent ionic requirements called out in

The Century 2000C aqueous cleaning system automatically controls process parameters

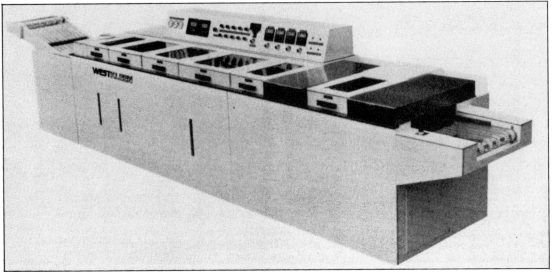

Available in Formulas II, III, IV and V, the Formula Series aqueous cleaners are modular in construction to customize for the needs of individual manufacturers

MIL-P-28809. The purity of soft water is not fixed but is a function of its mineral content. If the water ranges from half hard to hard, generally it can be safely treated and used for cleaning most electronic assemblies. While soft water contains ionic sodium in place of the insoluble heavier metals, its overall mineral content is not harmful as long as the water does not stagnate on the work.

Ultrasonic cleaning equipment

The implosion of cavitation voids induced by ultrasonic radiation produces mechanical agitation or scrubbing between parts and cleaning baths, and results in high-powered cleaning action.

The liquid medium selected for the bath greatly affects the cavitation obtained. Solvent cleaners generally have better wetting capabilities than aqueous solutions, but will only remove organic soils and particulates. After cleaning, boards are dry when withdrawn from the solvent solution. Alternatively, aqueous solutions, with the proper chemical additives, adequately wet board surfaces and also dissolve both organic and inorganic soils.

Traditional cleaning methods, especially of surface-mount components (SMCs), cannot be taken for granted. SMCs have a mounting distance of 0.001–0.005in. away from the

board surface compared to 0.010–0.015in. for axial lead and DIP-type components. As the popularity of SMCs increases, ultrasonic cleaning and high-powered cavitation may be seen as the process uniquely suited for removing contaminants from the underside of these components.

Serving as an inhibitor to the more widespread use of ultrasonics, some government specifications ban this cleaning technique. The fear is that cavitation causes resonances inside electronic devices, weakening or destroying wire bonds and interconnections. No hard evidence of this phenomenon appears to support this.

Ultrasonic agitation may be a controversial procedure, but John E. Hale and Warren R. Steinacher assert in a DuPont Freon Products Division paper that "the benefits of using it (ultrasonic cleaning) appear to be substantial."

The authors combined solvent boiling action with ultrasonic-induced cavitation and reported the following test results using two glass microscope slides, separated by a 0.003in. clearance and contaminated by solder flux in a controlled fashion: at 35°C cleaning time is one minute; at 40°C cleaning time is about half a minute.

Batch systems, synergistically combining

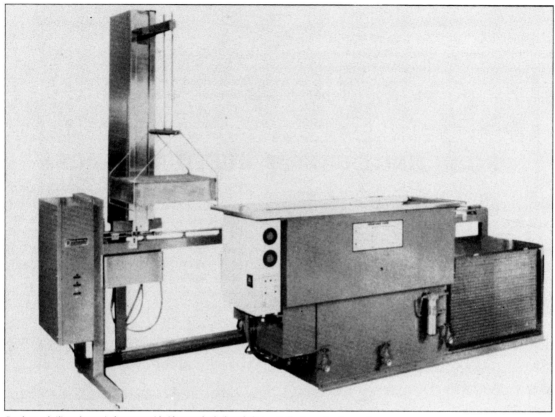

Batch-type, boiling ultrasonic degreaser with sidearm robot boils and ultrasonically cleans simultaneously

ultrasonic agitation with solvent boiling, are available in the marketplace. A machine from Delta Sonics Inc. is specifically designed for PCB assemblies with surface-mount devices having clearances of 0.001–0.005in. between device and board. Delta Sonics advises that the combined agitation technique is not recommended for other types of parts since the cavitation intensity is substantially reduced by the boiling action.

Cleanliness standards and measurement

Cleanliness standards for the PCB industry are partially defined in military specifications such as MIL-P-28809. These specifications, together with those generated by QC engineers from the industrial community, form the basis for the following guidelines: boards with a surface ionic contamination of 1.5μg

$NaCl/cm^2$ or less are virtually sterile; boards between $1.5μg/cm^2$ and $5.0μg/cm^2$ are clean enough for most high-quality applications; boards between $5μg/cm^2$ and $10μg/cm^2$ may be clean enough for some applications where high quality is not necessary; and boards over $10μg/cm^2$ are definitely not clean enough.

MIL-P-28809 specifies a simple measuring process useful as a production-line check on ionic contamination. The inspection process involves hand-spraying a test solution of 75% isopropanol/25% deionized water solution (by volume) onto the sample test board surface, collecting the washings and measuring their conductivity/restivity. A wash solution of resistivity greater than 2×10^6 $\Omega\text{-cm}^2$ is acceptable.

The development of more elaborate measuring systems has made the process less operator-dependent. Currently available systems typically employ several tanks of test solution in which the PCB assembly is immersed. As the test solution dissolves ionic contamination from the assembly, the solution is pumped to a conductivity cell. An output voltage is produced over the test time and then converted to a representative contamination level expressed in μg $NaCl/cm^2$.

A quantitative measurement of non-ionic (organic) contamination levels is both difficult and expensive. Generally, non-ionic contamination levels are indicated as pass-fail by measuring surface insulation resistance during accelerated environment tests at elevated temperatures and humidities.

PRODUCING QUALITY MULTILAYER PCBs

With increased usage of multilayered boards, it's time to look again at what is occuring at the manufacturing and design level. PCBs have become so commonplace since their inception that the main concern of electronic designers and packagers has been the chips affixed to the board, rather than the board design itself. However, there are still certain practices that must be followed and principles understood about board design and production in order to produce cost-effective electronic products. And, although much automation has been developed for producing PCBs, the human element is still critical.

Board design

Because of today's increased circuit density, one of the common myths in board design is that there are heat problems with multilayered boards. However, with digital ICs heat is not generally a factor, and power resistors and transistors are usually not mounted on multilayers. There are ways to dissipate heat if it does become an issue, such as using a heat sinking plane in the form of ground or power planes to dissipate heat away from heat-sensitive components.

In addition, a major benefit of multilayer boards is that the embedded ground plane and power plane become, in effect, an embedded capacitor. Thus there is distributed capacitance throughout the board, and integrated circuit decoupling capacitors are often not required. Another benefit involves high-speed logic, which requires a very clean signal-to-noise ratio. In a multilayered board, there is a very small potential drop across the ground plane, less than 0.1V. In double-sided boards this drop may be upwards of 0.3V.

There are many considerations in board design which affect the cost of manufacturing. The price difference between a four-layer board and a six-layer board is 30%, while the difference between a four-layer and an eight-layer is 70%. But this will change depending upon the width and space between the lines. To reduce a 0.010in. line and space to 0.008in., will increase the manufacturing cost anywhere from 5 to 10%. And the finer the line, the more problems with proper etching, voids, and registration.

Basically, the key factor is the electrical considerations versus the mechanical. With non-Schottky logic, switching speeds are low enough that boards do not require a ground plane. Thus multilayered is not necessary. However, with the advent of Schottky logic, multilayered becomes most advantageous.

CAD/CAM

At Integritek, many of Dibble's customers supply their own CAD designed board inputs and tapes for drilling. Dylon also uses CAD in designing boards. CAD/CAM allows the engineer to do what he was trained to do — engineer. Presently, many engineering functions are repetitive items that slow down the design function. As an example, making changes in engineering drawings is ordinarily a time-consuming job, but with the computer it becomes a fast and efficient task.

The increase in circuit complexity and the desire to use fewer boards per system, makes

First published in *Electronic Packaging & Production*,
Authors: Mike Friedman and Walter Berry, Integritek Inc.
© Reed Publishing (USA) Inc.

Monitoring and evaluating the etch characteristics is necessary in maintaining proper line widths as well as ensuring proper registration of inner layer patterns and uniform pad sizes for drilled holes

the use of high-density PCBs mandatory. High-density boards reduce packaging hardware requirements and provide the benefit of smaller backplanes. This results in savings in purchasing, inventory and parts handling.

CAD/CAM allows working smarter at the design end of the product. With the life of a product approaching one year or even less, it is necessary to move from design into production as soon as possible. Without CAD/CAM, a design entering into production could possibly be obsolete. For this reason Dylon has acquired a CAD system, the CDX-5000 from Cadnetix Corp.

The advantages of CAD/CAM are as follows:

- To realize the cost savings attainable by high-density boards, the quality of artwork must be exceptional, thus automatic equipment is the only way to ensure that the design can be transformed into a 'manufacturable' board.
- From initial design to the final photoplotting artwork, board quality is greatly upgraded, and board yield in the manufacturing process increases.
- On the manufacturing side, hole registration problems are eliminated because the system allows producing all manufacturing documents from the same database (artwork for all layers, padmaster, silkscreen and solder mask). Because of the accuracy in registration, hole break-out is a problem of the past.
- Screening and etching processes are more controllable because of the sharpness of line edges, unattainable when using hand taping/photographic artwork generation. Likewise, solder bridging is less likely to occur, and lifting of etch caused by shaving of pads will no longer be a problem.
- Layer-to-layer or side-to-side registration is virtually assured due to the system's ability in allowing up to ten levels of data to be entered simultaneously. Automatic insertion of components can be accomplished without concern for hole-to-hole location tolerances.
- Boards can go from design to production without first going through the old wire wrapping stage. This in itself saves a great deal of time.

PLASMA DESMEARING

As interconnection densities continue to increase in response to the drive for faster, smaller and more complex electronic systems, greater demands are being placed on multilayer PCB technology than ever before. More layers are being used and trace geometries are shrinking. As a consequence, contaminants on inner-layer conductors are more likely to hinder the formation of reliable interconnections.

Resin smear generated during drilling operations is one factor that can prevent the formation of a reliable plated-through hole connection to inner-layer conductors. Other foreign materials from the panel surface or on the drill bit can also contaminate inner-layer conductors during drilling.

During drilling the heat generated by friction between the drill bit and the panel can cause localized temperature rises that exceed

First published in *Electronic Packaging & Production*, October 1984, under the title 'Plasma desmearing assures reliable multilayer interconnections'. Author: Tom Dixon, Senior Editor.

the glass transition temperature of the resin. When this temperature is reached, the resin will flow and can be smeared along the inner surface of the hole wall. Unfortunately, this resin material coating, which is a dielectric, may be deposited on inner layer copper conductors where it can interfere with the reliable formation of electrical connections and degrade the mechanical strength of the through-hole connections.

One approach to the resin smear problem has been to eliminate the problem at its source: the drilling operation. Investigators experimenting with drill speeds and feed rates as well as drill-bit designs have been able to significantly lessen the drill smear problem.

This is apparently the plan followed by Japanese PCB manufacturers. Because the Japanese, as reported by one industry source, are relying primarily on process control of the drilling operation to prevent smearing, their use of desmearing process is limited. Advocates of desmearing point out that some of these apparent process control-related solutions to the resin smear problem are really the result of the use of laminates with higher glass transition temperatures.

The desmearing process, a post-drilling operation for mechanically or chemically removing the resin smear from the conductors, is viewed by many US manufacturers as additional insurance against running into interconnection reliability problems. They note that even one smear-related defect in a complex multilayer panel can cause it to be scrapped. Thus, the overall financial consequence of a single defect can be significant. There are also some military specifications that require some form of hole desmear and etchback. Etchback involves not only the removal of smear from the end inner-layer conductors in the hole, but also etching some of the laminate above and below the conductor at the hole's edge.

The use of desmearing techniques in multilayer PCB manufacturing is growing. Dr. William Loeb of Rose and Associates estimates that 17 million square feet of multilayer PCBs went through some form of desmearing process in 1980, of which approximately one million square feet were treated with plasma processing. For 1985, he projected that approximately 35 million square feet of multilayer laminate would be processed through some form of a desmearing process. Of this total, Loeb expected that between 7 and 8 million square feet would be desmeared with plasmas.

Several reasons are cited for the higher growth of plasma-desmearing process over the competing options. These include its favorable environmental impact, its controllability, its ability to work on newer laminate materials and the process's cleanliness.

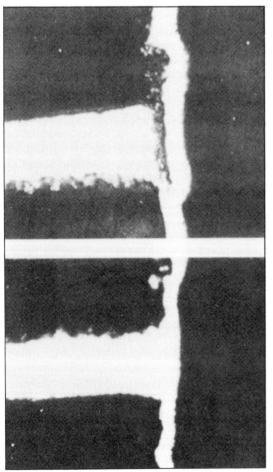

These cross-sections of through-holes in a multilayer PCB illustrate the differences in the connection which might be formed between the through-hole plating in an inner layer conductor for a hole that was not desmeared (top) and one that was plasma desmeared (bottom) (Courtesy of Branson International Plasma Corp.)

Removing resin smear

A number of desmearing techniques have been developed to remove resin smear from copper conductors in through-holes after drilling. There are mechanical methods such as air-water blast, ultrasonics and vapor honing. There are also wet chemical methods which etch away the smear. Frequently mentioned techniques, according to the IPC, are sulfuric acid, chromic acid, fluorosulfonic acid, mixtures of hydrochloric and sulfuric acid, alkaline permanganate, alkaline swell with chromic etch, and a number of proprietary processes such as Tetra-Etch which is used on Teflon multilayers.

Etching with plasmas is another possible means of removing resin smears. The use of plasma as a desmearing tool for through-holes in multilayer PCBs is only about a decade old. While the plasma process is a chemical process, it is normally referred to as a dry process since it involves gases rather than liquids, used in wet processes. It is only within the last few years that viable, production-oriented plasma production scale units have emerged on the market.

Interest in this method of smear removal appears to be on the rise. Several reasons are cited for this growing interest in the plasma process. One frequently mentioned is the environmental impact of the process, which proponents claim is quite favorable since plasmas do not present the waste disposal problem that the liquids do.

Plasmas

Plasmas are sometimes referred to as the fourth state of matter. A simplified definition of a plasma is a charged gas containing ions and electrons.

An external energy source must be present in order to sustain a plasma. Typically this would be an electric field which can act on charged particles. Electrons, with their small mass, receive most of the energy from the field. These fast-moving electrons collide with the gas atoms and raise them to an excited state. Eventually these excited atoms relax and emit light photons which are characteristic of the specific atom. This accounts for the

The chemical reaction shown in the figure was first proposed by Eugene Phillips as a possible explanation of the plasma desmearing process. It should be noted that the actual chemical reaction mechanism is probably far more complex. Some feel that the CF_2 acts more as a catalyst to the formation of oxygen radicals, which do most of the work

glow commonly associated with plasmas. The highly excited particles in the plasma react with organic materials, such as the laminate resins, and form smaller molecular units which are readily vaporized off the surface of the etched board.

The plasma desmearing process involves the conversion of stable gases into active metastable free radicals by exciting the gas with RF energy. The typical plasma desmearing system involves the mixing of tetrafluoromethane (CF_4) and oxygen (O_2) gases in a vacuum chamber where they are excited into a plasma state by an RF power supply. In this process the molecules are disassociated and ionized by electron impact. It has been theorized that these ionized molecules form an oxy-fluorine radical which impinges on the surface of the epoxy, and by virtue of their high energy content strip away atoms from the resin and combine with some of the free radicals to form gases such as HF, H_2O, and CO_2.

The variation in CF_4 and O_2 gas mixtures runs from around 90% oxygen and 10% tetrafluoromethane to 50% oxygen and 50% CF_4. A typical mixture would be between 20% and 30% CF_4, with the balance being oxygen.

Process advantages

Several reasons are cited for the desirability of plasma desmearing over conventional wet etching process. Some of the newer laminate materials, such as polyimide, cannot be etched by most of the conventional wet etch processes. On the other hand, proponents of plasma etching note that while there is some variation in the etch rate for various laminate materials, the plasma process has little difficulty etching the new materials.

Today's trend toward circuit boards with higher aspect ratios is another advantage frequently mentioned regarding plasma desmearing and etching. As through-hole diameters get smaller and boards get thicker, wet processes run into difficulties in penetrating into the centers of through-holes. Because plasma is a gas it has less difficulty penetrating the center of a hole, proponents note. However, wet process manufacturers have been responding to these problems by designing systems with better circulation and agitation to counter this problem.

A third favorable feature of plasma desmearing systems over conventional wet processing etching systems is that they present fewer environmental problems in regard to waste treatment and disposal. While some toxic gases are created in the plasma process, they are of a relatively low level and are easily diluted or treated. Wet chemistry-based systems require special waste treatment facilities and extensive record keeping for EPA.

Process controllability is one aspect of the plasma process which gives it an advantage over conventional wet chemistry processes. AT&T Technologies' Ray Rust explains that the plasma process typically takes around 45 minutes to etch away 0.006in. of material, while a concentrated sulfuric acid etch can etch away up to 0.008in. of material in 30 seconds with an additional 10 seconds resulting in an etch back of 0.001in. or more. This short process time with its narrow window makes precise control quite difficult.

The cleanliness of the through-hole after desmearing is generally better than that of most conventional wet processes, according to its advocates. The wet process may leave fluids in the through-hole which can dry if not completely removed, leaving residues which create difficulties in subsequent copper-plating operations.

The plasma process sometimes leaves an ash which is readily removable. Rust notes

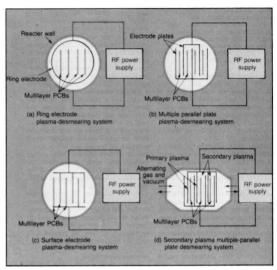

Uniformity of the plasma around the workpiece is crucial to achieve good results. Electrode design and placement, and chamber shape are among some of the factors that have been investigated to achieve a more uniform plasma process. The illustrations show some of the variations in electrode and chamber configurations that have evolved

that sometimes a small layer of fluoropolymers on the order of a few molecules in thickness, have been found on the surface after some plasma desmearing operations. This layer can present problems in subsequent plating operations if not removed; however, he explains that it can be easily removed by etching for only a few minutes with an oxygen plasma, since the CF_4 must be present for the polymer to form.

System design

A typical plasma desmearing system consists of a vacuum chamber, a vacuum pumping

This close-up picture shows the electrode configuration in one of Advanced Plasma Systems' units. Notice how the electrodes are configured in closely spaced pairs which are then placed at a much larger spacing to the adjacent pair

unit, an RF power supply, a gas source and some form of control circuitry. Inside the vacuum chamber can be found some form of racking into which panels are placed and electrodes which are connected to the power supply to generate plasma.

One of the key characteristics desired in a plasma desmearing unit is that the etch rate across the panels placed in the system be relatively uniform. To achieve uniform etching it is necessary to have a relatively uniform plasma. Electrode design and placement play a major role in obtaining such uniformity.

Early plasma system designs used a tubular construction with a ring electrode around the inner wall of the chamber to generate the plasma. This design had a severe uniformity problem because of the long path from the plasma-generating region and the shielding of one panel by another.

Some of this non-uniformity has been removed by adopting a parallel-plate electrode configuration. In this configuration, each multilayer PCB panel is placed between electrodes of opposite polarity that run parallel to the panel. These electrodes consist of solid metal plates, perforated metal sheets or metal posts. Some feel that this configuration does not give a completely uniform plasma etch and with further modifications could achieve even better uniformity.

One method developed by AT&T Bell Laboratories is to use the surface copper of the multilayer PCB panels which are being desmeared as the electrodes. In this scheme, adjacent panels are connected to opposite poles of the power supply. This configuration reduces the diffusion path and improves the power efficiency of its etching, according to its developers. This scheme still has some uniformity problems but these were alleviated by the addition of a dielectric perimeter around the edge of the panel.

Still another approach to achieving better uniformity can be found in units from Advanced Plasma Systems and Micro-Plate. These units place the panel to be etched between two sets of electrode pairs in what is known as the secondary plasma. On each side of the panel is a closely spaced positive and negative electrode. The theory behind this configuration, according to Advanced Plasma's Fazal Fazlin, is that in the center of the primary plasma, between the closely spaced electrodes, there is a null region where the plasma is not present. The secondary plasma has a more uniform plasma distribution since it consists of radicals escaping from the primary region and does not have a central null point. Another feature of these systems is that instead of continually supplying gases from one port and drawing it away from the other, these two ports on opposite sides of the chamber alternate between supplying gas and drawing vacuum. This counteracts non-uniform plasma distribution created by the depletion of the gas constituents which are consumed as they flow from the input to the exit ports.

Process concerns

A number of variables come into play in the plasma etching process and a careful balance must be struck to get optimum performance from the unit. A low gas pressure in the reactor vessel can improve the uniformity of the plasma, but if set too low it can result in too slow an etch rate. On the other hand, a higher gas pressure will increase the etch rate up to a point but degrade the plasma's uniformity beyond what is desired. If the pressure is set too high, the ions could collide with themselves and recombine, never reaching the surface to be etched.

Moisture levels both in the chamber and the panel itself can interfere with the process. Water is reported to hinder the plasma etching of polymers. If a system has been idle, it is general practice to run it unloaded for a cycle to bring the unit up to temperature, as well as drive out any moisture which might have condensed.

Different laminates have different affinities to absorb water, thus they can have dissimilar moisture behaviors. Micro-Plate's Rich Harrison notes that even laminates that are supposed to be the same behave differently, such as FR4 from different manufacturing locations and suppliers, reflecting differences in processing and environmental conditions.

Panel preheating is supposed to increase

the overall etch rate of panels. Most system suppliers recommend preheating to achieve a faster, more uniform, and repeatable etching.

Machine parameters which influence etching, and are therefore controlled, include chamber pressure, gas mixture, cycle time and power. The desirable power operating level can be influenced by the uniformity requirement, the number of panels in a load, or the number of holes in a panel.

CHIP-ON-BOARD

In most instances PCBs will go to drilling operations if used in conventional circuits or to adhesive or solder paste applications in preparation for surface mounted devices. In both instances, chips which are required in the circuit are connected by soldering to the printed circuit.

There are, however, applications where for economy of space chips may be connected directly to the circuit board without using chip packages. This approach is called chip-on-board (COB).

CHIP-ON-BOARD TECHNOLOGY

Chip-on-board (COB) technology involves attaching a semiconductor die directly to a PCB substrate with adhesive, wire bonding it to a circuit pattern already present on the board, and then encapsulating it. Chip-on-board is surface-mount technology (SMT) taken to its extreme, the difference between COB and SMT essentially being that COB usually involves high lead count, active devices and dispenses with the ceramic or molded plastic outer device packaging.

COB offers substantial savings in real estate, and, in addition, holds out the promise of a materials savings due to a radically simplified package, as well as reduced inventories and faster turnaround time.

Until now, COB has largely been restricted to use in simple and inexpensive consumer products, like digital watches, electronic game cartridges and calculators. As the market for electronic games has withered, COB has experienced a corresponding slowdown in usage. This situation, however, is expected to turn around in the future as we begin to see even more consumer products, as well as some industrial applications, sharing in the benefits of this technology.

COB in general

COB is not really new. Manufacturers in the hybrid business have for many years been mounting dice cut from wafers directly onto (ceramic) substrates, foregoing the use of a surrounding package for the device. In the world of PCB manufacturing, a conventional substrate takes the place of ceramic.

In practice, the production sequence for COB is fairly simple. First, a screen printer applies a pattern of adhesive dots, usually an epoxy, to the board to be used for die attachment. As an alternative to screen printing on the mounting adhesive, it can be applied by needle and syringe or by a pin transfer technique. As a further alternative, epoxy preforms may be used. Matching the

thermal coefficient of expansion between the part and the board is a critical consideration with this adhesive.

Next, a pick-and-place machine positions the bare dice onto the adhesive dots. As a potential option, the use of tape automated bonding on board (TAB-OB) is possible, although this is still a technology very much in its infancy. It allows the testing of chips before they are placed on the board.

Fran Dance at Qualitron adds that TAB-OB permits mass bonding of all the inner leads on the chip at one time. Similarly, all the outer leads can be mass bonded. He also notes that TAB-OB allows the user to reflow solder leads onto the board, rather than using wire bonding. This means a cheaper board since no gold plating is necessary and rework is easier.

Dance feels that greater growth of TAB-OB with COB is currently hindered by a chicken-and-egg situation: parts availability lowers demand, and demand is stunted by low parts availability. In this, COB with TAB-OB mirrors a situation that SMT is just beginning to outgrow.

After die placement, the adhesive is cured and (usually) an optical inspection is done.

Injection-molded substrate from Circuit-Wise Inc. (sold under the name MINT-PAC) with a built-in recess for an IC chip which is to be wire-bonded to the circuit traces. The finished product is called a CRIB (for Chip Recessed in Board)

First published in *Electronic Packaging & Production*, July 1985, under the title 'Chip-on-board alters the landscape of PC boards'.
Author: Robert Keeler, Associate Editor.
© 1985 Reed Publishing (USA) Inc.

This step is followed by wire-bonding: aluminum or gold wire is used to connect the bonding pads on the die to the pads on the PCB. After an electrical and/or optical inspection is performed, the dice are surrounded with a 'glob' of encapsulant material, either an epoxy or silicone-based compound; a plastic or metal lid might also be used.

If any further curing is necessary it is done at this time; finally, another inspection is performed. Test and inspection is essential throughout the COB assembly process, especially if the glob-top approach is used. This is because after final curing of the epoxy, rework is either impossible or very difficult; non-functional or unacceptable boards usually have to be discarded.

Wire bonding

Two kinds of wire are used for making interconnections in COB. In one case an aluminum wire is bonded to gold-plated bonding pads; in another gold wire is used. With aluminum wire, the method is to ultrasonically scrub the wire through the thin (5 to 15 millionths of an inch) plating of gold and weld it to the tin-nickel or nickel-plated layer below. The outer gold plating serves primarily to prevent passivation (oxidation).

As Dale Baird, marketing manager with Printed Circuit Builders notes, if the gold layer here is too thick, the wire won't be able to scrub through it and will bond to the gold instead, forming an amalgam of the two metals that is known as the 'purple plague'. This can ultimately lead to bond failure. Aluminum wires can be welded with either ultrasonic bonding or wedge bonding, which is a compression welding technique.

Gold wire bonding, on the other hand, needs a soft gold pad to bond the wire to, and

this must be thicker than in the case of aluminum wire bonding, since now the object is to scrub the wire into the gold metal itself to form a gold-to-gold bond. Aside from thermosonic, gold wire bonding sometimes uses a compression welding technique. Notes Printed Circuit Builder's Baird, since it is relatively easy to bond both gold and aluminum wires into silicon, most bonding to the die is done with a compression technique, like ball or wedge bonding.

Of the two principal methods of wire bonding for purposes of interconnecting dice to PCBs (thermosonic and ultrasonic), thermosonic bonding is more prevalent because it is faster. It uses both ultrasonic vibration and elevated temperatures to form the bonds. The interconnection is made with a gold wire in the diameter range 0.0007–0.002in., and the bonding machine often incorporates a ball-forming capability that flame-shapes the wire-end before the interconnection is made.

In ultrasonic bonding, by contrast, aluminum wire (similar diameter range as above) is attached to the aluminum bond pad on the die and to the lightly gold-plated bond pad on the board by vibrating the wire at up to 20kHz. A weld forms due to the heat generated by the vibration – the process does not require the use of elevated ambient temperatures. Machine-vision systems, which can be used throughout the COB assembly process, can be used to guide and inspect the wire-bonding process in both methods.

Protection for the die

After wire bonding is completed, the wired die is sealed to protect it from its environment, usually by means of an encapsulant based on an organic compound, such as an epoxy, although silicones have also been used. Silicones allegedly do not embrittle at low temperatures as epoxies do, and as Justin Bolger, a vice president and technical director of COB encapsulant research at Amicon points out, they offer superior resistance to cracking during temperature cycling. Nevertheless, most COB encapsulants are epoxy-based because epoxies offer better moisture permeability resistance. With regard to the glob-top, Bolger distills the situation as a

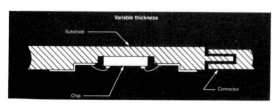

Illustrated is the capability that injected-molded substrates for chip-on-board products have for providing variable thicknesses and cross-sectional shapes

Category description	Units in millions	COB and TAB-OB percent penetration	PWB area in 1,000 sq ft
Communications			
Digital switch line FCN transmission	19.1	20	212.2
Channel end	6.2	15	51.7
Repeater	4.1	25	56.9
Telephone instrument	17.1	95	1,707.5
Pager	1.9	95	150.4
Portable cellular telephones*	18.0	98	2,205.0
Consumer			
Home computer	9.0	45	656.3
ROM cartridge	119.6	98	3,255.8
Color TV	8.0	5	55.6
Appliance control	14.3	95	1,413.1
Automotive control	11.1	98	757.6
Computer			
Personal computer	5.5	33	252.1
Handheld computer	1.0	98	72.3
Disk drive	20.2	90	1,497.5
Terminal	4.9	55	280.7
Smart card	25.0	100	1,041.7
Industrial/Instrumentation			
Digital display	3.8	98	206.9
Total			**13,873.3**

*New markets for COB and TAB-OB
Note: PWB area includes both rigid and flexible
Source: Marketbase International Technology Group, Palo Alto, CA

1990 Markets for COB and TAB-OB

trade-off between achieving resistance to thermal shock or moisture resistance.

Bolger feels that the glob-top is ideal for protecting gate arrays, and he also mentions quad packs as a likely application, although noting that device manufacturers unrealistically expect COB-mounted quad packs sealed with a glob top to pass reliability specifications written for DIPs.

In addition to liquid-dispensed epoxy, investigations are being carried out on epoxy pellets – basically a B-stage epoxy stamped out in a square shape and melted on top of the device. It offers improved esthetics. Whatever the choice of encapsulant, though notes Gary Stenerson, program manager at PCI, its formulation must include a consideration of a compatible temperature coefficient of expansion with the substrate.

The glob-top is not yet technically capable of providing a true hermetic seal, unable as it is to pass current helium leak tests governing standards on hermeticity. Organic compound-based encapsulants are permeable to moisture, a reality which means COB is still not ready for the military market.

As an alternative to the glob-top, it is possible to protect the die with a lid or cover of molded plastic or metal. These can be mounted into holes in the board and, after some preliminary testing is done on the chip and bonds, secured by ultrasonic welding, screwed down or simply attached by epoxy (coating or preform).

As Alan Berteaux of Valtronic points out, the great virtue of the plastic lid is that it is more easily removed than the glob when repair is needed. PCI's Gary Stenerson feels that encapsulation is the singlemost critical of all COB processing steps, whether the method is epoxy glob, pre-form or lid.

Advantages and disadvantages

COB offers many advantages, perhaps the most notable being that it saves space on the board, since the bulky outer package surrounding the die is eliminated. Moreover, the encapsulant can conform to the space available; thus it does not impose space constraints on the board, any more than the bare die alone would. In addition to space savings in the X and Y plane, COB also offers a lower profile height than even standard SMT.

By eliminating the need for a protective outer package, the cost of tooling that package is eliminated, as is the need to provide space in inventory storage for items related to the outer package. The need for a

A table-top view of the MEI 9010 automatic aluminum wedge bonder with automatic feed system. The 9010 is shown here bonding 4in. boards with two ICs on each board

lead frame is also eliminated.

Designers no longer are forced to match their circuit layout to the configuration of a given DIP pin-out – one which is possibly inconvenient. Order turnaround times grow much shorter, too, says Bob Luthi of TEAM (Technology Engineering and Manufacturing), because the COB assembly sequence overall takes less time than for the DIP/through-hole/solder sequence.

On the other hand, COB does have some limitations. For example, the outer protective covering in COB is unable to afford a true hermetic seal, although Ed Fuchs of the Photocircuits Division of Kollmorgen feels that the markets COB will likely serve in the future will not require hermeticity.

With COB, a typial chip is limited in power dissipation capacity (the capability ranges from 0.3–1.0W), because surface area for cooling is sacrificed when the hard DIP package is discarded. David Feindel of Mech-El Industries points out that CMOS, which draws relatively little power, is thus probably more suited to use with COB than bipolar or ECL devices.

COB is touted as highly economical, although this is only true if there is sufficient board production volume to justify the capital

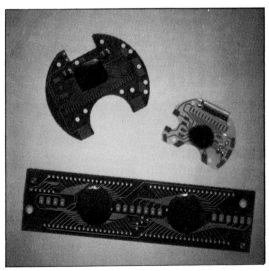

Epoxy-based blob-top encapsulants are used to protect the chips on the electronic watch modules (top) and optoelectronic circuit (below). (Courtesy of Emerson & Cuming)

expenditure necessary for COB production equipment. Also, there must be sufficient numbers of chips to be attached to the PCB before COB becomes more economical than the use of DIPs or even standard SMDs.

In the USA currently, the range of chip quantities that can be put onto a board with COB is from 1 to 7; however, as Len Distaso of Oki points out, the realistic limit is more like 3 to 4. One stateside manufacturer, however, claims to have seen the Japanese doing hundreds of chips (in this case 256k ROMs) per substrate with COB technology.

The reason for the limitation on chip quantity is that each die has some yield factor associated with it, originating in the quality of the wafer from which it came. For each new die attached to the COB board, the overall probability of a good board is found by multiplying the reliability figure for each successive die by the previous product.

Thus, for seven dice, each with a reliability of, say, 0.95, the overall board reliability, and therefore the yield, is 0.95^7 or only about 70%. In-process inspection and testing is clearly essential with COB. The above calculation does not take into consideration yield problems associated with factors not related to the die, either. Thus, the yield in the example above would realistically be factored down.

COB assembly is extremely sensitive to process control and substrate quality. The wire-bond size is miniscule compared to the size of the soldered joint of the through-hole conventional DIP board, and is therefore much less forgiving of out-of-adjustment processes. Also, surface contamination, roughness and geometry variations can easily contribute to a bad bond. The metallurgical quality of both the wire and the plated bond pads are other factors that must be kept under strict control.

Printed Circuit Builder's Dale Baird notes that, since one major motive for going to COB in the first place is a saving in real estate, care must be given to provide for component holes and pads that are small enough not to negate the space savings. Gold-plated boards are especially susceptible to scratching, pitting or denting. If these are present on a trace, they can lead to a bond failure. If a COB board fails final test, rework is difficult or impossible, and most often the product, with all its value added, must be discarded.

New applications

Until recently, COB has seen use only in inexpensive, simply constructed consumer products like digital watches, calculators and video-game cartridges. However, exciting new applications are predicted for it by industry watchers.

In future, we are likely to see COB turning up in consumer audio equipment and displays, and on expansion boards for minicomputers, microcomputers and their associated disk-drive assemblies. Bob Luthi of TEAM has suggested that COB may be used in the future in ROM and EEPROM memory cartridges used to replace diskettes as an attempt to avoid the problem of software piracy. He notes that 1 Mbyte ROM chips are just about to become available on a production basis.

TV and appliance controls, as well as industrial controls, will also make use of COB. In the area of telecommunications, COB will make a contribution in intelligent telephones and hand-held pagers.

The increase in speed afforded by COB will open the door to its use in high-speed, highly dense multichip modules. In the area of medical electronics, COB will appear in products requiring special miniaturization, like hearing aids, hand-held X-ray controls and heart defibrillators. Automotive controls will benefit from it too, and despite its limitation for the forseeable future in the area of hermeticity, even the military will begin to take a second look at COB for certain discardable items.

Graham Hall, president of Nepenthe, a firm which markets COB products for the Japanese company, Ibiden Ltd, points out two hot new areas for the 'package' approach to COB involving pin grid arrays (PGAs). One is products demanding high electrical or thermal performance, achieved by composite structures and perhaps inset copper slugs. The other, a low-cost alternative, involves a double-sided board and fitted pins.

The field of injection-molded substrates may lend itself to an easy marriage with COB, since both technologies require large volumes of production to cost-justify their use. Jack Mettler, president of Circuit-wise Inc., a firm which markets a molded substrate technology called Mint-Pac (for Molded Interconnect Package), developed in a joint venture with General Electric, notes that molded products offer several areas of compatibility with COB.

For example, integral standoffs can be molded right into the board to afford raised legs for the chip to rest on, facilitating cleaning and cooling. As an alternative, says Mettler, a recessed pocket can be molded-in to lower the chip's profile height. He notes that, for surface mount in general, the coefficient of expansion can be made close to that of copper with the proper choice of plastic resin formulation. Heat sinks, often necessary for surface-mounted devices, can be insert-molded right into the board.

CHIP-AND-WIRE TECHNOLOGY

There are several ways to package and assemble integrated circuits (ICs) on printed wiring boards. Yet, in the case of hybrid circuits, IC chips have been used almost exclusively in packages supplied by the IC manufacturer. During the past several years, however, bare IC chips have come to be used, both singly and in multiples, on printed wiring boards in what is referred to as chip-on-board (COB) technology.

First published in *Electronic Packaging & Production,* August 1985, under the title 'Chip and wire technology: The ultimate in surface mounting'.
Author: Gerald L. Ginsberg, Contributing Editor.
© 1985 Reed Publishing (USA) Inc.

Advantages of COB

Factors that lie behind the increasing interest in COB include the following:

- *Size* – The benefits of using COB all stem from the absence of the IC package. A wire-bonded chip takes up about one quarter of the area of a dual-in-line package (DIP) and is also more space efficient than leadless chip carrier (LCC) packages. Also, with a lower profile, COB can be used in applications not possible with the other packaging methods (e.g. in 'smart' credit cards).

- *Cost* – As mentioned, lower overall system cost is another factor behind the selection of COB technology. A related cost-saving consideration is due to the fact that the cost for making circuit interconnections is much less when made directly on the IC chip, as opposed to making them on a higher level packaging scheme. Therefore, the trend has been toward more sophisticated IC chips.

- *Semiconductor technology* – Trends in semiconductor technology have also paved the way to COB technology. Low-power CMOS devices, already taking a major share of the IC chip market, are ideal for the power-limited COB technology. Also, the trend toward custom and semicustom ICs should provide additional importance to COB.

- *Encapsulation* – Reliability has improved with the development of encapsulating materials that more closely match the thermal expansion properties of conventional printed wiring board materials. In the past, the expansion mismatch between the encapsulant (usually a 'glob' of epoxy) and the board would excessively stress the joint between the chip and the board. However, recent developments with improved sealants have reduced the significance of this problem, thereby making COB attractive for use in even more applications.

- *Automation* – Companies that have had only limited results competing in labor-intensive assembly markets may pay more attention to COB, since many, if not all, of the processes involved can be automated.

The three basic chip attachment and termination techniques or derivatives thereof, i.e. wire bonding (chip-and-wire), tape automated bonding (TAB), and controlled-collapse (flip chip) bonding, have been used for semiconductor packaging and in hybrid assemblies, for over 20 years in some cases. All of these techniques have been refined and improved considerably, to the point where they are suitable for COB applications.

Chip-and-wire technology

Chip-and-wire technology is commonly used in COB applications to make the majority of interconnections between the IC chip and the board. Unlike some of the other COB technologies, chip-and-wire requires the COB assembler to make two bonds per chip interconnection (one at the chip metallization and one at the board land).

By the time a chip reaches the interconnection stage of assembly, a considerable investment has been made in materials and labor; thus, high yields in these late process stages can result in greater cost savings than can be realized at earlier stages of the process. For this reason, COB manufacturers carefully follow precisely defined processing procedures and place stringent demands on the materials and equipment used in wire bonding.

Chip-and-wire technology has found usage in COB applications with a wide variation in size and complexity. For example, video-game cartridge packaging is entering a second generation of evolution – first generation cartridges consisted of plastic DIPs wave soldered onto printed wiring boards. The current generation is being built with direct COB construction that eliminates the first level of packaging. Further generations are expected to incorporate even more efficient and cost-effective approaches, such as TAB or flip-chip mounting.

Slightly more sophisticated COB packaging incorporates bubble memories. In these applications, the bubble substrate (chip) will be mounted on a small, but complex, multilayer board that will interconnect the chip and the I/O leads of the coils that are wrapped around it. The entire COB subassembly is then

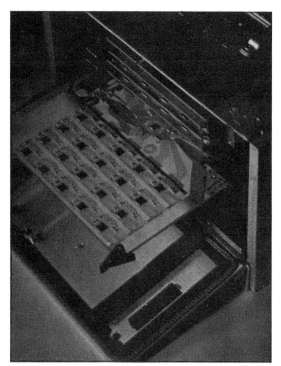

The high-density PWB shown is one of Hewlett-Packard's computer boards. It employs a gold-plated dielectric and uses HP's Thinstrate COB technology

enclosed with magnets in a shield to complete the assembly.

Perhaps the most sophisticated chip-and-wire COB application to date is the multichip 'Finstrate' assembly from Hewlett-Packard. This COB assembly contains 22 VLSI (very large scale integration) IC chips, which are interconnected by the Teflon-dielectric multi-layer board.

Wire bonding

The three basic types of wire bonding used in chip-and-wire technology are thermocom-pression, ultrasonic and thermosonic. They vary primarily with respect to the shape of the bond, the type of wire used, and their basic operating conditions (such as temperature and pressure).

Thermocompression wire bonding is one of the most frequently used bonding processes. The principle here is to join two metals using heat and pressure, but without melting. The elevated temperature (approximately 270°C) maintains the metals in an annealing state as they join in a molecular metallurgical bond. The process is quite involved, but, generally speaking, the softer the metals, the more readily they bond together. The basic method actually encompasses three different bonding processes – wedge, ball and stitch.

The oldest form of thermocompression bonding is known as wedge bonding. This technique requires two separate alignments (X and Y) and is rather slow. However, it is very useful with small diameter wires. It is also quite simple to produce two bonds with the same wire on the same land, thereby impro-ving joint reliability.

A technique more suited to higher bonding rates is known as thermocompression ball bonding. With this method, the wire to be bonded is fed through a capillary, usually heated to between 300° and 400°C. An open flame (or spark discharge) is then used to melt the end of the wire to form a small ball (only one alignment is necessary, thus allowing for faster bonding rates). The bonding tool subsequently forces the ball onto the bonding land and the bond is formed. Wire cut-off is achieved by a flame-off operation, that also produces the ball used in the next wire termination.

A technique frequently used as a comprom-ise between wedge and ball wire bonding is

Bonding process	A(min.)	B(min.)	C(max.)	D(min.)	E(min.)	D(max.)
Gold wire bonding						
Thermocompression	4	4	30	25	10	100
Thermosonic	4	4	15	40	10	100
Aluminum wire bonding						
Ultrasonic	2	4	30	25	10	100

Wire bonding features limits (0.001in)

thermocompression stitch bonding. In this technique, a cut-off arrangement is used in place of the flame-off. This allows for the use of gold and aluminum wire and smaller bonding areas. The cut-off process also forms the wire for the next stitch-bond operation.

Ultrasonic wire bonding is yet a different approach to bonding. It employs a rapid scrubbing or wiping motion in addition to pressure as a means of achieving the molecular bond. The scrubbing action removes any oxide films that might be present. It must be noted, though, that a slightly larger area of contact is necessary to accommodate the scrubbing action. Also, extreme care must be taken not to damage the chip during the ultrasonic bonding operation.

Because ultrasonic bonding can create bonds between a wide variety of dissimilar materials, it is an extremely flexible process. Thus, both gold and aluminum wire are compatible with this technique.

Void-free junctions are produced by ultrasonic wire bonding with relatively few foreign material inclusions, making it a desirable means of creating high-quality/low-resistance electrical junctions. Also, bonding dissimilar

metals at low temperatures eliminates, or greatly decreases, the formation of intermetallic compounds and allows bonds to be made in the immediate vicinity of temperature-sensitive components without adverse effects.

Finally, thermosonic wire bonding combines ultrasonic energy with the basic elements of thermocompression wire bonding. This technique relies on the vibrations created by the ultrasonic action to scrub the bond area to remove any oxide layers and also to create the heat for bonding. Combined with the pressure of the tool used to force the wire into the bonding land, this provides the final bond. Since bonding here can take place at temperatures around 120–150°C, thermosonic wire bonding can be done with low-temperature materials.

Board design considerations

The layout of a board for COB applications, as with conventional designs, begins with the positioning of components and the routing of conductors. A prime factor that should be kept in mind is that the exact placement of components is not always possible. Therefore, a good layout for high volume production is one that is relatively immune to the variations in part placement.

Ideally, discrete parts should be able to be mislocated by up to 0.010in. and rotated up to 10°. (Note, although the following information pertains specifically to chip-and-wire technology, similar considerations apply to the other COB technologies.)

Automatic wire/lead bonders require special layout considerations to optimize their operational efficiency. These include:

- Uniform wire/lead length.
- Preferred wire/lead lengths of from 0.060 to 0.080in.
- A narrow range of chip-to-board 'step-down'.
- A uniform and homogeneous bonding surface. Since an automatic machine is often not selective, it will not hunt for the best bonding spot on the substrate metalization.
- Conservative layout practices for the effective utilization of automatic equipment.

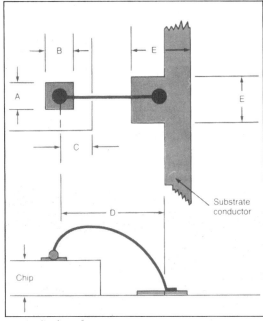

Wire bonding layout features

Guidelines	Reasons
Die-to-board attachment	
Chip attachment land should be at least 0.020 in. larger than the die size on all four sides.	To allow sufficient tolerances for the chip attachment adhesive and die placement during assembly.
Multichip-to-board attachment	
The location of multiple chips on the board should be equally spaced and on the same axis.	To simplify die attachment, wire/lead bonding, and encapsulant/ cover placement automation.
Lead bonding	
The bonding land on the board should be at least 0.020 in. from the chip attachment land.	To avoid bridging of the chip attachment adhesive.
The width of the wire/lead bonding land on the board should be at least 0.010 in. The bonding area should be at least 0.010 by 0.030 in.	To allow sufficient area for bond placement and rework.
Bonding wire/lead	
The length of the bonding wire/lead between the chip and board lands should not be greater than 0.100 in.	To minimize wire/lead sagging.
Spacing between adjacent wires/leads should be a minimum of 0.100 in.	To avoid shorts.
The tip of the bonding wire/lead should preferably have a square configuration.	To enable easier judgment of the reference points taken during automatic wire/lead bonding.
Solder mask	
The solder-mask opening should be at least 0.050 in. from the edge of the bonding lead.	To avoid bond placement on the solder mask.

General COB design guidelines

(The manual wire/lead bonding machine operator can often inspect and compensate while bonding, though the automatic machine cannot; it is precisely repetitious. Also, its bonding defects, though fewer, are nevertheless more difficult to detect.)

- When wire/lead bonds cross uncommon conductor metalization, the conductor metal should be covered with protective dielectric, such as solder mask, to prevent sagging wires/leads from causing shorts on the board. (Thermosonic wire bonding is particularly sensitive to wire sagging.)

Substrate materials and chip handlers

The selection of an appropriate board substrate material depends a great deal on the cost-tradeoffs associated with the end-product application and the type of COB technology being employed. For most low-cost applications where thermocompression wire bonding is not used, conventional printed wiring board materials are sufficient.

Where thermocompression wire bonding is employed, consideration must be given to the high temperatures associated with this process. When cost is an overriding consideration, localized bond-site charring and de-

lamination are often tolerated when conventional board materials are used.

However, when this is not allowable, special high-temperature epoxy/glass and polyimide/ glass board materials are often used. For special COB applications, the use of flexible printed wiring, conductive polymer, and molded thermoplastic board materials has also been considered.

Gold plating is usually used on the wire-bonding lands of the board. A major criterion for evaluating its suitability is bond yield. A proper gold wire bond should have a pull strength of from 8 to 10g for 0.001in. diameter wire, with failure occurring in the

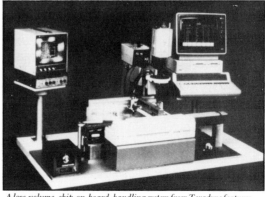

A low-volume, chip-on-board, handling system from Teradyne features computer controls and a video display

wire and not between the wire and bonding land (bond lift).

The chip-on-board assembly industry has now begun to closely follow similar manufacturing methods and procedures of those already well-established in the semiconductor and printed circuit industries. However, due to the relatively small size of this market and the associated geometric complexities, necessitating extremely accurate placement of chips, there has been a slower rate of progress in automating the COB chip handling operations.

With this in mind, various unique approaches have been taken to solve COB manufacturing problems with cost-effective solutions for the particular applications involved. A typical approach for low-volume applications consists of a computer-controlled pick-and-place machine for selecting devices from chip tray carriers, vibratory feeders and tape input.

The system orients and aligns devices on an automatic centering nest; it then accurately assembles them on the boards at preprogrammed locations. Such a system can also be equipped with provisions for die attachment epoxy dispensing, wafer frame or die tray loader elevators, pattern recognition for orienting devices and vibratory feeders.

Adhesives, wire and polymer

Depending on the COB wire/lead attachment method being used, it is often necessary to attach the chip to the board with an adhesive. Both gold and aluminum wire is used for COB applications. Gold wires are typically alloyed with small amounts of beryllium-copper to control grain growth during bonding. Aluminum wires, on the other hand, are typically alloyed with 1% silicon. In the case of gold, to achieve thermosonic wire bonding yields approaching 99.5%, the board plating should be a minimum of 0.000040in. (40μin.) of 99.99% pure soft gold deposited over 0.000150–0.000200in. (150–200μin.) of conventional nickel.

Also, the materials must be formulated specifically to meet the stringent demands of automatic wire bonding if bonding rates of up to four wires per second are to be achieved.

For these applications, it is particularly important that the plating be optically dense and that it consistently provides the smooth, flat bonding surfaces required to significantly increase yields and quality.

Aluminum (ultrasonic) wire bonding can be done reliably with board land plating, similar to that for gold-wire bonding. However, it is reasonable to expect to achieve bond strengths of only 3g minimum and automatic wire bonding rates of up to two wires per second.

Printed boards produced with polymer conductive materials on low-temperature substrates are now being made that are wire bondable. The application of a plated-metal surface over the printed thick-film polymer conductor pattern makes this possible. Polymers currently being used for this type of board construction include thermoset epoxy-silver conductors and non-noble nickel polymer.

Chip protection

Increasingly, many applications are requiring that the COB assembly protect the chips from the atmosphere, i.e. that the chips be sealed (hermetically, when practical). The means for doing this fall basically into two categories: glob-top coatings and lids.

In selecting the technique to be used for the seal, several important properties must be considered:

- *Sealing temperature and time* – The technique chosen must be able to form a seal at a low enough temperature and in a short enough time as to minimize the heating effects on the chip and board components.
- *Thermal expansion* – The thermal expansion of the sealing material/device should also closely match that of the board in order to minimize thermal stresses and, thus, maintain the integrity of the seal.
- *Hermeticity* – The seal must provide the degree of hermeticity required and maintain this level when exposed to the operating and storage environment of the equipment.
- *Cost* – The sealing technique must be cost-effective, not only with respect to material cost, but also with respect to application and replacement costs.

- *Repairability* – If it is necessary to be able to replace a chip or repair a wire/lead termination after the chip has been sealed, it is important that the sealing technique be such that it can be readily 'broken' and replaced.
- *Stability* – When coatings are used, they should be sufficiently stable that they do not tend to put excessive stress on the wire/lead bonds or the die attachment bond when exposed to the equipment operating and storage environments.

The final step in the chip-and-wire (and TAB) process is to protect the bare chip from environmental abuse. Depending on the environment and the degree of sophistication desired, this is usually done with either protective coatings or sealing lids.

Typical RTV dispersion coatings are room-temperature vulcanizing (hence the designation) one-component silicone-rubber coatings supplied as a xylene dispersion, with no mixing required. The curing process uses a crosslinking mechanism that generates methanol during cure. Once applied and exposed, the material vulcanizes by reaction with moisture from the air to form a soft, resilient elastomeric coating that will withstand long term exposure to temperatures as high as 250°C.

Special controlled technology can yield a silicone gel that affords the non-flowable permanence of a solid, but also gives the freedom from large mechanical and thermal stresses of a fluid. Chemically, a typical silicone gel is very similar to silicone fluids, but with just enough crosslinking to prevent separation of the individual polymer chains and give non-flow thermoset properties. The fully cured dielectric gel is a soft, jelly-like material that exhibits tenacious pressure-sensitive adhesion to virtually any substrate.

Epoxy coatings are also available for COB self-crowning or glob-top applications. Typical materials are two-compoent, liquid epoxy/anhydride systems that have been formulated for their superior thermal shock performance, substrate adhesion, moisture resistance, as well as glass-transistion temperatures in the range of from 165° to 180°C. Metal lids can be used to seal individual chip sites. For COB applications, soldering is generally used to attach the lid to the board.

The use of preforms is often the most convenient method for applying the solder, in addition to the solder coating on the board. The heat required to reflow the solder can be supplied by one of the soldering processes that is associated with conventional surface-mounting technology.

Type	Advantages	Limitations
Phenolics	Very high bond strength	Used mostly for structural applications; possibly corrosive; difficult to process at low temperatures.
Polyurethanes	Easy to rework	Not suitable for temperatures above 120 C; relatively high outgassing; some decomposition.
Polyamides	Easy to rework	High moisture absorption; high outgassing; variations in electrical insulation properties, especially when exposed to high humidity.
Polyimides	Very high temperature stability	High cure temperatures; requires solvents as vehicles.
Silicones	High temperature stability; easy to rework; high purity; low outgassing.	Moderate-to-poor bond strength; high coefficient of thermal expansion.
Epoxies	Some are easy to rework by thermomechanical means; some are low outgassers; easy to process; can be filled to 60-70 percent with a variety of conductive or non-conductive fillers.	Depending on type of curing agent used and degrees of cure, possible occurrence of outgassing, catalyst leaching, corrosivity.
Cyanoacrylates	Very rapid setting (≈10 s); gives very high initial bond strengths.	Bond strengths often degrade under moist or elevated temperature (≤150 C) conditions.

Various chip bonding adhesive types

TAB AND FLIP-CHIP TECHNOLOGY

A relatively new concept which appears to have a bright future in COB applications is known as tape automated bonding (TAB). This technique utilizes photoimaging/etching processes to produce fabricated conductors on a dielectric tape.

The most striking aspect of the TAB system is the carrier tape; it looks much like movie film because of its sprocket drive holes and comes in widths from 8 to 70mm. Windows are punched at specific locations and a thin (often only 0.0014in. thick) conductive foil is bonded to the tape, which is itself usually either Mylar or polyimide. A conductive pattern is then etched in the foil to give the desired interconnection circuitry with 'beam-type' leads that extend over the windows in the tape.

In subsequent processing, the beams are simultaneously bonded to the chip, which is precisely located under the window. The exact location of the chip with respect to the sprocket holes on the tape carrier permits the use of automated tape-handling equipment that accurately positions the individual chips for the subsequent processing operations. For COB applications these operations include testing, burn-in and mounting onto the board.

The initial stage of TAB wafer preparation is much the same as that for wire-bonded wafers. A pinhole-free silicon nitride passiva-tion layer, or in some cases silicon dioxide or polyimide, is deposited at low temperatures. The passivation is selectively removed, leaving a good portion of the aluminum land exposed. The chip is then at the state where it can be either presented for wire bonding or for additional TAB fabrication.

COB applications for TAB vary widely, depending on which of its main attributes are being exploited. For example, where low cost is the driving force in single-chip applications, assemblies such as a camera-flash firing unit have been produced in very large quantities.

Also, where an extremely low assembly profile is important, TAB is used to assemble 'smart' credit cards that are only 0.030in. thick. Such cards are expected to find wide usage in banking and telephone applications.

The need for small size and relatively low-cost military personnel 'dog tags' is also being satisfied by using TAB. In addition, the direct replacement of conventional through-hole and surface-mount printed wiring board assemblies with several TAB chips is being done in applications such as the camera exposure-control assemblies.

Bumped chip and bumped tape TAB

A chip slated for TAB can be processed in one of two ways: in the basic TAB approach, a barrier metal such as titanium-tungsten is deposited over both the exposed aluminum and over the passivation on the periphery of the land. This, plus the addition of 0.001in. high gold bumps, helps to ensure the reliability of the TAB connections to the tape during the inner lead bonding (ILB) process.

The bumps on the chip are electroplated onto the barrier metal at each land position. Copper bumps can also be used, and both copper and gold can be tin plated. This completely seals the chip, further enhancing TAB's reliability as a COB assembly process.

The other processing approach puts the bump on the tape rather than on the chip. This approach is known as BTAB. In either case, the bumps are necessary to elevate the etched tape conductors above the chip to prevent shorting of the leads.

This hybrid circuit makes use of multiple chip-on-board units assembled with tape automated bonding. (Courtesy of International Micro Industries (IMI))

First published in *Electronic Packaging & Production*, August 1985, under the title 'Chip-on board profits from TAB and flip-chip technology'.
Author: Gerald L. Ginsberg, Contributing Editor.
© 1985 Reed Publishing (USA) Inc.

An 80-lead, TAB-assembled display driver with chip resistors (Courtesy of IMI)

One of the most recent interconnection technologies is known as area TAB. In this approach circuit patterns are fabricated on the tape, allowing the chip designer to put the I/O lands at any position on the chip as opposed to being on the periphery of the chip. This reduces the signal path lengths required and thereby reduces signal delay.

Lead bonding

After the tape is fully fabricated, the next operation is to make the inner lead bonds (ILB) between the chip and the etched cantilevered conductors on the tape. The positioning for this operation is controlled by the holes on the tape. Monitoring the chip position is done using a television monitor or microscope.

A mass-bonding operation is then performed by thermocompression bonding of all interconnections at one time. Once the chips are bonded to the tape, the points of connection between frames are punched out to allow the individual chips to be electrically tested and, if required, burned-in.

The outer lead bonding (OLB) operation transfers the chip, with its leads, to the board. Outer lead bonding is a two-step process that is usually done sequentially on the same piece of equipment.

OLB equipment is predominately custom because of the wide variety of chips and boards involved. However, for a given combination of chip and board types, only the excising tool, thermode, lead-forming tool (if required), and work fixtures need be customized.

The first step in the OLB process is to excise the leaded chip from the carrier tape. The leaded chip is then transferred to the board in a way that precisely maintains the original relationship of the chip with respect to its tape sprocket holes. Thus, the chip will be placed on the board at a position that is accurately known.

Depending on the TAB mounting configuration, a secondary lead-forming operation may be included as a part of the process. When required, care must be taken to assure that the leads are shaped down to the board level without shorting them against the electrically active edge of the chip.

The OLB process can then be completed in one of several mass-bonding ways. The connection between chip and board can be done using either thermocompression bond-

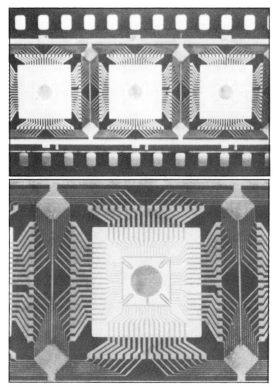

TAB tape has the appearance of more film, an effect which is enhanced by the sprocket holes (top); a close-up view of the same tape is also shown (bottom) (Courtesy of IMI)

ing, reflow soldering, or by the use of conductive adhesives.

TAB provides a means of mounting bare IC chips to printed boards, with the following advantages:

- The bonding areas on the chip can be hermetically sealed when a gold bump is constructed on the bonding land.
- Less gold is required for TAB than is needed for wire bonding.
- TAB provides a means for pretesting and 'burn-in' of the chips prior to their final mounting on the board.
- Bonding lands can be as small as 0.002in.^{2} on 0.004in. spacing.
- Input/output counts can be as high as 300 or more leads.
- Very low (0.030in.) profile COB assemblies can be made.
- TAB provides a method for automated (and robotic) simultaneous bonding of all the leads to the chip and board.
- TAB process chips require only a fraction of the board surface area required for mounting packaged chips.
- The rectangular TAB lead provides a lower inductance than does a round wire to enhance its use in high-speed applications.

Disadvantages associated with using the TAB process include the following:

- It requires specially designed equipment to match each application, both at the chip and the board interface.
- In general, TAB chips cannot be purchased economically in small quantities.
- At present, there is a lack of commercially available 'bumped' wafers or chips.
- TAB bonding and bumping equipment is rather sophisticated.

Flip-chip bonding

Flip-chip technology is based on the use of controlled-collapse solder bonding. The original flip-chip concept employed small, solder-coated copper balls sandwiched between the chip termination lands and the appropriate lands on the interconnecting substrate.

The resultant solder joints were made when

Close-up of a TAB module in a pocket calculator (Courtesy of IMI)

the unit was exposed to an elevated temperature. However, the handling and placement of the small-diameter balls was extremely difficult and the operation was costly.

In a more advanced technique, a raised metallic bump or lump, usually solder, is provided on the chip termination land. This is normally done on all lands of all of the chips while they are still in the large-wafer form. The individual chip is then aligned to the appropriate circuitry on the substrate and bonded in place using reflow soldering techniques. In this way, the interconnection bonds between the chip and the substrate are made simultaneously, reducing fabrication costs.

The use of flip-chips with controlled-collapse bonding has potential for very high assembly rates. For example, with only the use of simple optomechanical assists, manual chip-placement rates as high as 200 per hour have been achieved. This results in a very high assembly/bonding rate if a chip has many input/output terminations.

To date, the use of flip-chip devices has been limited to very special packaging applications. Whether or not these can be considered to be COB applications depends on the operative definition of 'board', since most of the applications pertain to the use of ceramic substrates.

In general, such applications have been referred to as 'hybrids'. With respect to conventional printed wiring boards, these

A 'smart' credit card makes use of this TAB module (Courtesy of IMI)

A conduction-cooled flip-chip assembly

techniques have not been applied to a significant degree.

Perhaps the most publicized application of these devices is that of the IBM thermal-conduction module (TCM). This very-sophisticated assembly can contain up to 133 flip-chips. As many as 20,000 logic circuits or 30,000 memory bits have been packaged on a single substrate in this technology.

The primary advantages of using flip-chip packaging are fast throughput times and efficient use of board area. Major disadvantages include an inability to visually inspect the assembled chips, limited availability of bumped chips, difficult flux removal, and thermal transfer complications.

A solution to the availability problem is found in an approach where the bump is produced as a part of the interconnecting substrate, rather than as a part of the chip. The COB assembler can use any of the devices that are available in bare-chip form.

There is no one best solution to the choice of an integrated circuit packaging technique. The most suitable packaging approach must be determined by a system analysis that considers the application, overall cost, manufacturing capabilities required, and degree of maintainability (modification and repair) needed. The risks necessary to achieve the desired results are also major considerations. These relate to design, procurement, and management acceptance when appropriate. It is to be expected that in several of these instances chip-on-board technology will be the best choice.

CHAPTER 3
COMPONENT INSERTION

All discussions of component insertion by automatic equipment must be viewed in the light of a management alternative – human assembly labor. The inherent flexibility, dexterity and sensory capability of the human worker set a competitive standard for automated placement equipment. When these human capabilities are enhanced by ergonomic workstations they will often prove the best approach to component insertion.

This chapter will span the range of component insertion for those components where lead wires and prongs or tabs penetrate the circuit board and have some degree of physical integrity with the board prior to soldering.

For the most part automatic insertion machinery is limited to component parts coming to the assembly floor in some pre-oriented form. These components may be attached to tape wound on reels. Other parts such as interconnect pins may literally be fabricated as part of the insertion operation. Other parts such as potentiometers, DIPs and transistors will come in magazine tubes already oriented for use in automatic insertion machinery.

There will always remain non-standard components that are normally purchased in bulk un-oriented condition. Most of these will continue to be handled on an ad hoc basis.

Automatic insertion of leaded components has reached a very high degree of maturity and reliability. Robots have found a higher degree of utilization in electronic assembly than in small parts mechanical assembly. This chapter gives special attention to this use of robots. It is further addressed in the chapter on SMT.

COMPONENT INSERTION EQUIPMENT

Today's electronic asembly tasks are much more demanding, but that is being effectively addressed by a vast array of new and/or improved machines and tools to make assembly far more efficient and effective than ever before. Insertion/placement rates can vary from a few hundred per hour with machine-assisted manual equipment to thousands of discrete component chips per minute with automatic equipment!

Mechanized component insertion is a mature technology now, with over 25 years of innovations and refinements to make it even better. And the makers of hand tools also have some very effective products that are worth considering. The solution may lie with either manual or fully automatic assembly, some-

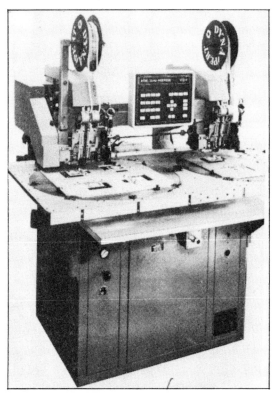

A microcomputer-controlled axial inserter from Dyna/Pert, the VCD-F, is said to have the highest productivity rates in the industry

First published in *Assembly Engineering*, March 1983.
Author: Terence Thompson, Editor.
© 1983 Hitchcock Publishing Co.

where in between, or a blend of the technologies. Selecting the right blend of assembly automation, however, does require some homework. The user should examine the many factors that impact final assembled product costs before making capital (or leasing/rental) outlays. The following should be considered:

- *Printed wiring board (PWB) volume.* Unusually high or low volumes tend to automatically exclude certain equipment options.

- *PWB configuration and tolerances.* Are the boards of conventional design or perhaps the newer, smaller hybrid substrate types? Are board/substrate sizes reasonably standardized or different for each product? How uniform and repeatable are overall dimensions, mounting holes and pads, and handling locating features?

- *Product mix.* Setup times are certainly a significant consideration regardless of assembly technique. What will it cost to make routine changeovers? Is one operator or machine expected to handle all possible component configurations?

- *Component mix.* A very interesting problem. Discrete only and mixed discrete/DIP integrated circuit (IC) application configurations are now commonplace. But what about the new discrete component chips (that can go on both sides of a PWB) or the new SMD (surface mounted device) technologies? Will you mix discrete conventional and chip components along with DIP or SMD type ICs? And don't forget the new leaded and leadless chip carriers either. The various chip and SMD and carrier devices can increase board densities 2:1 or 3:1. Are the components themselves compatible with both your handling techniques (static zaps many components) and/or the fluxing and soldering systems (sealed components are a must with certain fluxes and high densities preclude certain cleaning approaches)?

- *Component location/orientation.* With leaded components, hole patterns typically fall into the following categories: (1) all on one axis, all leads one span/distance, (2) all on one axis, two or more spans, (3) two axes, two or more spans or (4) any axis, many spans. Option (1) simplifies equipment selection but limits design freedom and (4) may result in a designer's dream and a production nightmare.

- *Component clearances.* Depending on how the component is held (by the leads, by the body edges or with a vacuum pickup probe on top of the body), with human hands/fingers, tools or insertion machine heads/robot grippers, they certainly require room to place and subsequently release the components. Higher densities are becoming very common, so watch clearances.

- *Product maturity.* With a new promising product, evaluate all of the alternatives but new equipment may have a marginal ROI (return on investment) for a mature product. Is the product on the declining side of the sales curve?

- *Pre-oriented parts.* Do you obtain components in loose form, on reels or rolls, in 'ammo' boxes or magazines/tubes? Will you need any pre-insertion processing such as sorting, orienting, lead forming and/or cutting, taping/sequencing, and electrical verification to determine that values, tolerances and polarity are correct? Remember, any incoming defective parts make the probability of assembly success marginal at best. Such processing is an integral part of many assembly machines.

- *PWB handling.* Will operators and/or machines transport boards/substrates between opertions? Is any buffer storage required?

- *Post-insertion considerations.* Do you need a turnkey vendor that can handle all of your requirements from incoming parts inspection through component insertion/placement, fluxing, soldering, cleaning, lead trimming and testing operations? And have you given adequate consideration to how rework/replacement of components will be accomplished if the product doesn't work?

- *CAD/CAM.* A familiar concept with very promising potential/actual applications in component insertion. Has your bare PWB fabricator shared the *X-Y* pad/hole coordinate information with those responsible for inserting/placing components on those same *X-Y* points? A shared database is worth considering. The other advantage that CAD/CAM offers is the built-in opportunity for convenient and rapid design/manufacturing interface prior to production commitments.

Automation

Automation or semi-automation, does offer tremendous potential for productivity improvement, lowering costs and improving quality – in many, but not all, situations. In some low-volume applications with highly skilled, efficient and dedicated employees, it might be difficult to eliminate enough direct and indirect labor costs to justify the equipment costs. Many machine-assisted manual operations are extremely efficient too, which

The A1-1000 automatic axial inserter from Amistar selects capacitors, resistors and diodes at random directly from reels or cartridges, thereby eliminating the traditional component sequencer

makes full automation a challenge.

The state of the art in highly automated component insertion machines is now looking more and more like those specialized IC chip placement machines used by the semiconductor manufacturers. It is more than a coincidence that some vendors offer both types of systems – pick your PWB density and they have the equipment.

Today's electronic assembly tasks are indeed more demanding. Manufacturers are faced with consumer demands for smaller, lighter, more versatile, highly reliable and, yes – less expensive products. This often means greater component density, more complex components, shorter product life cycles, high inventory costs for high value parts and shortages of skilled workers. And even if one is fortunate enough to have an adequate workforce, how do the operators muster the dexterity to precisely handle, sort, orient, insert or place the ever-smaller conventional discrete, integrated circuit and chip components into or onto the PWBs or other substrates? Imagine yourself with a pair of tweezers trying to pick up and place a common chip resistor that is less than 1/8in. long, 1/16in. wide and only a few thousandths of an inch thick. Don't panic, because there are solutions – in fact, too many to thoroughly discuss them all. What is covered, in overview and product round-up format, are some examples of problem-solving approaches to today's challenges.

Degrees of automation

Pragmatic solutions first require that you identify specific problems and opportunities, i.e. what do you want to accomplish in your board assembly program? Look at existing problem areas with product design, components, actual assembly or final testing to spot recurring problems that hamper overall quality and reliability while increasing rework costs. Automation may not be the answer if you haven't solved some basic producibility problems. If a task is difficult or impossible for a human, then machines may not be able to do it any better or at all. In fact, you may end up with computerizing problems. Make sure that both the parts and product are

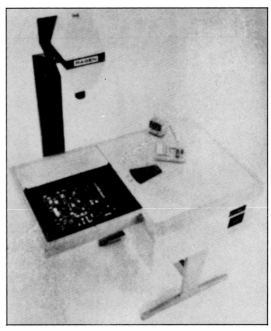

The 750 System from Ragen Precision Industries is a self-contained system that doesn't require off-line programming

designed for mechanized assembly. If not, try redesign to eliminate bottlenecks. And keep in mind that design for automatic assembly invariably pays off even in manual assembly because products designed for automation usually are much easier for human operators to assemble too. Semi-automation and full automation can lower manufacturing costs and increase both reliability and quality by improving the consistency of the insertion operation.

Programmed parts assembly

If one examines the published figures on manual component insertion, one discovers that rates generally range from 100 to 300/hour. When looking at semi-automatic programmed parts handling systems or component locating/delivering systems for low- to medium-volume operations, the rates are at least doubled. What then is a programmed parts handling system? It is a system designed to eliminate the causes of human error during component insertion in manual PWB assembly, and to reduce assembly time. This is accomplished by controlling both the presentation and sequence of components for the assembly operator. As a result, human error factors are reduced through lessened operator fatigue, elimination of ambiguous decision-making conditions, and simplified operation.

This semi-automatic solution falls partway between manual and fully automatic assembly methods. It uses an operator to insert parts into PWBs at up to 2000/hour rates while human error is minimized by using specially prepared parts dispensers, guiding lights and other aids.

The basic system components usually include a workstation containing a fixture that will accept and securely hold a board; sequential parts storage containers; a dispensing mechanism that presents prepared components to the operator; a light beam that indicates where to insert the component on a PWB and the controller containing the parts insertion program.

The system delivers only the correct components and they always appear at the same location thus encouraging a 'blind reach' technique that both saves time and reduces fatigue. Futhermore, visual aids, such as projected light beams, from above or below depending on the system, indicate the exact spot where a component is to be inserted. Projected colors, symbols and even notes can indicate component polarity and orientation plus any special mounting instructions. Also, automatic cut-and-clinch options can significantly reduce assembly time. It is estimated that by using a programmed parts handling system an operator can insert at least 300 to 600 parts/hour. These inserted parts can also be interconnect pins, jumpers or terminals in addition to the active and passive circuit devices.

Microprocessor (P) based control makes the systems work. The payoff is significant, in spite of the higher equipment costs when compared to conventional parts bins and manual assembly. A short payback period (usually less than a year), virtual elimination of errors, and relatively simple and fast production run changeover capabilities are but a few of the features being offered. The simple operation and short learning curve in turn is conducive to using less skilled operators which yields further savings.

Automatic component insertion/placement

You have several choices in overall automation concepts with some basic approaches being: (1) select individual machines based upon their merits regardless of source followed by in-house efforts to blend them into an efficient system; (2) go with a systems supplier; or (3) somewhere in between. The various approaches have strong points but the overriding consideration must be which route will quickly bring about the right degree of automation for your needs in an economically feasible manner.

An affordable building block approach makes sense if you've planned for future needs with both sound marketing inputs and an evaluation of current or emerging technological trends. Can you competitively build what the customer wants now or will want in a

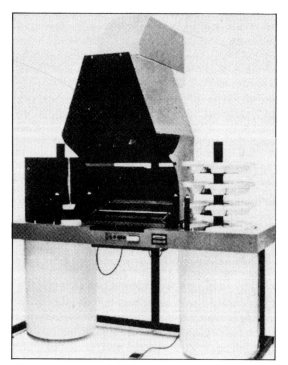

Electrovert's semi-automatic PWB assembly workstation projects component images precisely on the board, indicating where each component is to be inserted

few years? If 'now' means discrete parts assembly and 'in a few years' means SMD or chip carrier devices, then your equipment needs are simply different. One very positive aspect of equipment selection is that both the traditional and new microminiature insertion/placement assembly machines use low risk, proven technologies.

Computer control. The ubiquitous microprocessor generally provides all the standalone control needed by any inserter, sequencer, verifier, etc. Distributed processing allows every machine to operate or be programmed or diagnosed independently. And central computers with mass storage and supervisory control can run entire systems while supplying management information (MIS) or CAD/CAM data as required.

Fluxing, soldering and cleaning equipment. A natural outgrowth of insertion systems concepts includes the post-insertion fluxing, soldering (sometimes solder-cut-solder), cleaning and final functional testing of the assembly. Another vital aid to PWB reliability, i.e. solderability, can be virtually assured by pretinning boards and components – pretinning equipment is also available from systems vendors. Some offer all or part of the above now, others will undoubtedly offer them soon. If the components are in place, why not solder and have a complete assembly?

Systems. Universal Instruments Corp. was the first with a turnkey 'systems' approach for automated assembly of PWBs. The *Automated*

An automated Matsushita Electric factory with dozens of their component insertion, sequencing, testing, etc. machines operating as a total assembly system

Printed Circuit Assembly System concept is a step toward the creation of completely automated electronics manufacturing facilities. It builds on their proven 'Pass-Thru' system which includes equipment to both handle the boards and sequence, test and insert axial and radial lead components, as well as DIPs, SIPs and square wire pins. Also available are extremely versatile manual assembly and inspection capabilities incorporating a modular system for board handling, soldering, lead trimming, and cleaning operations. Recent additions such as the RHyMAS (Robotic Hybrid Microelectronic Assembly Systems) offer even more possibilities and assembly options.

Pass-Thru has a central Model 8222 Satellite Control System that is the librarian for all sequencing and insertion programs. It also includes an Executive Program with a full range of production, inventory, and management-control information. It takes boards from the stock room to insertion to inspection to soldering to cleaning to final test, etc. Ancillary systems such as sequencing and manual assembly are also provided. With the system, you begin with bare boards and end up with inserted axial, radial, DIP, wire pins and terminals which are soldered, cleaned and ready for transportation to any subsequent operations.

Dyna/Pert Div., Emhart Machinery Group, now has their *Flexible Assembly System* which combines standalone machines, a common transportation system (conveyor, carts, etc.), and the DBLU-F automatic board loader/

DuPont's Berg Electronics BP-195 automatic microprocessor-controlled pin-insertion machine stakes the full range of its standard pin and wrap post product line at a rate up to 8000/hour

unloader into a totally automated system offering maximum productivity. The DBLU-F automatically loads and unloads PWBs to all Dyna/Pert automatic insertion machines. It is fully modular and accepts a wide range of industry-compatible magazines. All machines are individually microprocessor controlled but can link to the DMSS 9000 Data Management System for program storage, information processing and inventory control data. Their wide array of insertion, sequencer, verifier and tester equipment can readily be included as parts of their Flexible Assembly System. And the new Modular Chip Place-ment System (MCPS) is available with 1 to 6 adhesive dispensing stations and matching component placement heads for a maximum output of 60,000/hour with both flat and cylindrical chips. Chips can be bulk fed or on 8mm tape.

Other vendors also offer systems with varying degrees of sophistication now. Obviously, the choice of any machine, system or groups of machines/systems for an auto-mated electronics assembly operation is de-pendent upon the user requirements. The best starting point in learning more about the capabilities is to simply ask the vendors.

WORKSTATIONS

The appearance of the electronics assembly plant is changing dramatically. Many indust-rial engineers believe that in just a few years the most noticeable line of demarcation between a factory's production floor and its office area will be the end of the carpet, says Bob Greendale of Ergo-Tech.

For many of these engineers, the ideal is to achieve in a production workstation what might be termed a 'manufacturing office'. Indeed, the inspiration for modularizing the production workplace originally grew out of design concepts developed in the business office environment where the open-plan scheme was based on fabric-covered, acousti-cally 'quiet' panel dividers.

Modular workstations for electronic manu-facturing and assembly, though influenced in design by office cubicles and furniture, are functionally direct descendants of the tradi-tional workbench. They are, however, much more sophisticated and offer more space-efficiency and productivity potential, due to the systems-oriented thinking which has gone into their design.

IBM, Hewlett-Packard, Tektronix, Texas Instruments, Xerox, General Electric, Digital Equipment Corp., Data General and Wang are just some of the major electronics manu-facturers which have found significant advan-tages in using workstation systems on their assembly floors. It was predicted (in 1986) that during the next five years as much as 40% of the market share now held by traditional workbenches would be surrendered to modu-lar workstations.

Modularity and change

Workstations are designed for just one task, whether it is stuffing and soldering a PCB, assembling a group of PCBs into a chassis, making a wire harness, etc. The person doing parts kitting will work at a different station than the one doing subassemblies, and that station in turn will not be the same as the one for final assembly (where the need for space is likely to be greater than in the previous two stages).

Says Ted Venti of Herman Miller, the term 'modularity' in reference to workstations thus pertains to the situation of having many different work activities going on. However, the term as used in the industry also suggests the ability to tear stations down and recon-figure them to meet changing needs.

Claims Jeff McMullin of Streator Inc., the great advantage to modularity in a production workstation system is the ability to deal with change, coming either in the form of new product introductions or existing product

First published in *Electronic Packaging & Production*, February 1986, under the title 'Systems design shapes production workstations'.
Author: Robert Keeler, Associate Editor.

revisions. Also, any necessary reconfiguration of the workspace can be carried out swiftly.

The work-surface area requirement is cut down due to the use of bins and shelves, associated with the modular station being 'vertical integrated'. Thus modular workstations are designed to carry shelving, drawers, bins, tote boxes and utility strips above the work surface. If parts can be stacked, yet still be accessible, it follows that square footage costs come down. This also has an impact on aisle space available and, ultimately, safety. Cluttering of work surfaces is reduced, as is congestion of aisles.

In some systems, shelves, drawers, bins, and utility bars can be attached to workstation frames at varying levels, and the frames in turn are interchangeable and easily reconfigurable. Tools and accessories, including power and pneumatic tools, hand tools, jigs, fixtures, hose arms, IC dispensers and tool balancers, as well as solder holders, document readers and many other items can be integrated into the work area.

Modular workstations can help reduce work-in-progress. They also encourage a much higher quality of worklife than available with traditional benches. This is seen in increased worker morale and pride in the work area, as well as lower rates of absenteeism, employee turnover and work errors.

Surprisingly, the issue of aesthetics is a fairly important purchase decision consideration among many system buyers. They want a 'high-tech look' in their workstations over and above functionality, since this in turn can build the confidence of their own customers in their capabilities. Jeff McMullin notes that this seems to be especially true with defense contractors. Observes Mike Christiansen of Isles Industries Inc., aesthetic design also encourages a sense of pride and professionalism in the worker.

Ergonomics

Ergonomics is the science concerned with adapting the workplace to the worker to achieve optimum efficiency and comfort. It thus has great importance in workstation design, where such basic comfort issues as avoidance of fatigue and backstrain induced by stretching, and proper utilization of a worker's cone of reach are critical. A related issue is height adjustability of chairs or station tabletops to accommodate variations in human dimensions. Other issues are proper lighting level, workstation surface coloration, line of sight, and focus. For years, many industrial engineers have unduly ignored these issues.

In particular, line of sight is a consideration since it has been determined that a straight-on, level field of view is not as comfortable for the eyes on a long-term basis as the case where the worker's eyes are about 12 to 15 degrees above the work. Placement of a scope or CRT screen above or below this angular range leads to eye fatigue.

Regarding lighting level, McMullin notes that just 10 to 15 years ago 200 foot-candles of illumination on the work surface was considered a desirable level for small parts assembly work. Today, it is known that this is too bright and leads to eye fatigue. Nevertheless, many plants still use it. Something in the range of 100 to 150 foot-candles is now considered preferable.

Notes Ergo-Tech's Greendale, certain environmental factors contribute especially to worker error. They include distractions (both visual and audible), the aforementioned inadequate lighting, and uncomfortable surroundings; also, inaccessability of utilities (power, air and vacuum) and improper placement of tools, parts and accessories. The latter encourage unnecessary reaching.

Human communication in a work environment is essential, having a great influence on productivity. However, the unrestricted communication typical of the bench is not desirable. Production seems to increase if workstation operators can talk to co-workers located to either side, but not facing straight ahead.

Parts delivery

The use of workstations does not mean that individuals will necessarily work any faster. However, it is possible to factor out much of the redundant non-productive human motion if thought is given to both intelligent layout and the way in which the workstation interfaces and communicates with the parts delivery system that supplies it with its 'raw materials'.

Studies on the use of older style production equipment like benches and parts-handling containers have shown that as much as 20% of direct labor time for an assembly worker is spent looking for the right component, getting set up, detrashing cardboard and vendor packages, and other non-productive effort. A modern, integrated workstation system, however, means that transfer of material from a general distribution point to the workstation becomes a simple transaction that only takes a short time. A related benefit is that work-in-progress does not have to be carried manually from workstation to workstation, being exposed along the way to risk of possible damage or contamination.

There are two general approaches to delivering parts to the workstation. The simpler and less expensive way is to distribute parts manually using some type of cart. The other, mechanized materials delivery, is substantially more conducive to workstation-operator productivity.

There are several aspects to automatic delivery. With an automated storage and retrieval system (ASRS) a robot (or 'crane') moves along a rail to pick parts out of a storage rack that amounts to a large closet. The robot can deliver parts directly to the worker, provided the workstation is located close to the UHS. Otherwise, a tote box is sent down a transporter to the station.

The carousel, rotating either vertically or horizontally, is somewhat less costly. Vertical

carousels are most useful when there are many kinds of small parts, claims Jeff McMullin. He points out that horizontal carousels, while somewhat less expensive, nevertheless suffer from a drawback seen in some dry cleaners' racks – they can cycle in one direction only, meaning a loss of speed in part recovery.

A transporter is basically an endless belt with push-out mechanisms along its course to divert tote boxes moving on the belt to waiting off-belt roller conveyors. When the workstation operator issues a signal to the dispatcher demanding more parts, a tote box is sent along the transporter belt and is automatically shunted aside at the worker's station. Thus, the operator can set a pace that is comfortable, while not feeling caught up in the drudgery of the old-fashioned asembly line.

Utilities and communications

Workstations are connected to the 'outside world' by a number of utilities. These include such lifelines as power outlets, compressed air and vacuum sources, and communications/computer interfaces.

Power service at a typical electronics assembly workstation is 115V, at 60Hz and 20A (sometimes 230V is also available). Compressed air at 100psi is usually present, with oil-bearing air in use for power screwdrivers, and oil-free air used for parts clean-off. Vacuum is used for solder removal. In special applications, such as wafer processing, a supply of compressed nitrogen may also be present at the workstation.

Electric current demand is fairly high at a production workstation because the station must support the use of such tools as soldering irons, desolderers, and heat guns, as well as a frequently present overhead task-lighting system and a lighted retractible overhead magnifier.

Herman Miller's Ted Venti says his firm's experience has been that networked information systems are more talked about today among their customers than even the subject of programmable automation when the topic of 'the factory of the future' comes up. He points to the great productivity enhancements

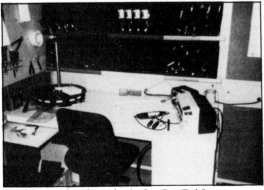

A modular manufacturing workstation from Ergo-Tech Systems

in manufacturing available if the workstation system is integrated electronically into a networked system.

Ergo-Tech's Greendale says that incorporating phone or intercom system wiring into workstations eliminates the need for operators to get up and leave their station to obtain information or materials. He adds that there is increasingly a trend to see customers put monitors in workstations to display all process steps, and some are adopting the use of barcodes to track part assembly progress.

Thwarting ESD

Electrostatic discharge (ESD) causes billions of dollars of expense in damaged devices to electronics manufacturers per year. If an electronic device is not protected from ESD damage at the workstation level and earlier steps, the assembly using the part can fail in the field. By that time, repair cost goes up tremendously. As Ted Venti notes, ESD is today a universal issue in electronic production workstation design. Those companies that cannot offer techniques to avoid or minimise the problem, he claims, will not even be serious contenders.

Workstation design can help to contain this problem with such features as static dissipative and conductive worksurface pads or special antistatic laminate surfacing for the

The West-Flex workstation from Westinghouse

station's desk top, grounded chairs and parts containers, and wrist-strap grounds.

Conductive tote boxes and part carts are other items in the arsenal against ESD. In an ideal scenario, kits or containers of parts that have been protected from ESD in previous parts storage stages are delivered to the workstation in static-protective containers, hung onto grounded rails on the station frame and made accessible to the worker who is grounded by a wrist strap to a single-point ground.

Static dissipative and conductive items should all be grounded to such a single ground point, agrees Mike Christiansen, but it should not be an electrical ground like the third prong hole of a wall socket. These can give the worker a shock if they are momentarily live when grounding out a charge on the casing of a power tool or electrical device. The preferred way, says Christiansen, is to use a separate ground like a water pipe or copper rod sunk into the earth.

Often a conductive floor is a necessary feature for the workstation too, especially if materials are constantly being transferred from one station to the next, or if there is much traffic in and out of a station. Air ionizers can be used to deionize a part or tool. Christiansen adds that many of his customers are adopting the use of a static monitoring system to alert the worker of a break in ground continuity with a visual or audible alarm.

Some available systems

Companies in the market for a workstation system should think in terms of an installation project timespan of four to six months. Before making the purchase, it is essential to have a concrete idea of company production needs and goals, and realistic expectations about the extent to which workstations can help to meet them. The following are some commercially available systems.

Advance Unit of Zero Corp. offers two different product lines: the Benchmaster includes four-legged-style electronic assembly benches, both with and without electrical options. The other line, the Universal Work Station (UWS), includes high production, ergonomically designed space-saving work systems having ESD-controlled components. UWS has an extensive accessory line to support the workstation, such as drawers, cabinets, lighting, chairs and parts bins. It features a slide-rail PCB push-along system, continuously adjustable in many angles ($-10°$ to $+70°$). Advance also offers a conveyor system, their modular Conveyor Line.

Ergo-Tech Systems offers modular manufacturing workstations with delivery of utilities and communications capabilities in mind. A channel which the firm dubbed a 'beltline raceway' houses and provides electrical, vacuum, air and communications outlets. Electrical utility service is based on a three-circuit, six-wire power distribution system. It allows for isolation for CRTs and communications equipment, and can energize up to 39 duplex receptacles (13 per 20A circuit). The system possesses a load-bearing capacity of approximately 400lb, and slotted support panels that accept panel-mounted components are available in a variety of sizes.

Herman Miller Inc. offers a comprehensive line called the Action Factory System that enables the user to integrate material handling and component kitting with the assembly operation. It includes, in addition to the workstation, carts, pallets and parts lockers, and permits a wide range of workstation types to be configured from a small ensemble of parts. The line includes a full range of static-protective items, including conductive

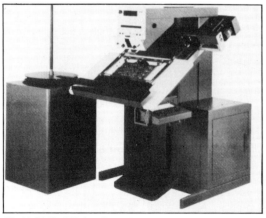

The Model CS-400B from Contact Systems is a component locator with an automatic cut-and-clinch feature

totes and parts containers, rails on which to hang the containers, work surfaces, shelves and a chair. A common ground serves all items.

The line of workstations from Isles Industries Inc. has static-dissipative work surfaces, and an ergonomically comfortable front edge for the worker to lean elbows on. Load-bearing capacity ranges from 500 to 1000lb, depending on whether the station is a cantilevered design or not. All foot rests, shelving, racks, lighting level and parts storage systems are adjustable. A central grounding point is present for ESD dissipative items. Electrical service is rated at 20A. Isles also sells a line of assembly conveyors, including chain rail conveyors to transport PCB pallets, conveyors with antistatic features and some with automatic tote take-off.

Streater THS (for Task Handling System) is a modular system with a $1\frac{1}{4}$in. spaced slotting system on which any number of functional accessories can be added at almost any height and width. Counters, lighting systems and other components are adjustable in the $1\frac{1}{4}$in. increments up and down to be compatible with either sit-down or stand-up operation, as well as with any type of automated handling equipment. Quick-disconnect fittings allow easy assembly. Air and electrical lines are concealed within workstation components. Trays are positionable at any level to suit an operator's arm length, sliding back and forth laterally for ease of use.

Westinghouse Furniture Systems' West-Flex is a flexible production system consisting of modular workstations, accessories, and material handling totes, subcontainers and distribution carts. Workstations are panel-based, with work surfaces, storage shelves and cabinets all adjustable in 1in. height increments on the panel. A patented panel-connector system allows quick rearrangement of workstation configurations to match specific operator tasks. There is an integral six-wire, three-circuit electrified panel present, as well as variable-intensity task lighting and air distribution systems. ESD protective materials are used throughout.

Wilfab Systems' line of workstations was designed from the beginning as production-

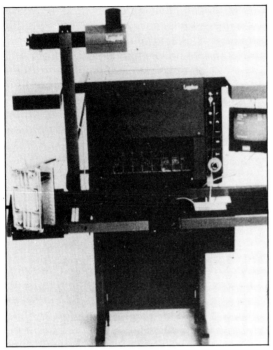

Universal Instrument's Model 4511B Logplace II component locator. A vacuum pen permits the placement of SMDs

oriented products, rather than being office environment stations with scaled-up capabilities. Structural frames are made of 10-gauge steel, shelves of 14-gauge steel and the binder bins are 18-gauge steel. The frame itself has a 2000lb rating (500lb on the work surface and 250lb on each shelf). A patented feature enables this modular system to be reconfigured with only an allen wrench as a tool. Drawers are hung from a rail, not screwed into the panels. A feature known as a Utility Raceway accommodates any number of electrical circuits, as well as telephone lines, compressed air, vacuum and even liquid lines.

Component-locator systems

A highly sophisticated type of workstation, that uses an electronic memory-guided light beam to direct a worker in manually assembling parts on a PCB, has carved out a niche in the workstation market. Known as component-locator systems or assembly directors, these products use a light beam, based on either a halogen, incandescent or laser source,

to indicate to the operator exactly where to insert the component leads.

With this approach, each board type will have its own component location program stored in memory. The programming here can involve a substantial amount of time, especially if lots being run change frequently, but once a board has a program it can be recalled quickly for subsequent runs.

Component-locator systems may include automatic drawer filling and automatic lead cut-and-clinch capabilities. Some versions make parts available at a constant height, so the operator will not have to stretch. The beam of light in some systems is also used to indicate the proper part.

Says Rory Stoner of Royonic Inc., the two greatest benefits of component-locator systems are a significant increase in production output (up to 60% greater than a manual station, he claims), and the banishment of the need for paper documentation or assembly instructions. Parts can be brought to the operator's reach automatically, and the likelihood of making a part-selection mistake is almost eliminated. Also, there is no need to rely on the operator's memory of the population scheme for the board.

Universal Instruments offers three systems in the component-locator field. These are the Logpoint, the Logplace and the Man-U-Sert. The Logpoint and Logplace are similar in design, with the Logpoint being primarily for through-hole application, while Logplace is aimed more at surface mounting. Both products feature microprocessor-controlled halogen light-beam guidance systems. Software allows the user to integrate the machines with a variety of systems, such as a host computer, CAD system or even an in-line repair station. The Man-U-Sert differs from Logpoint essentially in that it has cut-and-clinch.

Ragen Corp.'s Model 3600 computer-aided PCB assembly system incorporates a laser indicator, personal computer, rollaway component delivery system and cut-and-clinch function. The system features fully automatic, self-contained programming and automatic step-and-repeat for the simultaneous assembly of multiple board. The laser indicator can form shaped symbols that can be positioned anywhere on a PCB for indicating component placement. The personal computer is used to create new PCB programs, edit existing programs and to provide direction and control for assembly operations through direct access of program steps. The cut-and-clinch system moves beneath the board and can handle essentially any lead type or configuration.

Royonic Inc. features a magazine kitting system which allows setup of any new job in less than three minutes, including change of all parts, installation of the new kit and change of program. The pre-kittable magazine system allows the user to exchange 96 bins, installed in 12 magazines (eight bins in each magazine). Software allows rapid alterations in a program to accommodate engineering changes, updates and revisions, and it permits bins to be automatically renumbered. The entire program can be reconstructed, and the assembly program can be rotated. Multiple boards can be implemented in the program.

Contact Systems' Model CS-400B features automatic cut-and-clinch to eliminate the need to cut leads after wave soldering. This feature also helps secure components to the board, and prevents component spillage during handling or component pop-up during wave soldering. The CS-400B, designed for operator comfort, has an arm rest; an X-Y table ensures that insertion is always made at the same spot, reducing operator stretching and visual searching. The table is cycled by a switch in the footpad. Automatic part delivery is provided by the rotary bin system, which can be conveniently located alongside the work area. Programming can be done on either the built-in keyboard or on an off-line programming unit.

AUTOMATIC COMPONENT INSERTION/ PLACEMENT SYSTEMS

Efforts to automate PCB asembly ('stuffing') have progressed from hand-formed component leads and manual insertion guided by blueprint, to pre-formed leads and programmable assembly directors (still manual insertion), on up to fully automatic component insertion machines. The present state of the art involves either 'islands' of automation (groups of machines) bridged by sophisticated materials handling, or an array of machines set up for continuous in-line assembly. All types of components are now being sequentially inserted, including axial, radial, DIP, surface-mount (leadless), and odd-shaped components, the latter via special robotics.

Satisfying the user

Buyers of automatic component insertion

systems (and pick-and-place surface-mount systems) are demanding features aimed at maximizing the benefits of their investment.

Connectors and odd-shaped components, both leaded and surface mount, can be placed by this robotic manipulator – the Microplacer from MTI. Robotic systems are also used for inserting components used in low volume instead of tying up a dedicated insertion/placement system

First published in *Electronic Packaging & Production*, August 1984, under the title 'Automatic component insertion/placement systems ready for the automated factory'.
Author: Howard W. Markstein, Western Editor.
© 1984 Reed Publishing (USA) Inc.

Dyna/Pert's supervisor of product management, Tom Nash, says customers expect insertion/placement machines to provide consistent quality output, offer expansion in productive capability, and have data communications for tying into higher-level computer systems.

"Future developments will be in the area of on-line inspection of the assembly process by use of vision systems, as well as more automated methods of inventory loading and use of robotics," says Nash.

Dependability and reliability are also high on the customer's list. The ouput of the component insertion line is governed by the output of the weakest machine on the line. Therefore, reliability and ease of servicing are critical factors.

Amistar's vice president, R.E. Hogan, says that speed, automatic repair of misinserted components, lead protrusion inspection, automatic load/unload, and interfacing with CAD/CAM, are a customer's prime requirements.

One other requirement gaining momentum is that of inserting odd-shaped components. "The goal," says Hogan, "is to provide 100% automatic population of a PCB regardless of the type of components."

In-line PCB assembly is made possible by sequentially conveying boards through various insertion/placement machines. This Panasonic line includes axial- and radial-lead component inserters, DIP inserters, surface-mount placement and a robotic station for odd-shaped components

A modern in-line PCB assembly system is capable of inserting axial and radial leaded components as well as leadless surface-mount components. A robotic station inserts odd-shaped components.

Other features in demand by many customers are, for the short term, versatility and flexibility, and higher feeder counts for the long term. Stan Mayer, product manager at MTI, defines versatility as the ability to handle a broad range of parts and applications. Flexibility is ease of changeover to accommodate different products.

Mayer says customers are asking for very high feeder counts to provide the machine with an inventory of a wide variety of part types. He cites one military application where 270 different component types had to be available to the machine at all times. "Then, no matter what job was run," explains Mayer, "the program was plugged in and the machine could go to work – even though no single board had a requirement for more than 70 different parts."

Benefits of automation

Full automation of the component insertion/placement operation provides the user with benefits such as consistency of product quality and quantity, high throughput and productivity, and automatic inventory tracking. Additional benefits gained from these factors include reduced labor costs, reduced rework, and an improvement in the production process.

Tom Nash says automatic component insertion systems offer a consistent production volume, and by using on-line test and verification systems, allow the user to control the quality of output. "Automated systems are usually justified on this basis," says Nash.

Speed and quality are therefore considered the most important benefits offered by both automatic insertion of leaded components and surface-mount placement systems. Marshal Trayber, manager of TDK's automation systems group, emphasizes that the decision to automate component insertion should not be based upon labor savings – that is a given fact. Rather, the consideration should be how the automated equipment will affect process

Surface-mount technology is expected to diminish the role of leaded insertion systems. High-volume SMD placement is offered by the MCM-II system from Philips International. Widely used by Delco and others, the system can place 32,000 components per hour

Automatic axial-lead component insertion is the most mature form of automatic insertion. This Universal Instruments' system and others offered by a number of suppliers have insertion reliability yields of 99.7%

capability. The buyer should ask: "How will automation improve my production process and product quality?"

Recent improvements

Automated board handling, a wider range of insertable components and the increased availability of surface-mount pick-and-place systems represent the most significant recent advances in PCB assembly technology. Surface-mount systems are now offered over a broad price range and performance capability, from the large, high-volume systems costing from $400,000 to $850,000 depend-

ing upon options, to medium-volume systems, and on down to lower-volume units such as the Mamiya (Osawa), priced at under $17,000 and offered by a number of US distributors.

Some of the larger surface-mount systems have been improved to provide a recycling feature whereby boards with missing chips are automatically returned to the pick-and-place station. Most systems have also been expanded to accommodate a wider range of components, and some now incorporate an optical system to automatically correct for board misregistration.

Ken Yokozuka, product manager for Panasonic, says the most significant advance in surface-mount has been the increased availability of equipment. Panasonic alone now

This automatic axial-lead inserter, the AI-6448 from Amistar, has random access to 64 reels and 48 cartridges

offers 13 different pick-and-place models, in addition to auxiliary systems for solder paste, adhesive and reflow.

Leaded systems

Leaded automatic component insertion systems, such as axial, radial and DIP, represent a more mature technology. Axial-lead insertion is really comprised of a family of machines, such as reel packaging systems, sequencing systems (where applicable), component verifiers and the component insertion machine.

Bob Groening director of product marketing at Universal Instruments, says that axial-lead component inserters offer the highest

Automatic radial-lead component inserter from Universal Instruments includes processing radial-lead capacitors and TO-92 transistors

level of reliability and machine utilization of all other insertion/placement systems. Insertion reliability yields of 99.7% are common. Today's machines can insert 30,000 components per hour and feature high-speed rotary indexing, automatic board handling and simple programming. The newer machines will incorporate vision systems, barcode reading and interfaces to CAD/CAM.

Axial-lead inserters are now available that feature random, rather than sequenced, insertion. The components are selected randomly from reels right at the inserter, eliminating the need for a separate component-sequencing machine. This feature also provides repair capability by allowing a component type to be reinserted automatically. The penalty of a sequenceless machine is reduced insertion rate; however, rates of 9000 components per hour per head are achievable.

Automatic insertion of radial-leaded components is complicated by varied body sizes and lead diameters. These problems were overcome by handling all radial-lead components by the leads rather than the body, and by reel taping the leads at standardized spacing. Some components have been difficult to handle. For example, the TO-92 transistor with its three leads and 0.014in. lead diameter was not capable of being automatically inserted until a few years ago.

Automatic DIP inserters are designed to

This high-volume automatic radial-lead inserter from TDK can handle up to 40 different component types and features magazine load/unload

DIP inserters have evolved into high performance, high-feeder-count systems capable of handling a variety of DIPs including those on centers of 0.300, 0.400 and 0.600in. Dyna Pert's Model HPDI can process 4250 DIPs per hour

handle DIP lead row spacings of 0.300in., 0.400in. and 0.600in. The newer systems can also insert DIP sockets, DIPs into sockets, and mount decoupling capacitors.

John Pompea, president of Contact Systems, says his company has resisted the trend to offer machines handling the full DIP range. He claims that 90% of the DIPs used are 0.300in. Therefore, a simple 0.300in. DIP inserter can fill the needs of first-time users of automatic DIP insertion.

Odd-shaped components

Unlike the sophisticated 'working' robots that can be programmed for a variety of tasks, such as those developed by Seiko, Intelledex and IBM, robotics for component insertion and placement are highly specialized, although widely varied in design. Robotic insertion/placement systems can automatically insert headers, connectors, trimmer capacitors, potentiometers, ceramic filters, DIP switches, crystals, SIPs, LEDs, etc.

Robotic stations for odd-shaped component insertion do not generally offer higher speed than manual insertion. Cycle times are usually between 2 and 6 seconds. The major benefit is that of increased quality by providing consistency to the insertion operation.

Robotic assembly stations are also used for

'Cells' of automation consisting of an axial-lead inserter, radial-lead inserter/sequencer and a DIP inserter are tied together by a system of automatic guided vehicles, as the Universal Instruments' system shown here at a recent trade show

inserting conventional parts when requirements are for low volume. For example, a board having 100 axial- and radial-lead components and only one or two DIPs would use a robotic station for inserting the DIPs rather than tie up a dedicated DIP inserter.

Factory automation

Two approaches to factory automation involving component insertion and PCB assembly are in-line systems using the production-line concept where the output of one machine is a direct input to the next, or separate 'islands' or 'cells' of automation where the output from each cell can be programmed as to destination.

Jack Stellhorn, product manager for factory automation systems at Universal Instruments, says that the individual cells can be linked by conveyors or automatic guided vehicles (AGV) for transporting magazines

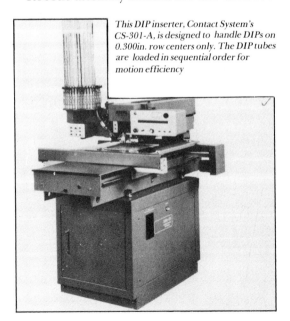

This DIP inserter, Contact System's CS-301-A, is designed to handle DIPs on 0.300in. row centers only. The DIP tubes are loaded in sequential order for motion efficiency

Robotics are employed for inserting odd-shaped components. This one is the Model 6511A from Universal Instruments, designed for electronic assembly operations

containing PCBs. After the boards pass through the cells for inserting axial, radial, DIP and surface-mount components, the boards can be transported to a conveyor where robotic systems and manual assembly stations insert the non-standard components.

After assembly, the PCBs are routed through an automatic vision system to detect missing components or leads not protruding from the underside of the board. The repair station precedes a computer-controlled wave soldering and cleaning cell.

This type of factory automation is controlled by a process host computer linked to a local process control host tied into the machine controllers through networking. The process host provides management data, downloads programs, controls work in process, schedules AGVs, and provides the basis for board tracking.

"This modular approach to factory automation allows users to start with basic machines without board handling," explains Stellhorn. "As production requirements grows, the need to provide more control over work in process increases, then the basic system can be easily transformed into an automated system."

According to Dyna/Pert, boards can be delivered to the production cells via magazines that can be adjusted for varied board sizes. The magazines hold from 20 to 60 boards each. Two approaches to board and machine loading are presently used. The first involves unloading the boards from one magazine at one end of the cell and loading into another at the other end of the cell. The second approach involves unloading and loading back into the same magazine at the same end of the cell. Each board in the second method is loaded back into the same slot from which it was retrieved, thereby maintaining work-in-process inventory control.

Are today's automatic component insertion/placement systems too sophisticated and costly to be affordable to any but the largest of companies? Apparently not. There are degrees of sophistication available, depending upon the user's requirements and production volumes. In essence, a broad mix of equipment is available to satisfy both the small and large user. The goal of most equipment suppliers is to offer automatic systems capable of being utilized by the widest range of companies and applications.

Suppliers say automatic component insertion and placement is not expensive when related to the benefits it provides. It is estimated that the payback period for an automated axial-lead component inserter is about one year. Suppliers also point out that since computers are expected to dominate production operations of the future, automatic component insertion systems will eventually dominate assembly operations.

SMD versus leaded

The widening use of surface-mount devices (SMD) is expected to eventually diminish the role of leaded component insertion equipment. Universal Instruments believes that any significant move toward SMD will take time, except for companies like Motorola, General Electric and Delco because of their need to reduce circuit size. Universal says that factors such as price of components, new capital equipment requirements, new process training curve and unknown long-term reliability effects of SMD connections will delay any swing toward surface mount until well into the second half of this decade.

Dyna/Pert views surface mount as an evolutionary technology that will, at its initial stages, act as a supplement to traditional automatic component insertion. In the future, however, it will become a competitive assembly technology.

Amistar says wide use of SMD will take a number of years because the components are still not cost-effective nor widely available in all types. PCBs already designed for standard components will also remain in the product lines of most companies for many years.

Fuji believes the present driving force toward surface mount is decreased circuit size. However, the driving force will soon be the low cost of components. As demand and production quantities increase, the cost of surface-mount components will significantly decrease. Fuji is developing a pick-and-place system featuring automatic optical alignment of chip-to-pad and integral laser soldering.

TDK says that SMD will definitely replace

axial-leaded components, while MTI predicts that leaded components will diminish but always have their place for certain applications. For example, SMD will make inroads in low-power applications but power supplies are expected to still use leaded components.

Contact Systems contends that surface mount will gain momentum when semiconductor companies begin producing chip carriers and small outline (SO) components in high volume. Chip resistors and capacitors alone will not swing the industry.

AUTOMATED PIN INSERTION

One of the most time-consuming operations in PCB assembly is that of actually mounting the components. For small-volume operations this is accomplished without difficulty, since traditional manual methods of insertion and soldering are very cost effective. However, when production reaches a certain level it is virtually mandatory for the manufacturer to carefully consider the realizable gains of automated processing.

The terminal pin, used to link double-sided boards and as probe test points, wire-wrapping posts, and connector parts, is a manually inserted PCB component that gives rise to many potential problems. As a loose piece it is small, awkward to handle and requires the use of a tool to fix it in place. In the header or wafer form, stock and procurement problems may arise when the number of rows and positions vary. Additionally, the plastic body may cause mounting and locating problems prior to, as well as during and after, the soldering process, thus necessitating additional operations. These include tape mounting and solder rework due to floating of the plastic housing. When using the pin in the bandolier form, the part is easier to stock and handle, but now the user is faced with either discarding or recycling the bandolier carrier.

In either case, further time and cost expenditures are added to the process. The terminal pin has, therefore, been subject to much development, both as a component and as part of the automated assembly process. In particular, the star form (upset) feature on the pin to ensure a good fit, and the

continuous feed of pre-stamped pins, have been vital developments on this component.

Assessments of PCB production assemblies often show that pin insertion is a prime target for cost benefits realized by introducing automation. In one recent study it was found that on a good day a skilled operator could insert an average of 600 pins into a PCB every hour. If production requirements demanded 150 boards per week, each containing 300 pins as a minimum, then two skilled operators would be able to reach this total in a five day, eight hours per day week.

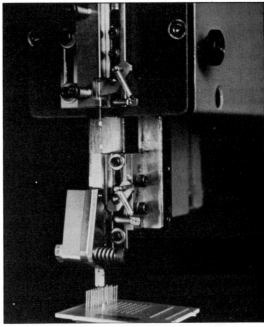

The Autopin II insertion tool module from Autosplice is shown inserting terminal pins. By coupling this machine module with a CNC positioning system, very high pin insertion rates can be achieved – up to 13,500 pins per hour

First published in *Electronic Packaging & Production*, December 1984, under the title 'Automated terminal pin insertion offers high throughput'.
Author: Brian Corner, Harwin Engineers SA.

However, output would only have to fall from 600 to 560 pin insertions per hour per operator for the target to be missed. The study results found that the average figure was, more often than not, only 450 pins per hour – in which case production falls to 120 boards per week. Either the operators must put in overtime or another operator must be brought in to help from time to time if production goals are to be met.

Since recent techniques in pin development and circuit board assembly have been concerned with higher and higher insertions, the current trend is to couple computer numerical controlled (CNC) positioning systems with insertion machines. This provides an increase in both production throughput and total cost effectiveness.

The demand for speed along with operator-friendly controls is being met by these CNC systems, which are well-suited to the task. They do not require technically trained personnel and at the same time are more efficient than skill-requiring operations and semi-automatic methods.

By automating the process, the two skilled operators can be replaced by one unskilled operator, freeing them to do other work. The unskilled operator can spend some time, in this application at least, minding other automatic machines as well as the pin-insertion machine. Thus, the continuous pre-stamped pin and automated insertion head with either the semi-automatic or fully automatic work-

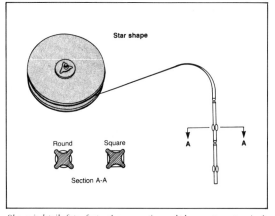

Shown is detail of star feature's cross-section and placement on a terminal pin for correct application sizing

positioning system is the optimum for automated assembly today.

Cost effectiveness

The continuous approach, together with the star-shaped pre-stamped terminal pin, offers reliability and repeatability while incurring low in-place setup costs, thus allowing increased cost and time efficiency.

Whereas the low- and medium-volume users may not need the sophistication of CNC machinery, it is a must for the high-volume user. Let us consider another case – this time a manufacturer whose requirements are 800 boards per week, each board having 250 pins. If manual methods were used here, 10 operators would be required to achieve the weekly target, assuming each could maintain an insertion rate of 20,000 pins per week (500 pins per hour). No user at this throughput level would opt for manual insertion, but using conventional bowl feed or semi-automatic pin-insertion machines it might be possible to achieve 1500 pins per hour per machine. This would require some four machines plus the same number of semiskilled operators.

With CNC systems of the type discussed, however, only one machine and one unskilled operator would be required. Again, the potential for cost savings is obvious. CNC machines have been expressly designed to meet the needs of the medium- to high-volume user, i.e. those who typically use two million pins or upwards per year. These pins are both relatively low in cost and highly cost effective, as the examples have shown. Reliability, ease of use, ease of maintenance, and flexibility in operation are some of the features of these machines. It has also been shown that medium- and low-volume manufacturers can benefit from this kind of equipment, particularly if expansion is planned at a later date.

However, these users would be wise to consider present and future requirements carefully to see if more modest machinery (which could be amortized over, perhaps, three years) would be a sounder, short-term investment. In that case, CNC machines

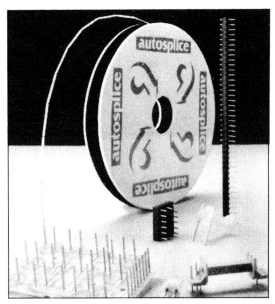

With Autosplice's continuous preformed pin approach, one reel of pins may hold up to 70,000 pieces, depending on pin size. A typical reel of 0.025in.² by 0.500in. long pins will yield 50,000 pieces

would be purchased later when fully cost justified. Present and future labor costs, inflation, cash flow, and many other factors have to be taken into account in this type of analysis.

Continuous preformed terminal pins

One of the many new developments in terminal pins is the continuous stamping process patented by the Autosplice Co., Division of General Staple Co. Inc. In this approach, continuous wire is stamped in such a manner that an hourglass shape is formed into it. The depression or notch made into a square wire has a pyramid configuration, while for round wire it is a hemisphere, with the notch accurately placed according to the pin length that is required. The continuous strip of pins is then layer-wound onto a supply reel.

Once the notch is formed, a nook is available for pickup and advancement of the material in a feeding mechanism. The insertion head is constructed for precise feeding and cutting of the pin material. The pin is cut at the center of the hourglass notch, producing a pyramidal contour on square pin tips

and hemispherical contour on round pin tips.

In operation, advancing and inserting of one pin pitch is accomplished in a 180ms insertion cycle. When the head is coupled with a CNC work-positioning package, automated insertion outputs of 13,500 pins per hour are attained.

As a further development on the pre-stamped continuous pin supply, a second simultaneous operation has been added to the stamping process. This was the introduction of the star form, or upset, to produce a low-stress press fit with the circuit board. The star is an integral part of today's high-speed automated process and offers greater cost savings and advantages than were available prior to its development.

The stamped star process produces a cross-sectional area which is greater than the pin diagonal or diameter – traditionally, about 30% greater for round pins and 50–70% greater for square pins, thus yielding a star size greater than the core size of the pin. The star is located along the pin length so that it is positioned precisely into the circuit board when inserted, leaving exact above- and below-board extensions for a given application. The feature is 0.060in. long and is inserted by machine into G-10 glass-epoxy board, typically, with plated-through holes. Holes are 0.006in. smaller than the star size, with the interference between the plated-through hole and the star producing a retention force of approximately 12lb.

The star/hole interference is advantageous since it permits the 'square peg to go into the round hole', so to speak, and it does so without seriously altering the hole geometry. In producing an equivalent retention with a standard square pin, a broaching effect upon insertion would alter the hole, producing a large-radius square form where the hole once was. Lands or pads and through-hole plating would possibly be destroyed. For round pins without a star, close tolerances on hole size are required, even when the retention force necessary to stay within specification is not as high.

Upon insertion of a sqare or round pin with star, hole geometry is minimally changed. Observation would show four areas – the edges of the star – that actually dig into the inside walls of the plated-through hole. Furthermore, since the core of the pin is anywhere from 30–70% smaller than the star, depending on type, large clearance spaces between the hole walls and the face of the pin occur. These allow free and unhampered reflow and wicking of solder.

AUTOMATED PLACEMENT TECHNIQUES

Large-scale manufacturers of hybrids stand to gain from recent advances in pick-and-place machinery introduced by Philips of The Netherlands. Based on technology developed for automatic placement of SMDs (surface-mount devices) on PCBs, the new system, the MCM IV, features a software-controlled simultaneous or sequential placement mode of operation. Other notable features include a multipurpose placement head and mechanically pressing a component onto its footprint without damaging the substrate.

The new system is specifically designed to work with standard 2 × 2in. ceramic substrates. The first machine, delivered to Philips Audio Division in Louvain, Belgium in April 1984, working 80 hours a week is turning out complete hybrids at the rate of 1.2 million units a year. General Motors' Delco Division, which assembles several million car radios a year, is just one of the US-based customers that has recently ordered the new system.

Special placement head

The key to system performance is a software-controlled placement head based on the surface-mount model. Standard versions of the machine can have either one or two placement heads, and in one machine program each can place as many as 992 components. Versions with more than two heads can be supplied on special order and, currently, systems with up to seven heads are being developed for a customer.

The central feature of each placement head is a movable beam carrying 32 pick-and-place pipettes that can pick up 32 components

Future improvements include a pattern recognition system to detect defective substrates. The new module will fit between the substrate loading module at left and the placement module seen at right. The tape reel magazine accepts standard 8mm tapes. The second placement module is available to accommodate 8, 12, 16 or 24mm tapes

simultaneously and place them on substrates either simultaneously or sequentially, depending on the substrate layout and the corresponding control software. Accepting tape-packaged components fed from supply reels (also controlled by software) the pipettes transport them to their programmed positions on the substrates, rotate them to required alignments, then press them down into solder-paste footprints screened in advance onto the substrates.

Four standard 2 × 2in. substrates at a time fit under the pipette beam, so that each substrate is served by a group of eight pipettes, each drawing from its own supply of components from a 32-reel magazine. The beam moves under servo control so that any one pipette can place its component anywhere on the substrate with an accuracy of 0.2mm in both X and Y directions. Incidentally, the

First published in *Electronic Packaging & Production*, May 1985, under the title 'Assembly machine for hybrids combines automated placement techniques'. Author: Nikita Andreiev, Editor.
© 1985 Reed Publishing (USA) Inc.

system imposes no artificial grid restrictions of its own.

The pipettes are spaced at a pitch of 10mm and the walking beam substrate transport mechanism can be programmed to advance the substrates 80, 160 and 320mm at a time. For substrates having only one circuit each, the transport mechanism is programmed to advance 80mm at a time so that successive groups of pipettes, working sequentially, can place up to 32 different types of components.

On substrates supporting two identical circuits, pipettes working in pairs can place up to 16 component types. On four circuit substrates, pipettes working in fours can place up to eight component types. The transport mechanism would then be programmed to advance 320mm at a step so that all four substrates would move as a unit into and out of the placement head working area.

In one machine cycle the pipettes can be charged up to 31 times. After that, all four of the substrates operated on in that cycle must give way to four new ones. If two placement heads are available, the transport mechanism automatically positions the first four under the second head.

To minimize unnecessary constraints on substrate layout, any component can be rotated from its pick-up alignment by 90°, 180° or 270°, with an accuracy of 2°. Since selective rotation of individual pipettes would be unduly complicated mechanically, all 32 pipettes are rotated together by the amount required for a specific component, but only those engaged in placing the component are lowered to the substrate.

The pipettes are designed to pick up components presented transversely in their tape blisters. The arrangement is normal orientation for most small components. Components available in the SOT23 package, however, are packaged lengthwise to the tape. The MCM IV offers two options in solving that problem. The first consists of replacing standard pipettes by ones that are prerotated; the second solution involves rotating standard pipettes by 90°, then selectively activating them in a second pick-up phase. Note, however, that this adds an extra operation and thus slows down the assembly cycle.

Multifunction pipettes

The basic design of the pick-and-place pipette combines a central tube with an external mechanical chuck or jaw. The function of the tube is three-fold: to hold the components by a vacuum during the last phase of placement action; to monitor the component presence throughout the placement cycle; and to press the component firmly onto its solder-paste footprint on the substrate. The chuck has equally important functions: it aligns the component with the X and Y axes of the machine regardless of the tolerance in the original angle of component presentation, and exercises a positive, mechanical grip during transport of the component from the pick-up to the placement position.

A pin rising beneath the tape lifts the component out of its blister and presents it to the central tube of the pipette, which at this stage is pulling a vacuum. The pin continues to support the component until the jaws of the chuck close on it. The pin then retracts and the pipette transports the component to its placement position on the substrate. Microphones mounted in turbulence chambers in each of the 32 vacuum lines monitor the presence of the components on the pipettes. If the component is not picked up on the first

In a close-up of substrates and placement pipettes notice the selective pipette activation in the third pipette from the right. Note the component reels in left foreground

attempt, the microphone in that vacuum line signals the corresponding tape feed to advance a new component into position and the pick-up is attempted again. If after two such attempts the pipette still fails to pick up the component, the machine cycle stops and the control panel alerts the operator.

An important feature of the pipettes is their action during the last phase of component placement. The system is designed to place components on solder-paste footprints whose limited adhesion properties must suffice to hold the component securely in place until the solder paste is reflowed. To assure satisfactory adhesion, yet avoid damaging the substrate, the system presses the component onto the footprint with a precise pulse of air.

Automation provisions

In addition to the placement and transport modules, tape reel magazines and control electronics, the machine includes loading and unloading modules, under software control, to ease incorporation into automatic assembly lines. The loading module, with a capacity of 600 substrates stored in cassettes, feeds bare substrates to the transport modules in step with the action of the walking beam. In the event of feeding errors, sensors incorporated in the transport module detect vacant positions and signal the control electronics to adjust the placement progress accordingly. The machine, including the control module, occupies less than 30ft^2 of floor space.

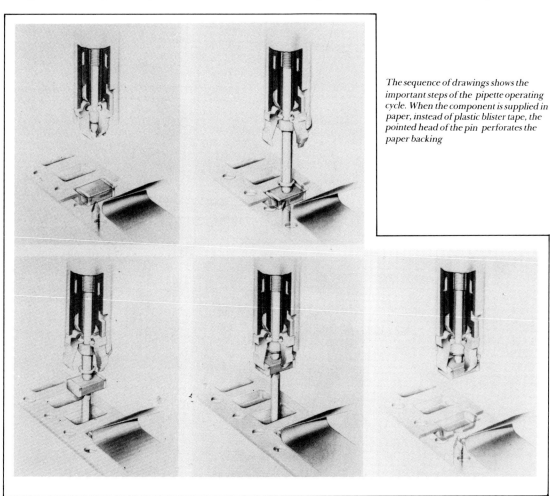

The sequence of drawings shows the important steps of the pipette operating cycle. When the component is supplied in paper, instead of plastic blister tape, the pointed head of the pin perforates the paper backing

ROBOTIC ASSEMBLY

In the assembly of discrete components prior to robotic assembly, two approaches were used. In high-volume applications, a circuit board carried in a pallet passed under a series of heads, each inserting a specific component at a specific location. Product change involved relocation to new positions. Yields are high, but downtime is extensive when going from one model to another.

On lower volumes, components are placed on reels in a specific sequence. The circuit board or boards are then moved under the fixed position insertion head by an X-Y coordinate positioning table.

Robots have the advantage of flexibility both in acquiring the components and in their insertion. Despite this flexibility, their utilization remains modest. This section examines the use of robots and also their limitations.

ASSEMBLY CHALLENGES

In recent years, electronics assembly has been touted as one of the major market and application areas for robots. A key argument has been the flexibility robots offer for placing or inserting odd-form components on PCBs. Close to 10 billion odd-forms will be assembled in 1990, according to market research.

Since 1980, the market for auto-assembly machines for PCB population has grown at over 20% each year, yet very few robots were actually installed on the assembly floor. Many reasons may explain the disappointing market penetration of robots for electronics assembly: over-optimistic projections, a 'high-tech US syndrome', and the complexity of electronics assembly processes are three issues whose understanding may help to boost the use of robots in this area.

Given a placement rate of 1000 odd-forms per hour, the total US market would support nearly 6500 flexible robots by the end of the decade. The market for odd-forms should not be expected to grow substantially, since the largest US companies have reduced their use of these parts in total component mixes to 6 or 7% due to board redesign. The introduction of new packages (e.g. surface-mount devices) is reducing the use of odd-forms even more. Low cost, dedicated robots could then enjoy a larger market, especially for DIP, SIP and selected SMDs.

The Japanese perspective

Contrasting approaches taken by Japan and the USA are the outcome of the differing views. In Japan, electronics assembly robots have been dedicated to the assembly of one type of package per robot, resulting in a low-cost machine and reduced tool-up. This has allowed the substitution of one robot for one manual assembler with a payback within 12 months. An approach that should not be overlooked is the tooling-up of dedicated, low-cost robots with a modular design. The Japanese have utilized this economical structure with success.

Deploying robots for dedicated assembly tasks has a two-fold impact: the robot and controller can cost as little as $15,000 to $20,000, and a fully tooled workcell can cost less than $50,000. Although robots are five to 10 times slower than dedicated auto-inserters, a robot-based assembler becomes competitive with higher priced machines in a low-volume environment. The potential market for electronics assembly robots is not restricted to odd-forms. It encompasses any environment in which one or two million per year of any component package are assembled.

Companies with in-house component-manufacturing capabilities can standardize packages, increasing the chances for successful introduction of robots: the same machine will insert different component part numbers packaged with the same physical dimensions. Assembly lines with several dozen dedicated robots are a feasible alternative for medium-volume or low-mix environments.

In the USA, robots are deemed to be flexible, not just pick-and-place mechanisms. This implies complex sensors, four or more axes, tight accuracy performances and ease of programming, which in turn create a need for high-level robot languages and associated computer power. The result is an expensive robot that must face two challenges: compet-

Four-axis articulated robot shown inserting a 40-pin DIP (Courtesy of Hitachi)

First published in *Electronic Packaging & Production*, March 1986, under the title 'Assembly challenges robots'.
Author: Charles-Henri Mangin, Ceeris International Inc.

download to individual microprocessors.

Users are realizing that assembly robots reach their full potential for flexibility when associated with a high-level off-line programming language. Several US vendors are developing the languages and computerized capabilities which flexibility implies. This trend is expected to be reinforced since more and more work center control languages are being developed.

Other considerations may affect the performance of a robot-based assembly cell. Feeding and component preparations required for electronics assembly are major concerns to users. Availability of a large number of different part numbers is key to the flexible use of a robot; a typical operation uses over 5000 part numbers. In many cases, component leads require straightening prior to insertion. If component preparation is performed off-line, it is possible to address a wider range of components with the robot and increase its flexibility. Axial, radial, DIP, odd-form and SMDs each require a separate forming device, meaning more than one work head will be needed. For this task, a tool-changing mechanism can be employed.

The component feeding technique is a concern due to the work area required to pick up components and the need to maximize the number of components available, increasing flexibility. The number of inputs is maximized when tape, magazine or vibratory methods are utilized, since these address most component sizes, styles or configurations. Limitation to one or more of these feeding methods will allow a reduced feeding area, benefiting cycle rates.

Safety in design also plays a role in the design of an assembly robot. A gantry configuration due to the constrained and well-defined work envelope has some advantage over jointed arms. This consideration must extend to the whole machine. Addressing machine requirements like safety will allow the application of electronic assembly robots to fulfil their market potential.

Electronic assembly robot-based machines can be constructed today using state-of-the-art US technology. Should US know-how not be incorporated into the design of turnkey assembly machines, the integration of truly flexible robots into electronics assembly machines and systems will continue to lag. Systems houses play a key role in designing and implementing unique products. Robot vendors are expected to enter the market with fully configured robot-based assembly machines. Robotic assembly, if given the opportunity, can prove to be inexpensive and reliable. But if the robot is to compete and win, it simply must be ready.

ROBOTS IN THE ELECTRICAL AND ELECTRONICS INDUSTRY

Although robots can handle a wide variety of tasks, they are still in their infancy in the electrical industry. The use of robots is still not being pushed hard enough in all sectors. There are various reasons for this:

- Uncertainties as to the economy of robots.
- A lack of robots with the accuracy, speed, and system interfaces that many applications demand.
- A user training backlog.

- A lack of experience of applications on the part of robot manufacturers supplying the electrical industry.
- Public statements of questionable authority as to the number of people made redundant by robots.
- Insufficient efforts to make production planners aware that the introduction of robots is not inconsistent with generally accepted social principles.
- A tendency, which dates back to the dawn of robotics, to ascribe human attributes to the machine; this is misleading because it

First published in *Assembly Automation*, February 1985.
Author: Gerhard Börnecke, Siemens AG.
© 1985 IFS Publications.

ing with existing auto-assembly machines, and tooling up so that it might perform its function on the assembly floor.

Axial, DIP, radial and SMC auto-assemblers are high-speed, low-cost devices. They possess a hidden level of flexibility, since they can be associated with an in-line or off-line sequencer allowing typically 200 different part numbers on line. With setup times between five and 30 minutes, these machines can easily be operated in an environment with batches as low as 10–20 boards per lot, corresponding to 5000 setups per year, which is a level of flexibility required for a just-in-time assembly operation typical of computer, instrumentation or avionics manufacturing with volumes as low as 150,000 PCBs/year.

Dedicated auto-assemblers are also becoming part of integrated systems. They are associated with automated material-handling equipment such as AGVs or conveyors. The only constraint for minimizing setup is to standardize the width of the boards or the use of snap-offs or fixtures.

Tool-up

Flexible robots on the assembly floor can perform their tasks only if they are properly tooled-up. Users can be ill-equipped to perform the task and reluctant to acquire a piece of equipment which cannot operate immediately to specification on the shop-floor. Robot manufacturers can expect no mercy from those in charge of running electronic assembly operations. These clients will only accept robots that are cost-justified and delivered as part of a fully operational system: a robot-based electronics assembly machine. For such a machine to meet the user's needs, a set of precise specifications is required for the robot itself, the component feeding mechanism and other ancillary equipment.

Specification for the future

The electronic assembly market in 1990 will be demanding. It is estimated that a four-axis robot will cost only $20,000. At this price, a fully automated and flexible robot-based machine will be price-competitive with semi-automatic and low-volume insertion and

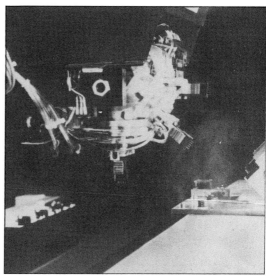

Fully tooled-up five-axis robot system capable of handling four different packages (Courtesy of Universal)

placement machines on the basis of a 12- to 18-month payback.

Basically, the work area of the machine of 1990 will be the same as it is now. Insertion rates are certainly expected to increase from an average of 1200 an hour to over 2000 an hour. Hole tolerances, lead diameters and straighteners, as well as positioning of the component in the gripper, will have an impact on the cycle rate. Insertion rate assumes correct placement on the first try, and that dispensing of adhesive for SMD operations is done simultaneously or at a separate station.

Repeatability will have to be improved, due to the shrinking geometry of devices. End-effectors, which today are hard-tooled, with chuck and turret, will have intelligent grippers with feedback capabilities. Some currently available equipment has a limited ability in this area. Although it is not yet firmly established, machine vision could be implemented in component selection and verification processes. At present, robot accuracies do not require a vision system mounted on the manipulator arm; proper feeding and board location will suffice.

Computer control, currently autonomous and decentralized with microprocessor and teach box being the norm, will acquire a more centralized form: supervisory computers will

gives rise to exaggerated expectations of robot performance.

In the past, when computers were first used for routine activities which human beings had been performing, no one thought of imitating humans. It was obvious that the human brain's faculty of associative thinking could not be duplicated. Yet in certain areas of robotics and related developments in sensors, it appears that forced attempts are being made to emulate human attributes.

In other words the robot is just as far from delving into a disorderly heap of parts as the computer is from synthesizing the human faculty of associative thinking.

By considering several aspects of robotics related to the electrical industry a number of points can be made which show how robotization can be accelerated. The subjects discussed are:

- Product variety.
- Material flow and data flow.
- Flexibility in production.

- Value added.
- Quality.
- Cost effectiveness of robots.

Product variety

Classical automation was concerned with production processes in which a limited number of activities were repeated frequently and items were manufactured in large quantities. It was against this background that numerous highly automated production lines came into being.

Very different approaches to automation and process control were taken in microelectronics. In the automation of test procedures, data processing played a decisive part.

In the electrical industry, however, there are many products which are not manufactured in large quantities, but in innumerable variants instead. These variants often require as much expenditure as completely new products. For this reason, recourse is often made to manual techniques.

The right solution to this problem of

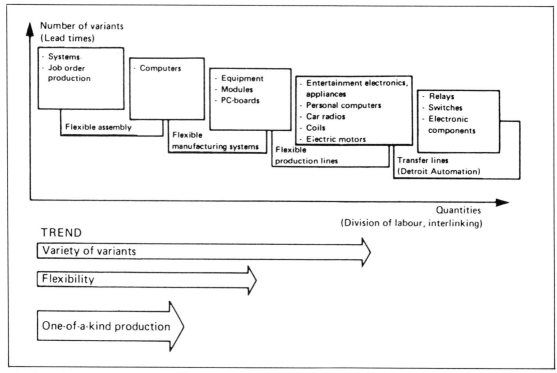

The advance of product variety and flexible automation into areas of high-volume production

product variants, which more often than not have to be manufactured manually, is substantial improvement of the flexibility of production. And here the robot can make a considerable contribution.

Material flow and data flow

In the electrical industry, the interlinking of production processes is gaining ground on a wide front. This is largely explained by the extremely short lead-times demanded by more and more manufacturers.

But interlinking is also important because it is a precondition of robotization. It makes sure that the right material is available in the right quantity at the right time and the right place and – what is particularly important and so often ignored – in the right position.

As any user of a computer knows, order must be maintained in the flow of data and the data itself must be available in a form suitable for the machine. The user of robots must do the same with his material. If flexible elements of production, and that includes robots in particular, are to be interlinked, both the material flow and the data flow supporting the production process must be automated.

Control of such production processes and the supply of supporting data generally take place within a four-level computer hierarchy.

This comprises, from top to bottom, a host computer with master data on planning and production, production control computers, production cell computers and computers built into the machinery, robot controls, and so on.

Many of today's robot controls are designed for autonomous operation only, i.e. they are still not equipped with standard interfaces via which they could be supplied with ready-made programs from a central program source. But this kind of data flow is essential, and the data must be supplied in absolute coordinates if adjustments are to be made on the spot without correction of control programs stored at the central location.

There must also be a facility for reprogramming from the production control computer when a different product variant is to be manufactured. Robot simulation with subse-

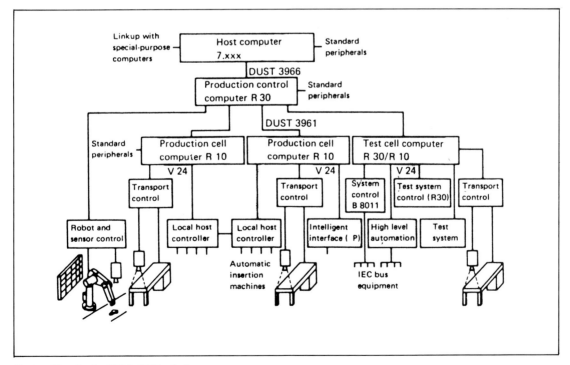

Computer hierarchy of an FMS for PCB production

quent program generation implies that robot controls likewise possess these capabilities.

The smaller the batch, the larger the quantity of data supporting the production process. In one-of-a-kind production, a data supply strategy is usually needed if the various computer levels are to be used with maximum economy.

Flexibility in production

The greatest obstacle to flexible manufacturing is thinking in terms of large batches, in terms of economies of scale which played an important part in automation in the past.

The term flexibility means the ability to cope with a lot size of one economically using the same production engineering and O&M techniques as for large batches.

Thanks to robots, this is no longer science fiction. Production of car radios at Blaupunkt is a prime example of this. Although Blaupunkt radios are manufactured in large

quantities, almost every single model requires a special housing because of differences in the cassette drives, controls, PCBs, and so on. The combined efforts of production and development engineers resulted in a system housing design which permits all variants to be produced on the same assembly line, even in lot sizes of one.

Commercially available robots and handling system components were used for this. Similar assembly systems are also being introduced at Siemens.

The essence of the robot's flexibility is its programmability. But in robot applications, programmability is only effective where very different activities follow one another and full advantage is taken of the robot's many degrees of freedom. This should be borne in mind when introducing robots; their appeal lies not in their speed, but in their flexibility.

Value added

Mere handling of a workpiece does not substantially increase the value of a product, because the product is not altered in any way. Yet most robots currently in service in the

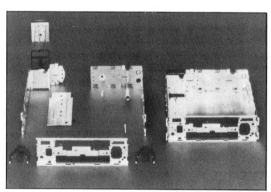

Car radio system housing

Car radio housing assembly

Television tube assembly line

electrical industry are used for handling. In this context, however, handling frequently involves far more than just loading and unloading operations: a number of interrogation processes are performed in parallel with it.

In most of these robot applications there is an orderly, defined flow of workpieces. In a smaller number of applications, the order of handling is generated in the robot's work area or by the robot itself.

Even in these more complicated applications, the value added is inconsiderable; production engineers do, of course, learn something from these handling robots. But robots should be used where they can make a substantial contribution to value added.

Quality

Creating quality in a product represents a substantial contribution to value added and an obvious challenge to the powers of the robot. Every instance of automation and hence every robot application that contributes to value added imply an improvement of quality. As Henry Ford pointed out in his memoirs: mechanization not only reduces the price to a fraction, but infinitely improves the product itself.

Areas of production in which quality problems exist should therefore be examined. It will be found that these are especially suitable for automation or for robotization in particular. High quality in manufacturing is in turn essential to the implementation of automated, high-speed, interlinked systems.

Cost effectiveness of robots

Profitability studies of automation projects which are largely based on comparing and contrasting the performance of human labor with that of the machine are not sufficient for a realistic appraisal. Wherever workshop inventories have an appreciable effect on costs, one includes potential reductions in lead-time with their calculated interest earnings in the economics appraisal.

As automation projects in general, and the application of robots in particular, lead to improved quality, the cost savings resulting from reduced after-sales warranty obligations are included in the profitability study.

The cost effectiveness of industrial robots can also be indirectly increased by their

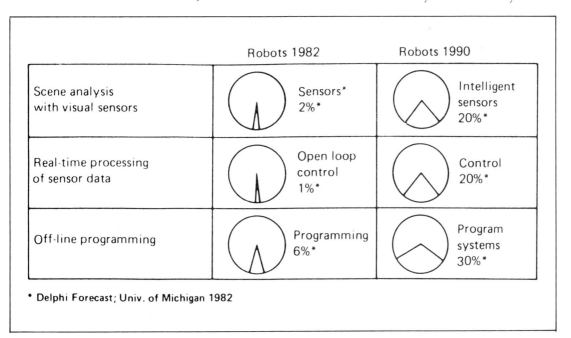

	Robots 1982	Robots 1990
Scene analysis with visual sensors	Sensors* 2%*	Intelligent sensors 20%*
Real-time processing of sensor data	Open loop control 1%*	Control 20%*
Off-line programming	Programming 6%*	Program systems 30%*

* Delphi Forecast; Univ. of Michigan 1982

Robotics: development trends and tasks

influence on product design. Several hand-books on robot-oriented design have been published, and their recommendations should be made known throughout industry.

If assembly procedures are too complex for robots, an attempt must be made to restructure the sequence of operations. Pre-assembly routines can often be rearranged so that they can be performed by a robot without intervening in the design.

As more and more robots are used for assembly automation, standardized robot peripheries must be created. This will permit application of existing solutions and improve economy.

Sensor technology currently focuses on processes involving tactile functions. As for sensors developed for cognitive functions, it should be remembered that it is always cheaper to maintain a defined material flow than to attempt to cope with a diversified one.

In the manufacturing context, cognitive sensors rarely contribute to value added. Special efforts must also be made to standardize sensor interfaces.

In applications where the special capabilities of robots are used to process workpieces for several production areas and the robots cannot be interlinked, technological islands will remain, but these should be supplied with programs and planning data from central sources of master data.

The future

At Siemens, the subject of 'the factory of the future' is being pursued with increasing intensity: robot applications are being probed in depth, and special attention is also being given to interlinking of process lines and to industrial data communications.

In the electrical industry, robots will at first be widely used for:

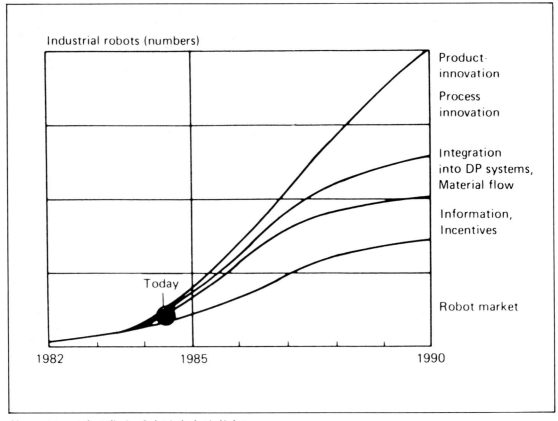

Measures to promote the application of robots in the electrical industry

- Flexible assembly lines.
- Production steps that increase value added and have so far been performed manually.
- Technologies of critical quality.
- As part of interlinked production systems.

Robots will not only attend to mass production, e.g. in entertainment electronics and domestic appliances, but also move into small-batch applications demanding greater flexibility.

A decision in favour of a defined flow of workpieces instead of gripper intelligence will speed up the introduction of robots. Just as computerization at first called for analysis and restructuring of the data flow, the first priority of robotization will be to master the flow of material.

The combination of computers and robots, i.e. the hardware and software of both instruments of automation, will open up new dimensions in productivity.

CONSISTENCY AND QUALITY USING ROBOTS

A robotic workcell may represent the first small step that leads to a manufacturer's giant leap toward an integrated, flexible manufacturing system (FMS).

Islands of automation, for instance, may be created – with robots inserting odd-form components or loading boards into automatic test equipment (ATE). Then the factory of the future will evolve with more computer-based processes and assembly islands, linked by local area networks and AGVs.

Today, robots perform a number of tasks for many manufacturers, centering around the production of circuits, their assembly and test. Parallel with a major thrust in operations at the PCB level, robots are also making significant contributions to enhanced productivity and product yield in highly competitive rigid disk drive manufacturing.

Components are inserted into PCBs with Universal Instrument's Vari-Cell robotic assembly system

Component insertion

Inserting odd-form or non-standard components in PCBs is one of the significant applications of robots in electronics assembly. These components, like connectors and PCB-mounted relays, are odd form in the sense that their various shapes and lower frequency of use on a board does not merit the use of hard automation like insertion machines for DIPs and axial-packaged components. The

First published in *Electronic Packaging & Production*, May 1985, under the title 'Robots lend consistency and quality to assembled products'.
Author: Ronald Pound, Senior Editor.

odd-form components are normally inserted manually.

Standard components like DIPs, too, may sometimes be inserted by robots. Where the manufacturer has low volume products, and needs flexibility to handle a large mix of products, a robotic approach may be more effective than hard automation with automatic insertion machines. Some products, in addition, may have boards with only a few ICs on them. Again, robotic insertion may provide the most effective approach in these cases.

A basic robot that returns repeatably to a point provides some component insertion capability. Higher levels of operation, however, provide decision-making capability within the insertion workcell and integration with other computer-based automation.

Robotic capabilities progress in complexity from repeatability, to accuracy, to intelligence and finally to communications, explains robot manufacturer Intelledex. This progression takes machines down a path beginning with a basic pick-and-place capability and ending with a sophisticated robot with all four capabilities.

Repeatability alone is sufficient for tasks such as inserting odd-form components, indicates Intelledex. Repeatability means the robot has the ability to return repeatedly to previously taught points. Teaching is done by positioning the robot at each desired point and recording the points' coordinates in the robot's memory.

With accuracy added to repeatability, the robot can locate a point on the basis of a programmed description, contained in software, of that point. Though still limited to tasks such as insertion of odd-form components, the operator is not required to teach the robot each point to which it must move.

In terms of robots' specifications, insertion of odd-shaped components is done with robots having a repeatability of ±0.001 to ±0.002in. in the horizontal plane. Since repeatability is better than accuracy, this latter parameter is in the order of ±0.002 to ±0.004in. for these robots.

For robots with the next level of development, intelligence, decisions can be made on the basis of data from sensors. This may be a tactile sense, with data from a force sensor, or vision, using a camera.

Tactile sense, for robots inserting components, can tell the robot if the component it is grasping has been seated in the holes of the PCB. If the force sensor indicates the component's leads have not moved freely into the holes, the robot may be programmed to act on this information by intiating a search routine. The robot can slightly move the component to find the holes. If the component's leads are bent, the robot may go through the search routine for a maximum, preprogrammed length of time. Generally, the larger a component's lead count, the longer the time period of the search routine.

A different sensing approach is used in Universal Instruments' Vari-Cell robotic assembly system. Their approach is to sense the travel of the robotic system's stinger that pushes the component into the board's holes. Full travel is an indication that the component has been inserted. Universal says full-travel sensing is inexpensive, is not affected by temperature changes or oil and dusty atmospheres, and does not require start-up stabilization (as with tactile sensors).

With the tactile sensing and the search routine approach, time may be spent in the search only to find that the component is not insertable. Universal thus emphasizes the idea of controlling the process. This means interacting with the component vendor and the manufacturer of the PCB to obtain consistent lead and hole geometries that can accommodate insertion at a rate of 600–800 components per hour. For example, with high pin count components, such as 64-pin DIPs, Universal indicates that high insertion yields will not be attained unless one works with the component leads.

The three capabilities – repeatability, accuracy, and intelligence – yield a pretty sophisticated robot for inserting components within its workcell. The next level of development, communications, allows it to interact with other automation in an integrated, flexible manufacturing system.

Robots' organization

Operating with a single workcell is one way to put a robot to work inserting components. With this approach, the robot is required to handle multiple component families.

A single workcell may also contain multiple robots. For instance, a couple of robots that insert components could be served by another robot that handles boards, perhaps taking them to a machine vision inspection station within the workcell.

On the other hand, robots may be organized into several workcells for inserting components. Each workcell is then dedicated to handling only one family of components, rather than a larger range of components. This arrangement may be selected to accommodate a higher volume of boards, or to handle a broad range of components that

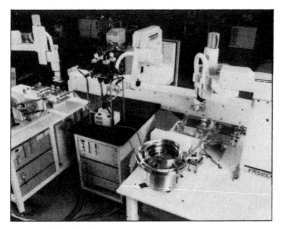

PCB assembly workcell uses two four-axis robots to insert components. The third machine, a six-axis robot mounted on a track, handles boards for the insertion robots and a vision system for inspection (Courtesy of Intelledex)

cannot be effectively inserted by a single robot end-effector. It could also simplify the arrangement of parts feeders, and facilitate their placement in a manner that increases throughput.

Combining the concepts of multiple workcells and multiple robots in a single workcell. Control Automation has configured an insertion system with eight of their assembly robots operating in five workcells. In one workcell, two robots insert SIPs of various sizes. In two other workcells, each with two robots, axial-leaded components and DIPs, respectively, are inserted. Workcells with one robot are dedicated to hardware such as stiffeners, and to connectors.

Grasping and placing

End-effectors are analogous to human hands and fingers. The ability to grasp components with a wide range of body shapes, sizes and body-to-lead orientations can be achieved by tool changes. That is, the end-effector may be changed automatically to match the component to be grasped. Using this approach, the order of components, and hence the order of insertion, must be matched to end-effector changes.

A primary tool with auxiliary tools describes the end-effector approach taken by Universal Instruments for their Vari-Cell system. The primary tool is designed with the intent of

handling the highest volume or the most complex components. It is custom designed based on the components a particular customer must insert. The primary tool becomes a part of the auxiliary tool.

For example, a primary tool may be designed to handle IC sockets, DIPs and box relays. These components are an extended family characterized by their horizontal shapes and similar widths. An auxiliary tool might be a plug that is picked up by the primary tool and then pushed into PCB-mounted connectors by the assembly system for subsequent insertion. In addition to the primary tool, one job could require up to five auxiliary tools.

Instead of tool changing, Intelledex takes the approach of using a programmable, indexing head supporting perhaps four end-effectors. If the range of components to be inserted exceeds the capability of this indexing head, additional workcells can be added. Intelledex's view is that tool changing impacts the production rate too greatly.

Control Automation, too, uses a multiposition gripper for its MiniSembler 2000 robot. With the FlexiGrip gripper, over 30 end-effector designs are available for inserting different classes of components. The gripper is a programmable device that changes end-effectors in less than a second. This eliminates complex mechanical changeover cycle time and accuracy problems associated with off-

Random insertion of components is performed with no change of end-effector in Chad Industries' ECA101 robotic system

line tool-changing techniques.

Chad Industries, however, developed its robotic insertion system with an end-effector that handles a range of components with no tool change or indexing. The approach is to have the end-effector compensate for variations in the body-to-lead orientation of irregularly shaped components. This allows 15 or more different components to be inserted in an average board randomly without the need for a specifically prescribed order or pattern of insertion.

Once a component has been grasped by the robot's end-effector, movement of the robot's arm takes it to the proper position and places it on the board.

Either a four-axis or a six-axis robot could be used for inserting components, according to Intelledex. To use a four-axis robot, both boards and components must be presented in a horizontal plane.

A six-axis robot is more expensive, continues Intelledex, but it allows a broader choice in how parts are presented to the robot. It is also much easier to retrofit into existing PCB assembly workstations, offsetting its increased costs for modification and fixturing.

A four-axis, DC servo robot will perform 90% of electronic assembly applications, according to Seiko Indstruments. Chad Industries' robotic insertion system, for example, uses a four-axis Seiko RT-3000 robot. This same robot model is used, as well, in

Chad's soldering system and liquid dispensing system (for solder masking and conformal coating masking).

Intelledex's Model 405, designed primarily for loading components, is also a four-axis robot. The Mini-Sembler 2000 robot, used in Control Automation's FlexCell assembly system, is also a four-axis machine.

The work envelope required, or the area in which the end-effector can operate, is relatively limited for insertion of components, in comparison to an application like loading PCBs into ATE. Loading boards into ATE may require a six- or seven-axis robot with a 10-foot-radius work envelope in a horizontal plane to reach board presentation and removal points, and any of the multiple testers in the workcell. On the other hand, insertion requires only a reach to relatively nearby parts feeders and over the area of a PCB. This may require a horizontal reach of about two feet.

In terms of speed, generally insertion robots' end-effectors may move around 55 or 60in./s, and insert components at a rate in the range of about 500–1000 per hour. A high-speed assembly robot, such as Unimation's Unimate Series 100, may move at a rate of 150in./s.

Hirata Corp. and Panasonic also have robotic insertion systems. The Hirata EPI-1 system uses that firm's AR-H assembly robot. It is suitable the firm says, for larger numbers of similar parts, such as DIP headers, connectors and electrolytic capacitors. Panasonic's Pana Robo AIKH robot, a four-axis machine, has an optional dual insertion head to handle a wide range of components.

Assembly tasks

Though component insertion represents a major thrust of robotic applications, other PCB assembly tasks, as well as testing, are being handled with robots. These include:

- Soldering (Microbot, Overseas Consulting & Sales, and Chad industries).
- Solder masking and conformal coating masking (Chad Industries and Microbot).
- Loading depaneling systems (Cen-corp, using an IBM 7535 robot).

A four-axis robot, Control Automation's Mini-Sembler 2000, is used in their FlexCell component insertion system. An idle station precedes the robot workspace to permit barcode identification and inspection of boards

- Loading PCB ATE (United States Robots, American Robots, Intelledex, and Zehntel).
- Processing dry-film coated panels (Hercules).
- Handling hybrid circuit substrates in a laser-trimming system (Teradyne).
- Adjusting and tuning PCB components (Anorad).

Rigid disks

While robotic applications have been developing in various PCB assembly and testing tasks, a parallel movement has been taking place in manufacturing rigid disk drives.

Intense competition characterizes the disk drive peripheral market. Thus, firms need to maintain productive assembly with a high product yield. The picture is complicated by the need to perform manufacturing operations in a clean room environment.

For manufacture of rigid disk drives, the key steps to getting out an assembled product have seen robots applied. They are: sputtering (where the disks are coated with magnetic and other materials), disk certification (testing of the disks), and the assembly of the drives.

In the sputtering operation, Phase 2 Automation's VS-100 robotic system, for example, provides automatic handling of disks at the input and output of vertical sputtering equipment. This system is Class 100 clean room compatible.

At the input station, a robot removes disks from cassettes and places the disks into

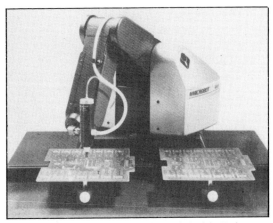

Solder-mask dispensing is performed with Microbot's Alpha II robot, a machine with ±0.015in. repeatability

pallets. Once loaded, the system moves the pallets into the entrance of the sputtering chamber, and the sputtering process is performed. Having passed through the sputtering chamber another robot unloads the disks from the pallets and places them into cassettes at the output station.

One of the key benefits of automating the loading and unloading of disks is that rigid-disk manufacturers increase their product yield significantly. This is primarily because human handling of materials is eliminated.

The end-effector, or gripper, handles disks by their edge to preclude damaging them. A dual gripper is standard. This provides higher throughput, compared to a single gripper design. The dual-gripper robot can pick up two disks from a cassette, then load them into a pallet.

Disk certifiers, the test equipment for rigid media, also use robots for loading. These certifiers write and read a test sequence. On the basis of the test, disks are sorted into as many as 10 grades, including one reject grade, depending on their quality.

In a certifier workcell, explains Applied Robotic Technologies, a robotic systems house, the disk arrive in cassettes. The robot takes a disk from the cassette and places it on the spindle of a certifier. Three or four certifiers may be served by a single robot.

An algorithm in the robot's controller uses data from the certifier to calculate the grade of the disk. Then, the robot places the disk in

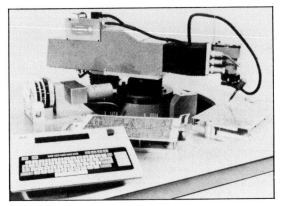

Soldering system using a four-axis robot (Courtesy of Chad Industries)

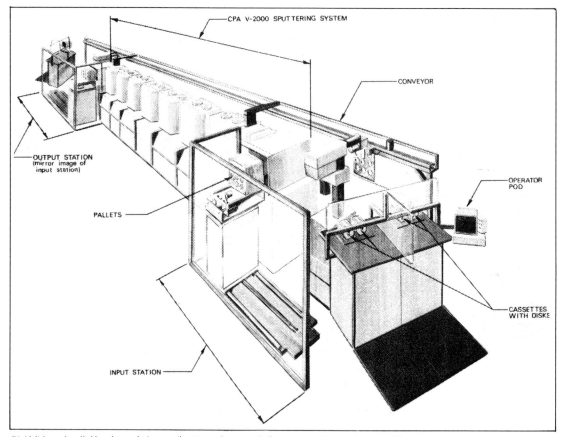

Rigid disks are handled by robots at the input and output stations to vertical sputtering equipment (Courtesy of Phase 2 Automation)

a cassette provided to collect disks of that grade. The controller also keeps track of the rate at which disks are accumulated by grade so empty cassettes can be ready to move into the proper position. If a disk is identified early in the test as a reject, data fed to the robot controller causes the certification process to be terminated for that particular disk.

Working in a Class 100 clean room, the robot's gripper handles disks by the edge so they will not be scratched or contaminated. A double-sided end-effector takes a disk off a certifier, then flips over to place a new disk it is holding onto the certifier's spindle.

A five-axis robot is normally required for a disk certification workcell, indicates Applied Robotic Technologies. The job could be done with a four-axis machine if it has a special wrist that can snap 90° to change the plane of the disk from vertical (coming out of the cassette) to horizontal (as it is put on the certifier's spindle).

Accuracy of 0.002 or 0.003in. is required of the robot. Thee is no more than about 0.005–0.010in. of freedom in getting disks onto the spindle.

Drive assembly

Automating the assembly of Winchester disk drives in a Class 100 clean room is also performed by robots.

GMF Robotics' A-1 robot is used by a manufacturer to assemble the spindle motor to the head-arm assembly backplate, install six disks and five disk spacer rings to the spindle hub, and install the top disk stack clamp ring. This robot has ±0.002in. repeatability.

On clean room models of the A-1 robot, potential contaminant sources are sealed and/or gasketed. Non-oxidizing paint is used

A disk-certifier workcell uses a robot to take disks from an incoming cassette, place them on the certifier's spindle and return graded disks to outgoing cassettes (Courtesy of Intelledex)

on the exterior surfaces. A filtered air return system removes particles from inside enclosures to prevent particle shedding.

The cylindrical coordinate robot is programmed to attach one of five end-effector mechanisms to its arm for executing its tasks. Tooling, positioned on a stand within the work envelope, includes two vacuum grippers, a mechanical manipulator and two screwdrivers.

Viewing their robotic system as a computer peripheral, Adaptive Intelligence's AARM product is also targeted for disk drive assembly as a primary market. Their robotic system, compatible with the IBM PC, incorporates tactile sensing.

A compliant gripper responds to the force signature during drive assembly. Tactile sensing is fast, compared to machine vision, and is

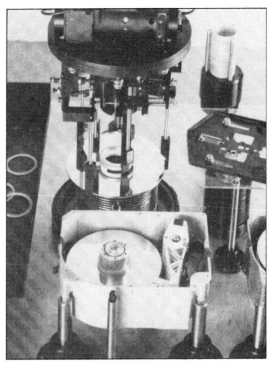

Force-sensitive, compliant gripper places disk and spacer together on $5^{1}/_{4}$ in. Winchester drive spindle (Courtesy of Adaptive Intelligence)

more practical for robot control, the firm says. The gripper moves at 40in./s, with repeatability of ±0.001in.

The machine design is based on the overhead gantry approach, rather than an articulated arm. Six-axis configuration is provided. The DC servo motor-driven X, Y and Z axes combine with the stepper motor-driven yaw, pitch and roll axes.

ROBOT HAND EXCHANGERS

Today's industrial robot can normally use only a single gripper or tool for manipulating materials or assembling products.

For example, a robot equipped with a gripper designed to handle minute components will generally not be able to handle the

First published in *Electronic Packaging & Production*, May 1985, under the title 'Robot's hand exchange expands its gripping capability'.
Author: Tom Petronis, Applied Robotics Inc.

circuit boards themselves. This severely limits the capability of a machine that could otherwise perform both tasks easily. This implies greater production cost. If the robot cannot handle the circuit board, either another automated device or a human operator must do so.

However, new developments in gripper technology have greatly expanded the capabilities of a single robot. Now, hand exchange systems permit virtually any robot on the market to overcome the limiations of being 'one-handed' by allowing them to quickly exchange one specialized hand or gripper for another. These systems allows a single robot to perform different tasks that previously required complex multitools, or even multiple robots.

Hand-exchanger design

A hand exchanger is basically a mechanical and electrical interface on the end of a robotic arm that allows it to exchange tools (hands) automatically during the work cycle – without human intervention – as its programming requires.

As an additional benefit, the most advanced hand exchangers integrate all the electrical, pneumatic, and coaxial lines that may be needed to control the tooling. So instead of connecting each tool to the robot control through bulky umbilical cords that crowd the work envelope, control signals are passed through the robot tool interface.

Most hand exchange systems require an interface half that attaches to the robot, and an interface half that attaches to the tool. The two interface halves mate to each other and complete the connection. Typically, a complete system would consist of one robot interface, one tool interface for each tool that the robot will be using, a connector to the robot controller, and a rack for the tools.

There are several methods by which the robot interface couples to the tool interface. Conventional designs use either pneumatic clamping, rotational screw clamping, or bayonet type mounts. All conventional methods require the robot to exert some type of coupling force or torque for the interface to link. Each of these methods also requires the robot to exert these same kind of forces to pull the interface apart during decoupling.

A method developed by Applied Robotics of Latham, NY, uses an air-driven mechanical linkage to lift the tool onto the robot arm without actually applying any torque or

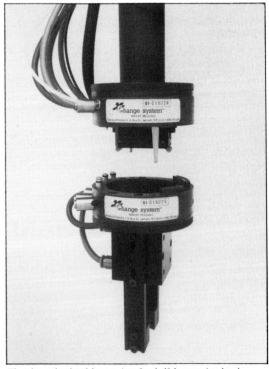

Hand exchanger on the end of a robot's arm can connect with a variety of tools placed in a rack, waiting for use

The robot and tool each have an interface half that mate in a hand exchange system

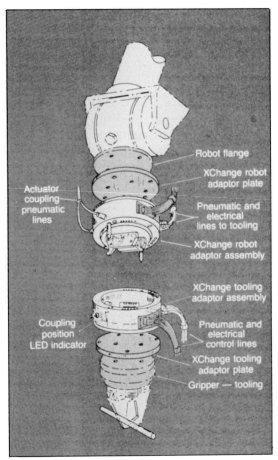

Air-driven mechanical linkage lifts a tool onto the robot's arm in Applied Robotics' XChange hand exchange system

The following labels appear in the diagram:
- Robot flange
- XChange robot adaptor plate
- Actuator coupling pneumatic lines
- Pneumatic and electrical lines to tooling
- XChange robot adaptor assembly
- XChange tooling adaptor assembly
- Coupling position LED indicator
- Pneumatic and electrical control lines
- XChange tooling adaptor plate
- Gripper — tooling

1. Using a jaw gripper, a robot takes an unpopulated circuit board from a rack and places it on the work surface.

2. Exchanging the jaw gripper for a small vacuum cup, the robot inserts a number of the smallest components into their proper locations.

3. Exchanging the vacuum cup for a different, specially designed jaw gripper, the robot then inserts custom components.

4. Exchanging the special gripper for a test probe, the robot tests sections of the board.

5. Finally, the robot switches back to the original gripper and exchanges the finished circuit board with a new one.

Without the ability to exchange tooling, this process would never have been possible without using a number of robots, complex tooling, or manual intervention.

A broad range of tools can be made available to a robot with a hand exchanger. according to A.J. Overton of Digital Equipment Corp. – in one application DEC has robots equipped with hand exchangers that use 14 different tools. In this assembly operation, the tools vary from simple screwdrivers to complex, custom tools designed for doing special jobs.

In another example of the versatility provided by a hand exchanger, one disk-drive manufacturer uses a clean-room version of a hand exchanger that allows a single robot to handle five tools in building 8in. Winchester disk drives.

Using a screwdriver, the GMF Robotics Class 100 A-1 robot first installs fasteners on the hub spindle. Exchanging the screwdriver for a vacuum gripper, the robot adds a disk and a disk spacer to the hub. With another vacuum gripper, it places a plastic disk spacer ring.

Once all six disks are in place on the hub, the robot exchanges the vacuum gripper for a spindle orientation tool which radially orients the assembly. After the orientation, the tool is switched for a low-torque screwdriver which installs fasteners to the clamp ring. Finally, the low-torque screwdriver is exchanged for a high-torque one, which tightens the assembly.

coupling forces. This method provides substantial benefits in certain cases because it allows light-duty robots to be used where a heavier, more costly robot would have been needed to apply enough force to pull the arm free when changing tools.

Once the tool is lifted into place on the arm, the robot performs the operations that tool is designed for, swings back over the tool rack and deposits the tool in its original position. Then it repeats the same kind of operation with the next tool.

Electronics assembly

Tool exchangers can be used in a wide variety of electronics manufacturing operations. A description of a typical application follows:

Class 100 clean room robot from GMF Robotics uses an Applied Robotics hand exchange system in the assembly of Winchester disk drives

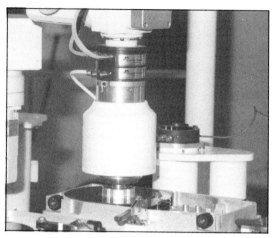

Tool for placing disks on a centering device before assembly is connected to a GMF Robotics A-1 robot by Applied Robotics' XChange system

Advantages

There are numerous advantages to using robotic hand exchangers. To begin with, the premise behind using a robot, instead of a hard automation system, is its flexibility. Robots give manufacturers the ability to automate production where the designs are complex or frequently changing.

Hand exchangers vastly increase robot flexibility. This is why DEC is using an exchange system. "We were trying to build a flexible system, and we wanted to be able to convert from one product to another on the fly," said Overton.

Robots are also a big advantage in situations where volumes are low. Certain applications may require only a few copies of a particular circuit board to be built once or twice a month. Instead of having to re-tool robots for this one board, a hand exchanger allows the right tools to remain in the rack until they are needed.

Dr. Leo Hanifin of Rensselaer Polytechnic Institute's Center for Manufacturing Productivity in Troy, NY, believes that hand exchangers may mean more than just adding flexibility to today's robots. He feels they will help expand the overall role of robots in manufacturing, particularly in electronics. "The ability for robots to automatically change hands without human intervention is a key to their cost effectiveness and, there-fore, will be critical to their increased levels of application," he said.

Price of flexibility

What hand exchangers really offer is flexibility, but not without some compromise – speed. It takes approximately two seconds for the robot to make the exchange. This time is not excessive, but it is two seconds longer than if two robots were used to do the job instead of one.

Of course, a second robot means a significant added cost (if there is enough work space to install a second robot in the first place). Also, adding a robot means two robots to maintain instead of one.

There is also the alternative of using a complex multitool that can do the jobs of several different tools. This will increase the production speed by eliminating the tool exchange.

However, multitools have two big problems. First, they are much more massive than single tools. This may require the use of a stronger, costlier robot. Second, multitools have a multitude of moving parts. This lowers the mean time between failure. When a multitool does fail, replacement requires costly downtime to remove it and put the new one on (assuming an extra, custom-made multitool happens to be in stock).

Applications

A hand exchanger can be found for most applications where robots are being used. This includes component insertion of PCB assembly lines and disk drive assembly operations, in clean rooms, radiation environments and even underwater. There are models to suit the needs of virtually every robot.

Now it comes down to: "Why haven't hand exchangers really caught on in the electronics industries?" There are several reasons.

First, many production environments are such high-volume operations that it is simpler to design and build a hard automation system, and not use robots at all.

Second, until recently, hand exchangers just could not withstand the demanding requirements of the medium-volume electronics production environment. The electrical connectors on most existing exchangers, using male/female connectors, require significant insertion force. Not only does this type of connector have a life expectancy of only approximately 500 cycles, it also sheds particles from its first cycle. This makes it unsuitable for any clean room application.

However, the Applied Robotics hand exchange, XChange (a registered trademark), is designed specifically for connector durability. This system uses spring-loaded pins that require no insertion force to make up to 31 electrical connections inside the interface. This particular connector is waranteed for 1 million cycles, and is rated for use in Class 100 clean rooms.

Perhaps the most dramatic point to be made about hand exchange systems is their price. A 40lb version, with interfaces for two different tools and a cable, costs about $6000 from a variety of manufacturers. Comparing this to the price of an additional robot, hand exchangers become a cost-effective choice.

VISION-GUIDED ASSEMBLY

Competitive forces moving manufacturers to reduce product cost and improve quality, and advances in computer processor and sensor technology are coupled to form the unifying idea for a young machine-vision industry. Like many new ventures drawing on the power of semiconductor technology, the machine vision industry was created by visionary entrepreneurs. Their products are now moving into assembly applications.

Machine vision is the use of a computer to acquire an image, determine what it is, and make a decision on the basis of that determination. The decision may result from an inspection process, or it may be the basis for subsequent control of a machine or robot. While providing vision for robots will be an important contribution of machine vision, process inspection for PCB and hybrid assembly, and control of wire bonding machines, are now common manufacturing needs where machine vision systems are ready to perform cost effectively, claim their manufacturers.

Sight for the blind

As might be expected, providing sight for a computer starts with a camera to view the scene of interest. The scene is transformed into an array of picture elements, or pixels. Following analog-to-digital conversion, each pixel has a value determined by its brightness and the image characterization scheme used. For example, in a system using only two brightness levels, black and white, a binary scheme assigns either 0 to 1 to a pixel, depending on its brightness relative to a threshold level. One now has the scene, represented by discrete numbers, in the computer.

Next, the computer compares the array of pixels, or some characteristic of the scene derived from processing the pixels, with reference scenes previously entered into the computer. With this processing, the computer determines which scene is being viewed, or

First published in *Electronic Packaging & Production*, May 1984, under the title 'Keen machine vision guides assembly operations'.
Author: Ronald Pound, Associate Editor.
© 1984 Reed Publishing (USA) Inc.

identifies any variations from a reference scene.

This is the process which enables the computer to determine, for example, what component is being viewed on a conveyor and how that component is oriented. The presence or absence of a component at a specified location on a PCB can also be determined. Based on the results of this process, the computer can issue control commands to a robot taking parts off a conveyor, or to a reject mechanism that removes faulty PCBs from a line.

Complex scenes may not be adequately characterized by the two levels of a binary image. A gray-scale system addresses this problem by providing more levels for the image. Each pixel receives a number based on its brightness in the gray scale.

The number of gray levels used varies widely in production floor machine vision systems, says Cognex, a supplier of inspection systems. Generally speaking the larger the number of gray levels used by the system, the closer the digital image approximates the scene as it appears to a human observer. Studies have shown, however, that 50 to 55 shades of gray are all the human eye can see. Thus 64 gray levels provide all necessary visual information. This is important in applications such as inspection of IC package print quality.

Prior to final processing, most systems convert the image to binary, claims Cognex. Their inspection system, however, uses 64 gray levels throughout all stages of image processing to maintain the visual information content.

Most machine vision systems start with a binary image by comparing the gray-scale image from the camera with a threshold to determine the binary value of a pixel, says Applied Intelligent Systems Inc. (AISI). Although the system may dynamically adjust the threshold, AISI claims that the technique fails when there is low contrast or non-uniform illumination across the scene. For example, low contrast between a capacitor and its hybrid substrate background means the pixels across the capacitor and the adjacent background will be close together in

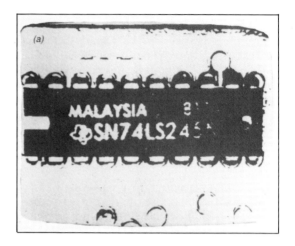

All visual information processed by the human eye is provided with 64 gray levels. These component inspection images are from (a) binary, (b) eight gray-level, and (c) 64 gray-level vision systems. (Courtesy of Cognex)

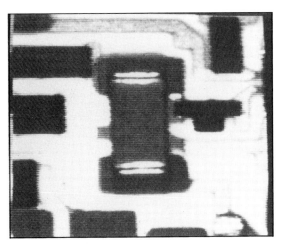

Low contrast between a capacitor and adjacent substrate background, complicated by circuit traces and screened resistors in the field of view, creates a difficult machine-vision problem. This capacitor image is from a 64 level gray-scale system (Courtesy of Applied Intelligent Systems Inc.)

their gray-scale value. Only a threshold level between these closely spaced gray levels will result in a binary image that distinguishes the capacitor from its background.

Once the pixels with the appropriate binary or gray-scale values are in the computer, the process to examine the image may be done with several algorithms.

Edge detectors look for transitions between levels, either binary or gray, as the computer moves from pixel to pixel across an image line. The recognition process in their Pixie-1000 system, says AISI, relies primarily upon edge detection, both magnitude and direction, with subsequent analysis of the relative placement of edges. The process caused the image from the camera to be transformed to a new image in which only the edges of objects in the scene are visible. An edge will become bright, with a brightness proportional to the contrast. Uniformly smooth regions will become black.

Edge detectors are useful for making measurements. In this case, the physical size that a pixel represents will have been calibrated. The number of pixels between transitions can then be expressed in physical distance between the edges that cause the level transitions.

Algorithms to extract features from a scene associate parameters with individual items in the scene. These parameters are measurable and include, for example, width, height, perimeter, and number of holes.

In template matching, or correlation, the computer searches for a match between the scene being viewed and reference images previously entered into the computer. To program a correlator requires only that the reference scenes be viewed by the camera and the resulting pixels be stored in the computer, says Richard Hubach, president of View Engineering. Correlators are useful for alignment of complex parts because of their ease in programming and capability to automatically search through a complete 360° of ratation without additional programming, notes Hubach. Template matching, however, requires precise placement of the part, says Machine Vision International.

For gray-scale analysis, a histogram can be created for the distribution of the pixel gray-scale level across the image. That is, the computer finds how many pixels there are for each gray-scale level.

John Fitts, president of Electro-Optical Information Systems (EOIS), says they have found there is undue emphasis placed on whether images are characterized by binary or gray levels, or whether they should be processed with correlation, histogram, or edge detection techniques. As algorithms become more complex, processing speeds go down and costs go up. Each job has to be looked at on an individual basis, says Fitts. EOIS maintains a large library of image-processing algorithms. These algorithms are placed together to form a complete image-processing system.

Machine eyes

Viewing scenes in the visual spectrum is done with standard vidicon tube or solid-state TV cameras, or with line-scan cameras. Incorporating charge coupled or charge-injection sensor devices, solid-state cameras offer several advantages over conventional cameras. Advantages include small size and weight, high sensitivity, and wide spectral and dynamic ranges.

With line-scan cameras, information is obtained from a linear array of detectors.

Such a camera, for example, may have 64 detector elements, or as many as 4048, notes John Fitts of EOIS. These cameras may be used in applications where parts are moving along a conveyor line. The camera scans the scene along the array direction, while the motion of the parts past the detector array provides the scan in the orthogonal direction.

To reposition the viewing scene for a standard camera, an *X-Y* mechanism may be used to move either the camera or the part being viewed. For example an *X-Y* positioner is used by AISI to move the camera to inspect the placement of over 120 surface-mounted components within a 20s period. Clever systems of mirrors and multiple cameras may also be used to view different aspects of a part without camera or table movement.

The accuracy of a machine vision system, or the physical distance in a scene that can be resolved by a single pixel, depends on the camera's field of view and the number of pixels available. Since a given camera has a fixed number of pixels in its field of view, decreasing the field of view decreases the distance that each individual pixel must represent. Currently the array size for standard cameras is up to 512×512 pixels. As an example, with a 1in. field of view, and the 512×512 pixel array, resolution is ± 0.002in.

Multiplexing several cameras to divide the scene into pieces with a smaller field of view can be used to maintain accuracy in some cases, says Richard Hubach of View Engineering. The same general concept can be implemented with an *X-Y* positioning mechanism to successively place a small field of view over parts of a complete image to be examined. The more pixels available, the less the requirement for translational mechanisms to reposition the camera over the electronic components, says John Fitts of EOIS.

Line-scan cameras with a very large number of detector elements, continues Fitt, have partially solved the resolution problem. These cameras, however, require an intense light source. In addition, vibration of the moving, non-fixtured part passing through the field of view effectively reduces the resolution, notes Fitts.

Illumination of the part to be viewed is a

Overhead coaxial and oblique lighting, with GE TN2200 charge-injection cameras, are used in the Interscan 1000 PCB inspection system. This system can inspect an 8 × 10in. PCB for the presence or absence of 40–240 components in 4–5 seconds (Courtesy of Control Automation)

crucial parameter. This was demonstrated by Larry Curtis, Intelledex's regional manager in Santa Clara, as he cast a shadow over a component on a PCB being inspected. The vision system reported the component missing.

Inspection of PCBs usually requires bright lighting with no shadows, says Integrated Automation. Built-in, overhead controlled lighting with dimmer features to adapt to any environment, and overhead charge-injection cameras in Control Automation's Interscan 1000 system illustrate one approach to illuminating and viewing PCBs under inspection.

Experimentation with part fixturing and lighting usually represents a significant portion of the algorithm development, says AISI. Generally their system works best when parts to be inspected can be fixtured within the field of view of the camera. Fixturing helps to simplify the description of the inspection problem and allows the lighting to be optimized. Their system was designed to avoid costly lighting requirements by using a gray-level processor which automatically accounts for non-uniform illumination.

Strobe light systems, with strobe synchronization, are used for the instantaneous image capture of components moving at high speed with extensive vibrational or rotational motion.

Machine communications

In order to operate, a system requires that someone tell the machine vision's computer what to look for, and once it has looked, what to do with the results of its examination.

Control Automation describes the setup of their Interscan 1000 system for PCB inspection as a teaching process. A good PCB is positioned in the field of view of the camera. The system learns the characteristics of this good board to use as a reference in the subsequent evaluation of production boards for the presence or absence of components. A menu-driven program leads the user through the setup routine. Component locations are identified and stored using a teaching mouse. A single command will tell the computer to learn what is being viewed. The user can enter names for each view. These names can be used with subsequent commands telling the computer what reference view to use at a specific point in the evaluation of production PCBs.

Downloading board characteristics from a host computer via an RS232C interface is an alternative to the teaching process. This communications capability will enable a user to interface a PCB CAD system with a machine vision inspection system, says Control Automation.

Parallel I/O ports can be used to integrate their vision systems with other production equipment, notes Control Automation. These ports can be used, for example, to activate devices such as solenoids, relays, and feeding mechanisms.

Machine vision for loaded-board inspection can operate in a closed-loop manufacturing system, says Cognex. The Cognex Checkpoint system, communicating with a host computer via a serial RS232C port, provides inspection results to a data base. On the basis of trend analyses, the host initiates corrective action for real-time quality control. Batch reports can also be generated for management's assessment of manufacturing performance in relation to quality goals. All analyses will be based on 100% inspection, at high production-line speeds, and free from subjective and casual evaluation by a human inspector.

Process inspection

About 40% of PCB faults found by in-circuit testing are presoldering faults, observes Stephen Denker, market manager at Cognex. Assembly errors such as missing and misoriented components, or bent leads which do not come through the board holes, are easier and less costly to correct before soldering. Cognex believes machine vision systems will be a powerful substitute for in-circuit ATE and, possibly, have an influence on the resources devoted to functional testing.

Vision inspection is a likely substitute for in-circuit testing of boards with surface-mounted devices, says Denker. Bed-of-nails fixturing for in-circuit testing of boards with surface-mounted components can be difficult and expensive. Often it may be nearly impossible to test at some nodes.

With mixed technology on a board, leaded and surface mount, automatic vision inspection will be a supplement to in-circuit testing. For boards populated only by surface-mounted devices, vision inspection with functional testing will be the approach used. Denker explains that placement of the surface-mount component is the crucial item. With a machine-vision inspection system, one can examine how the solder paste is screened, verify component placement, and view the results after solder reflow.

Current vision technology is adequate to handle boards with surface-mounted devices, says Denker. It will however, require some learning and experience at test sites because

Machine vision performs the inspection function at station six on Universal's computer controlled in-line system, which can completely populate a printed substrate automatically. Applied Intelligent Systems Inc. supplies the Pixie-1000 vision system used

of problems that may spring from the small size and close spacings on surface-mount components. Lighting, and setting up a system, needs more investigation, notes Denker.

Contrast is a major concern in inspection of boards with surface-mount devices, adds Control Automation. Component color in relation to the board color, and reflection of light, are crucial issues.

Inspection of hybrid circuits for the presence and correct positioning of components such as IC chips and capacitors is an application of machine vision, says AISI. With little contrast against the substrate background, and a complex scene with circuit traces and screened resistors in the field of view, there are some failures to detect capacitors even when they are present. On the other hand, IC chips, with a higher contrast, are detected with 100% accuracy, reports AISI. Inspection of 10 components per second is the rate claimed by AISI in this application.

A thick-film inspection system designed by Synthetic Vision Systems (SVC) will provide automated inspection at each level of the printing and firing process, from initial conductor application to final laser trim. Each layer of the circuit is inspected for flaws, voids, path width and spacing, and misalignment. This system has an optional communications link to allow interfacing with the user's CAD system.

Color and x-ray vision

Pseudo-color coding provides an aid for the operator to interpret selected views displayed on a CRT in SVS's thick-film inspection system. This technique assigns a color to a gray-scale level. Pseudo-color makes gray tones more perceptible to humans, explains Russell Petersen, vice president of marketing and services at SVS. The scene is not viewed with a color camera and the computer operates on gray-scale information until the pseudo-color display is generated.

A Color Discrimination Device (CDD) will, however, provide a color camera interface to gray-scale vision systems, says co-inventor Petersen of SVS. The CDD takes a composite video signal from a color camera, extracts hue information, and assigns different gray-scale values to the various hues, rather than to different brightness levels as in gray imaging of a scene. Using a composite video based on intensity, hue, and saturation offers speed and cost advantages over the chromatic red, green, and blue, or RGB, approach, claims Petersen. An RGB approach would require multiple buffering and a more expensive camera.

Reading color codes on components is only one of the advantages color viewing brings to PCB inspection, Petersen remarked. From his industry experience, he has concluded that the only reliable way to distinguish components from the board in difficult situations is with 3-D imaging using laser illumination or color. Detecting red and blue components on the green solder-mask background is a difficult inspection problem for gray-scale imaging to handle.

Gray-scale imaging cannot distinguish between solder, copper, or gold, continues

Real-time x-ray imaging is combined with machine-vision processing for automated inspection of multilayer PCBs (Courtesy of Penn Video)

Petersen. With color viewing of a board, one can handle the difficult vision problem of determining whether component leads have come through holes by using color discrimination of the solder-tinned leads and copper pads.

Imaging need not be done in the visible spectrum, whether gray or color. Machine vision, unlike the human, can see x-ray, infrared, and ultrasonic images when the appropriate illumination sources and sensors are used. The combination of real-time x-ray imaging and machine vision enables Penn Video's Videomet II to automatically inspect multilayer PCBs for registration parameters. Each board is x-ray scanned and the resulting images are computer analyzed to determine if pre-programmed registration limits are met. Inspection speeds are generally 15–20 seconds per board.

Machine control

Control of wire bonders used in the assembly of IC DIP and hybrid packages is a frequent application of machine vision. Wire bonders lay down 0.001in. gold wire from the IC chip to the DIP lead frame, with accuracy of 0.0005in. or better, at rates of four or five wires per second, says Richard Hubach of View Engineering.

Both a correlation and an edge-finder system may be required for some wire bonding applications, notes Hubach. Correlation is used to locate wires at the IC chip. The metalization on a device is not precisely positioned relative to the sawed edge. Thus, the metalization pattern itself must be used for alignment in the wire bonding machine,

rather than detection of the sawed edge.

The edge finder offers advantages in locating the position of the fingers on the lead frame. Each finger may be looked at individually, even though many may be in the field of view at the same time, continues Hubach.

Edge detectors are required when the position of both ends of the wire must be examined just prior to bonding. High lead-count packages and ceramic DIPs may require edge detectors as well as correlators to control the bonder, says Hubach. In high lead-count packages the wires must touch tiny fingers on the lead frame. For ceramic DIPs, flow of glass around the lead frame can cause misalignment between the fingers and wires.

The fundamental machine-vision problem is the same whether controlling bonders in DIP or in hybrid assembly, says Hubach. There is a difference though, in the programming required to take the bonder through the assembly process. With DIPs, a wire bonder is set up by the semiconductor manufacturer for a long run of one device with an unchanging chip pattern. Hybrids, on the other hand, have several different IC chips on the same circuit. Any one chip, supplied through more than one source, may vary in appearance. The

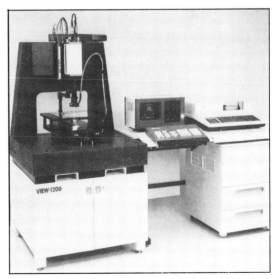

Control of a wire bonder is a typical application for the View 1101 vision system which uses correlation techniques for image processing (Courtesy of View Engineering)

wire bonder's vision must be programmed to call up reference views for all of these possibilities at the appropriate time.

Flexible robot control is provided in keyboard assembly by Octek's machine vision system. Keycaps are sent down a positoning chute that is intended to orient them with the correct side up. A camera attached to the top of the robot's arm checks the keycap for position and character. As the key is lifted by the robot, orientation is corrected, if necessary, and unacceptable keycaps are sent to a reject bin. The Octek vision system then gives coordinates to the robot for placement of the key on the assembly board.

Reducing the need for precise parts presentation lowers the cost of robotic automation. This is one of the benefits in using the 64 level gray-scale V-1000 robot vision system, says its manufacturer, Intelledex. Not limited to providing information for control of the robot, Intelledex says another benefit of its vision system is better quality control which results from inspection of parts just prior to robotic assembly.

The vision-system controller can function independently of the main controller. Thus, the robot can work while the vision system is locating and inspecting parts. An Intelledex robot can be fitted with vision by connecting the V-100 controller to the main controller, and plugging in a charge-coupled device or vidicon tube camera. A wrist-mounted camera on a robot may eliminate the need for multiple cameras.

Bin-picking capability, or the ability to

Machine vision guides a robot in keyboard assembly: (a) camera attached to the top of the robot's arm checks orientation of a keycap, (b) the character is identified as the keycap is lifted, and (c) the vision system gives the robot coordinates for placement of the keycap (Courtesy of Octek)

select randomly-oriented and jumbled parts from a container, is provided by Object Recognition Systems' i-bot vision system. This approach eliminates the need for feeders and cartridges which orient and place parts for industrial robots. The system combines both vision and tactile sensors for identification, acquisition, and manipulation of workpieces.

Interfacing with a Seiko D-Tran robot has been used by Machine Vision International to demonstrate the capabilities of its Genesis 2000 machine vision system. With 512×512 spatial resolution image, and a capability of 256 gray-scale levels, this system is intended for inspection as well as machine control and robot applications. Programming and editing are interactive, using prompts displayed on a CRT. Up to four development workstations are supported.

Looking ahead

The trends are evident that machine vision will become increasingly intelligent, coping with color and 3-D, and coordinating tactile sensing with vision in robot applications, states consultant Nello Zuech, president of Vision Systems International.

Future machine vision systems will feature larger image pixel arrays, faster processing speeds, more sophisticated gray-level processing, and easier programming, says AISI. Larger pixel arrays and faster processing speeds will give better resolution over a large field of view. This will be especially helpful in verifying the placement and quality of traces on PCBs.

Products designed specifically with the intent of using machine vision in their assembly will facilitate better image processing, says AISI. Coordination between product design, component procurement, and vision system operations should be directed at reducing the variability in appearance of parts, and eliminating components with colors that have little contrast with the circuit background.

ROBOTIC
APPLICATION

An interesting case history of robotic use now follows. In this chapter robots have been treated as insertion tools. Their increasing use in SMT assembly is discussed in Chapter 4 and their integration is discussed in Chapter 10.

The following paper addresses itself to non-standard insertion following standard insertion techniques such as axial lead insertion.

VIBRATORY INSERTION PROCESS: A NEW APPROACH TO NON-STANDARD COMPONENT INSERTION

AT&T Technologies Inc. has developed a robotic non-standard component inserter (RNSCI) capable of inserting a class of non-standard components into printed wiring boards. This system is based on a fundamentally new concept, called the vibratory insertion process (VIP). VIP's combination of high speed and high reliability has enabled the development of a machine which represents the state of the art in non-standard component insertion. VIP overcomes positional inaccuracies inherent in non-standard components, and reduces the need for high precision in the system's mechanical devices.

While commercial robotic machinery has demonstrated typical insertion times of 10–20 seconds per component, the RNSCI's cycle time is approximately three seconds. Insertion reliability in excess of 99.8% was realized under normal operating conditions. Systems design accommodates dozens of board codes and board dimensions, making it a versatile production tool. A high degree of automation has been achieved by interfacing a VIP insertion system with an automatic parts preparation stage. In the present embodiment, disordered, bulk parts enter the system, while boards leave the system with non-standard components automatically inserted. The machine can be modified to interface with a variety of production lines, ensuring its utility throughout many of the production facilities of AT&T.

While many components are presently automatically inserted into PCBs, there is a class of devices called non-standard components which generally still must be inserted by hand. Examples of non-standard components in use in AT&T Technologies' products are capacitor packs, mercury relays and transformers. These components are often bulky,

First published in *Robots 8 Conference Proceedings*.
Authors: Brian D. Hoffman, Steven H. Pollack and Barry Weissman, AT&T Technologies Inc.
© 1984 Society of Manufacturing Engineers.

making automatic feeding difficult. Additionally, non-standard components typically have poor lead-to-case tolerance, so that even if the device is held by its case, the location of the leads can vary enough to confound an automatic inserter. The leads themselves are often recessed from the part's edge on the part bottom. This makes it extremely difficult to design a gripper which can grab the leads for location (as is done on DIP inserters) and still not interfere with previously inserted components.

There is strong economic motivation for automating non-standard component insertion. Because the present process is labor-intensive, the cost of inserting non-standard components is very high. For example, on one assembly line, the cost of inserting a non-standard component is four times the cost of inserting a standard component. The required rate of component insertion can be very high as well. In some cases, a component must be inserted at least every four seconds.

Many technical problems have inhibited automating the non-standard component insertion process. Common techniques, such as dead reckoning, will not work because of the poor lead/case tolerance of the leads and the tight lead/hole clearances (required for high quality solder joints and board density considerations). More sophisticated techniques such as 'visual servoing' (looking through the board to 'see' the leads) are too slow (typically over 15 seconds per component) and expensive to implement. A technique which registers the leads with respect to the robot, through additional fixturing and process steps, can increase the cycle time. Additionally, this approach does not account for positional inaccuracies in the robot, board location, and hole location within the board.

The vibratory insertion process

A different approach has been successfully worked out at the Engineering Research

Center in Princeton, New Jersey, for a class of non-standard components. The class of components discussed in the remainder of this paper is 'rigid-leaded cubes', i.e. non-standard components that are basically cube-shaped and have leads stiff enough to resist bendng and crumpling under small forces (on the order of the weight of the component).

The basic concept is named the 'vibratory insertion process' (VIP). A programmable manipulator (robot) grabs the part by its outer case and brings the component down to the nominal insertion location on the board. This nominal position is such that if the holes, leads, robot, and mechanical fixturing were perfectly located, the leads would drop in through the hole centers. However, cumulative positional inaccuracies generally result in insertion failure.

Meanwhile, the board is vibrated with predetermined frequency and amplitude. Eventually, the board holes 'find' the component leads. When the holes and leads align, the component drops into the board. The technique is 'open loop' in that the frequency and amplitude is pre-set for all time (based on geometric considerations), and does not depend on the initial lead-hole misalignment in any way.

The main advantages of this approach are speed, insertion reliability, simplicity, and optimum equipment usage. The robot is asked to do what it can easily do (moving a component near the holes, which change from code to code), while mechanical aids rapidly bring the component into the holes.

To take full advantage of VIP, it is important to understand the process physics. Parameters such as vibration frequency, vibration amplitude, and robot characteristics interact during VIP. If parameters are chosen incorrectly, experience has shown that insertion reliability is unpredictable and unacceptably low. However, ERC's research efforts in VIP have resulted in process characterization and optimum parameter selection. Based on the studies, the full advantages of VIP have been realized for AT&T production.

The heart of the technique lies in creating relative motion between the part and the board. While board vibration was implemented, it is recognized that, alternatively, the gripper, or even the robot and gripper, can be made to oscillate, while the board is held stationary.

System capabilities and features

A robotic non-standard component inserter (RNSCI) has been developed which, using VIP, inserts non-standard components into printed wiring boards.

Compared to systems of similar reliability, the RNSCI offers state-of-the-art insertion speed. The RNSCI's cycle time is 3–4 seconds per component, with a 99.8% insertion success rate. From the time the component first contacts the board surface until the component inserts, the elapsed time is typically 0.25 seconds. This means that the actual insertion is so fast that part presentation limits the cycle. Thus VIP provides, with improved parts presentation, the potential for sub-second cycle times.

While other systems achieve reliability through high-precision/high-cost devices, the RNSCI achieves high reliability through a new approach (VIP), which greatly relaxes the system's positional accuracy requirements. The system can tolerate inaccuracy in the: robot, gripper, board acquisition, part acquisition, hole location within the board, leads (with respect to the outer case) of the non-standard component, and in teaching. The RNSCI has demonstrated excellent

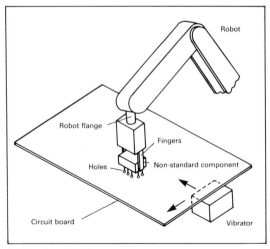

Basic hardware for VIP

insertion reliability, where system tolerances have exceeded ±0.030in. Furthermore, because the system can tolerate inaccuracies, no warm-up period is required. The machine is robust, in that it can operate reliably in a wide range of temperature and humidity conditions.

The system is intended to be easy to operate on the factory floor by featuring a high degree of automation. Non-standard components are taken from bulk storage and placed on a feeder plate. Boards are coarsely placed onto a tapered plate at the input station. It is the job of the machine, not the personnel, to orient the parts, trim and straighten the leads, test each component, acquire the board, and finally, insert the components automatically into the insertion sites. In the event of a problem, various diagnostic displays alert the operator with reasons for the difficulty. For trouble requiring prolonged debugging, parts of the machine can be selectively disabled, while the rest of the system continues to make product.

By making simple adjustments, the RNSCI can handle dozens of board codes, board sizes, and differing non-standard components. It typically takes under 10 minutes to perform all tooling and programming changes required to build a different board. This flexibility is essential in ensuring the RNSCI's utility, in the face of rapidly changing product, produced in lots of varying sizes.

Flexibility is also demonstrated in the modular design of the RNSCI. The machine consists of 'workstations', each capable of inserting several non-standard components during the board building cycle. Presently, three workstations link into the system's indexing conveyor. If more components must be accommodated, additional workstations can be readily integrated into the existing machine.

System architecture

The RNSCI consists of 'workstations', tied into a board indexing conveyor, which in turn is tied into a larger circuit board assembly line. Boards and parts are manually fed into the system, but, due to the special feeders, accurate material placement is unnecessary. A cycle begins with all boards simultaneously indexed by the conveyor. Each workstation receives a board, and accurately captures it from the conveyor. The captured board is vibrated, while a robot inserts non-standard components using VIP. Following the insertions, the board is released, allowing the next conveyor index, completing the cycle. For cycle time efficiency, board transit takes place during the robot's acquisition of a component. Following processing at the last workstation, the board is transferred to subsequent processing stations for insertion of components that cannot as yet be handled by the RNSCI. The boards then go to a wave soldering machine, and out of the system for test and shipment.

Control. The RNSCI has a distributed control system. Control responsibilities are divided

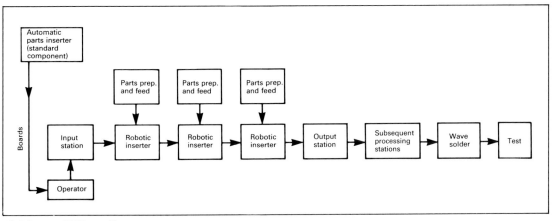

Block diagram of PWB assembly system

among the conveyor controller and the workstation controller. The workstation 'controller' is actually three separate controllers: the components preparation controller, the robot controller, and the local controller. These controllers operate asynchronously, and only communicate with each other at the end of their respective jobs. This operation allows for parallel processing of individual system tasks, minimizing cycle time.

A brief description of the control responsibilities will help to clarify the architecture. The conveyor controller is dedicated to indexing the automatic conveyor in synchronism with all board building activity. Compo-

nent preparation controllers ensure that bulk parts get individually processed and delivered to the robot every three seconds. The robot controller servos the robot, drives the workstation's pneumatics, and checks the station's sensors during insertion and board/part acquisition. The local controller allows manual or automatic operation of a workstation. In automatic mode, the local controller acts as an interface between the station's pneumatic valves and sensors, and the robot controller. In manual mode, the operator has control of the pneumatics for debugging and setup. The local controller also allows a 'bypass' mode, in which a disabled station can be electrically

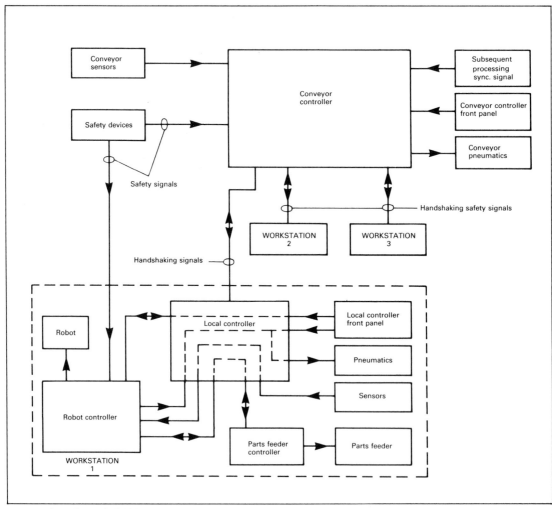

Overview of control system for RNSCI

taken 'off-line', so that board building can continue with fewer than three stations.

Workstation. The workstation is responsible for inserting non-standard components into printed wiring boards. Additionally, the station must provide prepared components, and must acquire a new board (from the conveyor) onto the vibration tooling (for VIP). These tasks are accomplished with a variety of devices, including a robot, a special-purpose gripper, vibration tooling, a parts preparation system, and controllers. Some of these devices are flexible (programmable) and some are not. This mix of automation promotes overall system efficiency. The dedicated devices, such as the vibration tooling, are optimized to perform unchanging or infrequently changing tasks. Additionally, these devices are specially designed to support the rapid insertion properties of VIP, removing the normally time-consuming task of hole/lead alignment from the (slower) flexible equipment.

Indexing conveyor. The system conveyor, which ties the workstations toether, has several requirements. Firstly, to maximize board building time, rapid station to station transfer was necessary. Secondly, conveyor adjustment was necessary to accommodate code changes due to different board dimensions.

(Both board length and width vary by several inches from code to code). Furthermore, production schedules require that any system changes be completed during operator breaks.

An indexing conveyor was used to accomplish the interstation transfers. The conveyor design chosen utilizes a pneumatic slide capable of translating all boards 18in. in 1.5 seconds. The slide transfers boards by pushing them from the rear edge as they ride in two edge siderails. All mechanism resides on only one side of the board, eliminating any inteference above or below.

Positional accuracy for final board registration was ensured with cam-activated board stops. Conveyor drive and stop mechanisms were protected from damage (due to potential system jams) with an overload release, analogous to an electrical circuit breaker. Under excessive force, this device decouples the drive cylinder from the transfer mechanism. This also provides for additional safety of personnel, in the event that someone interferes with conveyor motion.

Conveyor adjustment to accommodate board codes of different dimensions can be made in minutes with a set of hand cranks. With this adjustment, the system is capable of transferring printed wiring boards varying in length from 9 to 14in., and width from 5 to 12in.

CHAPTER 4
SURFACE-MOUNT TECHNOLOGY

The decade of the 1980s has seen a slow but ever-increasing trend to the use of surface-mounted components in electronic assembly.

This chapter will examine this trend, discuss its limitations and the challenges in assembly soldering and testing that its implementation will cause production engineers.

It is obviously a technology where capital investment can compete effectively with low-cost producers in third world countries. This will be done, however, by a methodical approach to mechanization, sound product design and an awareness of manufacturing problems.

SMT GROWTH

In the mid 1980s, surface-mount technology (SMT) has moved from the design and prototype stage to a full-fledged production technology in many US companies, complete with commercial production equipment, procedures and a full complement of surface-mountable components. In fact, over 1000 US electronics manufacturing locations are now using some form of surface-mount technology in preproduction or production. About 5% of all components used in the USA are surface-mountable, while in Japan (1984), nearly 50–60% of passive components and 5% of active components being assembled are surface-mounted.

Nearly 16,000 surface-mountable active device part numbers were available in 1985, compared with a universe of approximately 90,500 active device part numbers of similar electrical functions. Although only 17% of the component part number universe is available in surface-mountable form now, it is estimated that approximately 50–70% of the lower power memory, digital, microprocessor/microcomputer, digital, linear, audio/video, transistor and interface electrical functions are available in at least one surface-mountable (SM) package. With a few exceptions, multiple watt devices are not yet readily available in SM packages.

However, not all desired components are available in all of the desired SM packages – and in some cases not available in any SM package styles. This leads to circuit designs using both surface-mount and through-hole components. Furthermore, package standards have not kept pace, and the confusion and difficulty stemming from the lack of standard package geometries and standard shipment containers (tubes, sticks, tape and reels, etc.) have caused some would-be users to wait and see.

Statistics

In the last few years, the number of suppliers to the surface-mounting industry has grown dramatically. Passive-component suppliers offering surface-mountable components have gone from 44 in 1983 to 90 in 1985, an increase of 104%. Active-component suppliers offering surface-mountable components have gone from 36 in 1983 to 122 in 1985, an increase of 238%. Similarly, the number of commercial equipment suppliers and contract service houses has increased significantly, indicating the widespread commercial 'buying' interest in this assembly technology.

The net effect of all this commercial effort is a momentum which will economically and efficiently put surface-mount technology in the hands of all those companies who find a need for it. At this point in history, it is safe to say that surface-mount technology will most definitely become a major electronic packaging technology.

A striking statistic is the growth of surface-mount MOS component part numbers. The availability of MOS memory increased by 170% between 1984 and 1985, and MOS linear components have increased 225% in the same time frame. Overall, the availability of active (ICs and transistors) surface-mount component part numbers has increased by over 31%.

Ceramic leadless chip carriers and ceramic flat packs still dominate in terms of available part numbers. However, plastic leaded chip carriers and SO packages are now a major offering. Clearly, the early efforts by the military to standardize ceramic leadless chip carriers are now showing up as part of the wide product offerings by many suppliers. One notable observation is that there is now available a number of leaded ceramic chip carrier products – a real boon to those who do not want to deal with the thermal coefficient of expansion problems of leadless ceramic packages.

The major quantity of component part numbers is in the three I/O (transistors), 14–20 I/O (SSI glue chips), 24–32 I/O (memory devices), 40–44 I/O (LSI logic) and 68 I/O (microprocessors) devices. Very few

First published in *Electronic Packaging & Production*, January 1986, under the title 'Expense , availability direct SMT growth'.
Authors: Don Brown, Contributing Editor, and James Bracken, Anthony Manna and John Brasch, D. Brown Associates Inc.

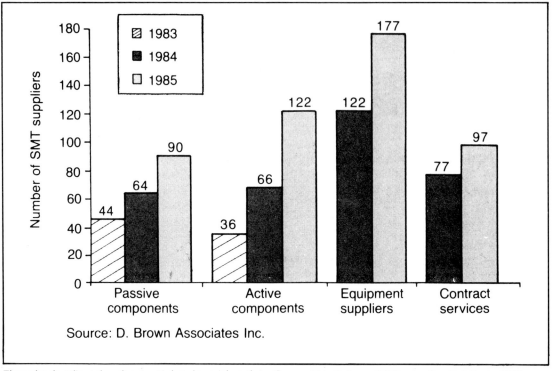

Source: D. Brown Associates Inc.

The number of suppliers to the surface-mount industry has varied over the past five years (Note: the figures of 1985 were gathered in September 1985)

part numbers of high I/O are commercially available. Based on this data, it seems clear that the largest functional use of surface-mount components is to replace the large number of logic glue and memory devices on a circuit, thereby reducing the overall area of the circuit assembly.

Type of component	1984*	1985**	Percent Change
Active components			
Bipolar memory	1,168	1,420	+ 19
MOS memory	166	449	170
MOS linear	47	153	225
MOS digital	2,053	3,070	47
Bipolar linear	761	1,113	44
Bipolar digital	4,443	5,540	15
Gate arrays	793	1,123	42
Microprocessor/support	428	732	71
Miscellaneous components	164	253	54
PNP transistors	489	603	18
NPN transistors	780	958	18
Unijunction transistors	0	5	500
Field effect transistors	97	164	69
Total	**11,389**	**15,583**	**+ 31**

*As of August 1984
**As of September 1985
Source: Surface Mounting Directory

Surface-mount component availability in the USA

Standards: The chaos factor

During the late 1970s and early 1980s, JEDEC successfully did what was never done before: developed and approved a 'standard' footprint for the not yet widely available surface-mountable leadless chip carrier package. Since then, the explosive growth of surface-mount technology has taken place worldwide. However, following this growth is an uncontrolled proliferation of 'non-standard' surface-mount package designs (not footprints) such as the SOIC, leadless and leaded chip carrier, as well as passive chip and other components. Still worse, in many cases manufacturers have used similar package designs, but have made some small changes, not obvious at first to the designer and user.

During the last decade, worldwide standards organizations have attempted to bring order to this problem, but efforts have sometimes been uncoordinated, incomplete, late and confusing to users and designers.

As a result, a situation can exist where the circuit designer is faced with the need to use a

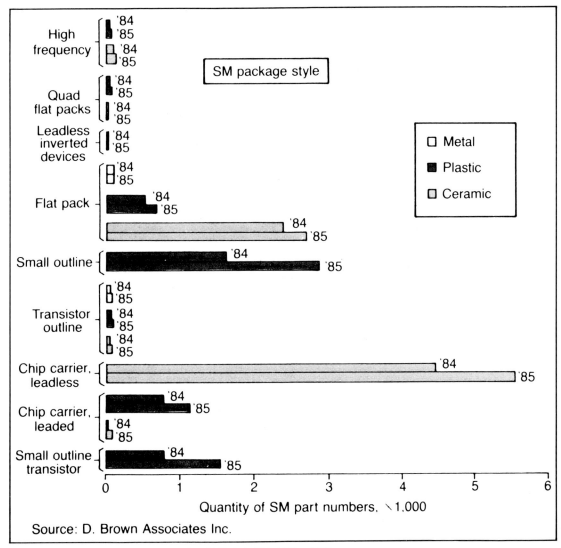

Surface-mount package style and package material availability in the USA, 1984 vs. 1985

surface-mount package from vendor A, second-sourced from vendor B with a different footprint and different physical package design – a very difficult logistics, PCB/substrate design and automation problem. Not only are the physical shapes of the packages different, but often, as in the case of a leaded package, the lead material, shape, thickness, stiffness and plating is different, resulting in different processing and assembly steps depending on which vendor's package is used.

The problem can be illustrated graphically.

Shown to scale are the differences among six major suppliers of SO-8 packages – one of the most 'standardized' packages available. The 'foot' of these commercially available SO packages ranges from a minimum of 0.010in. long to a maximum of 0.050in. The result is an almost impossible situation for the designer and the manufacturing manager, who must try to design and build a board/substrate to accommodate multiple package footprint designs so that the desired components can be multisourced.

Even if the designer can succeed in laying

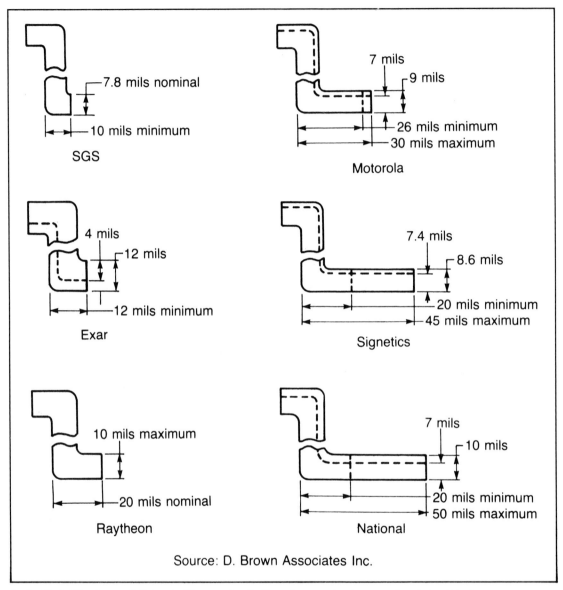

7.8 mils nominal
10 mils minimum
SGS

7 mils
9 mils
26 mils minimum
30 mils maximum
Motorola

4 mils
12 mils
12 mils minimum
Exar

7.4 mils
8.6 mils
20 mils minimum
45 mils maximum
Signetics

10 mils maximum
20 mils nominal
Raytheon

7 mils
10 mils
20 mils minimum
50 mils maximum
National

Source: D. Brown Associates Inc.

Lead profiles for SO packages available from six different suppliers. Profiles are drawn to scale to demonstrate the variation which exists

out the board to accommodate the various footprints, the manufacturing manager is faced with the problem of not being able to predict the field reliability of the solder joints, given a fixed amount of screened-on solder with varying lead geometries from various vendors, not to mention the possible machine setup variations for different package geometries. Now, how can one predict the field reliability if a production run starts with

SO packages from vendor A and switches, midproduction to vendor B's parts with a different lead geometry?

Research indicates that the lead shapes in the SO family of packages are relatively close in geometry compared to many other packages. The problem really gets interesting when quad flat packages with gull-shaped leads are considered.

Hopefully, the IEEE task force on compliant leads will resolve some of the reliability questions regarding the various J-shaped leads, by the thorough testing of a number of commercial offerings. The test data will then be available to component manufacturers, to help in choosing the lead geometry and material which will be most advantageous for their customer base.

A worldwide standard for surface-mount package geometries is absolutely necessary for the promise of surface-mount technology to be fulfilled. The way to achieve this goal is either by successful worldwide industry standards activities, with the full support of users and component manufacturers, or by the purchasing power of at least one (or a few) major users dictating the 'standard' package family and hence creating a standard by default.

What the component suppliers say

Because the suppliers of components are faced with a general slowing down in the industry, their optimism about SMT is evident. An overall growth of 31% in surface-mountable active component part numbers is an obvious sign. This growth is attributed to the increasing industry acceptance of surface mounting as a viable manufacturing process.

In high lead count devices such as 68 and 84 I/O standard cells, application-specific ICs and microprocessors, surface mounting has tremendous potential. Lea Schwartz, merchandising manager for NCR Microelectronics (Ft. Collins, MO), reported that NCR's shipment of components in surface-mountable packages will be up to 80% for the first quarter of 1986. Schwartz also reported that 50% of all high lead count devices were in a surface-mountable package and 65% of all semicustom and standard cell devices shipped from Fort Collins in the third quarter of 1985 were packaged in plastic chip carriers.

Jack Fuller, marketing manager of Amperex (Slatersville, RI) feels "the vendor base in discrete semiconductors has respectably introduced products vs. market requirements. Up until the beginning of 1986, the major concern was capacity. The small signal market, in terms of device types, is pretty well defined by their leaded counterparts. Timely

introduction of small signal devices in surface-mount packages as dictated by the market has, in my opinion, been rather good."

Some suppliers are faced with problems in converting their product lines to surface-mount form. Bo Matthy, manager of strategic product planning for Pulse Engineering, in San Diego, states that for his product, there is no industry-wide standard in the world for the type of package they need. He goes on to say that the surface-mountable packages they developed have gained acceptance among some of the major manufacturers. Matthy also states: "Whereas the package size for ICs has shrunk in SMT, it is not possible to do that for components such as pulse transformers and delay lines, while also maintaining high reliability. This is due to the physical size and internal structure of elements used to assemble them (such as wound cores, for example). Most users do not see this as an inconvenience because, in most applications, there are only a few such components on any board, and their size reduction would have only minimal effect."

Other component types faced with similar size-restriction problems are optoelectronic and high-power devices. For many optoelectronic devices, the die size is too large to convert a product offered in a dual in-line package to an equivalent SO package. For high-power devices, the package is too small to adequately dissipate the heat generated by the die. The manufacturer is therefore forced to do one of two things: either keep the same package as the conventional component and modify it (by clipping or bending the leads to enable the product to sit on the board), or design a custom package.

Matthy justifies the approach his company took by stating, "We believe that, whereas our package type and size may appear to be different from popular IC packages, our components can be assembled together with other SMT components on a board in one single process. That's where the real economy of SMT is."

"Reliability is an area where we (vendors) are perceived to be behind the customer (market) demand or expectations – i.e. zero defects, or at least PPM numbers that can be counted using only low 10s figures. Low PPM figures are achievable, but *only* when customer/vendor dialogue is ongoing, responsive and accurate, starting at the design level. Believe me, we are more than willing," states Jack Fuller of Amperex. This statement exemplifies the vendors' willingness to work with the market.

Fuller admits that there are problems in other surface-mount application areas as well. "Problem areas have been in tape and reel standards – every equipment manufacturer and contract supplier has been 'doing their own thing.' Now, it appears the EIA-proposed RS-481A, being driven primarily by Delco and IBM, is generally being accepted by the industry."

The overall view from the vendors of SMT components seems to be that the application problems are being solved one at a time. Some solutions come by necessity, others by market demand (e.g. tape and reel standard EIA RS-481A).

The status of the marketplace, as seen by the component vendor, is summarized by Gene Williams and Al Duvall of Rohm Corp.: "The move toward surface mount has continued to intensify over the past few months. The high-volume users in the automotive industry have been joined by a second wave of specifiers in the telecommunications, instrumentation and peripherals areas. A third wave, consisting of small- to medium-size users, is building. This third contingent of users is influenced by space savings and reliability as well as assembly cost savings. Pick-and-place equipment costs continue to drop and are now far less capital-intensive than auto sequencing, sorting, drilling, inserting and wave-soldering equipment required for leaded component use. This is encouraging acceptance of the technology."

Innovations are evident in packaging designs such as Micron Technology's use of 36 pins on a 68 pin J-leaded carrier to provide 0.100in. centers between (functional) contacts for its DRAMs. Their technique allows single tracing through the use of conventional PCB trace tolerances. By stacking ten 64k-bit chips into a single chip carrier, Micro achieves a memory density of 500k-bits/in.^{2} These clus-

ter DRAMs can be stacked for even greater memory density. Since all the 64k and 256k cluster DRAMs use the same pin-out arrangement, a board designed for the 64k × 8 or 64k × 9 cluster DRAMs may be upgraded to the

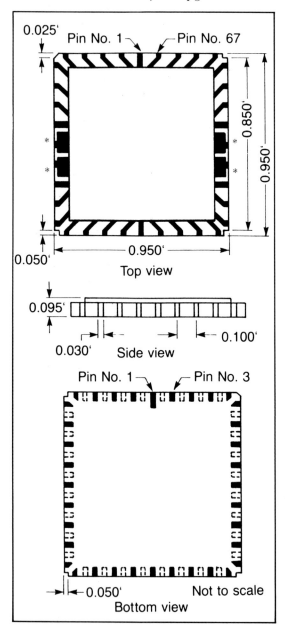

The 36-pin ceramic chip carrier from Micron Technology uses the standard 68-pin ceramic quad pack format with every other lead metallized, i.e. 36 pins on 0.100in. centers. Standard 68-pin sockets are suitable

256k, thus increasing memory density by a factor of four or eight (stacked cluster). The stacked module of two 256k × 9 clusters mounted on both sides of a circuit board enables the designer to achieve a density of nine megabits or one megabyte per square inch. Board changes or redesigns are not required to upgrade memory density.

The surface-mount indsutry is growing furiously with new products. Increased availability of components in surface-mount packages and the advantages SMT offers in size and automation move user and vendor in a spiraling upward race to enhance and perfect the technology. The future of SMT is exciting; its growth and acceptance is inevitable. The investment of vendors and pioneers in the industry guarantees a bright future for surface-mount technology.

A view from SMT users

Surface-mount technology has often been described as offering smaller size, higher performance, improved reliability and lower costs for electronic assemblies. In recent months, many users have come to realize that SMT is primarily a size-driven and electronic-performance technology, with short-term overall costs often being higher than its through-hole equivalent. If a given product design requires a smaller package or more electronic function in a given space, surface-mount technology offers a means to an end – but at a price.

REALIZING THE BENEFITS OF SMT

US electronic systems manufacturers are rapidly turning to surface-mount technology with the hope that it will be a more cost-efficient manufacturing alternative to conventional through-hole mounting. This change represents a major shift in electronic system manufacturing technology for a substantial portion of electronics firms. Major manufacturers such as IBM, AT&T and GM are making significant investments in surface-mount technology. Other firms, not wanting to fall behind, are rushing to design products using surface mount.

Why are electronic systems manufacturers stampeding to bring products with surface-mount technology to the market? Primarily it is that SMT holds forth to them the promise of being able to at least meet and hopefully beat foreign competition. This hope springs from the perception that SMT has advantages which will make it much more cost-effective than through-hole technology.

Among the commonly cited advantages of SMT are:

• The ability to readily automate the SMT manufacturing process.
• Increased functional density which allows smaller circuit boards or more functions within the same boundary constraints.
• A reduction of the number of interconnect layers necessary in multilayer boards.
• The use of shorter signal-path lengths, which can improve high-speed circuitry performance.

Because of the widespread media coverage given to SMT, these advantages are well-known. Yet the adoption of SMT is not without pitfalls and detours.

Unfortunately, the changeover to SMT involves a number of technical and logistic differences and problems that must be grappled with before SMT's benefits can be fully realized. Some of these are within the control of the user, while others are not. Differences in circuit constructions, manufacturing processes, components and materials must be addressed. New circuit design rules have to be formulated and used. Different environmental constraints exist which must be identified and coped with.

The dilemma

Selecting the proper course through the maze

First published in *Electronic Packaging & Production*, January 1985, under the title 'Unresolved issues hinder full realization of SMT's benefit'.
Author: Tom Dixon, Senior Editor.
© 1985 Reed Publishing (USA) Inc.

of alternate approaches for the neophyte to surface-mount technology can be agonizing and confusing. Even the knowledgeable can experience anxiety over whether or not they have made the correct choices. Unfortunately, the indvidual assigned the task of implementing surface-mount technology to produce a finished product is faced with a number of manufacturing alternatives from which to select and technical problems that must be resolved.

A number of questions must be answered in order to come up with the optimal solution. What are the technical requirements of the circuit to be built with SMT? Are there special electrical problems which must be considered? What special environmental constraints must the circuit be capable of meeting? How does the surface-mount manufacturing process differ from conventional through-hole mounting techniques? What is the availability of new production equipment?

If equipment such as a new component pick-and-place unit or a vapor-phase soldering system must be purchased, lead times of a year or longer may be encountered. This could force a decision to settle for a less capable system, or to make compromises in circuit construction which permit the use of existing equipment.

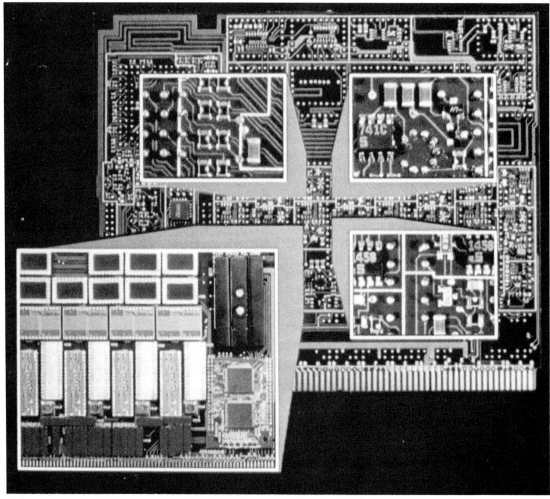

Surface-mount technology holds out the promise to the electronics industry of being able to compete with foreign products. However, before this promise can be fulfilled a number of technical and logistic issues must be resolved (Courtesy of Fairchild)

In addition there is the issue of component availability to be considered. While the number of surface-mount components being manufactured is growing rapidly, so is the demand for them. Lead times for certain types of surface-mount components can be quite long. A manufacturer can get into serious trouble if a product is designed for surface mount and parts shortage is encountered. Consequently, an option which might be explored is the mixed-technology concept.

The full potential of surface-mount technology will not be fully realizable until the 'glue' circuit components become readily available in surface-mount form, explains Jack Balde of Interconnection Decisions. Those components, according to Balde, include the MSI and SSI devices such as line drivers and hexidecimal convertors that tie the various VLSI components together and communicate with the outside world. Balde claims that such devices account for up to 98% of the total board space. With the short product-life cycles in today's electronics industry, it is not practical to use surface-mount packaging for such devices if they are not off-the-shelf items.

Three classifications have evolved which are used to categorize different surface-mount component assembly constructions. These classifications are Type I, Type II and Type III.

A Type I assembly uses surface-mount components exclusively. Assemblies which fit under this classification may have components attached only to the top side or to both the top and bottom side of the circuit board. This type of board construction permits the fullest utilization of the advantages of surface-mount technology.

Sometimes it is necessary to use both surface-mount and inserted components in an electronic assembly's construction. This construction is commonly referred to as the mixed-technology construction. The Type II classification for SMT assemblies is a mixed-technology concept. Included under this classification are assemblies with SMCs and through-hole components only on the board's top, SMCs and through-hole components on both the top and bottom of the board, and

SMCs on both the top and bottom of the board and through-hole components only on the top. These mixed-technology constructions can permit the manufacture of relatively high-density circuit assemblies by using SMCs while taking advantage of availability and cost breaks resulting from the use of through-hole mounted components. A drawback of this approch is the increased number of processing steps and manufacturing systems required.

A fourth type of mixed-technology construction, frequently referred to as a Type III assembly, uses only through-hole components on the board's top and only SMCs on its bottom. This technique offers some density advantages over through-hole technology and permits the utilization of existing insertion equipment, the one-step wave-soldering process, and readily available components. On the other hand, this technique has the worst circuit density of all the mixed-technology concepts since it does not make use of high density surface-mount ICs. Furthermore, it uses adhesives to hold the surface-mount components in place during wave soldering, eliminating the self-alignment of SMCs. Self-alignment commonly occurs with reflow soldering as a result of surface tension forcing off the molten solder.

The type of assembly construction that is finaly chosen will influence, if not dictate, what manufacturing processes will be used.

The adoption of surface-mount technology will result in the need to become familiar with a number of new manufacturing processes. Dynapert's Tom Nash notes that surface-mount technology involves more precisely controlled chemical processes than through-hole mounting. Instead of using bar solders and liquid fluxes, one more than likely will use solder pastes, which have a whole new set of key parameters. Wave soldering will more than likely be replaced by reflow soldering. This means that new soldering techniques, such as vapor phase or IR reflow, must be learned. Even if wave soldering can be used, the process will require such modifications as the use of a dual-wave configuration instead of the more familiar single wave.

Instead of using through-hole insertion

systems, automatic component placement equipment will be used. This will present a completely new set of problems which must be addressed. MCT's Fred Snyder notes that copper trace registration will become even more important than with through-hole mounting.

The close spacing between surface-mount components and the substrate to which they are attached increases the likelihood that flux residues will become trapped under them. Cleaning processes will have to be re-examined and possibly revamped or even replaced.

Component standards

Standardization of component packaging is another significant issue. The lack of uniformity in packages can complicate both circuit design and manufacturing. While efforts, led by JEDEC, are under way to develop standards for surface-mount component packages, much work still needs to be done. One of the main advantages cited for surface-mount technology is its adaptability to automation.

The lack of uniformity in component standards, coupled with the reluctance of some device manufacturers to conform to those existing standards, is complicating the automation process. Texas Instruments' Charles Hutchins notes that there is a need to develop standards that are acceptable to as many people as possible. Hutchins explains that so many compromises are made when standards are written that the resulting document often does not suit anybody's needs.

Component feeding techniques, using tape, were developed as a means of getting around the lack of uniformity in device packages. However, the lack of uniform tape standards has complicated the issue. Recently, the EIA released recommended standards for a variety of tape formats for component handling in an effort to resolve this problem.

Unfortunately, the packaging of surface-mount components in the reeled tape format tends to preclude testing and burning-in of components by the user. Texas Instruments' Charles Hutchins notes that a fundamental

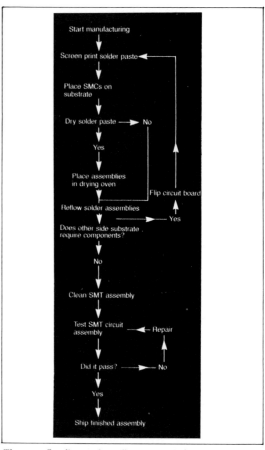

The process flow diagram shown illustrates a typical manufacturing sequence for a Type I surface-mount assembly. A Type I assembly uses surface-mount components exclusively

change in the component supplier and user relationship is needed to remedy this shortcoming. The end component user is going to have to rely more and more on the supplier to assure the quality and reliability of the components used. Thus the supplier, manufacturer, and user will have to establish a strong working relationship in which the user has total confidence in the supplier.

Solder-joint integrity

Probably the most significant consideration when adopting surface-mount technology is assuring the reliability of the interconnection between the components and the substrate. Surface-mount component attachment relies exclusively on the attaching medium, be it solder, the main technique used, or conductive adhesive, currently being investigated as an alternative to solder, in order to make the connection to the surface conductor. Through-hole mounting technology, on the other hand, also makes use of the clinched lead and solder in the plated through hole to assure a reliable connection. Thus the integrity of joints between the component, the attachment medium and the substrate surface is extremely critical for surface-mount technology.

The design of the solder joint plays a significant role in its ultimate reliability. ECR's Carmen Capillo notes that tests on leadless chip carriers show that changing the attachment land, which extends beyond the package's perimeter from 0.020in. to 0.040in., will significantly increase a solder joint's resistance to failure during thermal cycling. However, extending the land further than 0.040in. does not seem to provide much additional benefit, Capillo points out.

Because of the importance of forming a reliable solder joint, component and substrate solderability must be assured. Reliability is particularly important at this point, since defective solder joints are not easily observable or repairable. Planarity of component leads and flatness of the substrate are essential. It is generally agreed upon that lead planarity must not be worse than ±0.002in. Some feel that this is even too loose.

That it is critical to assure solder-joint integrity is reflected by the number of techniques being developed in an attempt to assure that joints will not fracture under the strains of expansion and contraction. These strains arise when components and substrates expand or contract at different rates. Ambient temperature changes, coupled with differences in the thermal coefficients of expansion of materials used for substrates and component packages, or temperature gradients within the assembly due to power cycling effects, can create strain on the solder joint.

The strain level associated with the expansion and contraction of small chip components such as resistors and capacitors, where the actual change in length is small, are normally rather low. However, as component

sizes increase above that of a 44 I/O leadless ceramic chip carrier, strains can increase

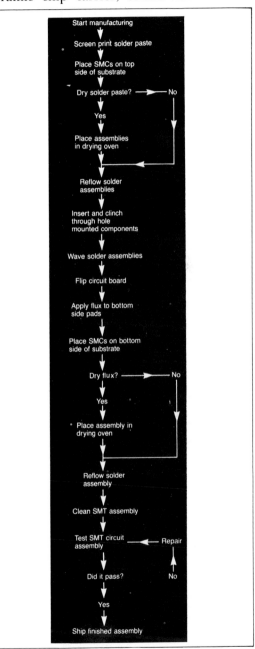

The Type II assembly manufacturing sequence illustrated in the diagram is for a circuit assembly with surface and through-hole mounted circuits on the top of the substrate and surface-mount components, including active devices, on the bottom. There are a number of other possible manufacturing sequences for Type II circuit assemblies. Which is most appropriate depends on the components used

substantially. These strains are especially pronounced if the substrate material is an FR-4, which has a thermal coefficient of expansion (TCE) more than double that of the ceramic.

Substrate solutions

One approach to the problem of minimizing strain on the solder joint is to use compatible materials, such as thick-film ceramic substrates for leadless ceramic chip carriers or glass-epoxy chip carriers to attach to FR-4 substrates, for both the component package and the substrate. In the former case one pays a cost penalty for the substrate; in the latter case one forfeits hermeticity of the package.

Over the last few years much time and

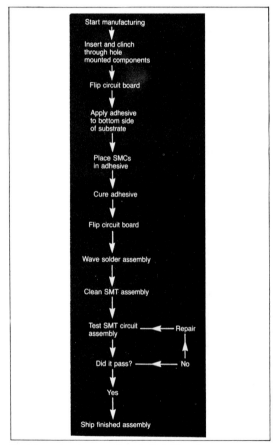

The Japanese have been manufacturing Type III surface-mount assemblies for a number of years. The process flow diagram shown above depicts a typical manufacturing sequence for a Type III assembly which has through-hole components on one side and surface-mount components on the other side

effort has been devoted to the research and development of alternate substrate materials to provide TCEs that closely match those of the ceramic chip carrier. Some have used fiber modifiers with low TCEs, which then combined with the proper amount of organic resins yield composites with low TCEs closely matching that of ceramic. Examples of such composites include epoxy-Kevlar, polyimide-Kevlar and polyimide-quartz.

Still another technique that has been tried to provide a substrate with a TCE closely matching that of a ceramic chip carrier is bonding a polyimide-glass or an epoxy-glass multilayer to a low TCE restraining core material. Typical of such materials are copper-Invar-copper, alloy 42, copper-molybdenum-copper, and copper graphite. Such restraining-core constructions usually require that laminate be bonded on both sides to form a balanced structure so that they will not warp or twist. Most restraining-core systems have a weight penalty.

Compliant interfaces

Still another approach has been the introduction of a compliant layer on the interface surface. This material is intended to absorb some of the strain. Unfortunately, all the laminates which have been reviewed are more costly than FR-4 glass-epoxy. Some are substantially more expensive.

Various techniques have been developed to increase the compliancy of the solder joint by increasing the solder standoff height between the component's bottom and the susbtrate's surface. Alpha Metals has been formulating solder pastes containing lead spheres which will not melt when the surrounding solder has melted, in an attempt to assure a minimum standoff. Multicore has added porcelain spheres into some solder-paste formulations to accomplish this. Raychem has developed solder columns supported by a helix of copper wire and contained within a dissolvable frame which holds the component up during reflow, thus assuring a high solder standoff.

In a paper presented at the 1984 IEPS Symposium, Hitachi revealed a fairly innovative approach to forming a controlled solder standoff height. The technique involved the deposition of large, high-temperature solder bumps underneath a flip-chip device. These large solder bumps hold the device up, causing the smaller I/O bumps with lower melting temperatures to form elongated pillar or hourglass-shaped solder joints.

Compliant leads

Relying on leads to act as compliant members between the substrate and the component is another possible solution to the TCE problem. Clips can be attached to chip carriers to hold them on the substrate and absorb strain. Post-molded, surface-mount IC packages, such as the SOIC with its gull-wing lead and the PLCC with its J-lead configuration, rely on the compliance of the lead to permit their use on FR-4 substrates.

Unfortunately, using a lead does not assure that there is enough compliance. The material from which the lead is made and the way in which the lead is formed and soldered can adversely affect its compliancy. Some manufacturers have continued to use alloy 42, which is fairly stiff, as a lead frame, while others have turned to more compliant copper alloys as leadframe material. Even improper soldering techniques can degrade lead compliancy. If excess solder fills in the bend of the gull-wing on an SOIC, the compliancy of that lead could be significantly lowered.

Balde pointed out, in a paper presented at IEPS, that some of the PLCC package device manufacturers anchor the end of the J-lead at its bottom. Others, he noted, were bending the leads around a rigid support along the bottom perimeter. Some were even doing both. All of these factors could potentially make them non-compliant members.

These packaging techniques have raised significant concerns over a lack of compliancy in leaded surface-mount components. As a consequence, a task force has been formed under the umbrella of the IEEE Computer Society to test the lead compliancy on PLCC packages supplied by a number of major device manufacturers. This test program, which is a cooperative venture involving a number of major component users and suppliers, is currently in progress at a number of sites throughout the country.

Because of the increased circuit density achievable with surface-mount attachment techniques, more circuit power can be concentrated in a smaller area. At the same time the exposed surface area of the device package is reduced, decreasing the surface area available for dissipating heat. Consequently, thermal management takes on increased significance.

Fortunately the reduction in package size also means that the surface-mount devices have shorter thermal paths. The thermal resistances (-jas) between the active junction and the ambient air of different surface-mount packages can become an issue. Device manufacturers are substituting copper alloys for the more traditional alloy 42 as a lead-frame material because of its improved thermal-dissipation.

Thermal-management problems can dictate whether leaded or leadless component assembly construction can be used. Sometimes the ceramic leadless chip carriers might have to be used even though hermeticity is not an issue, in order to take advantage of a shorter thermal path to the substrate.

Component reliability

Reduced component package sizes and variances from conventional through-hole mounting processing steps has raised con-

Pictured is a grouping of various leaded plastic IC packages designed for surface mounting. Concerns that the lead compliancy of these packages are not consistant from manufacturer to manufacturer is being investigated in a test program set up by a committee of IEEE's Computer Society

cerns about the reliability of surface-mounted components. Of particular concern to prospective surface-mount users are molded plastic devices.

National's Joseph Huljev notes that concerns over the reliability of PLCCs have, for the most part, tapered off. He attributes this to the growing recognition that the size of the PLCC package affords as much, if not more, protection to the chip than the standard plastic DIP. On the other hand, the reduced size of the SOIC over the DIP does raise some fear in prospective users. These concerns, he explains, would be justifiable if there had been no packaging changes from those used in standard DIP technology when the SOIC package was developed. However that is not the case, he claims. Huljev points out that there have been material changes, tightening of materials specifications and structural changes in the SOIC package that have improved its reliability. Changes in lead-frame design and plating have in effect made the path that moisture must travel larger than the width package.

Huljev offers the summary data from high-temperature bias humidity tests as an illustration of the reliability of the SOIC package. After 2000 and 4000 hours of testing, devices packaged in SOIC configurations under bias at 85°C and 85% relative humidity were as good, if not better than, the DIP. It was only after 6000 hours that the test results were slightly more favorable for the DIP. Huljev explains that this is much longer than the average component would likely be expected to operate under such hostile conditions.

Even with supportive data such as that presented, device reliability will still remain an issue; at least until historical data is collected. In addition, it must be recognized that components have process-related limitations, which if exceeded can create reliability problems. For instance, if an SOIC is exposed to a solder wave for longer than five seconds, stresses in the package could create cracks along the plastic-to-lead interface.

Testing and inspection

Two areas where surface-mount technology exhibits a significant difference from through-hole mounting technology are inspection and test. The increased density of the SMT circuitry significantly complicates the task of visual inspection. Even more troublesome is the fact that some of the devices, such as LCCCs and PLCCs, have solder joints which are formed underneath the package body, making visual inspection virtually impossible. Because of the extreme difficulty of visually inspecting solder joints it becomes essential that the solder process used for joining the circuit be extremely consistant and reliable.

The density of surface-mount component assembly complicates testing. Unlike through-hole mounting techniques which tend to space mounting pads on 0.100in. centers, most surface-mount pads are spaced on 0.050in. centers. This means that probe sizes have to be down-scaled to permit closer positioning. Test point pads on 0.100in. centers can be brought out from the lead attachment pads, however this detracts from the achievable circuit density. Further complicating testing is the fact that some surface-mount assemblies have SMCs on both the top and the bottom, generally eliminating the possibility of placing test pads on the substrate's bottom for components on the top.

Factron's David Clayman notes that involvement of the test group in the early design stages of surface-mount assemblies is more

Number of test hours completed	Percent failure	
	SOIC	DIP
2,000 hr	0.095	0.240
4,000 hr	0.336	0.350
6,000 hr	1.100	0.750
Test conditions: Components biased at 30V 85 C temperature 85 % relative humidity		

Source: National

Reliability test data comparison of the SOIC and DIP configuration

crucial than it is in the design of through-hole mount circuits. With through-hole circuits, the connection pad on the bottom side of the substrate is always available for probing. However, on surface-mount assemblies probing is normally done on the same side as the mounted component.

With many of the solder joints being formed under the body of the package, it is desirable to extend the component mounting pad beyond the package to provide a probing point. This eliminates the problem of trying to probe right on the component lead. Clayman notes that placing the test probe directly on the component lead can make an electrical connection on a cracked solder joint, resulting in the acceptance of a defective circuit.

DESIGNING RELIABILITY INTO SURFACE-MOUNT ASSEMBLIES

There are several major factors to consider when designing a surface-mount printed circuit board. First is consideration of the design type, which will fall under one of three categories: high density, moderate density or low density. Whenever possible, PCB low-density design guidelines should be used. There are several important reasons behind this; the two most important being manufacturing ability for zero defects and assembly reliability. Some other major design factors are component layout and orientation, conductor routing, land (pads) dimensions, solder mask, and substrate materials.

In each of the above factors, a thorough understanding of the many necessary conditions in surface-mount design has been extensively researched and tested under environmental and power-cycling conditions. Most of this research has been done and verified in the field by the military and aerospace industries which have used surface-mount assemblies since the 1960s for high-reliability electronic systems. It should be stressed that small changes in design, such as changes in land dimensions or in substrate materials, can make a huge difference in the assembly's performance or reliability. These differences can be as small as 0.005in. in land size or in the choice of an epoxy substrate vs. an epoxy-polyimide.

First published in *Electronic Packaging & Production*, July 1985, under the title 'How to design reliability into surface-mount assemblies'.
Author: Carmen Capillo, ECR Corp.

Component layout

Automatic machine mounting of surface-mount components such as SOICs, quad packs, SOTs and chips requires attention to

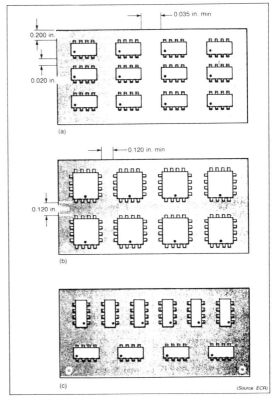

Preferred layouts for SOIC or quad packages. Dots represent pin 1 : (a) X-axis layout (SOICs), (b) pin 1 on quad packs all in the same axis, (c) X- and Y-axis layout

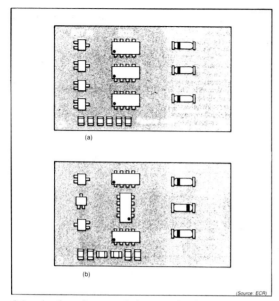

(a)

(b)

(Source: ECR)

Orientations for surface-mount devices: (a) preferred – one axis, one span and one polarity direction, (b) minimum acceptable – two axis, two polarity/pin directions and several spans

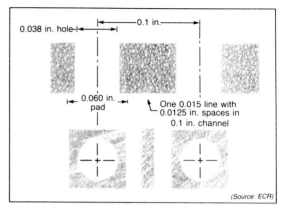

(Source: ECR)

One track, low-density routing dimensions

the orientation of these components during the layout stage. Optimizing component layout increases production efficiency and reduces costs.

The X-axis layout is certainly preferred over the others, since in this layout, the assembly can be soldered in several ways unlike the other axis layouts. All three types of layouts can be reflow soldered by infrared or vapor-phase technologies, whereas the dual-wave process can be used to solder the X-axis layout when certain process variables are optimized. When assembling quad packages with J-leads it is important that lead spacing between adjacent components should be 0.100in. minimum so as to allow visual solder-joint inspection and rework when necessary. This spacing is not as critical on SOIC packages in the leaded sides of the devices, where a 0.020in. spacing is recommended. At the non-leaded sides of the SOIC package, a 0.035in. minimum spacing between adjacent component bodies is recommended due to placement tolerances and extraneous epoxy material which sometimes extends from the component body due to the molding process of the package.

Another important concern is component

orientation. The preferred orientation occurs when components are laid out with one axis, one span, and one polarity direction. This minimizes machine placement movements such as head and table rotations. A component orientation which presents maximum machine head and table rotations can substantially reduce productivity. In some cases,

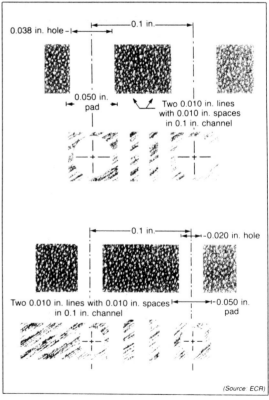

(Source: ECR)

Two track, moderate-density routing dimensions

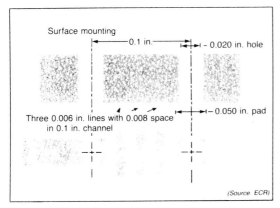

Three track, high-density routing dimensions

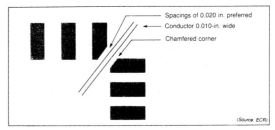

Chamfering corner land patterns for quad packs

certain SMD placement systems are unable to assemble a board with various conditions and can cause hand placement of some of the components.

Conductor routing

There are several areas of concern regarding conductor routing on a surface-mount printed circuit board that will have a large effect on manufacturing yields and reliability. Routing conductors between plated through holes and lands (termination pads) shall be the first major consideration. In low-density designs it is advantageous to avoid routing conductors between lands, since the width of the lands should be maximized and allow

insufficient spacing for routing. Low-density tolerances for routing a conductor between plated-through holes on 0.100in. grids are also illustrated. Here, a conductor of 0.015in. width can easily be routed between two 0.060in. diameter lands with 0.0125in. spaces. Moderate density or two-track circuit boards fall under this category, where two 0.010in. wide conductors are routed between 0.0160 lands with a reduced spacing of 0.010in. High density, or three-track designs, would allow a total of three conductors between plated-through holes with 0.006in. widths and 0.008in. spaces.

In even more complex or higher density designs on less than 0.100in. grid, pattern can be determined by the formula and illustration shown. Additional fabrication costs will be incurred due to increase in board fabrication complexity and possibly board assembly. Two important conditions must be met in routing

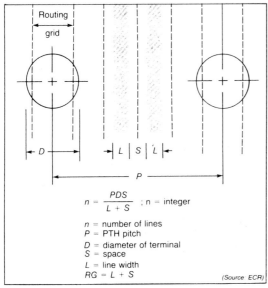

$$n = \frac{PDS}{L + S} \quad ; \ n = \text{integer}$$

n = number of lines
P = PTH pitch
D = diameter of terminal
S = space
L = line width
RG = L + S

(Source: ECR)

Formula for routing conductors between holes

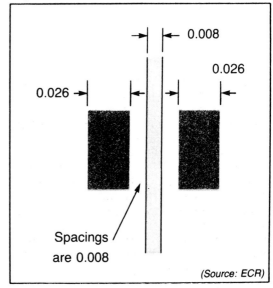

Routing conductors between lands

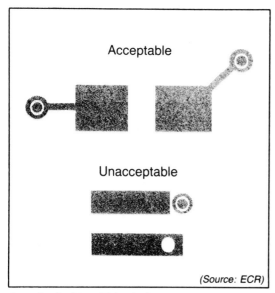

Connecting plated through holes to lands

conductors between surface-mount lands. The first is routing a conductor between the corners of quad-pack type land patterns. Here, chamfering the corners to allow optimum spacing can be illustrated. The other routing conditions are between lands. In general, land widths are normally 0.35in. (0.050in. centerlines), which do not allow sufficient room for conductor routing unless a 0.005/0.005 width and spacing is acceptable. Therefore, routing conductors between lands requires high-density designs in that the lands are reduced to a 0.026in. width with a conductor routed along the center. In such cases, optimum designs would leave all other

lands with 0.035 widths if no conductors are routed between them.

Also, another conductor routing concern is plated-through hole attachment to lands. In this case, to prevent component movement in the vapor phase or IR reflow process, a conductor 0.010in. wide by 0.020in. long (minimum) connecting land to PTH is preferred. For dual-wave soldering process this does not apply.

Land (pad) dimensions

Few aspects of surface-mount design are more important than land dimensions. Land dimensions obviously differ for the variety of SMDs that are available and being used in assemblies. Although there may be typical dimensions which are being used, it is important to use these tested land dimensions which offer optimum reliability with manufacturability a less important concern. From past research, with varying land designs a significant change in assembly reliability in regards to thermal, humidity, power, and vibration cycling testing has been observed for all types of components and their respective land patterns. Changes as small as 0.005in. have shown in a variety of cases from SOIC lands to LCC lands to make substantial differences in reliability. An example of such a case with 64, 36, and 16-pin hermetic leadless chip carriers is illustrated. Significant additional numbers of thermal cycles to failure (failure being any electrical or solder joint failure) can be gained by increasing the land extension beyond the

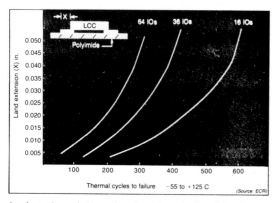

Land extension variations as they relate to thermal cycles to failure for ceramic leadless chip carriers

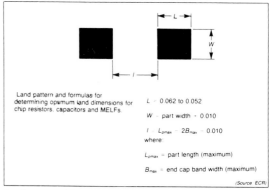

Land pattern and formulas for determining optimum land dimensions for chip resistors, capacitors and MELFs

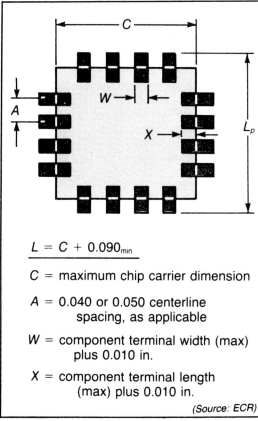

$$L = C + 0.090_{min}$$

C = maximum chip carrier dimension

A = 0.040 or 0.050 centerline spacing, as applicable

W = component terminal width (max) plus 0.010 in.

X = component terminal length (max) plus 0.010 in.

(Source: ECR)

Optimum land pattern dimensions for leadless plastic and ceramic chip carriers

edge of the device to a certain extent where additional land extension adds no greater reliability. This design/reliability relationship has been shown to exist for all types of land designs such as in J-lead quad packs, chip resistors/capacitors, and SOIC packages. In leaded devices, such as SOIC packages, the most sensitive design parameter is the land extension behind the heel of the lead. Land extension beyond the toe of the lead is less critical as it relates to reliability. Optimum and tested land dimensions are provided for some SMDs along with formulas which can be used for different size components.

Solder masks

Often one sees solder mask on termination lands and exposed conductors which prevents good solder joints and causes solder bridging. In order to determine the design criteria for solder masks, one should first determine the density of the PCB and assembly. For low- and some moderate-density designs, liquid solder mask can be a suitable material provided the designer understands that the screen printing of this material can range in accuracy of 0.008–0.010in. Therefore, when designs have conductors routing in between lands, liquid solder mask cannot be applied accurately enough to assure complete coverage of the conductor without some flow of solder mask onto the lands. Dry film solder masks or Probimer solder masks are certainly recommended for high-density and some moderate-density boards when conductors are routed between lands. The additional costs of these materials is easily made up due to the higher board and assembly yields as compared to the lower yields if liquid solder mask is used.

There are two types of solder mask windows which can be used in surface-mount designs. The 'gang' solder mask window can afford effective board masking and at the same time minimal chances for solder mask flow onto the lands. This type of 'gang' window should only be used when there are no conductors routed between lands.

The 'pocket' solder mask window should be used when conductors are routed between lands to prevent solder bridging upon reflow or dual-wave soldering and component lead

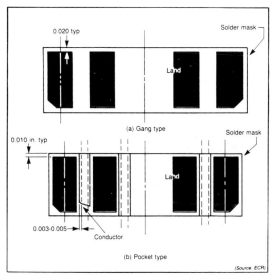

Solder mask windows

Solder mask flow characteristics: (a) high flow properties of screen-printed liquid solder mask (30X), (b) no-flow properties of photosensitive Probimer solder mask (30X). Artwork misregistered in this case

shorting to conductor. As illustrated, the tight tolerance requirements for this window would make it difficult to obtain with liquid solder mask and maintain acceptable conditions.

Substrate materials

The choice of substrate materials (sometimes called P/I, for packaging/interconnecting structures) can vary depending on the type of environment the assembly will see and also the types of devices being assembled. The substrate that one should use can also depend on the extent of reliability that is required for the assembly. For most commercial products, epoxy-fiberglass is being used for a supporting structure for leaded surface-mount ICs, chip resistors and capacitors, leadless resistor networks, leadless inductors and potentiometers. Other products that use epoxy substrates include, but are not limited to, automotive electronics, disk drives, exterior alarm systems, and cameras. When optimum designs are utilized along with quality manufacturing processes, epoxy substrates are suitable for most commercial products. When high-powered devices and leadless chip carriers are used, especially in extreme environmental conditions, more suitable substrates such as

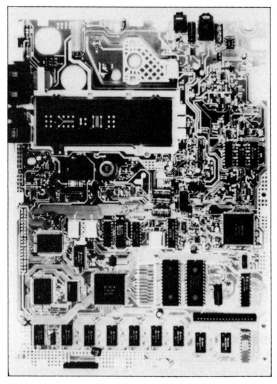

Portable computer assembly with a variety of SMDs and conventional devices (Courtesy of Convergent Technologies)

polyimides, copper-Invars, and polyimide epoxies must be considered.

The major concern of mounting surface-mount devices, especially leadless chip carriers, is the thermal expansion mismatch between the component package material and substrate material. This condition for conventional assemblies has not been an issue since leads are formed with sufficient 'stress relief' bends to avoid material mismatch stresses which can cause cracked solder joints. Much has been researched with these materials to determine their reliability difference with various assembly types and environmental changes.

Producing zero-defect products

In order to produce zero defects in manufacturing assembly, quality measures must start in the design. Poor and untested (unverified) designs can lead to catastrophic manufacturing problems and large product field returns. The surface-mount designs as specified herein have been thoroughly tested for moderate to harsh environmental conditions and utilized in the most sophisticated electronic systems. Some deviations from optimum design rules may be required. If such a case exists, then those involved should have an understanding as to what effect this design deviation has on reliability. Remember, in surface-mount technology as little as 0.005–0.010in. difference in design dimensions can have a remarkable change in reliability for better or worse.

DESIGN SOLUTIONS

Most manufacturers today hope to realize the full benefits of automation: increased yields, increased reliability, and reduced assembly times. To do this with DIP packages, layouts would have to be expanded by as much as 40% simply to accommodate the placement heads which close around the devices, crimp their leads and insert them through holes in the board. Placing the DIPs too close together would cause the placement head to bump devices already on the board. This problem does not exist with furace-mount devices (SMDs), which are much smaller than their DIP counterparts.

For example, a 14-pin small-outline (SO) package is 70% smaller than a 14-pin DIP, has a 70% lower profile, and saves 90% in total component weight. Furthermore, surface mounting enables the use of both sides of the PCB for mounting parts. Automatic pick-and-place machines use vacuum pencils to pick up the devices and place them on the appropriate spots on the PCB. The ends of these vacuum pencils are smaller than the parts they are lifting; therefore, there is no need to expand the board layout.

Because of the reduced inductance and capacitance of the SO package, device performance is less affected by package considerations than in a standard DIP. Parts generally appear to run faster in SO because of the reduced effect of package impedance.

Types of surface-mount packages

Packages for surface mounting have been around for many years. The ones discussed here are commercial SMD packages for integrated circuits; however, it is worth noting that there are SMD packages for resistors, transistors, capacitors, diodes, coils, and nearly everything else the modern PCB manufacturer needs for a total surface-mount assembly.

SMD packages for ICs are divided into two general categories: the SO package for low pin-count parts (eight through 28 leads) and the PLCC for the higher pin-count devices, or those 20- through 28-lead devices which have overlarge dies.

The SO package was developed by N.V. Philips and introduced to the Swiss watch industry in 1971. In 1977, Signetics pioneered

First published in *Electronic Packaging & Production,* January 1985, under the title 'Surface mount offers choice of design solutions'.
Authors: Mark Kastner, Signetics Corp. and Vern Solberg, Design Center for Surface Mount Technology.
© 1985 Reed Publishing (USA) Inc.

the domestic SO package by releasing several analog functions which were used primarily in the telecommunications and the hybrid industries. Until late 1981, only analog and CMOS parts were available. TTL was then made available in the SO package, and its acceptance and growth in the PCB industry has been explosive.

There is now a large variety of logic devices available in the SO package, including low-power Schottky (74LS), Schottky (74S), standard TTL (74), CMOS (HEF4000), high-speed CMOS (74HC/74HCT), and FAST (Fairchild Advanced Schottky TTL, a trademark of Fairchild Camera and Instrument Corp.) (74F). Signetics alone has over 350 logic and analog functions in SO packages.

The SO is manufactured in 0.150in. wide and 0.300in. wide versions. A 0.300in. version is called small outline – large (SOL) and offers a larger die cavity. Pin counts associated with these formats are shown in the table.

SO packages are industry standard, registered with the JEDEC JC11.3 committee, and multisourced. Some of the vendors of JEDEC SO packages are Signetics, Philips, Motorola, Texas Instruments, Siemens, Exar, SGS/ATES, RCA, Thomson CSF, Fairchild,

Pin count	SO (150 mil)	SOL (300 mil)
8	X	
14	X	
16	X	X
20		X
24		X
28		X

Note: There are two formats in 16 pins. When designing with a 16-pin SO package, check with the factory to ensure that the correct format will be supplied.

SO and SOL pin counts

Linear Technology and National Semiconductor.

Several Japanese vendors make packages similar to the SO package. None of them yet meet the JEDEC standards and most are different from each other. However, some do fit the same footprints as JEDEC packages.

A board using SMCs illustrates the variety of footprints. This board was stuffed by Philip's Model MCM III pick-and-place machine

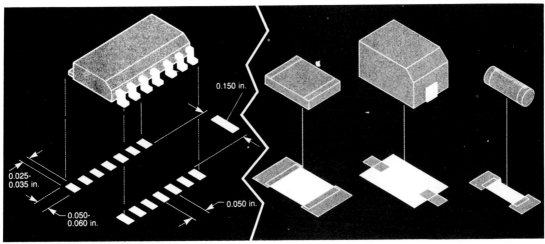

0.150 in.

0.025-
0.035 in.

0.050 in.

0.050-
0.060 in.

The recommended footprint for SO packages has a generous tolerance (Courtesy of NuGrafix Group Inc.)

Passive SMDs are supplied in many styles and sizes (Courtesy of NuGrafix Group Inc.)

Ceramic leadless chip carriers

The ceramic leadless chip carrier, a high-performance package for high-pin-count devices (such as memories, microprocessors, peripherals, and gate arrays) is used extensively in the military and other high-rel environments. However, because of cost and mechanical difficulties in mounting ceramic leadless devices to PCBs, the plastic leaded chip carrier was developed for commercial applications. The PLCC uses a lead frame with the leads bent down and under (J-hook); they can be soldered directly to footprints on a conventional PCB. This is not the case with the ceramic leadless chip carrier, largely due to the different thermal coefficients of expansion between the package and the glass-epoxy PCB. PLCC pin-counts include 20, 28, 44, 52, 68, and 84.

One other notable package form is the quad pack. This is a generic term for any package with leads on all four sides. It implies no standardization of lead spacing, lead bend, or pin-out. Normally, when people refer to quad packs, they mean the packages with the leads bent down and out on all four sides. Quad-pack ICs are supplied in several package sizes and lead spacing varies from one manufacturer to another. Only a few of these quad-pack formats are available in the 0.050in. lead spacing that is preferred by designers using CAD or grid layout systems. It is common to

see 0.04in., 1.0 and 0.08mm lead spacing on devices with 20 or more leads. The delicate leads are difficult to handle with automatic placement equipment without tedious hand alignment and touch-up after reflow solder.

Success begins with the design

Instead of being inserted into plated-through holes, SMDs are placed on metalized footprints etched onto the substrate surface. Designers will use a more refined set of rules for surface-mount PCBs because components can be spaced closer together with smaller contact spacing, making narrower conductor trace widths necessary.

A common signal conductor can be 0.010–0.012in. wide, and 0.015in. through 0.030in. is adequate for power and ground bussing. The footprint contact area suggested by vendors has a generous tolerance. For the SO, a rectangular pattern is used on 0.050in. spacing. The length of the pad is 0.050–0.060in. and the width can vary from 0.035in. down to 0.025in. or 0.020in.

The 0.025×0.050in. footprint pattern will work well using the grid placement system favored by most designers. The 0.012in. conductor width spaced at 0.025in. provides a reasonable 0.013in. air gap between traces.

However, if conductor traces are routed between contact pads, it will be necessary to neck-down the trace width to 0.008in. and still

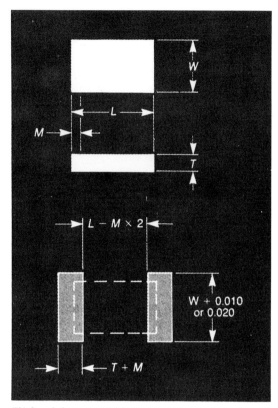

This formula for passive SMD footprints is applicable for both reflow and wave-solder processes (Courtesy of NuGrafix Group Inc.)

retain an equal air gap at each side. Because neck-down traces require additional time in both hand taping or CAD/photoplot generation of art masters, some compromises may be justified. By reducing the contact pad size to 0.020 × 0.050in., it is possible to route a consistent 0.010in. conductor trace width and still maintain the desired clearances.

Some PCB shops may not maintain the consistent quality necessary when using this fine-line approach over the entire board. It is important to discuss limitation and premium cost penalties with suppliers before making a full commitment to the 0.010in., and smaller, trace widths.

Another very important consideration is the thermal concentration caused by miniaturization. The same die is being used in the SMD as in the DIP, thus the power dissipated is the same. However, the smaller packages are being placed much closer together, concentrating the thermal energy.

The trade-offs between the increase in density and the concentration of thermal energy must be evaluated. These factors may influence the choice of PCB material, the number of layers and the thickness of the PCB. New methods to transfer heat from the package to the board and then away from the board should be considered by the designer.

Other factors, besides the placement system and soldering methods, are post-assembly cleaning inspection, test and the availability of parts in SMD packages. A first step is to list all the devices needed in a design and to determine which are available in SMD format. There are several cross reference lists available from design and assembly services as well as part lists from the various component vendors to aid the compilation. However, with the explosive growth of this market, none of these lists are current. It is a good practice to check with vendors because parts availability is growing daily.

Soldering methods

When choosing the type of footprint to use, it is very important that the designer consider the soldering method. Basically, there are two types of soldering in use today: flow soldering (wave, drag, or hot solder dip) and reflow soldering (vapor phase, infrared, thermal conduction through the PCB, and hot air).

The Japanese consumer electronics industry has had considerable experience using wave soldering with SMDs. They normally solder the PLCCs as an additional step after

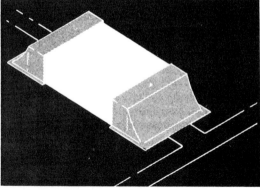

Shown is a typical footprint pattern for reflow soldering of passive SMDs. Solder paste holds components in position during the soldering process (Courtesy of NuGrafix Group Inc.)

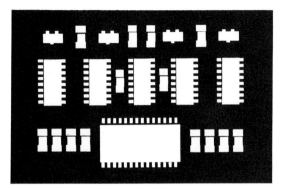

Passive SMDs and SO ICs should be mounted in one direction to take advantage of high-speed automated assembly (Courtesy of NuGrafix Group Inc.)

the discrete and passive devices are wave soldered (a secondary operation which adds cost to the product).

A form of reflow soldering should be considered if PLCCs will be used on the board due to the fact that wave soldering packages with leads on four sides usually result in solder bridges.

The SO, however, can be soldered using flow soldering methods. The devices must be attached to the PCB by means of an adhesive because the device side of the board will be facing down as it goes through the solder wave. The orientation of the part as it goes through the solder wave can play an important role in the elimination of bridges.

Experiments should be conducted by the user to determine the best footprints for use in a particular soldering system. Some users feel that narrower footprints help to reduce solder bridges. The Philips-sponsored SMD Technology Center in Milwaukee, WI, and other major users of these components have been experimenting with rounded footprints to reduce bridging during wave soldering. Both claim very good results.

Reflow soldering has been done for many years in the hybrid industry. A solder paste or solder cream is applied to the footprint prior to placement of the part. These pastes and creams contain tiny spheres of solder suspended in a carrier which contains the flux. As the substrate temperature is raised, the flux, solvents, and carriers are driven off and the solder liquifies.

Pastes and creams with various melting

points are available. As the liquid solder migrates to the metalized footprints, the surface tension is enough to move the leaded components. For SO packages, this can be an advantage because it acts as a self-positioning mechanism. However, it can be a problem for the smaller passive components if the solder paste is not printed on evenly. If there is an uneven amount of solder paste on one end of a two-lead part, the tension could tend to pull stronger on that end, thus lifting the other end from the board.

Passive components are supplied in many shapes and sizes. In an attempt to standardize and maintain material and process controls, components that will be available in the largest quantity and by more than one or two suppliers should be chosen.

Chip resistors and capacitors are usually furnished in rectangular packages. When specifying components, the value, size, voltage or wattage, and in the case of capacitors, the dielectric material should be listed. For general-purpose resistors, the most common size available is the 1206. This component is roughly 0.12in. long by 0.06in. wide and is rated at 1/8 or 1/4W. It is long enough to be handled easily in the assembly environment.

Many median-value ceramic capacitors are available in the 1206 package, but, more choices are open if the 1209 or 1210 size capacitor body is used. As with SO ICs,

Pin count	SO	SOL	PLCC
8	X		
14	X		
16	X	X	
18		X	X*
20		X	X
24		X	
28		X	X
44			X
52			X
68			X
84			X
* Rectangular			

PLCC package types overlap SO offerings

Note the difference in component size between a typical DIP and an SO package

footprint geometry is very important for each type of solder process employed.

Many variations of footprint patterns are possible. Many configurations are possible and should be tried on an experimental basis before making a large production run. In fact, first-time users of SMD are often advised by their more experienced colleagues to consider saving both time and development costs by seeking design and process consultants specializing in surface-mount technology.

Some typical footprints used in passive-component reflow soldering are illustrated. Note that the width of the footprint for the SO package varies from 0.025in. to 0.035in. Most users prefer the narrow footprint. The length of the print should be kept as short as possible

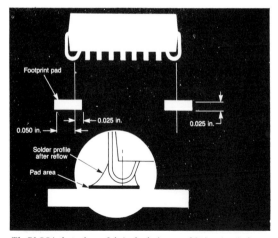

The PLCC is the package of choice for devices over 28 pins. A typical footprint design for the PLCC is shown (Courtesy of NuGrafix Group Inc.)

to prevent the part from swimming or sliding back and forth on the footprint while still allowing a good meniscus.

Another factor worth noting is that the footprint for a PLCC should not extend too far under the package, as this would promote solder bridges where they could not be seen during inspection. The footprint for the PLCC should extend out further from the package than the lead itself, to allow a good meniscus that will result in a strong, inspectable bond.

Careful placement of related components allows more effective use of a much smaller surface area. The interconnections that can be made on the substrate surface result in the elimination of feed-through holes. Reduction of these holes and their associated pad areas further increases the density of the layout, and reduces total board cost as well.

As indicated, the SO package has the same pinout on two parallel rows as the DIP package. Arranging related ICs in blocks or functional clusters with their associated discrete components can also help to maximize the use of the surface area.

Plastic leaded chip carriers

The smallest square PLCC is the 20-pin package. There are many reasons for this; the primary one is that below 20 pins, the package would be as thick as it is square, resulting in a cube which would be very difficult to handle in an automated environment. One vendor has a rectangular 18-pin PLCC and one has an 18-pin SOL package. There is some overlap in package types.

Many users have selected the SO format through 28 pins. There are many reasons for this. The SO is much smaller and lighter than the PLCC. The SOL, although a bit longer than the PLCC, still occupies about the same amount of board space.

Further, when using several packages and connecting them together, a given number of SO and SOL packages would take much less space than the same number of PLCCs, simply because of the interconnect geography.

Besides being smaller, the SO format is dual-in-line and has the same pinouts as those of a standad DIP (PLCC pinouts vary between devices as well as between manufacturers). The SO format is easier to handle and is much easier to visually inspect.

For devices over 28 pins, the PLCC is the package of choice, largely because it can hold much larger die than any of the SO formats.

Producing boards with SMDs

In the early days of PCB technology when plated-through holes were not possible, designers were forced to carefully plan component arrangements and connections. Using experience and ingenuity, they were able to eliminate crossovers while reducing the need for unwanted jumpers. With the advent of plated through holes and multilayer boards, the exclusive use of single-sided boards was eliminated. Using the single-sided concept, the techniques used to interconnect the SMD are as important as the footprint patterns. As noted before, the contact pads, on 0.050in. centers, range between 0.025in. and 0.035in. in width.

If a feed-through hole is to be added on the pad itself, two points must be taken into consideration. First, the hole diameter selected must allow for a reasonable location tolerance. A 0.010–0.015in. diameter plated-through hole in 0.062in. thick FR4 material may increase the cost of your PCB. Second, unless the feed-through hole on the footprint area is either plugged or masked, in a reflow soldering situation the solder will tend to migrate away from the IC contact, resulting in a poor solder joint.

It is more desirable to add a separate pad for via or feed-through requirements. To provide for the routing of conductor traces while ensuring an acceptable air gap, a 0.035–0.035in. square pad may be chosen for these feed-through holes. The square con-

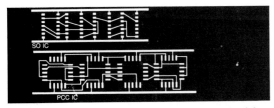

A plan for layout and component placement of SMDs on a single-sided board is illustrated (Courtesy of NuGrafix Group Inc.)

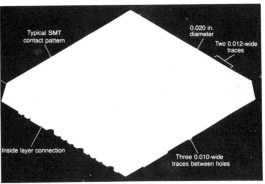

In this layout, plated-through holes have been used in conjunction with SMD contact patterns (Courtesy of NuGrafix Group Inc.)

figuration will furnish more than enough metal in the diagonal corners to compensate for the reduced annual cross section at the sides of the square. The 0.035–0.037in. square feed-through pad can be spaced at 0.050in. when necessary or on the more traditional 0.100in. pad. With this spacing it is possible to route two 0.012in. wide conductor traces between pads, something only possible before with costly multilayer designs using plated-through hole technology.

The feed-through pad is then connected to the component contact area with a narrow trace. This trace reduces migration of the solder paste during the reflow process. To further reduce migration of the liquid solder, application of solder mask coating over surface areas not requiring solder is recommended. This coating is applied with a wet screen process or photographically as a dry film and will act as a dam to contain solder to the contact area.

When using reflow soldering, the trace

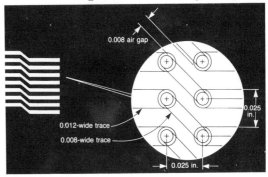

Offset stepping requires necking-down at the point of direction change to maintain the desired air gap

should be about half the width of the footprint pattern. As noted before, the signal-carrying conductors are generally 0.012–0.015in. wide. Supply voltages are carried on wider traces. When running traces between the device leads, it will be necessary to reduce the width to about 0.008in., which provides an 0.008in. gap between the trace and the edge of the pads when using 0.025in. pads.

Because the SMDs are so much smaller than their leaded counterparts, the scale of the layout should be considered. On larger boards with a mix of SMDs and leaded devices, a 2:1 scale may be adequate. More complex layouts can be designed at 4:1 scale with excellent results. The larger scale will make it possible to increase density while assuring accuracy. If designing with a CAD system, accuracy and desnity can both be increased by increasing the grid resolution.

Routing conductor traces requires careful planning. It is customary to use a 90° or 45° angle when traces must divert from a continuous line. Offset stepping several 0.012in. wide conductor traces on 0.025in. spacing will require necking-down at the point of direction change to maintain the desired air gap. The start and stop points of photoplotter aperture rungs must be carefully executed to reduce the chance of overlay and shorting.

If outside services are used for digitizing or photoplotting, requirements for accuracy should first be discussed. Some compromises may have to be made to ensure quality and control costs. Preparing art masters on Mylar using precision tape products and preprinted footprint patterns may afford more flexibility during your entry into surface-mount technology. Changes can be made easily, and economical photo reduction processes will provide high-quality working film. The technique used to prepare working film is a choice generally influenced by inhouse capability or services available in your region.

CHOOSING SMT AND SETTING UP A FACILITY

As surface-mount technology (SMT) continues to burgeon within the realm of electronics assembly, many companies face two problems: first, deciding what benefits a move to SMT might provide; and second, understanding the major considerations involved in setting up an SMT production line. The former is by far the easier task, while the latter requires expertise and, if implemented, a significant financial investment.

The benefits of SMT have been widely publicized. Electrically, SMT offers a reduction in circuit parasitics, IR drops, switching speed delay and EMI. Mechanically, the advantages are reduced size and lighter weight, or increased functional density. SMT design can be applied to existing products as well as new products in an effort to attain some or all of these advantages. For example, how SMT was applied in the redesign of an existing product is shown in the photograph. The conversion was not for mechanical reasons but specifically to improve electrical performance by eliminating a number of 'piggy-back' assemblies and leaded components in the 'hairpin' configuration. The SMT board illustrates the advantages of increased integration and the use of smaller package sizes such as SOTs and SOICs. (In fact, the primary advantages gained by SMT are a result of smaller component sizes, and only secondarily by the feature of surface mounting – optimized by mounting components on both sides of a board.)

Nicholas Hughes, manufacturing support manager at Tektronix, explains that the PCB shown here is a center-frequency control board used in a Tektronix spectrum analyzer. The original design exhibited noise problems and cross talk at certain frequencies. "The SMT version operates clean," says Hughes.

The latest SMT report by Electronic Trend Publications (ETP) says a practical approach

First published in *Electronic Packaging & Production*, July 1986.
Author: Howard W. Markstein, Western Editor.
© 1986 Reed Publishing (USA) Inc.

in deciding upon SMT is to first separate the product from the process. The initial analysis should determine if SMT will directly enhance the product and if it will provide some secondary advantages. Intangible items must also be considered, such as the effect on the product's competitiveness in the marketplace. The report states that the results of the analysis may differ for newly designed products, compared to existing products considered for SMT conversion.

For existing products that are well into distribution, exhibit satisfactory performance and have achieved success, the obvious answer is to leave well enough alone. However, the ETP report says certain factors should also be considered, such as: Does the market need a revitalized product? Is the competition increasing? Can SMT offer a decided marketing advantage? In essence, an affirmative answer to the question, "Will a conversion to SMT result in an improved product with added features and be available at or near the same price?," should decide in favor of SMT and thus maintain the product's competitiveness.

For new products, the ETP report says that the primary considerations for not using SMT are products manufactured in small quantity and those that require excessive hand assembly, such as power supplied with bulky and odd-shaped components. The present depth of the SMT marketplace ensures the feasibility of producing a design using surface-mount devices (SMDs). If in certain specialized circuits SMDs are not available, then the design should be considered for either a Type II (SMDs and through-hole components on either side of the PCB) or Type III (through-hole on top, SMDs on bottom) assembly.

Once the decision for SMT has been made – based upon an analysis of the anticipated improvements to the product – the next step is to acquire an SMT production capability.

Preparing for SMT

Education in SMT is a major prerequisite for planning and setting up a production line. Most of the larger equipment suppliers offer some kind of SMT training program. However, there are also companies that specialize in SMT education and offer seminars and formal courses on all aspects of SMT. For example, the Educational Division of AWI (Sunnyvale, CA) provides intensive seminars on SMT subjects tailored for managers, design engineers and manufacturing engineers. The SMD Technology Center (Milwaukee, WI) offers similar services.

The learning experience will also be enhanced by attending trade shows and SMT technical sessions, consulting with equipment

Surface-mount technology provides many mechanical and electrical benefits. Here, the original PCB assembly at right contains 'piggyback' and 'hairpin' inserted discrete components. The result: less than perfect performance at high frequencies. At left is the redesigned version showing a partial conversion to SMT. The result is improved electrical performance (Courtesy of Tektronix)

suppliers and professional consultants, and by visiting existing SMT installations. The most direct way of attaining expertise in planning for SMT is to hire a few experienced key personnel. For those not intent on setting up their own line, turnkey systems are available from specialists who supply all the equipment and training necessary for setting up an SMT production line.

Planning the facility

Once there is sufficient training and/or background in SMT available within the company, the time has arrived for making some critical decisions. Engineers involved in SMT activities at Tektronix advise that there are two major actions necessary in ensuring the proper impetus at the very beginning of the program. First, there must be full upper management commitment and financial support. Second, assigned responsibilities must be made to key individuals for each segment of the process, such as PCB design, assembly, test and purchasing. Management support

Responsibility	Personnel
Process development (solders, soldering, component attach, cleaning)	2
Components (selection, evaluation)	1
Pick and place equipment	1
Test and inspection equipment	1
PC boards (design rules, materials)	1
Thermal management	1
Program manager and project leaders	2
Design tools (subcontract)*	—
System designers*	—
Prototyping	2
Manufacturing	8

*Systems designers and software personnel

SMT staffing requirements

and efficiently organized engineering are the essentials for success in setting up an SMT production capability.

Tektronix says the two other activities that must then be addressed are equipment selection and the selection and training of the SMT manufacturing staff. An engineering/manufacturing/purchasing facilities team can be organized to select the proper equipment and to establish the overall facility. A manufacturing staff can then be assembled. Note the assigned areas for specialists (pick-and-place, test, etc.), in addition to general manufacturing personnel. This gives specified individuals responsibility for each critical operation.

Equipment and costs

Equipment selection decisions are based upon the type of SMT assemblies to be processed (all SMD or mixed), PCB size ranges, number of SMDs per assembly, and production output requirements. The basic equipment needed for manufacturing includes: screen printer for solder paste, drying oven, pick-and-place system, IR or vapor-phase soldering system, and a cleaning unit. For mixed technology using both leaded components and SMDs, the SMT line will require an adhesive-dispensing system for SMD attachment and a dual wave wave-woldering system. Adhesive dispensing can also be integral to the pick-and-place unit.

Generally, the pick-and-place operation is the gating factor in production volume flow. These units are classified as to the number of SMDs placed per hour, as opposed to the other equipment rated by the number of boards processed per hour. Pick-and-place systems have matured such that there is now a broad choice for satisfying a user's requirements for speed and variety of SMDs accommodated. There are now about 25 suppliers of pick-and-place equipment. The SMT Technology Center has in-depth literature and spectrum charts available that compare the latest pick-and-place systems as per placement capability and cost.

Quad Systems (Horsham, PA) says a major factor when setting up an SMT facility is to purchase equipment that is appropriate for present and future needs. Production re-

Equipment	Cost ($)
Solder screener	60K
Pick and place	200K
Soldering system	90K
Board cleaning	60K
Visual inspection (automatic)	120K
Board tester (manual)	100K
Miscellaneous (repair, etc.)	100K
Total	**730K**

(Source: Tektronix)

SMT production-line equipment costs

quirements, both current and future, must be analyzed to avoid being locked into equipment that is limited in scope. Quad also emphasizes the importance of having a long-range overview of plans for automation and factory integration, especially from the standpoint of off-line programming and CAD data download.

Whatever equipment is purchased, Tektronix advises that initial production be restricted to processing simple circuits having low density. A manual placement capability should also be maintained for backup and repair. As process control improves, boards of medium or higher densities can be assembled.

Cost estimates for setting up a prototype laboratory SMT capability range from $175,000 to $200,000. The equipment required would be of the batch type for small-lot processing.

Cost for a production facility will run in accordance with production volume requirements that dictate the choice of equipment. For example, pick-and-place systems can range from a 6000 SMD/h system from Universal Instruments for just over $100,000, on up to a 300,000 SMD/h system from Philips for over $1 million.

Tektronix estimates the capital resources required for a medium-volume facility to be in excess of $700,000.

As for floor space, AWI estimates a prototype facility as requiring 1500–2500ft^2, a low-volume facility up to 5000ft^2, medium-volume 5000–30,000ft^2, and a high-volume facility 30,000ft^2 and over.

Computer simulation

Setting up an SMT facility can be assisted by computer simulation, whereby the facility requirements are determined by the type of SMT assemblies to be manufactured and the production volume requirements. A special software program then allows the computer to calculate placement speed as to SMDs per hour (or SMDs mixed with leaded devices), the type of equipment needed, its cost, and percent utilization for varying production volumes. The Confacs Group, Tempe, AZ, provides such services, as do others, including AWI.

Most sources estimate the time span to set up an SMT facility and production capability as from 18 months to two years. Tektronix says that if the *total* program is approved, the development of a facility can be completed in six months and in full production in another six months. However, approvals usually come in incremental states based upon milestone reviews. Realistically, the process often requires 18 months to two years.

A subcontractor operation

Soletron, a San Jose-based assembly subcontractor, has developed an SMT capability consisting of chip placement, soldering, touch-up and visual inspection. Tomoyasu

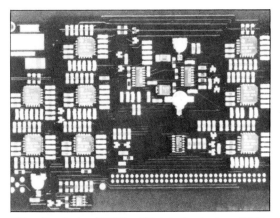

This medium-density, all SMT board illustrates the advantages of SMT, such as reduced component sizes, short lead lengths for better electrical performance, and the capability of mounting SMDs on both sides (Courtesy of Tektronix)

Computer input		Computer output		
Boards/month	100	Panels/hour	3	
Number of components		Components/hour		
Top		Top		
SOIC	25	SOIC	67	
SMD R&C	15	SMD R&C	40	
DIP	5	DIP	13	
Axial	3	Axial	8	
Manual	2	Manual	5	
Bottom		Bottom		
SOIC	0	SOIC	0	
SMD R&C	60	SMD R&C	160	
Days/month operating	5	Combined		
Hours/day	7.5	SOIC	67	
Board size	2 × 4	SMD R&C	200	
Manual insert rate/hour	300	Equipment cost and		
Number of boards/panel	1	percent utilized		
		Screening	1,200	20%
		Pick & place	25,000	50%
		Bake	500	25%
		Reflow	1,600	40%
		Clean	100	40%
		Misc.	100	100%
		Total equip. cost	30,300	
		Utilized cost	11,000	
		Product cost/year	65,000	

Computer simulation of prototype SMT production facility

Yoshikawa, SMT engineering manager, explains that the choice of a pick-and-place system is the most important factor in setting up an SMT production line. The speed of the chip placer regulates the volume of assembly for the entire line.

Soletron has incorporated two main types of pick-and-place machines in its assembly line: a high-speed system for high-volume production of a single design, and a low-speed system for lower production volumes and fast changeover. A Celma SMT-83 (Mamiya) is also available for very small lots.

"Our high-speed machine, a Fuji FCP-60, can handle up to 14,400 SMDs per hour,

while the low-speed machine, an Excellon MC-30, can handle up to 2200 SMDs per hour," says Yoshikawa. Soletron presently has one Fuji FCP-60 and three Excellon MC-30s.

Another aspect of SMT setup – expansion – is vividly illustrated by Soletron's planned additions to its present SMT capability. Two different TDK high-speed placement systems are on order and the plans are to also acquire a number of medium-speed systems (4000–6000 SMDs/h) to fill the gap between the high-speed Fuji and TDK machines and the Excellon MC-30.

While deferring to disclose specific TDK model numbers, Yoshikawa says the capabilities of one of the TDK machines are similar to the Fuji FCP-60, but it can handle 12mm as well as 8mm tape.

Soletron's other equipment, completing the SMT production line, includes an HTC IL-12 vapor-phase reflow system, Sensbey LR-300FM dual-wave wave solderer, Corpane B318RE inline cleaner, two Royonic assembly stations for manual insertion requirements, Baron MRR-10 vapor degreaser, and a Blue-M OV-490A2 oven.

Component availability/standards

Component availability is obviously an important factor in running an SMT production line. According to Electronic Trend Publica-

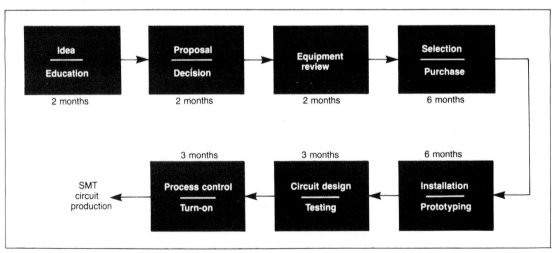

Although a full SMT production capability can be set up in less than one year, the need for program reviews and general red-tape procedures usually stretch this period to one and one-half years (Courtesy of Tektronix)

tions' SMT report, the often-mentioned 'limited availability' of SMDs is rapidly changing. Where previously there were only about 10% of the leaded component types and values available in surface-mount packages, up to 50% will eventually be available in a span of three to four years.

For the subcontract SMT operation at Soletron, Yoshikawa says that the SMDs are supplied by the customer in most cases. "However, many SMDs are still rare components in our industry," he says, "and we sometimes have to delay operation because of component shortages from the customer."

Since many customers prefer supplying their own SMDs, Yoshikawa reveals that parts packaging (bulk, reel tape and tube) can also be a serious problem for machine loading. Many high speed pick-and-place machines have restrictions on parts packaging, such as acceptance only of reel tape, while most medium- to low-speed machines accept a wider variety of packaged parts. The result is that, when a small volume of parts is supplied in bulk form, it restricts using the high-speed machines. Yoshikawa emphasizes that attention to parts packaging is an important aspect of setting up and running an SMT production line.

The ETP report says that problems with component availability are not caused by lack of standards but because too many standards exist. There are over a dozen IC packaging standards competing to succeed the DIP as the standard surface-mount IC package. The report contends that this forces pick-and-place machine suppliers to design equipment capable of handling all package types. This results in the marketing of sophisticated equipment having replacement heads or manipulators capable of handling the various packages, at a high machine cost to the purchaser. The alternative is to manually place the variety of components that a standard manipulator would not handle.

Avoiding problems

Tektronix says the best way to avoid problems in setting up an SMT facility is to initially make it a joint venture between engineering and manufacturing. Both should be involved in choosing equipment, setting up the line and in establishing the type of product design to be processed. Tektronix also advises that the basic necessities are to have a well-planned and well-supported project, a single point of accountability for process management, and a single point of accountability for program management.

Once the line is set up, those involved in the production of SMT assemblies agree that if problems arise during actual production operations it is most often caused by poor incoming material. Problems with component lead materials, platings or coatings, coupled with unsatisfactory PCB or substrate soldering surfaces, are sure to play havoc with production yield. SMT facilities are advised to deal with only a limited number of vendors – all of proven dependability – and purchase only the higher quality components. Likewise, PCB surfaces must be processed and protected to provide and maintain solderability. Loading inferior products in at the front end is the surest way to defeat the initial purpose of setting up an SMT capability.

Poor PCB design is another factor often cited as causing problems in SMT production operations. Too high a component density, crowded layouts and inefficient orientation of components can cause problems such as interference with the placement head, voids in soldering, and inspection difficulties. The best PCB layouts are produced when the

SMT facilities should be equipped with both high-speed and low-speed pick-and-place systems. The high-speed system here can place 14,400 SMDs per hour and has integral adhesive dispensing (Courtesy of Soletron)

design group works closely with manufacturing and is aware of all aspects of the SMT production process.

Carmen Capillo, vice president of R&D at Engineering Circuit Research, San Jose, CA, remarked in a paper given at the recent NEPCON Northwest that pad sizes, component spacing, tooling tolerances and dimensioning for solder-mask application all interact to grossly affect SMT production yields, Capillo claims the only way to achieve high yields is by employing proven board-design layout techniques. This requires extensive research and development and considerable hands-on experience. His advice: "PCB layout design can make or break an SMT production operation. Do not accept anyone's design standards or advice on SMT board design unless they can show convincing evidence that the design has been successfully employed and verified in an SMT production environment."

SMT ADVANTAGES

Surface-mount technology (SMT) is an old technology that has suddenly captured the limelight in electronics manufacturing. Its beginnings can be traced back to over 20 years ago when chip resistors and capacitors were first soldered to thick-film hybrid circuits on ceramic substrates. Today, SMT's popularity has been accelerated by the development of the chip carrier, the availability of small outline (SO and SOT) semiconductor device packages, a widening choice of automated production equipment, and the major driving force: the requirement for higher density packaging.

Why surface mount?

SMT, as applied to PCB technology, offers significant advantages over that of using through-hole mounted components. The primary advantage is space savings. SMT assemblies are smaller because of reduced component package sizes, and the density of interconnections can be increased because area-consuming through-holes are not required for each component lead. Depending upon the type of circuit and the components used, volume reductions of up to 10:1 can be achieved, although 3:1 and 5:1 are usually the more realistic ratios. The packaging density capable with SMT allows shorter interconnection paths and correspondingly better high-frequency operation.

SMT is not expected to be a panacea for all electronic circuitry. Power supplies, for example, having large capacitors and transformers would not benefit from a conversion to SMT. Most power supplies can be accommodated on a single-sided board, and many of the bulky components require the mechanical support afforded by a through-hole, clinched lead.

Assemblies are varied

A company desiring to acquire an SMT capability must initially define the degree and form of surface-mount device (SMD) integration most appropriate for its products. SMT takes many forms, such as conversion to:

- Conventional ceramic thick-film hybrid with both discrete SMDs and screened components.
- Ceramic thick-film interconnect with all components in SMD form and mounted on one or both sides.
- PCB with mixed through-hole components and SMDs on the top side.
- PCB with mixed through-hole components and SMDs on both sides.
- PCB with both through-hole components and SMDs on top and only SMDs on bottom.
- PCB with only through-hole components on top and SMDs on bottom.

First published in *Electronic Packaging & Production*, May 1985, under the title 'Automation thrives as companies strive for SMT capability'.
Author: Howard W. Markstein, Western Editor.

- PCB with all SMDs on one or both sides.

(*Note:* The PCBs listed here may also incorporate polymer thick-film circuitry.)

Implementing any of the above assembly types is dependent upon the availability of components having a surface-mountable configuration. The most familiar SMDs are SOICs, which are small-outline DIPs with leads bent outward at 90° for surface attachment; SOTs, which are small-outline transistor packages having leads also bent at 90°; MELF diodes; chip capacitors and resistors; and J-leaded and leadless chip carriers, both plastic and hermetic. These more common package types are now being complemented by the availability of inductors, potentiometers, carbon film resistors and electrolytic capacitors, all with surface-mountable leads.

SMD attachment

Wave soldering and solder reflow are the two basic techniques used for soldering SMD PCB assemblies, while hybrids are confined to the latter. Wave soldering requires that the SMDs first be attached to the board via an adhesive, then passed through the solder waves. The adhesive is applied just prior to or in conjunction with the pick-and-place operation. A dual-wave system has proved to offer the most reliable soldered connections for SMDs. Since the body of the SMD passes through the wave, 'shadowing' effects can result in poor or missed solder joints. Defects have been minimized by using a highly turbulent first wave to ensure complete coverage and wetting, followed by a laminar wave to remove excess solder.

Reflow soldering requires the screening of solder paste prior to pick and place, thereby temporarily securing the SMDs by its tacky surface. The solder reflow operation can be performed in a belt furnance, an IR system or by vapor phase.

In general, wave soldering is preferred for processing mixed through-hole leaded components and SMDs. Reflow, especially vapor phase, is advantageous for assemblies dominated by SMDs and mounted on both sides.

The basic equipment required for automated SMT production consists of a pick-and-place system, solder-paste screen printer, drying oven, reflow system and cleaning equipment. Mixed through-hole and SMD assemblies will require the addition of an adhesive dispenser, curing oven, automatic leaded component inserters, robotic odd-shape component inserter, and a dual-wave soldering system.

Acquiring SMT capability

Two ways of entering the realm of SMT are by in-house studies, equipment purchases and production line setup, with or without outside consultation, or by the acquisition of a complete turnkey system from an outside source. Both methods can be preceded by initial in-house studies using small laboratory systems for product prototyping and for gaining experience in designing for SMT.

Panasonic's applications engineer, Greg Stumpo, says that the major problem in setting up an SMT facility is poor or unrealistic planning. Success is dependent upon adequately defining the project time frame. He estimates that companies should allow up to 1½ years to establish an effective SMT capability. Other sources claim this period may extend as long as two or three years. The

---

- Space savings
  —Smaller device packages
  —Shorter leads
  —Shorter interconnections
  —Smaller volume (3:1 to 10:1)
- Smaller PC boards
  —Vias replace lead holes
  —Fewer circuit layers
- Lower cost in high quantity
- Increased automation capability
- More reliable construction
- Better high-frequency operation
- Vibration and shock resistant

SMT advantages

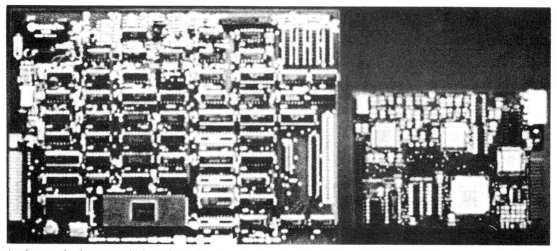

An advantage of surface-mount technology is that it reduces the required board size. Here, the Western Digital WD1000-85, a storage controller board (left) using through-hole technology has 176 components. The SMT version (right) is smaller and has only 86 components

list of factors involved provides an indication that these estimates are within reason. In essence, acquiring the necessary equipment and personnel does not constitute an SMT capability. *Productivity* constitutes a capability.

Companies preparing to acquire their own SMT capability can expect to consume significant time for research, equipment evaluation, integrating the production line, personnel training, and converting or designing new products for SMT.

Robert A. Reynolds, chip carrier business manager for Texas Instruments, avises that hardware decisions will be based on factors such as desired throughput and PCB complexity.

"The selection of a pick-and-place system is impacted by PCB size, number of different SMDs to be placed, and production volume requirements," says Reynolds. "At the same time, however," Reynolds warns, "decisions must also be made as to allowable cost, maintenance requirements and quality of performance."

Panasonic's Greg Stumpo says determining the equipment required must be preceded by defining the process. The PCB component mix will dictate the differing magnitude of the processes required, such as pick and place, leaded-component insertion, robotics for odd-form components, and wave solder or reflow. He says an obvious approach is to

define the most extensive process required for the company's product line, then survey the available equipment market for systems meeting these requirements, while also keeping an open mind to future growth and expansion of SMT.

As for maintainability, Stumpo says this is an issue often misunderstood by equipment purchasers. He says, "You can expect a machine manufacturer's support during installation and startup, and your engineers and

Four types of SOICs in 8- and 20-lead packages await placement from the tape carrier to the PCB approaching at right. The pick-and-place system shown has dual heads, one for adhesive application and the other for SMD placement (Courtesy of Panasonic)

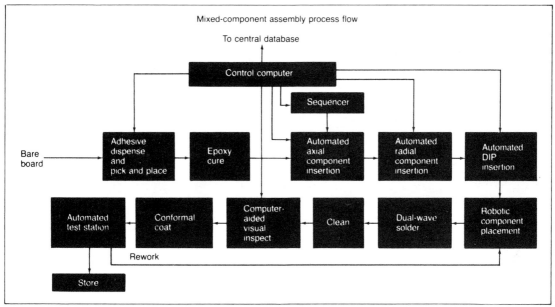

This production flow diagram is for a modern automated facility assembling PCBs containing both surface-mount and through-hole components (Courtesy of General Electric)

maintenance personnel should receive formalized training. However, small problems will arise and adjustments must be made even on the best of machines, and these are the responsibility of the user."

Staffing of an SMT facility will vary according to the size of the SMT production line and the degree of automation. The do-it-yourself method of implementing an SMT capability can be supported by the services of outside consultants well-versed in surface-mount technology, some more theoretical, others with 'hands-on' experience. Simulation modeling and computer studies (The Confacs Group, Tempe, AZ) are also available to determine SMT facility requirements for producing specific types of SMT products at desired production capacities.

Vendors provide support

SMT equipment suppliers, especially pick-and-place system vendors (Universal Instruments, Dyna/Pert, etc.), offer assistance or advice in setting up an SMT facility. As expected, all generally tout their own systems, complemented by a choice from other available support equipment. Quad Systems, Horsham, PA, for example, recommends the choice of equipment for a solder-paste screening and vapor-phase reflow SMT operation.

Panasonic has opened its Factory Automation Technology Center in Elk Grove, IL,

---

The following is an example of the possible types of equipment required by an SMT facility set up for solder-paste screening and vapor-phase reflow.

Screen printers for varied board sizes
• Weltek Model 44 (4 by 5 in. print)
• deHaart Model AOL15 (10 by 12 in. print)
• Forslund Model 2430 (20 by 22 in. print)

Auxiliary equipment
• Solder thickness probe: Engineered Technical Products, Model MG-400SW
• Digital scale
• Viscometer: Brookfield, Model RVT
• Hand solder dispenser: Electron Fusion Devices, Model 1000D
• Blue M oven, POM-336-E, floor model

• Blue M oven, table model

Placement
• Quad Systems, Model 34 pick-and-place

Soldering
• HTC, Model 1416 vapor-phase reflow (14 by 17 in. max. board size)
• HTC, Model IL6 (6 in. max. board width)
• Rework station: NuConcept Model 200A
• Defluxer: Branson, Model PSD-1216-W

Inspection equipment
• Ring-light magnifiers
• Stereo microscopes, 10-40X
• Profile projector
• Metallurgical microscope: Nikon, Model AFM with Polaroid camera
• Lead puller
• Stereo microscope: Wild Heerbrugg, Model M5 with camera

SMT facility requirements

• Product requirements	• Engineering training
• Equipment and machines	• Production personnel training
• Materials and components	• Product design training
• Personnel requirements	• Quality control
• Facility design	• Process implementation, fixtures, etc.
• Management training	• Documentation

(J. T. Stimy)

Factors involved in SMT facility setup

Personnel	Duties
Quality control manager	Schedule and direct incoming test and in-process QC personnel.
Process control engineers	Process control of assembly, soldering and conformal coating.
Test engineer	Maintain and update test software and fixturing. Assist in both incoming an d board test problems.
Manufacturing engineer	Provide assembly aids and custom fixutres. Assist in assembly equipment pro blems. Maintain and update assembly equipment software.
Component engineer	Monitor incoming components. Interface with vendors and resolve problems.
Shop supervisor	Schedule and direct first-level incoming test, stocking, kitting, and dispatch personnel.
Manufacturing engineering manager	Supervise maintenance and engineers (test, manufacturing, component and process control).
Clerks	Dispatcher for transporter. Clerk-typist.
(General Electric)	

Staffing requirements for an SMT facility

offering equipment demonstrations and consultation services for setting up an SMT production facility. North American Philips offers similar services through its SMD Technology branch in Milwaukee, WI.

Universal Instruments, Binghamton, NY, has developed an in-house SMT capability for both design and production purposes. A customer's board designs can be converted from through hole to SMT by Universal's CAD system, then produced and assembled on its SMT production line. In effect, the customer's learning curve is shortened, and equipment purchases can be preceded by actual first-hand operating experience.

The turnkey approach

The alternative to assembling an SMT facility via the piece-part approach is to purchase a turnkey system. J.T. Stimy, Inc., Trevose, PA, says it can supply a basic turnkey system and provide a company SMT capability in three months. More complex SMT capabilities can be acquired in six months.

Paul J. Polcsan, J.T. Stimy vice president, says a turnkey system is a pre-engineered, pre-designed SMT manufacturing system complete in all its elements. The company also provides the required instruction for engineering, production, production support and management personnel.

J.T. Stimy president, John Stimadorakis, views the pre-packaged technology of a turnkey system as a solution to the problems of fragmented equipment sources and the excessive time needed for a company to develop its own SMT capability. Stimadorakis believes that, eventually, the majority of SMT equipment sales will be to turnkey systems suppliers, who will then offer a total manufacturing capability.

Turnkey systems can be supplied in varying degrees of capability. J.T. Stimy, for example, supplies equipment packages ranging from simple manual placement stations rated at 500 component placements an hour, on up to automated in-line SMT production systems of either 2000 or 4000 component placements an hour and handling boards up to 16 × 18in. The turnkey systems include solder printing, IR or vapor-phase soldering, and batch or in-line cleaning.

A typical OEM SMT line

Numerous original equipment manufacturers (OEMs) have set up SMT production lines, including IBM, Sperry, Delco, Texas Instru-

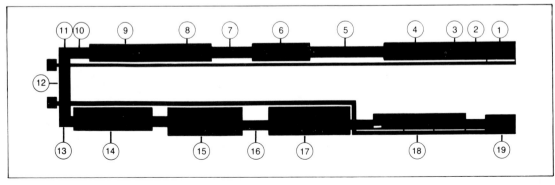

SMT production line at Western Digital (key—see pp. 161-162)

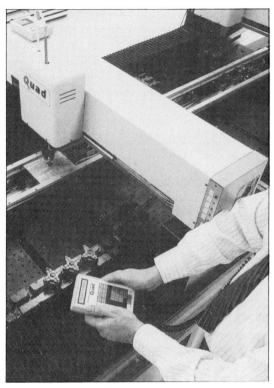

An operator uses a hand-held unit for programming the Quad IC placement system at Western Digital

ments, General Electric, Martin Marietta, Honeywell, RCA, Raytheon, Litton and Western Digital. The accompanying figure shows the SMT production line at Western Digital, Irvine, CA, (the ninth fastest growing US company, as per Electronic Business, January 15, 1985).

Western Digital, Irvine, CA, has assembled its own SMT production line geared for maximum automation and high-speed output. The U-shaped line and the equipment chosen by Western Digital is described as follows:

1. *TSM-4 Sanyo PCB loader.* The loader consists of a loading platform and an automatic feeder mechanism. Each magazine on the platform will hold 250 PCBs (bare) or 50 loaded boards. The automatic feeder will feed boards stacked (250) or cassette indexed (50) onto the conveyor belt. When a magazine is empty it will be unloaded, and a full magazine is then loaded without interrupting production.

2. *TPM-30 Sanyo screen printer.* The printer screens solder paste in a controlled layer onto the pads of the PCB. The printer accepts the board on the conveyor, aligns and holds the PCB to a vacuum plate. After the board is positioned under the screen, solder paste is applied and the PCB is transferred to the conveyor. While one board is being screened, another board is being aligned.

3. *TDM-20 Sanyo epoxy dispenser.* The epoxy dispenser consists of a programmable table and a controlled dispensing syringe. Both table and syringe are controlled by the main program that lists the X-Y coordinates and dispense time for each placement.

4. *TCM-40 Sanyo chip mounter.* The chip mounter consists of a programmable X-Y table, a revolving chip placement head, and a bi-directional reel track. Placement positions and reel selection are set in the main program. All programs can be written at the console or downloaded to the system from the MBC-225 (Sanyo minicomputer). Speed is about 9000 placements per hour with 99.9% reliability.

5. *Conveyor, Dynapace 303.* This system is a variable speed, adjustable-track conveyor that can be controlled by a microprocessor. The conveyor runs continuously.

6. *QS-34 Quad IC Placement (3).* Each unit has an X-Y programmable chuck that will pick up an IC and place it on the PCB. The board is received on a belt transport and locked into position. Programs can be easily written and process programs modified with the hand-held programmer.

Technicians check out the SMT production line at Western Digital. System in foreground is the Sanyo chip placer. Quad IC placement stations can be seen at rear

A technician monitors the control panel of the Sanyo chip placement system as the unit places SMDs at a speed of 0.4s/component

7. *Conveyor, Dynapace 303.*

8. *TFM-1 Sanyo UV cure and IR oven.* This unit uses UV for curing epoxy and has an IR oven to reflow solder, if necessary. The continuous conveyor has an adjustable track width and variable speed.

9. *SMD-718 Vitronics IR reflow.* The IR reflow unit has a continuous conveyor, preheat chamber and main oven. PCBs enter the preheat zone to stabilize the board temperature, then continue to the main oven where the solder paste will reach critical temperature and reflow.

10. *Conveyor, Dynapace 303-VB.* A continuous conveyor system similar to the 303 but uses vibratory brushes for moving the PCB.

11. *Angle transfer table, Dynapace 3131.* The unit will move the PCB from the conveyor onto its own conveyor belt, rotate the base to deliver the PCB to any programmed direction, then return to the original position.

12. *Conveyor, Dynapace 303.*

13. *Angle transfer table, Dynapace 3131.*

14. *Conveyor, Dynapace 303-3X.* This conveyor is a modified 303 to allow personnel to perform any hand placement if needed.

15. *G-12 Sensbey wave solderer.* The PCB first passes over a fluxer, then over a preheater, onto a scrubbing solder wave and to the smooth wave, then to a cooling section to reduce the PCB temperature.

16. *Conveyor, Dynapace 303.*

17. *WP1211 Corpane solvent cleaner.* This unit is a freon solvent cleaner. The boards rest on a continuous conveyor and pass through four spray heads.

18. *Dual-level conveyor, Dynapace 5000.* The dual-level conveyor is a continuous belt system. Boards will be received at one level, inspected, and good boards placed on the second conveyor for unloading.

19. *TZM-3 Sanyo unloader.* This unit consists of an automatic unloading system and a magazine load/unload platform. Each magazine will hold 200 finished boards. Magazines are automatically positioned and ready to receive PCBs. A board from the conveyor moves into the index slot in the magazine, and the magazine then indexes up to receive the next board. When full, the magazine is unloaded and an empty magazine is positioned automatically.

Western Digital's SMT line begins with solder-paste screening and an epoxy-dispense operation followed by pick-and-place using a Sanyo system for placing the smaller chip components. Larger chip components, especially the 68- and 84-lead plastic chip carriers used by Western Digital, are placed by a Quad pick-and-place system. Soldering is performed by a Vitronics IR reflow unit and a Sensbey dual wave wave-soldering system. A Corpane cleaning system completes the major production equipment comprising the line. The equipment can process PCB assemblies from 2 × 2in. to 10 × 13in.

Joseph Baia, vice chairman of the board for Western Digital, says that the SMT line can process an 85-component PCB approximately every 30 seconds, assembled and soldered. The placing operation is the Sanyo pick-and-place system which has a cycle time of 0.4 seconds per chip.

Baia says Western Digital set up its own SMT capability with only minimal formal outside consultation. Combined with its extensive semiconductor packaging expertise, most of the knowledge and information was acquired by attending SMT seminars, discussions with vendors, visits to other SMT facilities and hiring experienced key personnel. The timespan for developing the SMT capability was 1½ years.

"To be successful with SMT requires putting priorities at the front end," instructs Baia. "Since processing operations and workmanship are now controlled by automation, the focus should be on engineering, board layout, built-in testability and the use of quality material. My advice is to deal with a limited number of proven vendors, specify high-quality parts, and don't let price be the sole determinant."

PICK-AND-PLACE MACHINES

Machines placing surface-mount components on a printed circuit board meet a staggering cast of device package styles and sizes, not to mention that they come in contact with them at the rate of 2000 to 500,000 times per hour. There are chip resistors and capacitors, metal electrode face (MELF) resistors, mini MELFs, small-outline transistors (SOTs) in around 25 different sizes, small-outline ICs (SOICs), leaded and leadless chip carriers, and flatpacks. All of these demand the accuracy from a pick-and-place machine to accommodate their 0.050in. down to 0.020in. pad or lead spacing.

Programming the machine

A single pick-and-place machine may be called on to fetch from tape reels or other feeders a variety of surface-mount devices (SMDs). These devices are placed by the machine's head over most of the area of a PCB, perhaps up to the order of 18 × 18in. With the trend to microprocessor or desktop computer control of manufacturing equipment, giving instructions to a pick-and-place machine on the sequence of actions for performing this assembly process translates into programming the unit.

Walk-through teaching, keyboard terminal entry, and host computer entry are the three programming modes for pick-and-place mchines. A specific model may have the capability to be programmed in one or more of the modes.

In the teach mode, the programmer, with a hand-held portable terminal, walks a machine head through the required sequence of steps. The terminal unit, connected to the machine, has keys for specific actions the unit can perform. As the programmer activates the proper key at each location, and the machine's electronics note the coordinates, a program for placing the SMDs takes shape in the unit's memory.

In direct entry, a keyboard terminal is used to enter both the actions to be taken in the assembly sequence and the coordinates where they are to be performed.

The remote computer mode allows an assembly program to be developed on a host computer off-line from the pick-and-place machine, then downloaded over a communications link. Frequently the information transfer is through an RS232C interface.

A special case of this mode is when the remote computer is a computer-aided design system. In this case, data on component style and coordinates on the PCB, developed in the board's automated design process, is directly transferred from the CAD system to the pick-and-place machine. Similar to producing automatic insertion tapes from a CAD database, the CAD vendor has customarily been the point where post-processing programs have been developed to match the CAD data format with that required by the specific assembly machine.

DynaPert's MPS118, for example, is controlled by a group of microcomputers with progamming through a hand-held control unit. Approximately 900 unique placement positions can be programmed into the unit's RAM. Placement pressure levels are preset, and the programmer has only to select for each device one of the three levels available. The unit's software provides for operation of

First published in *Electronic Packaging & Production*, January 1985, under the title 'Pick-and-place faces a variety of SMCs'
Author: Ronald Pound, Senior Editor.

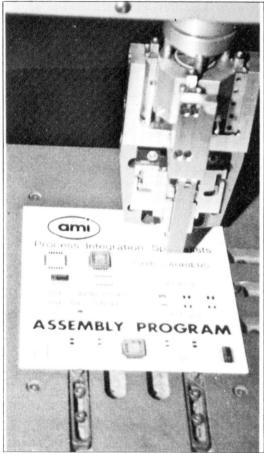

Chip carriers, small-outline ICs (SOICs), chip capacitors and resistors, and small-outline transistors (SOT-23) have been placed on this demonstration board by AMI's Stage II Mikronipulator pick-and-place machine

the machine functions, program editing, machine diagnostics, and basic management information.

With machine control through an on-board keypad, DynaPert's MPS500 model has RAM capacity that allows 4000 unique placement positions to be programmed. An optional programming console includes a desktop microcomputer with CRT, keyboard, dual floppy disks, and an integrated digitizing tablet. Programs written on the console, or other host computer systems, are downloaded to the placement machine.

Two files are used in this program generation. The first, a component library, supports many placement programs. It describes the components in each of the feeder positions, their orientation, and the placement force required. A second file is generated for each specific PCB. It contains the coordinates for placement and the orientation of components on the board.

Capable of all three programming modes, AMI's Stage II Mikronipulator uses an RS232 interface when transferring data from a CAD system. Placement routines can also be generated on Micro Component Technology's MCT 6000 placement system via walk-through, direct coordinate entry, or CAD/CAM interface.

Component feeders

Programming a pick-and-place machine involves not only where a component is to be placed on a board, but also where the machine head is to find the component to pick up. Tape and reel feeders, stick feeders, matrix trays, and feeders for bulk component are the sources where the machine's head picks up SMDs to be placed on a board.

Feeding components to the machine on tape has the advantage of ease of handling. It is, however, difficult to 100% test components, and not all components are available on tape from vendors. Bulk components, on the other hand, are easier to test and less costly, but more difficult to handle.

Standards exist for 8mm tape, which will

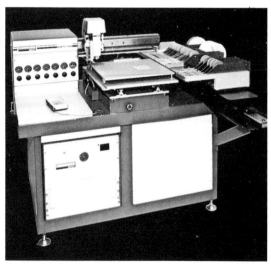

Component feeders, placement head, and controls (from right to left) are elements of the DynaPert MPS 118 machine

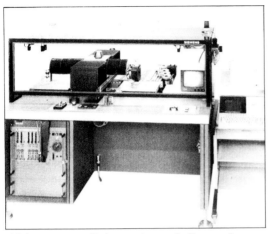

A desktop computer enables data entry in one mode of programming pick-and-place machines (Courtesy of Siemens)

handle discrete, passive chip components and SOTs. Available in 7in. reels as well as 14in. reels, the larger size may become more popular. About 12,000 components can be held on the larger size reel, and 3000 on the smaller.

Tape may be plastic with embossed cavities that hold the individual devices. A sealing tape secures the devices in the cavities. While this is the emerging method in the USA, another approach is paper tape with adhesive to secure the component.

Furthermore, there are options as to how a component is oriented on the tape. On a SOT-23 component, for example, the base and emitter leads emerge from one side of the device package, and the collector lead from

Placement routines can be generated through a CAD/CAM interface, direct coordinate entry, or machine walk-through with the MCT 6000 system

the opposite. Rotation by 180° provides the second of two fundamentally different orientations in which the component may appear on the tape.

SOICs and chip carriers are too large for current tape formats and must therefore be handled in sticks or tubes. As the number of components per stick is limited to a relatively small number, this approach may not be attractive from a production view. An Electronic Industries Association (EIA) committee has proposed standards for 12, 16, 24, 32, 44, and 56mm tape. Larger formats will enable components with 1½in. dimensions to be handled on tape.

Universal Instruments' Model 4712A Onserter surface-mounted component placement system, for instance, uses vacuum extraction to take a component from 8mm tape. A rotating magazine assembly, which is a two-tiered carousel, has the capacity to hold up to 32 input tape reels on each tier. This translates to a capability for handling up to 64 different components per machine. In addition, modular flexibility allows a user to expand this capacity to 256 different components.

As a stand-alone unit, Universal's Model 4621A Omniplace SMC placement system has a feeder input area that will handle up to 40 different components on 8mm or 12mm tape. This is expandable to 192 components.

Amistar's SM-1000 chip placement machine, to be unveiled at NEPCON West, will accept rectangular chips, SOT-23s, and similar devices on 8mm tape. The unit will have input stations for 64 7in. reels, expandable to 128 input stations.

Feeders to supply components are mounted in three blocks of 20 units on each side of DynaPert's MPS 500, allowing a total of 120 TF8-type tape feeders.

Component verification

That the tool has actually picked up a component from the feeders, and a component of correct size and value at that, may be verified prior to attempted placement.

With Sanyo's TCM-30 and TCM-40 chip mounting machines, for instance, if a chip is not of the correct size, or is turned incorrectly,

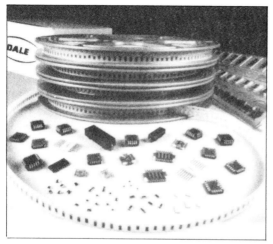

Tape reels hold surface-mount resistor components. The variety of other device package styles are all resistive components (Courtesy of Dale Electronics)

the vacuum head bypasses the placement point, and drops the chip in a collection receptacle. Automatically the machine picks up another chip and places it in the location previously bypassed.

Universal's Onserter placement system, according to the company, eliminates the two primary causes of board failure associated with surface-mount components. These are the inability of incoming inspection to test components prior to mounting and small chip size, making visual inspection during assembly difficult.

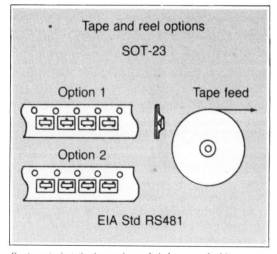

Component orientation is an option on the industry-standard 8mm tape and reel format (Courtesy of Motorola Semiconductor Products)

Universal's machine provides a method of chip verification in its basic system, thus reducing the need for repair work. Each chip can be verified for value, leakage, opens, shorts, and orientation. If a chip is bad it is automatically rejected and another taken from the feeder.

Placement process

One machine, required to run, for instance, at a placement rate of 5000 components per hour, and to place a range of component sizes from smaller discrete chips and SOTs through larger SOICs and chip carriers, need not be limited to one head and one tool to do the job.

Dual-head machines enable the unit to place a component with one head, while it picks up the next component with the other head. Also, multiple heads may be used so that adhesive for holding the SMD during wave soldering, can be applied by one head on one board, while another head is placing components on a board to which adhesive was just previously applied.

Using automatic selection, different pick-and-place tools for the head can be taken from a tool bank, depending on the size of the next component to be placed. Rotating turret configurations with multiple heads are also an approach to providing multiplicity that increases placement speed and the range of parts that can be handled.

With multiple pick-up heads, Amistar's SM-1000, for example will be capable of placing up to 14,500 chips per hour. Dual heads on DynaPert's MPS 500 contribute to placement at 6000 components per hour, under laboratory test conditions. The placement rate in actual practice will be affected by the length of movements associated with pick-up from the feeders and placement on the board, and by the number of tool changes for placement of a mix of devices.

On DynaPert's dual-head machine a tool bank contains four different tools on each of the two different component feeder carriages. Tools to handle the required range of SMD sizes may be distributed between the two banks.

A turret-style application head on Universal's Onserter system utilizes vacuum pick-up nozzles located at 90° intervals. As the head indexes, the nozzles extend to perform their particular function. An adhesive applicator head, functioning in step with the component placement head, deposits a dot of adhesive at each programmed placement location. This unit places 6000 components per hour.

Automatic insertion

Not many PCBs with surface-mount devices (SMDs) are populated by these devices alone – and this condition will persist for quite some time. Siemens, the German electronics giant, cited a forecast that indicates it may be 10 years before at least half of all components will be surface mounted.

As manufacturers make the transition from through-hole assembly to surface mount, they will generally find that limited surface-mount component availability will force them to mix through-hole components with SMDs on the same board. Some manufacturers may, as an alternative, construct small PCB assemblies supporting only SMDs. Then, as a SIP-like device, mount that SMD board on a larger one with through-hole components.

Odd-form components, too, whether through-hole or surface-mount, will have to be accommodated on any style of board by manual or robotic assembly.

There are two basic configurations of boards with both types of components. The industry is now calling these assemblies Types II and III. With SMDs of all styles, including chip carriers and small-outline ICs (SOICs), on top or bottom, and through-hole components on the top side of the board, Type II is the predominant form of SMD assembly in the USA today. Type III assemblies, popular in consumer electronic equipment from Japan, have through-hole components on the

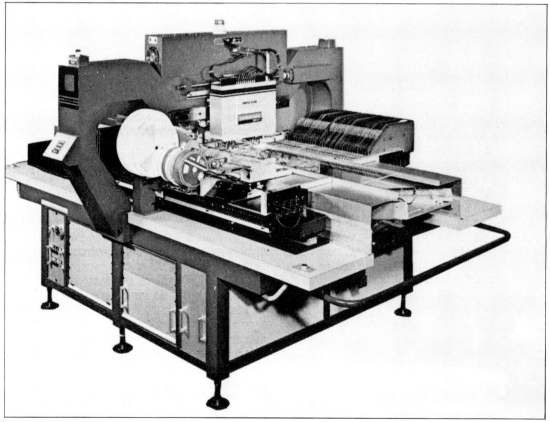

Dual heads on DynaPert's MPS 500 system take components from feeders on each side of the machine

Component parameters are verified before placement by Universal Instruments' Model 4712A Onserter system. A carousel component dispensing module is in the section on the right, and an elevator is to the left left

top side and discrete, passive chip components and small-outline transistors (SOTs) on the bottom side of the board.

Type I is a board populated only with SMDs, and represents a small portion of surface-mount assemblies today.

In terms of assembly machinery, mixing component types on a single board means insertion machines, as well as pick-and-place machines, will be required in a manufacturing line. A Type II, mixed-technology board

Adehsive is applied to one board, while components are placed on another by Universal Instruments' Omniplace machine

might pass through an assembly process as follows:

- Screen print solder paste on pads for SMDs to be placed on the top side of the board.
- Place SMDs on the top side with a pick-and-place machine.
- Solder SMDs in a reflow soldering machine.

Affiliated Manufacturers Inc. (AMI)
 – Stage II Mikronipulator. 7 in. by 8 in. PC boards
 – In-line machine. Places 32 k components/hr (cph)
Amistar
 – SM-1000. 14.5 k cph, 16 in. by 16 in. area max.
Anorad
 – Available 2Q, 1985. To handle 12 in. by 16 in. PCBs
DynaPert
 – MPS 118. Places 2 k cph, 18 in. by 18 in. area max.
 – MPS 500. Places 6 k cph, 18 in. by 14 in. area max.
 – MPS 100TU and TW. Placement in 4 in. by 4 in. area
 – MPS 100TC. Placement in 6 in. by 4 in. area
Excellon Micronetics
 – MC 30. Places 2.8 k cph, 12 in. by 12 in. area
Fuji America
 – Compatible with 8-mm tape and reel format
Henry Mann
 – Vacuum head, separate centering station
Micro Component Technology (MCT)
 – MCT 6000. Places 6 k cph in 12 in. by 18 in. area
MTI Corp.
 – Microplacer. Hybrids and PCBs to 14 in. by 24 in.
Panasonic
 – Panasert M2. Two adhesive, two component heads
 – Panasert MM. High-volume placement, 576 k cph
Philips
 – MCM I through IV. MCM III places at 552 k cph
Quad Systems
 – QS34 system. Communicates through data network
Sanyo
 – TCM-30. Chip mounter from 8-mm tape
 – TCM-40. Chip mounter from 12-mm tape
Siemens
 – Modular system. Components handled include MELFs
 – Programmed machine. Uses a desktop computer
TDK Corp.
 – SM2. In-line system for boards to 12 in. by 15 in.
Teledyne TAC
 – HAS-1000. For hybrids and small PC boards
Universal Instruments
 – RHyMAS II. Expanded 8 in. by 10 in. placement area
 – In-line RHyMAS. Up to 20 stations
 – Onserter. Verifies chip and SOT components on-line
 – Omniplace. Component verification an option
 – Logplace. Manual placement machine
Zevatech
 – Range of models. From 2.4 k to 10 k cph

Pick-and-place machines for SMD assembly on PCBs

- Insert through-hole components on the top with DIP insertion, axial insertion, and radial insertion machines.
- Insert odd form, through-hole components manually or with a robot.
- Flip the board over (bottom side is now turned upward).
- Apply adhesive (which can be done with the pick-and-place machine) to the locations where SMDs will be placed on the bottom side.
- Place discrete chip and SOT SMDs on the bottom side with a pick-and-place machine.
- Flip the board over (bottom side is now turned downward).

- Solder bottom side (SMDs and through-hole leads) with a dual-wave soldering machine.

It is apparent that pick-and-place machines must have the capability to be integrated into a larger assembly scheme of considerable complexity. Conveyors, automatic board loaders, PCB magazines, board inverters, and perhaps automatically guided transporters may link multiple pick-and-place machines with multiple automatic insertion machines, as well as soldering and cleaning equipment, in the assembly of a single Type II board.

SURFACE-MOUNT ASSEMBLY USING MACHINE VISION

Successful surface-mount pick-and-place depends on a number of operations orchestrated in such a manner as to ensure an end product, often manufactured in large volume, which functionally and cosmetically conforms to rigid specifications. Solder paste or adhesive is first laid down on the circuit board, or sometimes on the component itself, the surface-mount device is picked from its container and placed precisely at its predetermined address, and the assembly, with parts tenuously secured by viscous paste or glue, is soldered to provide permanent electrical and mechanical bonding. Prior to soldering, an inspection process verifies placement; while following final assembly, test assures functionality. Equally important are cleaning operations and through-hole insertion for mixed-technology boards.

Every technology is finally legitimatized when it evolves its own jargon. So it is with SMT (surface-mount technology). Not all frivolous, the terms and the language permit effective communications between designers and engineers, between manufacturers and

distributors, and between sellers and users.

Phil Zarrow of Excellon Automation shares some of the language of SMT and SMA (surace-mount assembly). SMT printed circuit boards come in three types. Type I covers circuits which utilize surface-mount devices exclusively, assembled on top, bottom, or both of single- or double-sided boards. Type II circuits use a combination of surface-mount and through-hole components; the SMCs may be on top of the board (inserted component side), or top and bottom (solder side). Type III defines circuits which use through-hole components on the top side of the board and SMCs exclusively on the bottom side.

Surface-mount components are available in a variety of packaging formats including: bulk; tape on reel (8, 12, 16, 24, 32, 44, and 56mm); magazine (metal and polycarbonate); stick or tube (SOICs and PLCCs); waffle pack; and cartridge.

Placement machines have been traditionally classified by their mechanical functions into four categories. In-line placement, the first of the four, refers to the progression of the substrate past a fixed-position placement station. The station places a single device onto a fixed location on the substrate.

First published in *Electronic Packaging & Production*, January 1986, under the title 'SMA promotes automation and boosts machine vision'.
Author: S. Leonard Spitz, East Coast Editor.
© 1986 Reed Publishing (USA) Inc.

Used in a high-reliability control panel, this circuit board features automatically placed SMCs, including chip capacitors and resistors (Courtesy of Kyocera)

Sequential placement, the second category, defines the action involved when software-controlled single or dual heads pick up the MSC from each feeder, or feed sequence, and place it on a substrate. On some machines, the head moves in both *X* and *Y* directions, while others employ a transport table to move the substrate in one or both directions.

In simultaneous placement, the third of the four categories, multiple heads transfer an entire array of devices onto the substrate at one time. A board, or part of a board, is completed in a single operation.

The final category, sequential/simultaneous placement, identifies a software-controlled system in which an *X-Y* table passes under multiple heads, each placing a single component. Heads may be fired simultaneously.

A user may focus more effectively on comparable pick-and-place machines by identifying the level of production. Accordingly, placement machines may be sorted by volume vs. time, as shown in the table.

While the above values are neither absolute nor representative of the most important criteria in selecting a pick-and-place machine, the ranges correspond well with equipment in the surface-mount assembly marketplace.

The heart of the placement system

The pick-and-place machine is the central organ, the heart of the surface-mount placement system, pumping components onto printed circuit boards, transported and fed so that trajectories intersect, and then quickly moving each assembly on its way to the next workstation.

Basically, automatic SMA machines are comprised of several subsystems: a conveying and presentation mechanism for the surface-

Designed for moderate placement speeds, the MPS-500 is an extremely flexible pick-and-place machine (Courtesy of Dynapert-Precima)

Level	Placement/ hour	Seconds/ placement
I	<2,000	>1.8
II	2,000 to 3,000	1.8-1.2
III	3,000 to 10,000	1.2-0.36
IV	10,000 to 20,000	0.36-0.18
V	>50,000	<0.072

Placement rates

mount boards; a mechanism for feeding components packaged in various formats to the placement head; an X, Y and 0 carriage to precisely locate the component positioning head with respect to the device's address; and a computer or microprocessor for programmable control of selection, sequencing and placement operations. Peripherals such as adhesive or solder-paste dispensers, multiple placement heads, and computer power provide machine enhancements. Mechanical configuration, flexibility, throughput, degree of automation, and board type accepted distinguish one pick-and-place machine from another.

North American Philips offers the MCM modular design placement system, permitting customization to the user's needs by the addition of appropriate modules. Available stations include: hardware-programmable multiple placement modules capable of placing up to 32 components simultaneously in 2.5 seconds; choice of manual or automated substrate load/unload modules; substrate transport modules, off-line tape holder loading modules; and pin-transfer glue modules.

The pick-and-place, multi-jawed pipette is central to the MCM placement module and serves as an example of sophisticated mechanical engineering in support of surface-mount technology. Initially, a surface-mount component is eased off the reeled tape by a pin which supports it against the pipette. Vacuum in the pipette monitors the SMC's presence; the pin provides support until the jaws clamp onto and firmly hold the device.

As they close on the SMC, the jaws establish a center line that provides a reference for placement, eliminating the influence of packaging tolerances on placement accuracy. The vacuum in the pipette holds the SMC while it is being lowered into position on the substrate. When the pipette withdraws, the absence of a vacuum indicates that placement has been completed.

Dynapert-Precima's MPS-500 represents another level of machine sophistication. Classified as a moderate speed but flexible pick-and-place system, the MPS-500 was introduced to the SMA marketplace about three years ago and has been continuously upgraded through several iterations. Its cycle rate of 6000 components per hour qualify it for the moderate-speed designation, and its ability to feed components from 8, 12, 16, 24, 32, and 44mm tapes, as well as from vibratory feeders and sticks, helps to support its claim to flexibility. The wide variety of component types, which includes small chips, SOTs, SOIC, and leaded/leadless chip carriers, can be intermixed and processed in one production run.

Two pick-and-place heads move alternately from feeder to board. Components are accurately centered with the tweezer jaw assembly. Full 360° placement rotation of the pick-and-place head is a standard machine feature and is programmable in 1.5° steps. Tool banks containing four different pickup and placement tools are located on both sides of the machine. Tools for the specified range of component sizes are preprogrammed and

The three-bit turret head on the System-4100 in-line machine places SMCs and dispenses epoxy for wave-soldering applications (Courtesy of Henry Mann)

Autovision provides accurate, real-time SMA inspection coupled with ongoing statistical process control (Courtesy of Automatix)

automatically selected in accordance with production demands.

Controlled by a group of integrated micro-computers, all primary machine functions are activated through a simple on-board telephone-type keypad mounted beneath a mini-CRT. Machine operation, basic program editing, machine diagnostics, and management data are available through the keyboard. The standard RAM memory allows up to 4000 placement positions to be programmed. Several programs may be stored simultaneously on the machine's floppy disk. Without options, but including operational and diagnostic software, the average selling price for the MPS-500 is pegged at $175,000 (1986).

While users have the right to expect capital equipment valued at hundreds of thousands of dollars to operate at the state of the art with advanced engineering, such expectations may be unrealistic when considering surface-mount placement systems proceed at almost an order of magnitude less.

Due for formal introduction in 1986, System-4100 from Henry Mann Inc. could well be an exception at a base price of less than $25,000, assuming it can satisfy the user's particular requirements of throughput, precision, and flexibility. Capable of inputting components from 8, 12, and 16mm tape, vibratory bulk feeders, sticks, and waffle packs, the 4100 handles over 80 different SMDs. These include resistor and capacitor chips, SOTs and SOICs. A three-bit turret head dedicated to specific component sizes places SMCs at 1500 devices per hour on an average 8 × 10in. board, and can also dispense epoxy for subsequent wave-soldering applications.

The eye of the placement system

Visual inspection is an extremely time-consuming, expensive task. It is tedious, and because acuity, concentration and judgment vary, objective and consistent results cannot be assured. Machine vision 'sees,' enhanced by the speed, accuracy, and consistency of computer power. Viewing the printed circuit board, an optical system takes in information which is transformed into machine code that can be handled by the computer's hardware and software. Microprocessors extract relevant data such as defects and measurements, comparing and analyzing the information for instant decision-making.

Multiple-width tapes accommodating SMCs from discretes through large ICs are stacked in reels, ready for high-speed placement (Courtesy of Signetics)

Manufacturer/model	Specifications
Amistar Corp. SM 1000	• High-speed dedicated machine. • Placement rate: 14,400/h. • Placement accuracy: 0.004 in true position. • PCB size: 15 in. by 15 in. in automatic feed version. • Handles 64 8-mm tape reels.
Automatix Inc. Microsert	• Data-driven software features a menu-driven user interface. • Placement rate: 3,000 chips/h. • PCB size: 18 in. by 18 in. • Accommodates PLCC, CLCC and 25-mil devices. • Component feeders include tape, stick and tube.
Celmacs SMI-85 placement-systems	• Low investment entry systems.
Dynapert-Precima MPS-111	• Designed for prototype production. • Placement rate: 2,000/h. • Placement area: 18 in. by 18 in.
MPS-318	• Placement rate: 4,300/h. • Placement area: 18 in. by 18 in. • Automatic feeder capacity of 60 8-mm tape reels and SOIC sticks. • Placement accuracy: ±0.008 in. • Vacuum pickup and programmable four-jaw tweezer centering.
Excellon Automation	• To program, system is stepped through desired sequence and instructed to remember each step. • Handles chip, SOICs, DIPs, and LIDs. • Applies solder paste directly onto SMCs. • Placement area: 18 in. by 18 in.
Fuji America Corp. CP-90	• Assembles up to 90 kinds of SMD chips in any mix and order. • Placement rate: 14,400/h. • Built-in squeegee puts adhesive across bottom of each component prior to placement.
Heller Industries Inc. Chiplacer	• Semiautomatic placement system. • Designed for low to medium volume. • Placement rate: 800-1,000/h typical. • Placement area: 9.8 in. by 13 in. • Position accuracy: 0.002 in.
Henry Mann Inc. FMS	• Low cost, top-of-the-line system. • Uses multiple placement heads, each dedicated to a specific SMD type. • PCB size: 18 in. by 18 in. • Vision system option available.
4100	• Lower cost, in-line stand-alone unit. • Component feeding includes bulk, tape and sticks. • Placement rate: 1,500/h. • PCB size: 8 in. by 10 in.
4083	• Lowest cost, bench-top stand-alone system.
Micro Component Technology Inc. MCT 6000	• Automatic board-handling standard. • Handles EIA/JEDEC standardized SMCs. • Placement rate: 6,000/h. • PCB size: 12 in. by 18 in. • Placement accuracy: 0.002 in R true position.
North American Philips Co. MCM II	• Sequential/simultaneous, software-programmable placement. • PCB size: 8.27 in. by 12.6 in. • Placement rate: 32,000/h. • Placement accuracy: ±0.008 in., 0 = ±2 degrees. • 32-channel tape feed.
Panasonic Industrial Co. Pansert Series MR	• Placement rate: 10,000/h. • 100 component input stations. • Component feeders include 8- and 12-mm tape. • Component types: chips, SOTs, MELFs.
Siemens Components Inc. System HS 180	• Designed as a flexible modular system of moderate speed to complement the MS 72. • Placement rate: 9,000 to 12,000/h. • PCB size: 18 in. by 18 in. • Handles chips, SOICs and DILs on 8-, 12-, 16- and 24-mm tape, also bulk and sticks.
TDK Corp. of America FM-3	• Handles 80 to 120 components/board. • Chips formatted on 8-mm tape. • Placement rate: 0.3 s/SMC. • Mounting directions: 0, 90, −90, 180 degrees.
Universal Instruments Corp. 4712B Onserter II	• Cycle rate: up to 12,000/h. • Placement accuracy: ±0.008 in R true position. • Placement area: 16 in. by 16 in. • Adhesive or solder paste can be preapplied.
RHyMAS II/E robotic assembly	• Standard tooling processes square components from 0.030 in. to 0.188 in. on substrate sizes up to 7 in. by 10 in. • Placement rate: 1,600-2,400/h.
Zevatech Inc. (1986 generation)	• Modular design, highly flexible, moderate volume. • PCB size: 24 in. by 24 in. • Package size: 1-mm sq to 34-mm sq (0.04 in. to 1.34 in.). • Component feeders include tape, stick and bulk. • Placement rate: 2,400/h.

SMA pick-and-place machines

Many surface-mount assembly pick-and-place machines come with vision built into the placement system, or can be fitted with an add-on vision system. These eyes check for chip placement in real time, calculate actual chip coordinate locations, monitor component polarity, identify the presence of glue and excess glue, and continuously report failure conditions for correction.

According to James L. Field, product manager at Automatix Inc., "Vision is used to avoid or reduce rework time. Repeating errors can be detected before many bad boards are built. Reporting the location and nature of non-repeating errors, such as a single chip out of place, also reduces repair time. In either case," Field adds, "the vision system builds statistical records of board quality, providing insights into long-term production problems."

The Automatix vision system, Autovision, works on-line at surface-mount assembly speeds, and can either be integrated into existing equipment or can be used as a standalone system at any point in the assembly cycle. Autovision's inspection capabilities include verifying the presence or absence of SMCs; monitoring the position and orientation of SMCs; identifying the polarity of SOICs and SOT-23s; determining the presence or absence of solder paste, as well as the solder-paste area; and monitoring the presence of excess glue before and after chip placement.

The jargon of SMT clearly identifies the many format types of packaging by which surface-mount components may be presented to the feed mechanisms of pick-and-place machines. These are tubes, sticks, magazines, bulk, and tape on reels. Positive indexing and proper orientation at high speed tend to preclude or at least inhibit gravity or vibratory feed systems. Manufacturers of placement equipment favor tape on reels, as this format enables machines to reach very high cycle speeds and production volume. However, one impediment to the universal use of tape has been the lack of standardization.

Mark Kastner, surface-mount device marketing manager for Signetics Corp., provided a historical perspective. "In 1984, board assemblers, equipment makers, and IC vendors, under the auspices of the Electronic Industries Association (EIA), worked together on a common standard for IC tape on reel. The result was the EIA proposed specification RS-481A, "EIA Standard for Taping of Surface Mounted Components for Automated Placement." This specification standardizes tapes in sizes ranging from 8mm through 56mm, and covers SMC IC packages including SO, SOL, and PLCC.

"The tape is made up of a PVC, carbon-filled carrier tape, molded with cavities to carry components individually, and a clear polyester cover tape, heat-sealed to the carrier tape to hold components in place after the cavities are loaded."

Planning for SMT: A case study

Best known for its power supplies, Computer Products is a medium-sized company with gross annual sales of $100 million. The Real Time Products Division accounts for about $28 million. Real Time designs and makes process simulation and process control systems for industrial users, including paper mills and automotive paint-spray operations. The company manufactures some 55,000 boards per year, of more than 450 basic designs. Typical kit sizes on the production floor number between 5 and 15 board assemblies, with a low of one and an infrequent high of 50. A commonly used through-hole board measures 9.5 × 11in. and carries over 300 components. In 1986, the company expected to go to 14 × 14in. boards, each having as many as 2000 components.

In October of 1983, surface-mount technology and surface-mount assembly seemed worthy of investigration. The events that followed chronical one company's march toward adopting this state-of-the-art process:
- *January 1984.* A preliminary task force is formed, made up of representsatives from development and manufacturing engineering, materials, quality control and operations management.
- *March 1984.* Made entirely with SMCs and displayed at an industry trade show, a competitive Japanese product introduces more pressure for change.

- *April 1984.* Led by the director of advanced technology, a more focused team emerges from the task force. A development engineer and a reliability engineer are assigned full time to the project.
- *June 1984.* The team visits users and vendors in the industry for guidance. Texas Instruments, North American Philips, and AWI prove to be particularly helpful, providing SMT information and direction.
- *September 1984.* An in-house training program is initiated, complemented by college curricula.
- *January 1985.* The equipment procurement cycle begins with a visit to NEPCON West to evaluate alternative systems.

- *June 1, 1985.* The SMA facility is turned on. Laid out in an area of less than 1000ft^2 the equipment purchased includes a deHaart AOL-15 screen printer, an Excellon Micronetics MC-30 surface-mount pick-and-place machine, and a Model CVP-SS-1B inline vapor-phase soldering system from Centech Corp. A Series 2 Comparascope, made by Vision Engineering, aids and enhances the manual inspection process. The total capital investment amounted to $275,000, and has enabled the shop to routinely apply surface-mount assembly techniques to 4.5 × 9.5in. standard control boards made up of two layers and 50 SMCs. Projections for the first year of operation called for 335 to 400 units per week.

SMD PLACEMENT USING MACHINE VISION

VLSI allows designers to pack more circuitry onto their chips, but as chip complexity increases, so does the number of I/O lines needed to access that added complexity. To increase the number of terminations per device while maintaining the smallest mechanical package, vendors are reducing the lead spacing on surface-mount components from 0.050in. to 0.025in., and in some cases, even to 0.010in. As lead spacing shrinks, the requirements for accurate automatic devices placement become more stringent.

Placement machines fall into several categories: high-speed machines that place small discrete chips; slower, but more precise machines for placing PLCCs and CLCCs; and multifunction machines that do both. Today's placement machines rely purely on mechanical accuracy and precision for board assembly. However, for 0.025in. active SMDs, and large 0.050in. active devices, tolerance stackups make mechanically based placement machines unreliable. To make the move to these larger devices, users will turn to placement machines that incorporate additional sensor capabilities, such as machine vision.

Placement accuracy

The ability of a mechanically based assembly system to reliably place a component is governed by mechanical tolerances in the assembly machine, the board, the components that the machine places, and the fixture that holds the board. Taken together, these tolerances define a tolerance stackup that determines the range of components an assembly system can reliably place.

In contrast to through-hole technology, where tolerances are relatively large (0.010–0.012in.), high pin-count PLCCs and CLCCs present a more difficult placement problem. Because the leads are so closely spaced, both pad and lead size are small, demanding a much more accurate placement. To avoid solder bridging between adjacent leads, solder pads are typically made no more than 25% larger than the lead. Normal workmanship standards require at least 75% of the lead to be on the pad.

Error budget

To determine if a placement system can accurately place components with acceptable

First published in *Electronic Packaging & Production*, January 1986.
Authors: Jim Field, John Payne and Chris Cullen, Automatix Inc.
© 1986 Reed Publishing (USA) Inc.

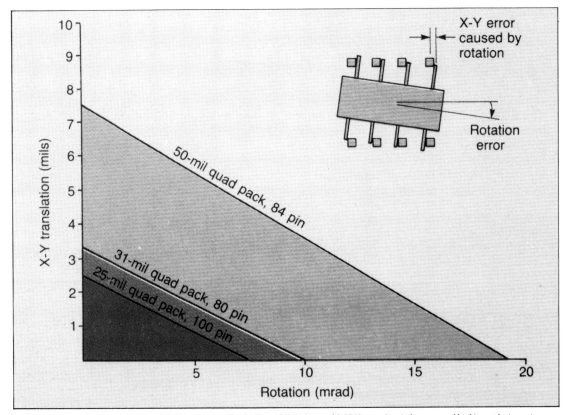

The accuracy required for a typical through-hole device, as well as SMDs with 0.050in. and 0.025in. spacing, is demonstrated by this graph. Assuming that a placement process incurs zero rotational error, a 0.007in. translational stackup can be sustained and still make automatic device placement of 0.050in., 84-pin devices possible. However, adding a rotational error may bring this device out of the shaded area

reliability, the system's error budget must be calculated. For SMD placement, there are three fundamental error sources: machine, board, and component. The table below lists the error sources and tolerance stackup for both the through-hole and SMD assembly processes. For each error source, two components are listed; an *X-Y* translational error gives a measure of how far the center of a lead will be displaced from the center of its pad as a result of a particular error. The rotational error gives a measure of how much a component will be rotated relative to the pads.

The first significant source of error in the placement process is the placement machine itself. Two errors that define machine performance are repeatability and accuracy. Repeatability (or repeat accuracy) is the ability of a placement machine to move to a previously taught point on the board. Accura-

cy is the ability of a machine to move to a point on the board calculated from data only (not previously taught).

Most commercial machines specify repeatability only, which is typically 0.001in. Using a machine which relies on repeatability requires the user to go through the arduous process of

	Through hole		SMD w/o vision		SMD w/ vision	
	Translation (in.)	Rotation (mrad)	Translation (in.)	Rotation (mrad)	Translation (in.)	Rotation (mrad)
Machine						
Repeatability	0.001	NA	0.001	NA	0.001	NA
Accuracy	0.005	NA	0.005	NA	NA	NA
Fixturing	0.002	0.1-0.5	0.002	0.1-0.5	NA	NA
Tooling	0.001	1.0	0.001	1.0	NA	NA
Board						
Mech. features	0.002	0.1-0.5	0.002	0.1-0.55	NA	NA
Artwork	NA	NA	0.001	--	0.001	NA
Artwork to mech. features	NA	0.2-0.6	0.006	0.2-0.6	NA	NA
Distortion	0.002	NA	0.002	NA	0.002	NA
Component						
Lead distortion	0.004	1-2	0.001-0.004	0.5-2.0	NA	NA

17 mrad (milliradians) = 1 degree
1-mrad error at 1-in. radius = 0.001 in.

Comparison of manufacturing tolerances

Designed with flexibility and precision in mind, MICROSERT, the SMD assembly system from Automatix, uses machine vision in two ways. One camera, mounted on the robot arm, registers board artwork. A second camera, mounted below the table, provides component lead inspection and registration. The Autovision controller then directs component placement, carefully matching component leads to board artwork

teaching every point on each new board. In addition, it is not possible for an operator to reliably teach a point to within the repeatability spec of the machine.

Inaccuracies in the placement machine contribute significantly to the total system placement error, but the ability of a system to reliably place a component depends on more than the machine's ability to move accurately to a given point in space. Errors in board and component manufacture and alignment cause the board's actual pad locations to differ from the location programmed in the placement machine. Thus, though the placement arm moves precisely to a given location, that location may be incorrect.

Another source of machine error is inaccurate board fixturing. This occurs when the board is locked in place in the assembly machine using mechanical features on the board. During this process the board may not be accurately located. Burrs or jagged edges on the boards may prevent perfect seating on the tooling or shot pins.

Centering repeatability error is another major source of component misalignment. These errors are a function of the tolerance of the jaws that grasp the component. Typically, a placement arm's jaw can center a component to within 0.001in. Wear of the mechanical components can, however, eventually lead to greater errors.

Printed circuit board errors

Once the board is clamped, errors on the board itself contribute to system inaccuracy. These board errors include mechanical-feature location error, artwork error, artwork to mechanical-feature registration error, and board distortion. Mechanical-feature error is caused by drill or mill machine inaccuracy.

Artwork error is caused by thermal expansion of the artwork during the printing process, or by poor etching process control. During the milling, drilling or etching process, improper fixturing causes errors in artwork to mechanical-feature registration.

During board manufacture and handling, it is possible for the board to develop local or global distortion errors. An example of a global error would be thermal expansion or contraction of the substrate material. Local errors are commonly caused by severe warping or incorrect handling. In many cases, the board distortion errors can be eliminated by more careful process control and by selecting different materials. A strict climate-controlled environment, for example, could eliminate thermal distortion errors. However, such control measures can increase the board cost.

Component errors

Manufacturing errors may cause a component's leads to be distorted with respect to its body. Since the body is used by the placement arm to locate the component, any lead distortion will result in misalignment of leads to pads even if the body is placed correctly. Typically, the X-Y translational error that results from lead misalignment is 0.004in.

Tolerance budget

Each of the three major error sources discussed thus far acts to chip away at the assembly process' tolerance budget. If the total stackup exceeds the budget, then reliable interconnection to the substrate cannot be guaranteed. In the through-hole placement process, the translational stackup is 0.010in. and the budget is 0.012in. Consequently, reliable through-hole insertion is feasible using mechanically based placement systems. The same applies for passive SMDs where the translational stackup is typically 0.012in. and the budget is commonly 0.025–0.30in. Typical values for individual error components in the tolerance stackups for different processes are also given in the table.

With 0.050in. and 0.025in. active devices however, translational tolerance stackups approach, and sometimes exceed, the tolerance budget. The point at which automatic device placement becomes unreliable for large active devices depends on the total error contributed by both the rotational and translational components.

Sensors close the loop

To make automatic placement of large 0.050in. and 0.025in. devices feasible, sensing systems can be incorporated within the placement machine to correct for board and component misalignment. Machine vision is one such sensor for surface-mount assembly. Using vision and the appropriate software, a placement machine can compensate for mechanical erros in the assembly process. A vision-based placement system detects the salient features in a board and then uses them to establish the exact location of the artwork. This bypasses the previously described board and board fixturing errors.

By examining the exact position of the component leads, the vision system enables the arm to accurately place the device. If the leads are out of tolerance, the part will be rejected, resulting in a higher product yield than is possible without machine vision.

THE DEMANDS OF SMT

SMT puts stringent demands on quality – in concerning components, placement and testing. These cannot be ignored.

HOT-AIR LEVELING

For the past few years, there has been a growing trend to accept the use of surface-mount devices on printed wiring boards in the electronics industry. Due to the difficulty of cleaning flux residues from under these low-profile components, the use of mild rosin fluxes has been adopted. For very simple applications, where single-sided boards may be used, the copper tracks are covered with solder mask to reduce the probability of bridges during wave soldering on double-wave systems.

Because of the rapid introduction of this technology, the bare copper boards arrive 'just-in-time,' and therefore solderability preservation has not been a problem using mild rosin fluxes. However, for boards requiring double-sided circuitry with plated through holes, the use of solder mask over tin/lead electroplate is not recommended.

Recent studies undertaken by the military in North America recommend that boards with solder mask over tin/lead electroplate not be used. The use of hot-air leveling to protect solderability of printed circuit boards for surface-mounted devices is advised to prevent wrinkling of the solder mask, to permit improved cleaning results and to ensure solderability for an unlimited time. Since nothing solders better than solder, SMD components on such boards can be soldered at higher process speeds, reducing the leaching of metalization from chip components.

Zero-defect soldering

As the electronics industry matures, it is becoming more important to control and reduce total costs in order to supply the customer with a product of higher quality at an overall lower product cost. Solderability committees and zero-defect soldering programs are being established in many firms to resolve problems at the source and to control all stages of the assembly process, in an attempt to reduce production costs and maximize profits.

Poor solderability of printed wiring boards is estimated to cause 50% of the solder defects; 20% are caused by components. Economic conditions over the past few years have forced the industry to become more aware of the cost of defective solder joints. Instead of doing it right the first time, the industry has generally accepted visual inspection and hand touch-up as part of regular production costs and has not identified this area as over-cost. More mature companies are saving enormous amounts of money by directing their energy positively: personnel are retrained to accept only those parts which are solderable, thereby eliminating visual inspection in many cases.

Maynard Eaves of Hewlett-Packard states that printed circuit board quality is a major concern in the electronics industry. Quality constitutes the largest impact on a product's cost and opportunity to reduce cost. A $20 board can escalate to as much as $5000 by the time a failure is discovered – a failure that could have been detected in fabrication.

Specifications for boards should be clearly defined at the start to establish and preserve solderability. Instead of continuing to accept boards of inferior quality, the industry is beginning to realize the large-scale financial advantages of paying a little more for quality printed wiring boards which have been hot-air leveled.

Several board fabricators have totally removed their tin/lead electroplating lines and replaced them with machines which apply pure solder of 63/37 eutectic alloy composition for optimum solderability retention. In some cases, specifications call for 100% of the boards to have solder mask applied over bare copper, followed by hot-air leveling. Jack Fellman describes a 30% increase in the number of boards being hot-air leveled during 1983. He mentions that there are reports which claim that boards produced in hot-air leveling equipment may reach 50–60% of the market share.

In the past few years, many technical

First published in *Electronic Packaging & Production*, February 1986, under the title 'Hot-air leveling excels for surface-mount boards'.
Author: Donald A. Elliot, Electrovert Ltd.

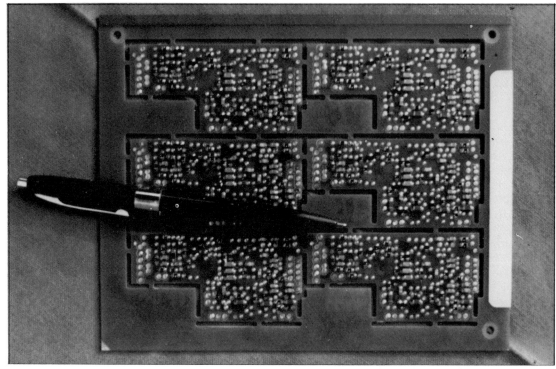

Boards for SMD applications will be smaller due to the smaller size components and because components can be mounted on both sides of the board. For easier handling, boards will be fabricated, assembled and soldered in single-panel form before breaking apart. Each small board here measures 1.3 × 2.6in.

articles ahve been written on the topic of surface-mounted devices. At the Printed Circuit World Convention III, held in Washington in May 1984, almost all the technical discussions touched on this topic. However, very little has been written on the subject of hot-air leveling of printed wiring boards for this application.

SMD assemblies

James Hall was one of the first to identify three basic types of assemblies. The assemblies are classified as follows:

- *Type I – Total surface mounting.* This was pioneered by the military and aerospace industries. All components are of the surface-mount type. Assemblies contain all types of components including SOICs and chip carriers. Circuitry is of the high-density type on single-sided and double-sided boards.
- *Type II – Mixed technology.* Assembly consists

of both leaded components and SMDs on double-sided boards with plated through holes.
- *Type III – Underside attachment.* Small chip resistors, chip capacitors, SOTs and, in some cases, SOICs, are found on the underside of conventional single-sided boards or double-sided boards with plated-through holes.

For Type I assemblies, vapor-phase reflow soldering is the predominant technique, although infrared is also used. Products employing Type II technology are generally soldered by a combination of wave soldering using the dual-wave principle and one of the above reflow methods for the more complex SMDs. Type III circuits are most often soldered by the dual-wave principle.

Cleanliness levels

In conventional electronic assemblies, decisions on flux activity were generally dictated

The dual-wave principle is the primary method used for wave-soldering boards of the Type II and Type III surface-mount classifications

by the solderability of the boards and components. If solderability was poor, a more aggressive flux was often selected to overcome the problem. Afterward, intensive measures were undertaken to ensure these flux residues were adequately removed. With assemblies containing SMDs, the choice of flux cannot be selected in the same way.

Interesting discussions are underway in attempts to establish the best cleaning method after soldering assemblies containing surface-mount devices. Most of the industry has adopted mild rosin fluxes. Many are concerned as to how the solvent will clean the flux residues out from under these low-profile components. Uncertainty of final cleanliness dictates this as being the safer approach.

To guarantee sound joints when mild fluxes are specified, the boards and components must be extremely solderable. After soldering, a superficial cleaning to remove the excess sticky rosin flux residues is considered sufficient. If some of the mild rosin flux

residues should be left under the components, no damage is done. Therefore, in order to have an extremely solderable assembly, the majority of boards for these applications will be hot-air leveled.

Disadvantage of bare copper

Most single-sided printed circuit boards are of the print-and-etch variety, leaving no protection for long-term solderability. This may be fine for boards which arrive 'just-in-time'. For example, if the board were etched yesterday afternoon and delivered, according to Japanese practice, directly to the assembly area by the vendor before 10 a.m. and wave soldered before noon, then copper boards of this type would be most cost-effective. However, until this practice is implemented, it is advisable to add some means of preserving solderability.

While roller-tinning and high-pressure hot oil spray systems afford a means to this end, the thickness results are less than the solder-

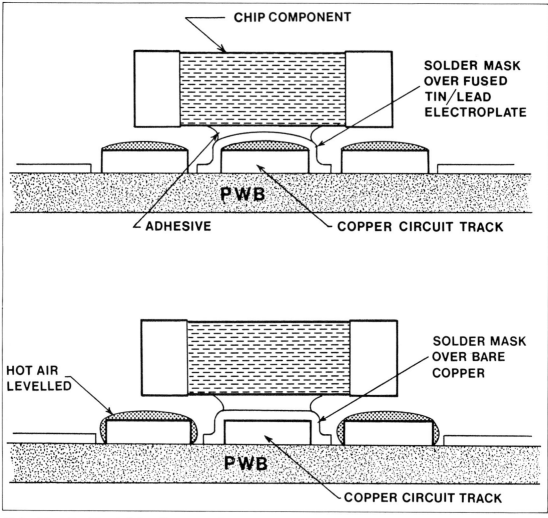

The hot-air leveled board shown in the lower illustration is preferred in order to avoid the risk of having the SMD move or wash away during wave soldering, as may occur in the case shown above

coating thickness produced in a hot-air leveling system. The thicker coating increases confidence in the solderability protection. Extremely thin coatings, produced by other means, may be only as thick as the intermetallic compound, or the coating may have been applied over oxidized copper which will not appear dewetted until the components are soldered.

Problems with tin/lead

Dewetting occurs when the electroplate is too thick or when it is plated over oxidized copper. Rough and gritty solder indicates that the composition of the plating is far from the 63/37 eutectic composition which he describes as giving the best results. He reports on the wide variations of electroplated composition from seven vendors. When fusing the tin lead electroplate, the tin from the electroplate reacts with the copper to form the intermetallic compound, leaving the remaining electroplate lead-rich. Other reports indicate that the difficulty in controlling the additives and brighteners in the tin/lead electroplating system dictates a double pass through the

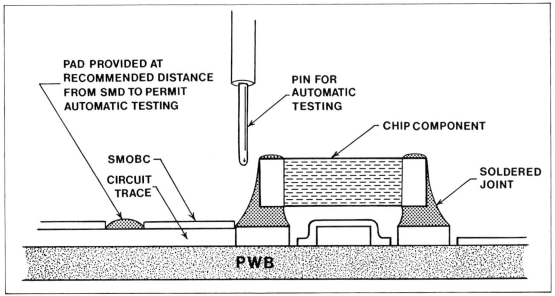

PAD PROVIDED AT RECOMMENDED DISTANCE FROM SMD TO PERMIT AUTOMATIC TESTING

PIN FOR AUTOMATIC TESTING

CHIP COMPONENT

SMOBC

CIRCUIT TRACE

SOLDERED JOINT

PWB

As shown here, the shape and small size of solder joints present test-pin difficulties. To meet automatic testing requirements, board designs will incorporate test pads at a distance from each SMD

reflow system to eliminate dewetting resulting from outgassing of these additives.

Wrinkled solder mask

It is common practice to apply solder mask on printed wiring boards to eliminate the possibility of solder shorts on printd wiring assemblies during wave soldering. On assemblies which require double-sided circuitry and plated-through holes, the most common technique is to use tin/lead electroplate as an etch resist.

However, far too many in the industry still have specifications which permit their board fabricator to apply solder resist over the electroplated circuit tracks. The solder mask wrinkles as the electroplate under the mask melts during wave soldering, IR reflow or vapor-phase reflow soldering. This causes problems of flux entrapment and does not guarantee total elimination of bridges, especially on narrow circuit tracks when the electroplate is thick and the overhang is excessive.

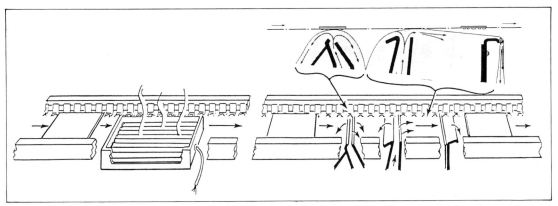

High-intensity preheaters and solder tanks with dual pumps are being retrofitted to many existing machines. The nozzle on the left provides a turbulent jet of solder, penetrating all points on the board to ensure proper wetting of the circuit pads and the terminals of the SMDs. The second nozzle provides a smooth laminar wave building upon the solder deposited by the first wave

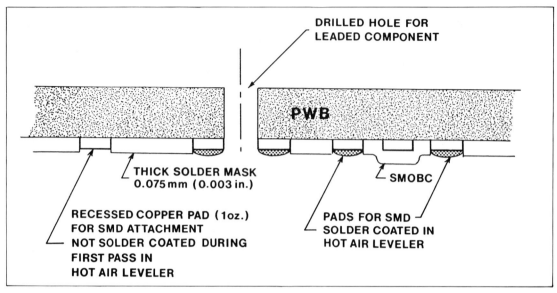

When thick solder mask is applied to copper boards containing pads for SMT applications, there is a possibility that a number of pads may not be solder coated during the first pass through the hot-air leveler

In Type II and Type III assemblies, the surface-mount component is epoxy bonded to the bottom of the board. If the droplet of epoxy is deposited onto solder mask over a tin/lead electroplated circuit track, the epoxy and the component may move or flow away when the heat of the solder-joining operation wrinkles the solder mask and causes the solder plate to flow.

Donna Sanger studied boards for military applications where solder mask was applied over tin/lead electroplate after fusing. Results were compared with those obtained on boards where the solder mask had been applied over bare copper, followed by hot-air leveling for solderability protection. She concludes that solder mask should not be used over a tin/lead plating due to wrinkling, and the entrapped flux residues can lead to degradation of the assembly in service. It is difficult to ensure that conformal coating will adequately protect the assembly if it is applied over loose, flaking or wrinkled solder mask.

These disadvantages can be overcome by hot-air leveling double-sided boards where solder mask has been applied over the bare copper. The advantages are noted as follows:

- Nothing solders better than pure eutectic solder.

- With an adequate solder-coating thickness, solderability is guaranteed and shelf-life is limitless.

- Since the exposed metal surfaces are solder coated in the hot-air leveler, very mild fluxes may be used in the solder-joining operation. This provides higher surface insulation resistance after soldering and cleaning the assembly.

- During hot-air leveling, the board is immersed in hot solder for approximately four seconds. During this time, the intermetallic compound is formed. A portion of this time is devoted to permitting the heat from the solder bath to provide the activation energy required to ensure the formation of the intermetallic bond between the tin in the solder alloy and the copper circuits. This means that the soldering of the SMDs to the board can take place more rapidly. This puts less heat into the components and reduces the leaching effect.

- Solder mask over bare copper will not wrinkle. As a result, higher levels of cleanliness can be assured, and conformal coatings will ensure better encapsulation.

This typical hot-air leveler is of the vertical type used for application of a hot-dipped solder coating to optimize the solderability of circuit boards

Hot-air leveling for SMDs

For the board fabricator who is already familiar with the details of hot-air leveling and who is successfully producing high-quality conventional boards, there are very few surprises when applying this technique to boards for SMD applications. There will be many minor changes that do not affect the process. However, the need to apply high quality, liquid-type solder mask for proper registration and the need to prevent bleeding of the mask onto areas to be solder-coated still require the same attention to detail. Also, the need to have chemically brightened copper just prior to hot-air leveling is equally essential. Boards for SMD applications are essentially similar to conventional boards, ranging from the single-sided types with no holes all the way to complex multilayer boards.

There are several factors which are im-

mediately apparent on boards for surface-mount technology. Initially, boards will be much smaller. Therefore, for easier handling throughout all stages of production, the design will specify multiple step-and-repeat patterns on the same panel where, in the final stages, each individual small board will be broken apart. Due to the microminiaturization that is possible in SMD applications, electronic designers will add more features and options to the circuitry and, where potential field failures are costly, may also add redundant circuits on the same board. Therefore, boards for SMD applications will eventually get larger, if space permits, depending on each particular application.

Board flatness will be an important factor in the proper application of the components. Warpage specifications will become tighter. Also, with small breakaway boards, routing or punching between adjacent small boards should retain sufficient rigidity. Methods will be developed to retain flatness during the mounting, assembly and joining of SMDs on flexible circuit and thin paper boards during processing.

The most notable characteristic of boards for SMD applications is the many small square or rectangular pads where the tiny components are to be joined to the printed wiring board. Also, there may be exposed pads between the components to permit bed-of-nails testing to be undertaken after assembly.

Some boards will be designed without any holes; therefore, fabrication costs will drop. On double-sided boards, for Type II boards, the plated-through holes will be of conventional size, but there will be more of the small interconnecting via holes. For these assemblies, specifications may call for the solder mask to cover the entire pad and via hole on both sides of the board. In this case, it will be essential to ensure the continuity and long term integrity of the copper plating in the via holes. Alternatively, the small hole can be solder coated during hot-air leveling. Depending on the hole diameter, board thickness and the operating parameters during processing in the hot-air leveler, these small holes may be left filled with solder to ensure reliable continuity between the bottom and top side of the board.

The solder alloy used to hot-air level printed wiring boards for SMD applications will most probably be the same 63/37 tin/lead eutectic alloy that is presently used for conventional boards. This is based on the use of solder creams which minimize leaching and, in the case of wave soldering, by using the recommended process parameters of increased preheat temperatures and the dual-wave principle. The majority of producers of Type II and Type III assembly configurations are wave soldering their boards with 63/37 solder.

With reference to board cleanliness after fabrication, board manufacturers may be asked to meet cleanliness specifications with the same degree of importance as solderability specifications. If corrosive residues are left on the board from the fabrication process, it will not be removed from under the SMDs after they are soldered to the board surface.

Thick solder mask

On boards for SMD applications, a new problem has arisen in a very specific case. It has been observed that when very thick solder mask, measuring 0.003in. thick is applied on printed wiring boards with 1oz copper thickness measuring 0.0014in. thick, there is a high probability that a small percentage of the bare copper pads will not be coated with solder. By processing such boards a second time and, in some cases, a third time, all pads become solder coated. This eliminates solderability of the copper as a variable.

This occurrence is observed on both horizontal and vertical hot-air leveling machines. Due to the fact that the solder mask is much thicker than the copper, this geometry and the small pad size as well as machine process variables combine to form this problem.

On horizontal systems, the problem has been addressed by improving cleanliness and maintenance of the equipment, ensuring that the copper is truly clean prior to processing and by passing the panels through the system twice.

To solve this specific problem in vertical hot-air levelers, flux manufacturers are looking at modification of existing fluxes to be able to process such panels in one pass. On an experimental basis, it has been demonstrated that diluting the flux with alcohol permits such boards to be processed in one pass. However, this is not recommended as a safe production method due to the flammability of alcohol. Until this problem is resolved, the easiest solution is to specify thinner solder maks.

In summary, the industry is maturing. Saving a few pennies on the cost of cheaper boards which, during assembly and soldering, may result in the need for costly inspection, touch up and repair can no longer be tolerated. Therefore, the industry will demand that its boards and components be solderable. To eliminate 50% of the cost of inspection and touch up after soldering, only the highest quality boards should be specified. This is the primary reason for electronics manufacturers specifying printed wiring boards which have solder mask applied over bare copper followed by hot-air leveling.

TESTABILITY CIRCUIT AND BOARD ACCESS PROBLEMS

There is a significant need today for improved design for testability and testing

First published in *Electronic Packaging & Production*, January 1986, under the title 'Testability circuit solves SMT board access problems'.
Author: Jon Turino, Logical Solutions Technology Inc.
© 1986 Reed Publishing (USA) Inc.

techniques, tools, and methods throughout the entire electronics manufacturing industry. This need is driven by increased product complexity and speeds, new packaging and assembly methods (e.g. surface-mount technology) and ever-increasing costs for automa-

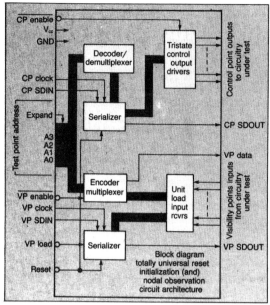

The T-Circuit permits controllability of inputs to circuit elements in the functional circuitry and observability of logic states at the circuit outputs

tic test equipment and test programs at both the device level and the printed circuit board level.

In an effort to satisfy the specific need for a testability approach to surface-mount PCBs, a (patent pending) circuit has been developed which can be installed right on the PCB to render it testable with functional or limited access in-circuit testing techniques. Designated the T-Circuit, its specific hardware implementations can provide, in minimal real estate and impact on product parts cost and reliability, all of the visibility and control point interfaces that are required between the internal nodes of complex electronic circuit assemblies.

These include situations where spring probe fixturing technology is not applicable, specifically in the areas of high volume, high complexity SMT PCBs which must be tested with functional automatic test equipment (ATE) to verify the performance of those circuits. Major firms that have already expressed serious interest in evaluating the T-Circuit include Apple, Siemens, TRW, DEC, Motorola, Magnavox, Rohm, Grumman, Philips, and several others.

The T-Circuit chip set has been protyped

using commercially available components and has been proven to function. It allows the user to select the optimum amount of parallel and serially accessed visibility and control nodes to achieve the optimum tradeoff between added circuits and test/troubleshooting time. It furthermore involves the addition of far less circuitry to a unit under test at a given level of testability than other existing testability methods.

The initial chip set, and the many variations of it that will be developed in terms of pin-outs, fabrication technologies, and interface configurations, can be used as a collection of high-technology business productivity improvement devices. Users can design these directly into their own products to reduce testing and trouble-shooting costs during product manufacture and service.

Disadvantages of past approaches

There have been a great many approaches described in the literature regarding methods to enhance the testability of electronic circuits. Structured schemes, such as level sensitive scan design, random access scan, and built-in logic block observation, require complete interaction between the functional circuitry and the testability circuitry. They have thus been limited to applications where the large investment in CAD tools and personnel training is economically justifiable.

Non-structured schemes, e.g. the random addition of test points and control points to

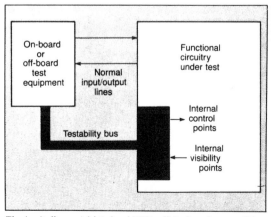

The circuit allows serial data-based ATE to efficiently evaluate parallel logic states within the functional combinatorial and sequential circuitry

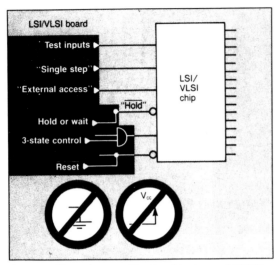

A summary of the usual names and functions of the pins on an LSI/VLSI device that should be controllable

logic circuits, suffer from the disadvantage that it is usually not possible to standardize the testability interface in a way that provides a common test interface. The most flexible of such schemes to date, the scan set approach, requires less interference with the logic circuitry to which it is connected than other methods, and has the ability to provide a relatively standard interface (scan data in, scan data out, clock, etc.). Even so, it suffers from the very long test pattern lengths that result when it is applied to very large scale circuits.

The T-Circuit overcomes these limitations by using a combination of parallel (random access) and serial techniques to provide complete flexibility in selecting visibility and control points, without interference with system functional circuitry. Further, it entails no special personnel training, CAD tools, or other complications.

The interface bus between the unit under test (UUT) and the ATE, when implemented with the T-Circuit, is expandable up to many thousands of control and visibility points. It also requires only one additional line to double the number of control or visibility points required. These can be operated independently from the bus interface at any time in either a serial or parallel fashion. The bus is technology-independent and is there-

fore compatible with semiconductor technology arising from virtually any fabrication process (e.g. CMOS, TTL, ECL).

The T-Circuit can moreover be applied at the device, board, or system level. From the standpoint of pin-out, it must be admitted, it is not very efficient on devices of 40 pins or less in its raw form. This is because up to 20 testability bus pins are typically needed to gain adequate access into the UUT. For devices using pin-grid array or other high fan-out connection techniques, however, the testability bus provides a relatively economical set of additional pins that make the test interface to any device standard. With some clever multiplexing, the device's normal input/output pins can be converted to testability bus pins under the control of one or two test mode pins.

At the printed circuit board and system levels, the circuit provides a very economical means of gaining all required visibility and control points for effective logic simulation, test pattern generation, and built-in test. The features of the circuitry utilized to implement the UUT/ATE interface also provide for on-line built-in self-test with minimal additional circuitry – usually less than 1% additional logic for the T-Circuit circuitry itself.

Diagnosis down to the component level from the board interface, and to the board

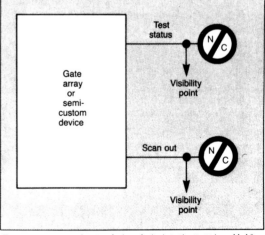

For gate arrays or semicustom devices, device input/output pins added for test should be connected to visibility point inputs to allow rapid fault isolation right to the faulty devices, instead of leaving them unconnected as many suppliers do

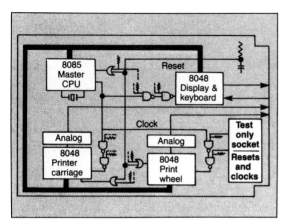

The difficult-to-test board has an average product cost of $124.75, while the more testable one costs an average of $116.17 – a saving of over $8 per board gained by adding parts cost to achieve testability

level from the system bus interface, is easily implemented using the testability bus. In fact, the T-Circuit architecture makes it feasible to achieve component-level fault isolation from the system interface.

If boards are implemented with this interface scheme, all boards can be tested from the same interface, thus reducing the need for unique electronic interfaces. New boards using surface-mount devices (SMDs) will lend themselves to new ATE architectures that can take full advantage of the increased visibility and control. The same interface and ATE architecture can be used to test entire systems, as well as the boards and subsystems that make up the functional system under test itself.

Description of the circuit

The T-Circuit is applicable to both synchronous and asynchronous circuitry (or any combination of circuit types), and it is usable both for standalone operator-initiated testing or for on-line monitored built-in test. It accomplishes this without undue hampering of the normal logic design function, and with minimal impact on system functional reliability.

The circuit is made up of two major sections – the control point section and the visibility point section. These sections may be interconnected as shown and produced as a single device, or they may be implemented as

separate devices, combining the appropriate signals at the board or system level.

The control point section includes a serializer subsystem which can receive serial input data for later parallel application to the control points selected within the functional circuitry to which the T-Circuit is applied. This serializer may also be loaded in parallel from the test point address inputs via the decoder/demultiplexer subsystem. All control point signals are applied to the circuitry under test via tri-state control output drivers so that there is no effect from the T-Circuit until it is activated.

The visibility point section consists of a serializer subsystem which can output serially the parallel data received through the unit load input receivers connected to visibility points within the functional circuitry. Visibility point data can also be presented at any time by inputting the proper address to the test point address inputs. The test point address inputs, via the encoder/multiplexer subsystem, can also access the unit load input receivers and present the (tri-state capability) visibility point data output line.

Both the control point section and the

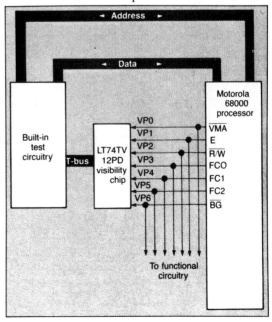

The built-in test circuitry on the LT74TV12PD visibility circuit can look at the critical visibility points on the Motorola 68000 processor by addressing them and reading them through the testability bus

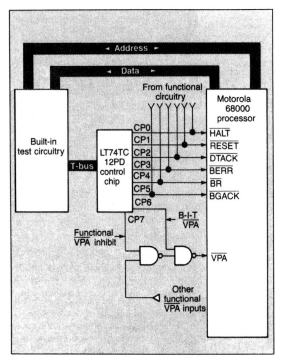

The same Motorola 68000 processor, this time with the control points identified. The built-in test circuitry can control the operation of the processor at any time via the LT74TC12PD IC. If the built-in test circuitry does not need to interfere, it simply tri-states the testability chip

depending on the circuit under test and level of implementation. By comparison, the T-Circuit will typically require well under 1% added circuitry.

Strictly serial techniques result in long test patterns and long test times. The combination parallel/serial architecture of the T-Circuit grants the user the flexibility to trade-off number of patterns and test time with the size of the testability bus. Twenty-five pins on the bus allow for 8192 visibility and control points, all accessible either serially or in parallel.

All structured approaches have significant impact on circuit design. The T-Circuit has minimal impact on circuit design. In most cases, for control purposes, an occasional extra input must be provided to a gate, and standard testability practices (e.g. not connected 'sets' and 'resets' hard to power or ground rails) must be followed. Other than standard testability practices, there is minimal or no impact on the circuit design with the T-Circuit.

Impact on testing

To determine whether or not the T-Circuit would achieve the objectives planned for it in terms of lowering programming, fixturing,

visibility point section are equipped with serial data inputs and outputs so that circuits may be cascaded if desired. To expand the number of control and visibility points, it is only necessary to connect the serial data outputs from the first circuit to the serial data inputs on the next.

The T-Circuit takes only a few hundred gates for accessing up to 32 visibility and control points. The gate count goes up linearly with the number of control and visibility points desired. Thus, it becomes quite easy to assess the optimum trade-off between the number of visibility and control points to be implemented vs. the costs associated with generation of the desired test pattern, testing, and fault isolation.

How does the T-Circuit compare with well-known structured and unstructured testability approaches? The following points give some indication.

LSSD and similar approaches take anywhere from 2 to 20% of circuit real estate,

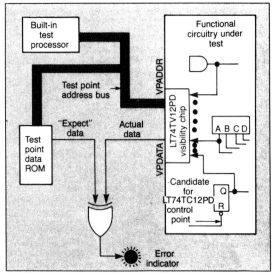

The built-in test processor is able to periodically compare actual data from the unit under-test's internal mode with expected data stored in ROM. If a mismatch occurs, indicating a failure, the processor might activate an error light or take other action

testing and trouble-shooting costs (at both board level and product level), a commercial ATE company was asked to run a logic simulation on two typical boards, first without and then with an extra two dozen control and visibility points.

The results, based on boards from two major electronics manufacturers, were as follows:

- Programming time reduction: 20% minimum, 50% typical.
- Number of test patterns reduction: 15% minimum, 100% typical, 300% maximum.
- Percent fault coverage increase: 95% typical vs. a current 75–90%.
- Trouble-shooting time decrease: 25–30% minimum, 50% typical.

These figures are in line with real-world experience. More conservative figures have been used in calculating the cost differences cited than those shown earlier.

A testability bus, implemented with the architecture of the T-circuit, provides the first universally implementable standard testability interface for commercial and aerospace/military electronic devices, boards, and systems. It holds the potential for drastically reducing system built-in test design costs, and for drastically reducing test engineering costs in all aspects of test program generation and logic simulation. This last includes go/no-go testing and fault isolation, particularly with respect to assemblies utilizing SMT.

Such a circuit potentially provides the most viable means for major electronics manufacturers to improve productivity within their business environments, with minimum cost and effort. T-Circuit utilization is expandable to any degree of size to handle any degree of system complexity, and provides total testing interface compatibility between any or all of the electronic devices, subassemblies, and systems designed to incorporate it. The circuit itself is easily implemented as a standard cell for gate array, custom, and semicustom LSI and VLSI chip designs.

It is believed that companies which are faced with choosing between spending as much as $40,000 for an application-specific product function that cannot be purchased off the shelf, or spending $40,000 for a producibility function that can be purchased off the shelf, will opt for their own product functional designs and will purchase the producibility features from an off-the-shelf vendor.

In the industrial marketplace, the incentive to the customer in using the T-Circuit is achieving lower overall business costs. Customers will find that by adding incrementally to parts costs in their products (e.g. the price of the T-Circuit chip set), they will reap a large savings in test equipment, test programming, test execution, and fault isolation costs. This will more than pay for the extra parts cost.

CONDUCTIVE EPOXY FOR SMT SOLDER REPLACEMENT

Solder paste may not be familiar turf, even now, for some amid substantial changes that loom with the anticipated surface-mount technology revolution. Undaunted, some manufacturers of electrical-conductive epoxy, following on their success in hybrid and IC die attach, are looking to replace solder paste with

conductive epoxy for attachment of surface-mount components to PCBs.

Some electronic OEMs are evaluating conductive epoxy as a solder-paste replacement. Such a bold move, from solder paste to conductive epoxy, certainly demands thorough evaluation of the processing, reliability, and cost issues involved.

Solder problems

The premier question is: 'What's wrong with

First published in *Electronic Packaging & Production*, February 1985, under the title 'Conductive epoxy is tested for SMT solder replacement'.
Author: Ronald Pound, Senior Editor.
© 1985 Reed Publishing (USA) Inc.

Surface-mount components (such as these from Philips) are now attached to PCBs using solder paste, but conductive epoxy between the component leads and circuit pads is an alternative being tested

solder?' Differential thermal expansion between leadless, ceramic chip carriers and their epoxy-fiberglass substrate is, by now, a familiar problem in the application of SMT. Differential expansion leads to stress on solder joints attaching the chip carrier to the PCB, and subsequent cracking of the rigid solder may occur.

The differential expansion problem, according to models presented in the literature, is due to differences between the thermal coefficient of expansion for a ceramic package and its plastic substrate. Even with the same thermal coefficients of expansion, differences in temperature between component and board can introduce differential expansion.

Since the problem is created by the interacting system of chip carrier, solder joint, and substrate, the solution may spring from sources other than the method of attachment. In fact, matching the substrate expansion to that of the chip carrier, or matching the carrier to the substrate by using plastic packages where permitted, are in the forefront of solutions proposed. Compliant leaded components may also be used instead of leadless components. None of these solutions directly involve the issue of changing the attachment material.

Solutions impacting the attachment material include not only the conductive-epoxy alternative, but also retaining solder in a form that resists cracking. Greater solder-joint height reduces the stress induced by differential expansion. Indium alloy solders may provide greater fatigue resistance than the common 63/37 Sn/Pb solders.

Streamlined assembly becomes apparent when the assembly process involved with reflow soldering is compared to that of conductive epoxy. Proponents of conductive adhesives are pointing to a process with five steps, rather than nine with reflow soldering.

Just as important is the nature, in contrast to the number, of solder reflow processing steps that lead to assembly concerns. Vapor-phase soldering, a leading contender for surface-mount assembly, requires components that can withstand its high temperature of 210–265°C. Subsequent cleaning of solder-flux contaminants is generally required.

Cleaning of soldered assemblies involves cost and manufacturing floor space for capital equipment. There is a concern, too, that with surface-mount components perhaps as little as 0.003in. off the board, the cleaning medium may not be able to reach entrapped flux contamination. Solvent cleaning, rather than aqueous cleaning, offers a medium with lower surface tension. There are, however, associated costs and processing control issues involved with the purchase and safe handling of chlorinated and fluorinated solvents.

Formation of solder balls and leaching from component and circuitry metals, particularly of silver, are other solder problems that are cited. The use of 62/36/2 Sn/Pb/Ag solder paste, however, appears to remove the elimination of leaching as a potential advantage of silver-loaded conductive epoxy.

As might be surmised, the differential thermal expansion problem with subsequent

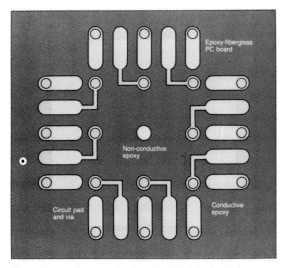

Conductive epoxy is screen printed onto circuit pads for electrical attachment of the surface-mount component

cracking of solder joints, and overall processing cost and complexity of reflow soldering with solder paste, are translated into potential advantages for conductive epoxy.

Conductive epoxy processing

Epoxy manufacturers make their products electrically conductive by loading them with metals, silver in particular. Using epoxy in attaching surface-mount components to PCBs involves screen printing the conductive epoxy onto circuit pads, placing the component on the pads, and curing the epoxy.

The epoxy is screen printed onto circuit board pads to a thickness of 0.002in., utilizing a 200 mesh stainless-steel screen with 0.0018in. emulsion thickness, said Dick Estes, quality control manager for Epoxy Technology, describing the use of their EPO-TEK H20E. Similar screen-printing parameters are recommended by Electro-Science Laboratories for their ESL-1900 silver conductive adhesive.

Designed to be screen printed, Amicon's C-770-4 conductive epoxy has a viscosity of 100,000 centipoises. This is the area where one generally wants the viscosity for screen printing, notes Mark Edson, Amicon's product manager for conductive products.

In addition to electrical-conductive epoxy screen printed onto each pad, a dot of non-electrical-conductive adhesive may have to be dispensed to provide mechanical strength, and perhaps thermal conductivity, between the component and its substrate.

Epoxies that provide the highest bond strength are the most rigid, says Edson. If a board is going to be highly flexed, or subjected to a lot of vibration, the conductive epoxy would have to be a more elastomeric material with a higher modulus of elasticity. The more elasticity in the adhesive, however, the lower the bond strength. Thus, non-conductive adhesive may have to be used in conjunction with the conductive epoxy in such applications.

In some cases, like the smaller chip capacitors and chip resistors, conductive adhesive alone might be used, providing both electrical contact and mechanical strength. The larger the component, however, the more consideration must be given to providing strength with a non-conductive adhesive, Edson indicates.

Components must be precisely placed on their circuit pads, particularly if they have a large number of leads. With small, closely spaced dots of conductive epoxy, one cannot afford anything but precision placement that leaves the epoxy unsmeared, remarks Edson. If a 64-pin component is placed crooked, and it is twisted to correct its position, the chances are that a short will be created by smearing the conductive epoxy. Solder paste, on the other hand, when it is smeared, tends to flow back and gather on the pads when it is reflowed, in effect unsmearing itself.

Pick-and-place machines provide sufficiently precise placement. Smaller manufacturing operations which may use manual or semi-automatic placement could have epoxy-smearing problems.

Thermoset epoxy is cured, depending on the particular material, by heating in convection ovens, by exposure to infrared or ultraviolet radiation, or by heat released in condensation of hot vapor on a cooler circuit assembly. Curing times are on the order of a few minutes to an hour, depending on the curing equipment and epoxy parameters. High-strength materials cure at 150–180°C, while lower strength, highly elastomeric materials cure at 80–160°C.

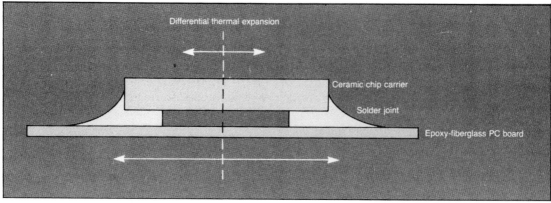

Differential thermal expansion is created between a ceramic chip carrier and epoxy-fiberglass PCB, resulting in cracked solder joints

Normal cure time for conductive epoxy in an air-circulating oven is about an hour, says Joe Vaccaro, marketing manager for electrical/electronic liquid systems at Furane Products. Curing times can be reduced by using infrared. With a typical adhesive that cures in about an hour in an oven, IR can reduce the time to about seven minutes, remarks Furane's product development supervisor for microelectronics, Ron Hunadi.

Most people will be using convection ovens or IR, Vaccaro says. There are no conductive adhesives that can be completely cured with UV alone. Its penetration is limited in metal-filled materials.

Pivotal attractions

Solder problems contrasted with claimed conductive-epoxy attributes for surface-mount assembly is the basis for the list of epoxy advantages cited by some of the materials' manufacturers. Reducing the number of processing steps and providing a more compliant attachment between surface-mount components and their substrate are the major attractions.

Five processing steps, instead of nine as with solder paste, will result in cost reductions that will more than offset the higher cost of conductive epoxy, some of its manufacturers claim. Not only are cleaning and attachment material reflow steps eliminated, but board solder plating and component lead solder tinning are also unnecessary for component attachment.

In fact, observes Amicon's Mark Edson, one does not want component leads to be tinned when epoxy is used. While epoxy will stick to virtually any substrate, two materials can present adhesion problems – tin and lead, the components of solder. Conductive epoxy will work with lightly oxidized surfaces.

There are two purposes for using conductive epoxy, according to Edson. It can be used with heat-sensitive components that cannot withstand the temperature of vapor-phase soldering; and it eliminates a large portion of the resources associated with a soldering operation.

Elimination of soldering and cleaning equipment is the main advantage of going to conductive adhesive. The greatest interest is where manufacturers are looking ahead to totally surface-mount boards, says Edson. If soldering is still a requirement, the advantage of using conductive epoxy is not as great.

Mixed-technology boards, with through-hole components on one side and surface-mount components on both sides, would present problems for the conductive epoxy approach, though it could conceivably be done. If the surface-mount components were attached first, the conductive epoxy could withstand wave soldering and cleaning with solvents once the through-hole components are inserted, says Furane's Vaccaro.

Solving the differential thermal expansion problem, and thus cracking of solder, involves both the rigidity and thermal coefficient of expansion of the attachment material, accord-

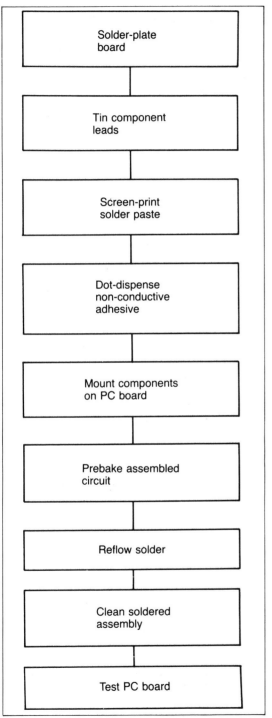

Reflow soldering of surface-mount assemblies requires nine steps for board and component preparation, and attachment of the components (Courtesy of Furane Products)

ing to epoxy manufacturers. Models of the process, however, account for the problem in terms of the difference in the thermal coefficients of expansion of a ceramic package and its plastic substrate, and rigidity of the shorter solder joint alone. As the plastic substrate expands more than the ceramic package in response to heat input, the rigid solder joint cracks.

Even the most rigid epoxy still has more flexibility, measured by the modulus of elasticity, than does solder, says Edson. With epoxy, too, one has the freedom to use the flexibility that is needed for the application.

In addition, metal-loaded epoxy has a thermal coefficient of expansion more closely matched to that of an epoxy-fiberglass PCB than does solder, claim some epoxy adhesive manufacturers. However, Herb Kraus, con-

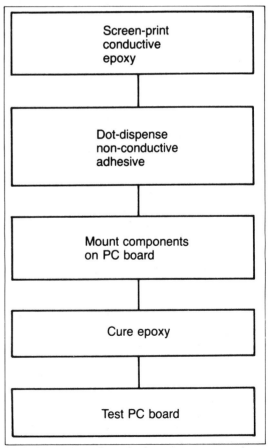

Conductive epoxy for attachment of surface-mount components requires five steps in the manufacturing process (Courtesy of Furane Products)

sultant at Ablestik Laboratories, an adhesive manufacturer, states that there are few conductive epoxies below 40ppm/°C.

With epoxy the molecular weight is built up, but a true crystalline structure is not developed, like the process in cooling solder. Cross-link density normally is not disturbed by vibration, unlike the crystalline structure of solder when it is cooling, according to Vaccaro. Resulting defects in the solder joint's internal structure weaken its structural integrity.

Epoxy issues

Not free from its own application problems, some tough-minded evaluation of the issues is in order before walking out on solder paste and taking the step to conductive epoxy.

Silver migration springs to mind as a potentially severe problem when silver-loaded epoxy is in close proximity to other conductors, and electrified in a high-moisture environment. Dendrites of silver can quickly form between the epoxy and the conductors, creating a short, according to Ablestik's Kraus.

Surface-mount assemblies are typically unprotected and may be exposed to high-moisture environments in service, says Kraus. To minimize the possibility of silver migration, the assembly would need an effective conformal coating. This will add cost and complicate repair of assemblies, he observed.

While elimination of solder tinning is claimed as a cost savings, Kraus says that contact surfaces must be silver plated, or preferably gold plated, to provide compatibility with silver epoxy.

It is like a plating tank when silver is subjected to heat, humidity, and bias, remarks Funrane's Vaccaro. The silver tends to be pulled out of the adhesive and redeposited.

Fillers other than silver are being evaluated at Furane. Every metal migrates under the right conditions, explains Ron Hunadi, it is just a question of how fast. Copper, one metal being considered as a filler, migrates slower than silver, and is a lot less expensive than gold, another metal with slower migration.

The issue of providing acceptable conductivity with fillers such as specially treated

Property	Material	
	Conductive epoxy	Solder
Volume resistivity Ω-cm	0.001	0.000015
Thermal coefficient of expansion (ppm/C)	15-20	25
Modulus of elasticity	Flexible	Rigid
Bond	Chemical	Metallurgical
Processing steps	5	9
Source: Amicon		

Material properties

copper, or even in nickel-filled systems, is viewed optimistically by Furane. A lot of people in the industry are realizing that they can use materials with higher volume resistivity in many applications, states Vaccaro.

While at times there is a requirement for a specific resistivity, frequently people cannot state what is really needed. They play it safe by asking for the best conducting system available, according to the Furane spokesmen. For the conductive-epoxy user, a large safety factor may provide security if they use the adhesive close to its expiration date, since the resistivity can change with time.

The possible acceptability of higher resistivity materials paves the way for fillers other than silver. Use of fillers such as copper could aid in the solution of metal migration problems.

Another critical factor in the plating tank analogy of silver migration is that there must be an electrolyte present in the system, observes Bob Plewnarz, Western district sales manager for Furane. This electrolyte is formed from the ionic species that contaminate the silver-filled epoxy. By creating an

Material	Thermal coefficient of expansion (ppm/C)
Ceramic chip carrier	5-7
63/37 Sn/Pb solder	25
Conductive epoxy	15-20
Epoxy-fiberglass board	12-16

Thermal properties

epoxy with a majority of the ionic species removed, the electrolyte tends to be eliminated. Silver migration, though not precluded, is greatly reduced in high-purity adhesive systems, says Plewnarz.

The success of silver-filled epoxies in hybrid microelectronics does not alone provide a sufficient basis for extending the approach to surface-mount components on PCBs, indicates Herb Kraus of Ablestik. Conductive-epoxy attachment of leaded devices may present electrical reliability problems. Wire bonds in hybrids are never made with conductive epoxy, he says.

According to Kraus, the epoxy resin tends to selectively form a resin-rich, insulating layer around a wire. An electrical joint could exhibit high resistance, either immeidately or eventually. This problem can be expected with silver-epoxy-attached leads, he cautions.

Moisture resistance has not been a problem when the adhesive is used for chip attach on hybrids, because they are sealed in a package, Kraus indicates. Surface-mount assemblies, however, are exposed. Thus, it must be assured that the silver epoxy can withstand exposure to high-moisture environments for prolonged periods without failure at the adhesive-to-metal interface.

Concepts tested as an alternative to 63/37 Sn/Pb for attaching drop-in components onto the circuit traces of a microwave PCB in a stripline configuration have been reported by Texas Instruments. These components required surface mounting, with the solder joint providing both structural integrity and electrical interconnection. Sn/Pb solder joints cracked under temperature cycling.

In their evaluation, TI indicated that conductive epoxy, one of six alternatives considered, failed in temperature testing. They selected indium alloy solder to use for the assembly.

THE REMOVAL AND REPLACEMENT OF SMCs

Although surface-mount components (SMCs) have been used for about two decades, safe removal and replacement is a comparatively recent development. Factors which have generated this new emphasis include: increased packaging density; high frequencies and speeds; closer pitch lead-to-lead spacing; leadless carrier utilization; proliferation of device configurations without standardization of size; and one and two conductors between terminal areas.

Because it is easier to mount and remove, the dual in-line package (DIP) superseded the original SMC, the flat pack. However, as packaging density became more important, the high-density SMC has returned to prominence. As a result, the difficulties inherent in removing and replacing surface-mount components have become a growing concern within the industry.

First published in *Electronic Packaging & Production*, January 1986.
Author: Linus Wallgren, Pace Inc.
© 1986 Reed Publishing (USA) Inc.

Today circuit assemblies can cost thousands of dollars. Obviously, it is bad economics to scrap such an assembly because of a failed component.

Assembly design

Often the assembly design will not easily permit components to be safely removed. Since there can be no difference between a new assembly and a reworked one, proper component and board design must consider the repair and replacement option so when exercized it produces no adverse effects on the PCB, the adjacent components or the new component.

To avoid overheating over areas when removing a component, heat should be applied only to the component's terminations. Brute force heating can melt solder, scorch boards and damage components. To minimize expansion stresses during this rework, some compliance between components and boards should be incorporated into the design.

Because heat during operation is a concomitant of high-density SMC placement, designers are mounting heat sinks on some 'hot' components to protect others. The use of these sinks must be carefully considered so they do not shroud SMC terminations in case of rework. In applying a heat sink to a specific component, the sink is sized to remove undesirable heat from the component during normal operations. However, the heat sink should not be so oversized as to prevent the effective heating of the solder joints for the removal and replacement of the component.

Another technique for removing heat generated by high-wattage components is to heat sink them by interconnecting to the PCB with thermally conductive adhesives. Thermal adhesives may require extra heat during component removal, which mechanical-type adhesives with a low structural strength are more easily removed. The best structural adhesive is one that can be dissolved and washed away during cleaning operations after reflow soldering.

The ideal board design for reworking has thermally identical terminations to SMCs. Thus, no terminations require more heat than others. All electrical connections to the SMC should be made through the smallest conductor possible. The entire ground plane must not serve as a land for an SMC. Instead, the ground plane should be connected to the SMC land through a small circuit track to isolate the SMC termination thermally but not electrically.

Advice to designers: consider the need for reworking when planning packaging density. Adequate space between components is important. Conductors should not be exposed between lands and excessively large lands. In general, the exposed metallized portion of the board must be kept to a minimum.

Removing and replacing SMCs

The criteria for ideal removal and replacement of SMCs include:

- No damage to removed components.
- No damage to adjacent components.
- No damage to the PCB.
- Non-hazardous operation.
- Minimum skill required.
- Minimum time to effect rework.
- System capable of both removal and replacement.
- Low cost.

Emphasizing its comparative position with respect to other factors, cost is listed last in the hierarchy of criteria. The true guide to selecting a rework technique and appropriate equipment is the acceptability of the repair and the performance of the final assembly. The capability of the repair system must match the criteria defining desired quality.

Moving from the ideal to the acceptable, listing criteria is the first step in SMC removal. Each item on the list increases the cost of the system, conversely each one omitted may reduce rework acceptability later. To make the reworked assembly indistinguishable from a new one, four basic steps are required for the removal process: preparation, heating, removal and cleaning.

Preparation is the most important step in the process. There must be clear access to solder joints. All contamination must be removed from the SMC area with solvent, abrasives or heat. Components or heat sinks that obstruct access should be removed and stored for later reinstallation.

Adjacent components that may be damaged during the heat applications need protection, an inherent feature of sophisticated removal

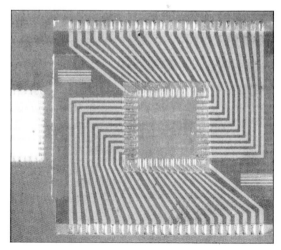

Thermally uniform foil at land terminations creates an ideal board design for effective rework

systems. Quality removal equipment heats only the solder joints of the subject SMC, eliminating the need for thermally shielding other components and the board with protective devices.

The last stage of the preparation process calls for coating the solder joints of the SMC with a suitable flux to assure proper remelt of the solder. The assembly is then secured for the application of heat. Almost any device that holds the board securely will suffice.

Heating is the second step in the removal process. Getting heat to the solder joints is easy; keeping it from other areas is difficult.

Indiscriminate heating of SMC assemblies can cause damage while selective heating control represents the key to successful component removal and replacement. Some ceramics currently being used as board substrates may crack when heat is applied to small portions of the board. To reduce thermal shock, the assembly should be preheated to a temperature just below solder melt temperatures.

Heat is transferred to solder joints by conduction, radiation and convection. One of

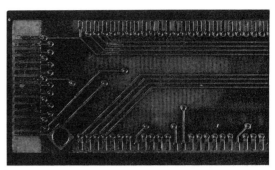

Overheating during an attempt to remove a leaded chip carrier damages the PCB substrate

the most common conductive devices is the soldering iron, but it is difficult to remove some leadless devices with conventional hand-held equipment. Even special-design soldering irons may have drawbacks.

Placing the assembly over a hot plate heats the entire board by conduction and melts all the solder joints. The failed components can then be picked off with tweezers. This technque makes the entire assembly very fragile when the joints are molten and is not suitable for assemblies with components on both sides.

Several forms of radiant energy (ultrasonic, induction, microwave) can be destructive to the assembly, but laser and infrared energy offer some promise. Laser heating presents cost problems and may be practical only in high-volume applications. Infrared, however, is enjoying increased use in this application.

Convective heat is the most common heating technique for reworking SMT boards. Hot gas directed precisely at solder joints melts them without exposing the rest of the assembly to heat damage, eliminating many problems associated with other methods.

Removal follows heating in the rework process. Using any of the heating methods, the simplest removal process is to pick off the faulty component with tweezers or with a vacuum pickup. Another, not uncommon method melts all the solder joints on the board, after which the board is rapped sharply and all the components are removed at once. This procedure, used only where the PCB is the costliest part of the assembly, requires that all the components be replaced. However, the machine-positioned, operator-

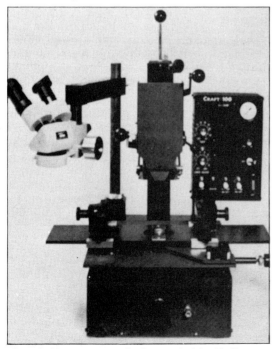

The Craft 100 system removes and replaces most leaded and leadless carriers in less than 20 seconds

An SMC is placed in a retractible locator and then positioned onto the circuit board with a vacuum pickup

controlled vacuum pickup is the most preferred for an acceptable reworked assembly.

Cleaning is the last step in the rework process before component replacement. After the component is removed, any residue, excess solder, flux or other contaminant must be cleaned. If significant time elapses between the removal of the SMC and its replacement, the exposed area of the assembly must be covered with a protective coating, and the board should be packaged and stored until the rework is completed.

Replacing the SMC

The procedure for replacing a faulty component is the same as for placing a new component in a new assembly. To replace a component, there must be clear access to its solder joints. There can be no obstruction of any kind to jeopardize the quality of the reworked assembly. As with removal, if there are any components, heat sinks or mechanical obstructions prohibiting access to the SMC, they must be removed and stored for rein-

stallation after the SMC has been replaced.

All contamination and protective coatings must be cleaned from the area; any adjacent components susceptible to damage during the heat application need to be protected. Finally, the solder joints of the circuit board and the SMC require pretinning and fluxing to assure clean and proper reflow of the solder.

Component placement is a delicate step. Before the solder joints can be safely melted, the components must be accurately positioned and firmly held on to the securely mounted PCB. Tweezers and adhesives are usually unsatisfactory for placing the holding components. The method of choice is machine-positioning, under operator control. Having few drawbacks, this technique offers the most cost-effective man/machine combination for SMC placement because the operator needs no special dexterity to position and hold the component.

Heating procedures are the same as for those in the removal stage. Before the actual placement, the assembly is reheated to slightly below solder melt temperature to avoid thermal shock. As in the removal process, the most common and effective method of heating when replacing an SMC utilizes hot gas aimed precisely at the solder joints. This technique permits mechanical manipulation of the component during heating and allows the operator to safely view and control the process.

Finally, the assembly is cleaned to remove any flux residue and contamination. Components, heat sinks, other parts and protective coatings that had been removed are then replaced.

CHAPTER 5

HYBRID CIRCUITS

Increasingly the complexity of demands on electronic devices must be
met by utilizing several different assembly techniques on a single
circuit board. The different requirements for power and logic functions
in ever smaller spaces has created a constantly growing demand for
hybrid circuits.

NEW APPLICATIONS

While recent years have not seen spectacular technological change or burgeoning growth for the hybrid industry, most signs point towards the steady influx of new applications for this mature and surprisingly popular packaging approach. Although most hybrids continue to be built with the 'standard' manufacturing method – noble metal thick film and wire bonding – other techniques also have their adherents. There appears to be no end in sight to the industry's traditional diversity. Moreover, the Silicon Valley doomsayers' predictions of the demise of the hybrid under the onslaught of VLSI show absolutely no sign of coming true – the hybrid sector is big, stable and healthy, perhaps more so than ever before.

The sheer size of the hybrid industry is a surprise to many who stereotype hybrids as a niche technology. It was estimated (1984) that the total number of hybrid shops in the USA was 700, approximately 40% of which were captive operations devoted to supplying other activities of their parent organizations. About 60% were merchant shops selling standard and custom parts in the open marketplace.

Application strongholds

One area where hybrids are widely applied is in medical electronics, where long-term reliability is crucial, along with dense circuitry and the need for special substrate shapes to fit odd-shaped packages. There are other, subtler advantages to hybrids in medical applications, particularly those involving implantation. R. Michel Zilberstein of Raytheon, Quincy, MA, has pointed out several of them. "Owing to the low level – or in some cases, total absence – of organic materials in the assembly, and the care exercised in manufacture, final products are contaminant-free. This guarantees freedom from internal corrosion which would degrade performance and reliability," he said.

Raytheon engineering laboratory manager Zilberstein also emphasized that hybrid technology is a low-risk route for medical applications because of the military's extensive use of hybrids. The resulting widespread adherence to stringen specifications like MIL-Q-9858A, MIL-STD-883B and MIL-M-38510 has created a cadre of experienced, highly reliable hybrid shops.

Micro-Rel of Tempe, AZ, is a supplier of hybrid assemblies for medical euipment, with most of its output going to parent company Medtronic, Inc., of Minneapolis, MN. To build the custom hybrid, Micro-Rel used its own in-house multilayer co-fired technology combined with surface-mounted chip carriers and chip discrete components on both sides of the substrate.

Significantly, one must look away from the main substrate to the internal connections of the chip carriers to find conventional wire bonding – Micro-Rel's effort has as much in common with surface-mount PC technology as it does with traditional thick-film hybrids.

Another application that has often found a hybrid solution is the fast-moving computer industry's need for higher memory density. Rather than sit on their hands waiting for the semiconductor suppliers to let the next generation of memory ICs out of the laboratory, computer makers have often turned to modular memory packages, like those produced by Electronic Designs of Hopkinton, MA.

This custom hybrid from Micro-Rel combines a multilayer co-fired structure with double-sided surface mounting

First published in *Electronic Packaging & Production*, November 1984, under the title 'Hybrids '84 – A mature industry moving into new applications'. Authors: Don Brown and Martin Freedman, D. Brown Associates Inc. © 1984 Reed Publishing (USA) Inc.

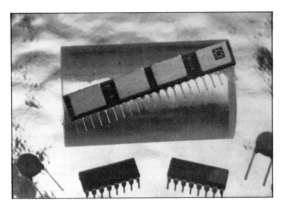

Hybrid memory modules, like this one from Electronic Designs Inc.,
provide computer makers with four times the density of DIP devices

Like the Micro-Rel custom hybrid, EDI's SIP and DIP DRAM modules use surface mounting, chip carriers and chip discrete components, with nary a wire bond in sight except inside the chip carriers. The module concept was born out of a need for more memory in less PC real estate and illustrates an important principle that will keep hybrids, or functionally equivalent technologies, viable for the foreseeable future.

Even though the modules were originally designed to overcome the density limitations of 64K DRAMs in DIP packages, the same concept has been applied to the latest 256K chips. In other words, whatever density and functionality is achieved in silicon can be improved upon by building that silicon into a hybrid or hybrid-equivalent assembly.

Surface mount – a mixed blessing?

Hybrid shops were using surface-mount solder attachment a long time before anyone thought of applying the technique to organic PC substrates. The application of surface mounting to PCB has affected the hybrid sector in two major ways. The good news is that PC surface mounting has produced an explosion in surface-mount component availability – 21,000 parts in 1984 compared to 9000 in 1983 – which the hybrid industry can partake of. Unusual items like inductors and crystals are now obtainable in chip form, along with an unprecedented number of packaged surface-mount ICs. The bad news is that the hybrid industry has all but ignored

the opportunity to move into PC surface mount and may well lose some of its formerly exclusive market as a result.

The reason for this situation is that surface mount, properly applied, can bring the density of a PC assembly squarely into the traditional hybrid ballpark, usually at lower cost. An example of this is the SIP memory module – similar to EDI's, but using plastic chip carriers and different pinout – developed by office-system giant Wang Laboratories. The design was initially executed in thick film on alumina, but was also on the drawing board as a PC assembly with the same form factor, at a lower cost.

There is a subcontracting industry growing up around such surface-mount PC assemblies, and those assemblies are the functional equivalent of a hybrid. They also represent an opportunity for hybrid houses who, after all, pioneered surface mounting, to expand and diversify. Unfortunately, few have done so as yet.

Cost matters

As befits a mature technology, the costs involved in hybrid production have been falling as yields increase and manufacturing techniques are refined. Jim Angeloni, director of operations at Natel Engineering Co., Simi Valley, CA, was emphatically bullish on the effects of this improved cost-effectiveness. "The most significant trend in hybrid circuit technology is that of lowering the process cost so as to make hybrids attractive to the industrial marketplace," he said.

Speculating on the future of hybrid packaging, Angeloni added, "As hybrids move into the industrial arena, metal packages will give way to ceramic . . . and possibly in the future, some form of plastic." He sees surface mounting on PCBs as an opportunity for hybrid makers to participate as component suppliers, and as a clue to the shape of hybrids to come. "Hybrids will become available in a variety of surface-mount packages," he predicts.

Cost-effectiveness also figures in a recently introduced hybrid product from Dynamic Measurements Corp. of Winchester, MA. While complex signal and data conversion

These voltage-to-frequency (VF) converters from Dynamic Measurements Corp. are cost-effective enough to compete head to head with monolithic ICs

functions are often assigned to hybrids rather than monolithic ICs on the basis of pure performance, DMA's 3815 family of voltage-to-frequency (V/F) converters can claim lower cost as well. The closest equivalent monolithic ICs require external components to perform at the hybrid's level, and these additional components make them more expensive. Director of marketing Bill Schrom says the company is "... well along in developing other data conversion hybrids that can be similarly positioned in existing monolithic functions."

Applications in which hybrids are chosen over monolithic ICs on the basis of cost are few and far between, however. Hybrids are generally specified where they provide performance not otherwise obtainable, like data conversion. This demanding and crucial task is difficult to implement monolithically because it inherently requires mixed linear and digital circuit functions, as well as 'tweaking' of resistor values – all strong points of the hybrid approach.

A/D converters

Analog-to-digital (A/D) converters are perhaps the closest thing to a 'glamour' product that the hybrid industry has. These sophisticated devices are found almost everywhere the very digital world of computer

technology meets the very analog realms of industrial, scientific and aerospace instrumentation. Although the requirements vary widely with the application – speed and accuracy being the major trade-offs – the demands on a hybrid analog-to-digital (A/D) converter are never easy to meet. Fortunately, hybrid technology allows the designer of the device to pick the very best monolithic IC dice and combine them with high accuracy, laser-trimmed resistive elements to optimize performance for the application.

In the ADC803 A/D converter Burr-Brown of Tucson, AZ, has plainly gone for conversion speed as a design goal. Although built on a conventional thick-film-on-alumina substrate, maximum resistor accuracy is obtained by putting all critical resistors in a thin-film-on-sapphire resistor network, which is wire bonded to the substrate as if it were an IC die. Other key components include a monolithic

The major feature of this analog-to-digital (A/D) converter from Burr-Brown is its 1.5 µs conversion speed

This Hybrid Systems D/A converter boasts 16-bit linearity and resolution, with CMOS hold power dissipation to less than a watt

12-bit digital-to-analog (D/A), 12-bit successive approximation register (SAR), and a precision op amp used as a comparator. The performance payoff is a blistering 1.5 microsecond conversion time and ±0.015% full-scale linearity over a −25° to +85°C operating temperature range.

Where accuracy is the name of the game, Hybrid Systems Corp. of Billerica, MA, has a real scrutinizer in the HS 9516-6 with ±0.0008% linearity error and 16-bit resolution – that means one part in 65,536. Hybrid Systems has kept power dissipation down to 900mW typical – about 50% lower than most competitive hybrids – by using a CMOS D/A converter die. As is typical of high-precision hybrid data converters, 'no missing codes' is specified over the entire operating temperature range, which in the case of the HS 9516 series is 0–70°C.

A refinement

Increasing hybrid sophistication is not restricted to the devices – the manufacturing process itself is also being improved on an ongoing basis. Although surface mount is an important thrust for the hybrid industry and TAB technology has its appeal, wire bonding remains the most popular method to connect semiconductor dice to hybrid circuits.

The work of John F. Graves of the Communications Division of Bendix Corp., Baltimore, MD, is an example of how painstaking research can significantly improve the performance of an 'old' technique. Graves discovered that contamination found in the first few angstroms of the thick-film bonding areas was the factor that limited the pull strength of thermosonic wire bonds. Although a brute force approach of crushing through or otherwise dispersing the contaminants produced a passable bond in many cases, optimum results require removal of the contamination before bonding. The method Graves arrived at, plasma processing with argon gas, improved the bond strength for 0.001in. gold wire from 6.5 to 9.5 grams, an improvement of over 31% that was achieved with no physical or electrical ill effects.

A major arrival

Perhaps the most significant event recently for the hybrid industry was the entry of Tektronix Hybrid Components Operation, Beaverton, OR, into the merchant market. The versatile Tektronix shop has been one of the most capable and widely respected captive houses in the world for over 15 years, and has played a major role in the success of the company's popular test and measurement equipment.

A look at what is going on at Tektronix gives a quick overview of some of the

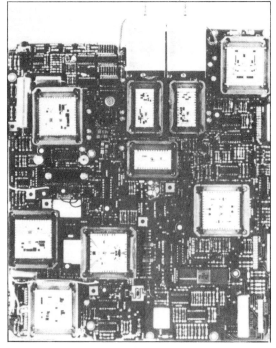

The popular Tektronix 2400 series of oscilloscopes uses 11 major custom hybrids to enhance its performance

directions the hybrid industry will be pursuing next year and beyond. On the captive side, there is a strong belief at Tektronix that custom hybrids are an essential ingredient in meeting future performance requirements. "The 2400 series of oscilloscopes has 11 major custom hybrids in it," according to director of marketing Peter Huntsinger. "They replace thousands of discrete parts, making the instrument lighter, more reliable and easier to manufacture."

Tektronix has also implemented some Japanese methods, including Just-in-time (JIT) techniques in its chip-and-wire lines. They claim a 97% yield for 1984, with improvements expected next year. The company is also a CAD leader, having developed its own software, called HCAD, which aids in custom hybrid design. The hybrid operation is heavily automated including screen printers, furnaces, pick-and-place equipment, laser trimming, substrate scribers and test gear.

Huntsinger also pointed out some major challenges. "The future in hybrids is higher density and this leads to multilayer thick- and thin-film hybrids. We see this as the major trend . . . to meet density requirements and improve performance." To do this will require attention to issues like line and space definition and thermal management – a big part of the ongoing evolution of hybrid technology.

Continued growth

The hybrid industry has established itself as a mature packaging technology. The growth of surface-mount technology has had the benefit of providing both a much larger quantity of miniature component types and specialized assembly equipment. Continued growth of the hybrid industry is assured as the electronics industry strives for smaller, more reliable designs for a multitude of tasks that are not readily achievable in monolithic semiconductor form because of performance or time-to-market constraints.

NEW MATERIALS AND EQUIPMENT

Hybrid microelectronics is growing at a rate exceeding that of both the overall economy at large and the electronics industry in general. Estimates of the average growth rate in real dollars range from 14 to 20%.

Part of the disparity in the estimates is caused by the difficulty in defining a hybrid circuit. Some definitions include resistor networks, while others include single integrated circuits mounted in DIP packages fabricated by one of the film technologies. The definition in the ISHM Design Guidelines, and the one used here, requires that a hybrid circuit include at least two components, one of which is active, and an interconnection pattern made by a film technology. Passive components may be integrated on the circuit or mounted as separate chip components.

First published in *Electronic Packaging & Production*, August 1984, under the title 'Hybrids take the lead in electronics industry growth'.
Author: Jerry E. Sergent, Contributing Editor.
© 1984 Reed Publishing (USA) Inc.

While the field of hybrid microelectronics has fundamentally changed very little in the past few years, it has been marked by dramatic improvements in materials, by advances in processing equipment which allow greater precision and repeatability and which greatly increase throughput, by the implementation of new applications which were not earlier thought feasible, by the automation of certain phases of the design process, and, perhaps most importantly, by an increased awareness on the part of hybrid designers and users of the characteristics and capabilities of the technology.

Some phases of the technology, such as polymer thick films and copper thick-film conductors, have been somewhat dormant in recent years and are experiencing a rebirth, while other phases, such as high temperature materials, are of more recent vintage and are still in the developmental stage.

Epoxy adhesives

Both conductive and non-conductive epoxy materials have been used as adhesives in hybrid microcircuits for years, but not without problems. The outgassing products of many of the epoxies used to date (so-called 'second-generation epoxies') contain potentially corrosive elements such as ammonia, boron, trifluoride, methane, chlorine, sodium and potassium. When combined with water (which is also an outgassing product), these materials can be highly detrimental to hybrid microcircuits.

When water combines with chlorine, potassium, or sodium, an ionizing path is created which can accelerate silver or gold migration. Boron trifluoride, when combined with water, attacks the aluminum metalization on semiconductor devices. Further, these epoxies provoke severe allergic skin reactions in certain individuals.

Recently, new epoxy materials have been developed which minimize or eliminate most of these problems. Referred to as 'third-generation epoxies', these epoxies have a chlorine conent of below 10ppm with correspondingly low sodium content and are manufactured with non-toxic resins.

Released on May 10, 1984, National Security Agency Specification NSA77-25B incorporates a number of changes from 77-25A. One of the most significant changes is the establishment of the cure time by maximizing the glass transition temperature. Previously, the cure time was twice the manufacturer's recommended time, which still sometimes resulted in epoxy which was insufficiently cured. While still retaining some of the features of the second-generation epoxies, this specification, along with the corresponding NSAS specifica-

tion, will result in improved results with epoxy adhesives.

Polyimide adhesives

While still not covered by an official specification, as is the case with epoxy adhesives, polyimide adhesives are playing an increasingly important role in the hybrid microelectronics technology, particularly in the area of high-temperature applications. Epoxy adhesives are inherently limited in the maximum operating temperature due to their chemical composition. At elevated temperatures, epoxy materials begin to emit outgassing components, and the shear strength drops dramatically at temperatures above the glass transition temperature. Polyimide adhesives are being used in high-temperature applications with considerable success.

These materials do not contain chlorine compounds and therefore have a very low ion content. Their major weakness is that they are solvent-based. This necessitates a two-stage curing process for many materials. When used to bond large-area devices, such as substrates, voids may be left at the interface. In addition, the pot life is somewhat limited due to solvent evaporation.

Particle gettering materials

Since the latest revision of MIL-M-38510E was released, officially allowing the use of organic materials inside a hermetically sealed hybrid circuit, the use of particle gettering materials has been on the upswing. These materials, basically uncured silicone materials, are dispensed inside the lid prior to the sealing operation. The circuit is then subjected to low-level vibration to trap loose particles in the silicone. These materials remain tacky at temperatures well below $-55°C$ and do not release any corrosive outgassing products at high temperatures. Use of gettering materials allows virtually 100% of the circuits to pass the PIND test and also allows particles generated during the life of the circuit to be entrapped. Since no conformal coatings are used, rework and repair is greatly facilitated, and the overall cost is much reduced.

Material	Supplier	ppm Cl	ppm Na	Generation
H20E	Epotek	170	30	Second
CT-4042-5	Amicon	130	20	Second
826-1H	Ablestik	175	19	Second
C-850-6	Amicon	160	12	Second
CRM-1020	Sumitomo	8	5	Third
EM4000	Hitachi	10	2	Third
C-860-3	Amicon	10	5	Third
C-869	Amicon	8	4	Third
843-1	Ablestik	12	5	Third

Properties of second- and third-generation epoxies

Polymer thick-film materials

Polymer thick-film (PTF) technology seems destined to grow at a rapid rate in the near future. The impetus for this growth is due to several reasons:

- The development of a directly solderable conductor. With proper techniques, components may be directly soldered to a PTF conductor without plating.
- The development of PTF resistor systems with electrical and stability properties which exceed those of ordinary carbon-film resistors.
- The increasing use of surface-mount technology (SMT). The capability of fabricating multilayer structures with PTF on printed circuit boards makes SMT and PTF a very attractive combination in terms of size and cost when compared to multilayer PCBs with a multitude of plated-through holes.

The prime conductor material is silver, with palladium-silver alloys used in some applications for improved leach resistance and increased resistance to silver migration. Some work is continuing on copper conductor materials, with the Japanese appearing to have the edge at the present time.

The prime methods of curing are by resistance heating, infrared (IR) curing, and vapor phase curing. Ultraviolet (UV) curable PTF materials are still somewhat in the future because a thicker film is required in most applications.

Because of their superior wear resistance and contact noise properties, PTF resistor materials have been used for some time to manufacture potentiometers. Work on multilayer structures has been somewhat more recent, but several large companies are doing

extensive developmental work for internal use and several others are in the process of developing facilities for manufacturing custom PTF circuits for outside consumption.

Copper thick-film materials

After a period of minimal activity, copper thick-film systems are making a dynamic comeback, with the catalysts for this activity being the development of improved dielectric and resistor systems. Multilayer circuits for digital applications have been fabricated in moderate quantities, and resistor networks in high volumes have been available for a number of years. In recent months, several companies have begun manufacturing analog circuits with both active and passive components with performance characteristics comparable to noble metal systems.

All of these applications to date have been in commercial systems. The use of copper thick-film systems in military applications is probably some time in the future due to the possibility of corrosion of the conductors. The overall cost differential of using copper as opposed to gold in a military circuit is not sufficiently great at this time to warrant serious consideration. As the use of surface-mount technology increases, however, copper may be considered due to its solderability and leach resistance.

Micro-optocouplers

Optocouplers have been used for some time in applications where a high degree of electrical isolation is required. While optocouplers for hybrid circuits have been available, they have necessitated a wire-bonding opera-

Two-way radio tone processor microcircuit made by General Electric Co., with the DuPont copper thick-film materials system

Sheet resistivity (Ω/□)	Carbon composition (ppm/C)	Carbon film (ppm/C)	PTF (ppm/C)
100	−300	−240	−150
1	−500	−260	−100
10	−800	−400	−125
100	−1000	−500	−200
1 Meg	−2500	−700	−500
10 Meg	−3000	−1300	−800

Comparison of carbon composition, carbon film, and PTF resistors

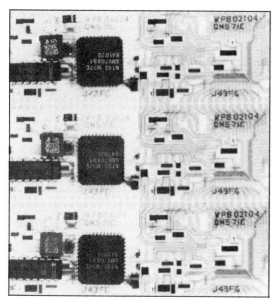

This 4 × 4in. ceramic substrate contains six individual copper thick-film hybrid circuits which are used in Northern Telecom telephone desk sets. The left side shows the circuits populated with components while the right side reveals the copper thick-film

Size	0.090 in. by 0.090 in.
Input to output voltage	±1 kV
Collector-emitter voltage	35 V
Emitter-collector voltage	7 V
Continuous collector current	40 mA
Diode reverse voltage	3 V
Diode continuous current	20 mA
Rise time (VCC = 12 V, 1C - 1.0 mA, RL = 100Ω)	15 μsec

Properties of TRW OPR1015 optocoupler

tion which increased the area required for use. TRW Optoelectronics Division in Carrollton, TX has just announced the development of an optocoupler which is compatible with surface-mounting techniques. These devices may be soldered or epoxy mounted to a hybrid circuit and are usable over the entire military temperature range from −55° to +125°C.

Automation

While individual pieces of hybrid fabrication equipment can be classified as automated, the concept of a totally automated manufacturing line is still somewhere in the future. The reasons for this are varied.

Most hybrid equipment manufacturers are

Micro-optocoupler for surface-mounting application was developed by TRW Optoelectronics Division

highly specialized and make equipment for only one or two steps in the process. Little progress has been made in standardizing tooling, fixturing or interface equipment. As a result, the circuits must be manually transported from one operation to the next, and the equipment must be individually programmed. This is the most common problem mentioned by those who have attempted to automate.

The manufacturing process of hybrid circuits depends largely on the application, and there is no such thing as a 'standard processing sequence'. This further complicates the interfacing problem when a wide variety of circuits is to be manufactured.

Automated equipment is expensive compared to manual equipment, especially when pattern recognition is included. Adding to the complication is the fact that it is difficult to automate only a portion of the manufacturing process. For example, to use automatic wire bonders to their maximum extent, the semiconductor chips must be mounted with a high degree of accuracy, requiring either elaborate fixturing or a pick-and-place machine.

It must be noted that an increased production rate is not the only reason for automation. Many companies are automating to improve reliability and consistency and to decrease visual inspection time and rework.

One aspect of automation which can be singly incorporated is visual inspection. Currently available equipment can detect voids in thick- and thin-film circuit boards, determine whether all components are present and properly oriented, and if all the wire bonds are present and properly located. While not completely eliminating the need for manual inspection, automatic inspection can minimize the time required and improve the efficiency.

Computer-aided design

The utilization of computer-aided design (CAD) techniques is increasing rapidly but has not quite advanced to the point where a multilayer circuit can be designed automatically. Interactive design, however, is readily accomplished with a considerable savings in time and with more optimum design.

Most CAD systems have the following features:

- *Automatic checking.* With the schematic entered into the system, the CAD system can perform a point-to-point continuity check of the layout.
- *Component boundaries.* The computer prevents the designer from placing a component within a predetermined distance from another component, to prevent interference or to allow room for wire bonds.
- *Parts list.* A complete parts list can be created. The area of each print can be calculated, and the amount of paste required can be determined from an algorithm.
- *Visual aids.* The visual aids needed for fabricating the circuit can be created by the computer directly from the layout.

As automated equipment becomes more standardized, networking with the CAD system as the central core unit will increase. The CAD systems will be directly interfaced with such equipment as wire bonders to program the bonding pattern directly from the layout.

Industry trends

Military applications are expanding in a number of areas, including radiation-hardened circuits, power circuits, communications circuits, and display circuits. This circuit manufactured by Litton Data Systems, combines LEDs with multilayer thick-film wiring boards to create a high resolution, high-contrast display. By cross-linking the 12 individual circuits, exceptional detail can be attained over a large surface area.

New materials systems and improved pack-

Display panel, manufactured by Litton Data Systems, is fabricated with 12 thick-film hybrid microcircuits

aging techniques have extended the ambient temperature at which hybrid circuits can operate. By carefully matching expansion coefficients and selecting bonding materials to avoid the formation of intermetallic compounds, chip-and-wire hybrid circuits can be made to operate at temperatures in excess of 200°C.

Similar efforts have also extended the power-handling capabilities of hybrid circuits. Such innovative techniques as heat pipes and copper-based packages dramatically lower the thermal resistance of the system, allowing more heat to be dissipated without excessively raising the operating temperature.

The companies which manufacture communications circuits have been among the leaders in advancing the hybrid technology. The requirements for this industry are complex circuits at low cost and high volume. To achieve these conflicting requirements, the technology has been improved in terms of both materials and equipment. For example, these companies have been among the pioneers in advancing copper thick-film circuits.

Periodically, new developments occur which prompt a new obituary for the chip-and-wire hybrid circuit. One of the latest of these is the use of CAD systems to design custom integrated circuits.

It is likely that the chip-and-wire hybrid will remain a part of the hybrid designer's repertoire for some time to come. Custom ICs simply cannot keep pace with the frequent changes during the initial design phase. By the time a design is sufficiently mature to warrant a custom IC, the program is frequently already completed. Further, analog circuits are difficult to customize because they require precision components.

An increase in the use of semicustom gate arrays for complex digital circuits is highly probable. They can frequently be designed in the same amount of time and for the same development cost as a hybrid circuit when the application is right, with a corresponding decrease in production cost and complexity, and an increase in reliability.

INTEGRAL SUBSTRATE PACKAGE TECHNOLOGY

Integral substrate package technology has been in existence for many years, although it is rarely used in large-scale hybrid circuit production. It offers the hybrid circuit manufacturer a method for producing high reliability, hermetically sealed circuits that possess many advantages over circuits mounted in plug-in, one piece, drawn packages.

Further development in ISP technology is expected to: reduce the cost of the ring frame to less than 50% of that of a standard flatpack metal package; minimize the tooling costs for non-standard packages; improve lead times for tooled-up and non-standard packages; and offer ring frames at realistic costs for prototyping.

Already, by modifying and improving existing equipment, it has been shown that bottomless ring frames and flatpacks can be produced at less than 60% of the cost of existing standard packages. It is also possible to produce ring frames in many shapes without charging tool-up costs and at reduced lead times.

Adhesives eliminated

A detailed study performed by British Telecom observed that there are three major deficiencies in the performance of hermetically sealed hybrid circuits during thermal overstress. These are component degradation, wire-bond degradation, and outgassing from adhesives.

The area that caused most failures was hybrid environmental pollution. It was analyzed that, under thermal overstressing, large quantities of ammonia and water vapor were evolved, in fact, three times the amount regarded as tolerable for the reliable operation of the semiconductor components. A number of epoxy adhesives were examined and all were shown to evolve various degrees of pollution. This posed the question, which merits serious consideration, as to why adhesives should be used at all? They obviously do contaminate the package interior, differing only in degree of pollution. It would make sense to use alternative technologies that still offer reliable hermetic sealing.

The integral substrate package (ISP) provides a reliable hermetic seal using the substrate as the structural base of the package. An overglaze insulates the output metalization from the Kovar bottomless frame

First published in *Electronic Packaging & Production*, August 1984, under the title 'Integral substrate package simplifies hybrid assembly'.
Author: A.W. Koszykowski, AWK & Associates.
© 1984 Reed Publishing (USA) Inc.

The integral substrate package (ISP) can eliminate the need for a substrate adhesive, and use of chip carriers can also eliminate die-attach adhesives. Both are lower-cost solutions to the reliability problem.

Original ISP construction consisted of low-temperature glazing a ceramic frame onto a pre-printed and fired substrate which thus formed the base of the package and onto which lead clips were subsequently soldered. After trimming of resistors and placement of components, the circuit was tested. Final lid sealing was accomplished by soldering a metalized ceramic lid to the top of the metalized ceramic frame.

More recently, the widely accepted sealing technique consists of using a stepped Kovar lid for attachment to a metal package. The current desire to construct an ISP using a metal frame and lid sealing in a similar manner is understandable. In this case, the bottomless ring frame made from an alloy, typically Kovar or its equivalent, is mounted onto a metalized track on the overglazed circuit, using solder cream. This enables the mounting and positioning of all add-on components directly onto the solder cream or preform, and then solder reflowing both the ring frame and components simultaneously. Any flux residue present can easily be removed at this stage. The package is now ready for any additional die or wire bonding requirements.

Advantages

At this stage the package can be 100% tested before finally attaching the metal lid. These lids are typically welded directly to the right frame using parallel seam welding equipment. Laser welding has also proved satisfactory.

Since there are no glass/metal seals in the plain ring frame, the possibility of poor hermeticity is reduced.

The ISP hybrid circuit weighs 50% less than a similar circuit mounted in a plug-in, one piece, drawn package. This offers a considerable advantage, especially for military applications.

It is possible to achieve less than five millibars of water vapor pressure within an ISP. This is due to the absence of expoxy normally used for mounting the substrate to the base of the metal package.

Wire-bonding failures are minimized because there are no posts for wire bonding the I/O connections.

All heat generated within a metal package with an epoxy-mounted circuit must be dissipated via the wirebonding pillars and posts. However, since the substrate in the ISP is the base of the package, any heat generated will easily dissipate within the substrate, soldered leads and heat sink.

Since heat from the ISP will dissipate more efficiently, an ISP circuit can handle more power.

As for thermal expansion, the ring frames are available in Kovar or alloy 42, so that there is a close match between the ring frame and ceramic.

CHAPTER 6

SOLDERING AND CLEANING

The most commonly used method of bonding electronic components, both passive and active, to printed circuit boards is soldering. We will not consider in this book wire bonding and resistive welding techniques more commonly associated with thick-film manufacturing processes.

The quality of the soldered joint will probably prove to be the most challenging task of the production engineer. Assuming good control of incoming boards and components, soldering techniques will determine both yield on first-time functional testing and ultimate product life.

This chapter contains a series of articles concerning soldering, cleaning and desoldering for rework. These techniques are varied and even contradictory. It seems useful to start with a brief overview, followed by an in-depth review of the processes available beyond manual soldering.

SOLDER/CLEANING, TEST AND REWORK

Wave soldering has been the mainstay of electronics mass soldering for a long time. And while it remains very useful for the attachment of surface-mount components, other methods are gaining popularity. Several types of reflow soldering, infrared, vapor phase and laser and robot soldering, as well as dual-wave soldering, are replacing the original wave soldering process. And at the same time, development work is being done that may make conductive epoxies competitive.

Surface-mount components can be used both as the only type of component on a circuit board and in various combinations with through-hole components. Different soldering techniques are used for each. With surface-mount components, some form of reflow soldering is often used, although wave soldering may also be used; in mixed assemblies, wave soldering is usually used. With mixed assemblies sometimes both are used sequentially. Less common in both cases is the use of robotic soldering or laser soldering, both of which, in effect, replace a human operator.

Wave soldering

If wave soldering is to be used, an adhesive is first dispensed at each component location. The components are placed and the adhesive cured, at an elevated temperature if required. The board is now ready for wave soldering.

A dual-wave soldering system is most often used to solder surface-mount devices. The first, turbulent, wave aids in the proper wetting of closely spaced components, while the second, smooth, wave reduces the chances of bridging. At least one manufacturer (HOLLIS) uses a hot-air knife, a sharply defined blast of super-heated air, to remove any remaining bridges.

Improvement of the wave-soldering process and wave-soldering equipment seems to be currently an evolutionary process. Wave soldering is capable of excellent yields, at least with lead through-hole components, if prop-

First published in *Assembly Engineering*, May 1985, under the title 'Solder/cleaning, test and rework in electronics assembly'.
© 1985 Hitchcock Publishing Co.

erly maintained and operated. And dual-wave systems with or without an auxiliary air knife are successful with many types of surface-mount components.

Reflow soldering

Reflow soldering techniques commonly used are IR (infrared) and vapor phase. Laser reflow soldering systems are available though not widely used.

IR reflow. In infrared reflow soldering, absorption of radiant energy heats the materials in the heat zone. The air in the heating path is not heated; there is no heat transfer by direct physical contact.

The PC assemblies first pass into a preheat zone. Relatively gentle heating of the solder paste is necessary to drive off volatile solvents, etc., without spattering and to avoid thermal shock to the board laminate. After an appropriate time in preheat (from seconds to minutes depending upon the nature of the materials involved), the assembly is raised to reflow temperature for a few seconds and then cooled.

Vapor-phase soldering. Vapor-phase soldering or condensation soldering offers precise temperature control through its ability to guarantee that the soldered assembly never reaches a temperature above that of the boiling of reflow fluid (usually about 215°C).

In vapor-phase reflow soldering, the circuit assembly is immersed in the vapor of a heavy, inert boiling liquid. The vapor condenses on the relatively cold assembly and gives up its heat in doing so. This raises the temperature of the circuit assembly to reflow temperature.

A suitable vapor-phase soldering fluid must have a boiling point minimum of 20°C above the reflow temperature of the solder, be non-flammable, non-toxic, have high vapor density, must leave no residue and must not decompose in use. The fluids are expensive. Major suppliers are 3M, Montedison, Air Products and ISC Chemicals Ltd. Production rate, in-line, vapor-phase solderers require rather high-power heaters, (up to 40kW or

Wave soldering is still the most widely used form of mass soldering. Initially used with leaded components, it is now being used successfully with surface-mount components also

more) to keep the liquid boiling when condensate returns to normal at a rapid rate. Power and fluid conservation features in recent equipment are well worthwhile.

Laser reflow soldering. Another reflow alternative coming into use is laser reflow soldering, using a robot to direct the beam to each solder joint in turn individually, usually as a replacement for manual soldering.

The advantage of laser reflow soldering is the ability to accurately limit the heat at the joint to that necessary to reflow the solder and no more. Where other automatic reflow methods usually heat the whole assembly and manual soldering can easily result in localized overheating, laser soldering can apply heat to an area as small as 0.005in. in diameter for 200–300 miliseconds. A laser soldering system can solder up to four joints per second and achieve work rates approaching those of mass soldering under some conditions.

A laser soldering system, once it is programmed, requires only unskilled supervision and has very low power requirements and operating costs.

Robotic soldering

A robotic soldering system can replace manual soldering of discrete points on a PCB and it can do it with a speed, accuracy and consistency a human operator cannot match for long. As with laser soldering, a major advantage is that only localized heating of the solder joint is necessary. Robotic soldering can be used where components could not survive the

Many capacitors are extremely heat-sensitive; because the heat input of laser soldering is both very localized and short in duration, components such as this can be reliably soldered without damage. In this application, the circuit board solder pads are screened with solder paste and the component connecting pads are tinned. Flux is applied locally and the laser beam is directed at a 45° angle to impinge on both the component pad and circuit board pad simultaneously (Courtesy of Coherent General)

temperature of wave soldering or vapor phase or IR reflow soldering. It can be used to solder temperature sensitive components after wave or reflow soldering has soldered all others or it can automatically solder the entire PCB. A robot system can solder 900–1200 solder joints per hour and can solder wire to connectors and other joints that reflow soldering would find difficult or even impossible, at a cost similar to that of laser soldering systems.

Typically the system can be programmed on-line or off-line. Parameters under control include, first of course, position of the solder joint, and then also, time at that position, tip pressure, amount of solder fed to the joint and iron temperature.

Solders and solder pastes

Solder pastes, occasionally called solder creams sare commonly used in surface-mount

Chad Industries robot soldering system can solder up to 20 joints per minute. Time, temperature and solder feed are all controllable. The system can solder joints that wave soldering could not. It can solder after heat-sensitive parts have been mounted on the PCB

component reflow soldering. These pastes consist of fine particles of solder suspended in a medium of such consistency as to allow easy dispensing yet not change characteristics while in storage.

The alloys in general use in solder pastes are similar to those in common use: 60% tin/40% lead, 63% tin/37% lead; or where leaching of silver or pladium silver component metalization is a problem: 62% tin/36% lead/2% silver. High lead content solders are sometimes used to attach surface-mount components for wave soldering and indium alloys are used where thermal problems are present.

Metal makes up to 80–90% by weight of the solder cream. The remainder of the cream is the flux/vehicle. Some progress is being made in increasing the amount of metal further; this can result in less residue when cleaning and a stiffer mixture which is less prone to flow or spread.

Major components of the flux/vehicle include rosin, activator, the thixotropic (thickening) agent, and solvents. The rosin and activator comprise the flux, the thixotropic agent combines with the metal powder and the rosin to set the viscosity of the paste. The viscosity must be sufficient to hold solids in suspension yet allow easy screening. It should also prevent separation in storage; there is a school of thought that, however, recommends always mixing solder paste shortly before use. The solvents take rosin into solution and make the paste spreadable. After application, which may be by screening or stencil or various types of spot dispensers, the paste is cured to hold the component during soldering and drive off most of the solvent which otherwise will boil during soldering and could cause spattering.

Fluxing

Fluxes are a critical factor in soldering whether in wave soldering or in the reflow of solder pastes. Flux must remove oxides and prevent their formation during soldering. It also promotes wetting. Five types of fluxes are commonly used for electronic soldering. Type R flux contains only water white rosin as a fluxing agent. It is the least active flux used. Type RMA (rosin mildly activated) and RA (rosin activated) are successively more active. A small amount of a non-ionic halogen is added to RMA; type RA contains an ionic halogen; type RSA is similar to RA and contains more halogen. Type OA uses organic acid as an activator. Both RSA and OA are corrosive.

Defluxing

While the cleaning of boards soldered using types R and RMA fluxes is considered optional, those using types RA, RSA and OA must be cleaned. The residues of these fluxes is corrosive and conductive. It can cause joint and conductor damage and leakage between conductors. In the presence of high humidity, dendretic growth, the ionic migration of conductive material across the space between conductors may occur. In addition, the residue of all fluxes may interfere with the visual inspection of solder joints and may interfere with the adhesion of coatings or encapsulants and the contact of test fixture probes.

Cleaning must remove flux from all areas of a PCB. The dense populating of surface-mount boards, and the minute clearance between the components and the board, contributes to difficult cleaning.

Competition has arisen between chlorinated and fluorinated solvents and aqueous, water-based, cleaning solutions. Aqueous cleaners with suitable detergent clean away rosin-based fluxes well. Organic acids dissolve in water and wash away with the detergent and rosin. Chlorinated and fluorinated solvents are necessary to dissolve the synthetic activators in types RA and RSA fluxes.

The washing sequence for aqueous cleaners usually begins with a spray wash to remove bulk contaminants, followed by a high-pressure detergent wash and a two-stage rinse. Drying may involve hot air blown at the board with sufficient force to carry away the solids remaining in the rinse water. A most effective method of solvent cleaning is vapor degreasing followed by a water rinse. A high-quality cleaning sequence may involve degreasing, spray, and degreasing again and possibly a de-ionized water rinse and then drying.

The advantage of solvents is that they dissolve residues that aqueous cleaners can-

not. Aqueous cleaners have the advantage of low cost, low toxicity and recyclability.

Solderability

Achieving and maintaining solderability of PCBs and components is of prime importance in surface-mount technology. The solder joints are very small in area and close together and the number of solder joints per PCB is increasing sharply. In addition, as the joints become smaller and begin to disappear beneath the component, and inspection becomes more difficult, it becomes more important to be able to have confidence in what cannot be easily seen.

The major cause of defective joints is the solderability of the parts being joined; touch-up rarely fixes things, its results are only cosmetic.

Bare PCBs are coated with solder mask with windows in it where solder joints must be formed. In these areas the copper can oxidize. Some vendors apply a rosin flux-type coating on board as short-term protection. Solder plating is more durable for long-term storage. Nevertheless solder too begins to oxidize immediately upon plating. A dry nitrogen atmosphere has been suggested for the storage of critical components.

Conductive epoxies

Conductive epoxies have been used for years in the hybrid and semiconductor areas of the industry, but are not commonly used to assemble PCBs. Conductive epoxies have the potential to neatly sidestep some of the problems associated with soldering. They cure at relatively low temperatures, require no fluxing and consequently no stringent cleaning process; they are not as rigid as solder and can tolerate the differences in thermal characteristics between the components and the circuit board and overall the assembly process would be shorter. The material cost of conductive epoxy connections is higher, about three times the cost of solder and the connections are of higher resistance.

Conductive epoxy technology for surface-mount applications is under development; additive process, surface-mount component boards may soon be an alternative.

SOLDER QUALITY DEFINITION

Before an in-depth examination of soldering and cleaning techniques, it is worthwhile examining the classification of solder joints and their more common defects.

SOLDER JOINT ACCEPTABILITY

Whether military standards are appropriate for use in commercial applications or not, they are the harbingers to industry specifications. When the IPC (Institute for Interconnecting and Packaging Electronic Circuits) rewrote its solder specification, the new spec was written with the intention of releasing it with military sanction. Because the IPC's spec had to conform with stringent military specifications in order to be pre-sanctioned, it was of necessity essentially the same as the military's, including such requirements as solder-joint brightness, joint smoothness, and plated through-hole (PTH) solder fill.

Solder-joint brightness is an issue important in all solder processes, particularly in microelectronics and surface-mount applications, where solder paste reflowed onto chip carriers and other chip components may have a more irregular and dull surface than that resulting from wave soldering.

The IPC is currently reviewing its position on the solder-joint brightness issue. The printed minutes of the IPC annual meeting in May 1984 contain these statements: "There is a major need to define acceptance criteria. Bright and shiny is not necessarily a good QC evaluation."

This new position was taken by the IPC when it was seen that surface-mounted chip carriers reflowed in solder paste could not always have bright joints. It has been demonstrated, after billions of operational hours, that controlling metal quality prior to the soldering operation is the only way to ensure solder-metal quality. Functional testing, not cosmetic inspection for brightness, is proof of the product's reliability.

While newer surface-mount processes such as infrared or vapor-phase solder-paste reflow may result in dull, uneven surfaces, the joints are nonetheless reliable. Western Electric has patented a process which deliberately renders a dull solder surface. The lack-luster surface, they claim, makes the joints more readily inspectable. Western Electric manu-

factures 240,000 printed wiring boards (PWBs) per day and solders over one billion joints per year. Their product life is estimated to be 40 years.

A few of the typical alloys used by designers to achieve special solder interconnections are shown in the accompanying table. They are all metallurgically acceptable solder alloys used throughout industry and the military. The surfaces of these alloys, by their nature, vary in brightness. In fact, some of them are quite dull.

During soldering with a eutectic solder, the alloy may pick up other metals (such as copper) from the PWB. Within reasonably broad limits this is acceptable. Solder-joint brightness and smoothness may be affected, but it can be seen that use of the above alloys dictates the addition of particular elements to achieve a solder-joint objective. When a metal (Au, Ag, Cu) is added for special alloy purposes, it is called an additive, but when added via the soldering process some consider it a contaminant.

Pure 63 Sn 37 Pb is bright. This pure composition cannot, in practice, be absolutely maintained. As 63/37 deviates even slightly from eutectic the brightness and smoothness change. While the light reflection off the surface changes with a change in smoothness, the joint may still be metallurgically acceptable. Joints should be expected to be dull to one degree or another.

Understanding brightness variations

Solder surfaces, by the nature of metal

Alloy	Application
50 Pb 50 In	Ductile, die attachment and general circuit assembly
80 AU 20 Sn	Die attachment
63 Sn 37 Pb	General circuit assembly; eutectic
5 Sn 95 Pb	High melting point
95 Sn 5 Ag	Minimize leaching on components for surface mount
42 Sn 58 Bi	Low-temperature eutectic
Sn PB Cu	Fine coil wire soldering
Sn Pb Sb	Antimony strengthens solder (also darkens)

Alloys used for special applications

First published in *Electronic Packaging & Production*, February 1986, under the title 'Redefining solder joint acceptability'.
Author: Joe Keller, Joe Keller Associates Inc.
© 1986 Reed Publishing (USA) Inc.

Dull, uneven crystalline surface facets of a well-wetted joint do not enter into the process of determining solder-joint reliability. The gritty structure at the fillet edge is a normal phase separation part of solder solidification while the joint is forming (Courtesy of Hollis Automation)

solidification, cannot always be smooth; for this reason they vary in the extent to which light is reflected back to the eyes. Solder shrinks upon solidification. Since a large amount of solder (500lb) in a wave-soldering machine cannot be maintained as pure eutectic, there will be alloy variations. This will result in joint solid-to-liquid phase variations during solidification.

As solder crystals (dendrites) nucleate and grow from the melt upon cooling, a point is reached where insufficient volume is available as space for growth. Solder grains then 'collide' at the grain boundary, become 'upset', and form less than a smooth surface. When viewed with a scanning electron microscope, this surface is sometimes seen as a mesa-like structure or other irregular surface of grains or crystal facets that work to diffuse incident light and create a perceived dull-to-bright surface.

Vapor phase, infrared, or conductive-belt reflow methods heat the entire PWB. This causes the solder joints to cool slowly. The large grains result in a solder joint that diffuses light and therefore has a dull appearance. Moreover, the organic residues from

solder paste float on the molten solder, causing surface roughness. When the paste alloy is indium or high in lead content, the joints will be duller than a 63/37 alloy.

During the solidification process, lead or tin dendrites precipitate or separate out of solution. Lead-rich phases form. This causes a natural non-uniform rate of cooling with subsequent variations in the rate at which solid phases within the liquid are formed, and results in a surface that varies in smoothness and light reflection. Joint geometry conditions are created wherein the liquid phase flows off the solid phase to form a gritty or bump surface, usually near the thinner portion of the fillet, away from the conductor lead and close to the PWB copper conductor. This surface condition does not affect reliability; it is only cosmetic.

Specifications regarding solder-joint brightness, indeed, the entire issue of physical metallurgy of solder solidification and solder acceptance criteria, deserves fresh consideration. At the IPC meeting in September 1983, John C. Mather of Collins made the following statement: "Rough and gritty solder is indicative of contaminated solder and/or a slowly cooled solder joint whose composition is quite

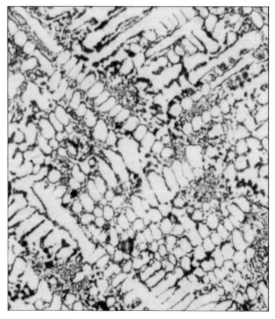

70 Sn/30 Pb. Dentrites of tin-rich solid solution (light), in a matrix of tin-lead eutectic (Courtesy of Kester Solder)

63 Sn/37 Pb. Globules of lead-rich solid solution (dark), some of which show a slightly dendritic structure, in matrix tin (Courtesy of Kester Solder)

far away from eutectic. Our customers will not accept rough and gritty solder joints, and neither would I."

The cause for rough and gritty solder is related to off-eutectic solder and cooling. But Mather's statement, designating off-eutectic solder as 'contamination' begs further discussion.

The alloy 5 Sn 95 Pb used for resistor lead coatings is clearly off the 63 Sn 37 Pb eutectic, but that does not make it the wrong alloy to use. An excellent off-eutectic high-temperature solder material, 5 Sn 95 Pb is used in industry to a very large extent. Sixty/forty solder, as compared to 63/37, should not be viewed as solder contaminated with 3% lead.

There is a strong technical case in support of the theory that solder coatings may vary substantially in alloy composition and still wet reliably when solder coating thickness is maintained. In fact, their sophisticated presentation, which includes a 3-D illustration of percent dewetting, alloy percent, tin deviation, aging, and solder thickness, clearly shows that solder composition may actually vary by 40% without dewetting. Even after accelerated age testing, solder joints with off-eutectic composition were found to be reliable.

The importance is placed on solder coat thickness and it is recommended that, since solder thickness will vary on all PWBs (edges of conductors, 'knees' of PTHs, etc.), an alloy range of 58–75% tin will still provide an additional solderability margin. As 63% Sn/37% Pb eutectic drifts in tin percent between 58 and 75%, tin-rich dendrites form, the physics of solder solidification is affected, surface characteristics and solder-joint brightness are also affected, but solder-joint integrity remains intact.

Solder-joint appearance

John Hagge confirms that solder-joint reliability is defined by wetting at the interface of the solder below the solder surface and surfaces being joined, and not by the external surface of the solder. While joint interfaces cannot be seen by inspectors, dewetted surfaces may be found by airknife wave-solder stress testing. Thick multilayer wave-soldered PWBs, as well as vapor-phase and infrared-reflowed PWBs, will always cool slowly. Minor phase separations in the alloy resulting from this slow cooling may produce dull, grainy or rough joints, but will not affect reliability.

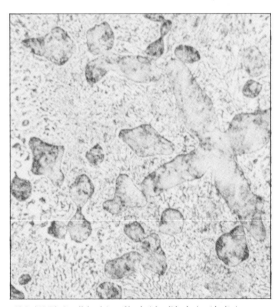

60 Sn/40 Pb. Small dendrites of lead-rich solid solution (dark), in a matrix of globular tin-lead eutectic finer than that shown in the previous figure (Courtesy of Kester Solder)

During an open forum at an IPC meeting in the spring of 1975, James Woodward of TRW's Defense Division (Redondo Beach, CA) commented on his studies in solder-joint formation and joint reliability. His specimens were chip lap and surface-mount lap joints on chip resistors, barium titanate chip capacitors, and other surface-mount hybrid components mounted on ceramic and glass-epoxy PWBs. His conclusion was that "visual appearance does not indicate solder-joint reliability."

Thermal-cycling and vibration tests performed by Joe Keller at Martin Marietta showed uneven, dull joints to be perfectly acceptable from a functional point of view. His research, combined with Woodward's data, resulted in a waiver permitting Martin Marietta to use dull joints on military hardware.

Solders are defined by ASTM standards as:

- *60/40 hypereutectic.* Small dendrites of lead-rich solid solution (dark areas) in a matrix of globular Sn/Pb eutectic (finer than that seen in 63/37).
- *63/37 eutectic.* Lead-rich solution (dark), slightly dendrite in a matrix of Sn.
- *70/30 hypoeutectic.* Dendrites of Sn/Pb solid solution (light) in a matrix of Sn/Pb eutectic.

All of the three solders above show a different dendrite freeze pattern, but are perfectly acceptable as solder metal. Dull joints represent a deviation from eutectic – not a deviation from quality. As the solder mass varies from the thin fillet at the PWB conductor pad or land to the bulk solder in PTHs, compositional inhomogeneities from differential freeze rates will exist within a solder joint. This condition can be expected and is not harmful. Bringing the solder closer to eutectic changes brightness but not joint reliability. Solder does not have to be eutectic to be good. Indeed, off-eutectic solders have their own qualities and are often selected for a particular characteristic as part of the design decision.

Faster conveyor speeds (over 8ft/min) in oil-intermix wave soldering prevent heat from penetrating into the PWB polymer and offer surface protection to the molten metal. This causes the solder to freeze faster. The joint will be brighter but not better.

Smaller PWB lands are conductive to bulkier solder-joint forms which tend to freeze freer of grainy portions. The larger pads allow solder to flow out into thin fillet portions. As the solder volume decreases during solidification, the shrinkage effect pulls the liquid off the solid crystals until the dendritic structure penetrates through the receding liquid solder, producing a grainy, pimply appearance. While a larger pad design is one method that will result in brighter joints for those who believe that 'bright is right', the reliability of the dull, grainy joint is just as high, in spite of its appearance.

WAVE SOLDER DEFECTS

Touch-up of defects after wave soldering is a nearly universal operation. Most people regard defects as an inevitable part of the wave-soldering process.

To reduce the amount of time spent in touch-up, wave-soldering machines have been made more sophisticated. All the major manufacturers of wave-soldering equipment have added computer control to their

First published in *Electronic Packaging & Production*, February 1986.
Author: J. Gordon Davy. Westinghouse Defense Electronics Center.
© 1986 Reed Publishing (USA) Inc.

machines to achieve better process control.

While automation has helped, the touch-up rate on PWAs is still high. The reason is not inadequate technology, but inadequate training. It has been pointed out that the touch-up rate often greatly exceeds the defect rate; and that the reason for the discrepancy is basically a lack of an objective standard that is comprehensive. Several military standards and specifications give lists of defects, but no distinction is made as to whether the soldering is performed manually or by machine, and no two lists are the same.

Defect list

The existing lists of solder defects are of two types. One type shows the appearance (by line drawing or photograph); the other has no visual aids, but offers a long list of possible causes for each – some of them highly unlikely.

A standard has been developed at the Westinghouse Defense Electronics Center that overcomes these limitations. The 12 defects cited in the standard are asserted to be a complete list of all wave-solder defects. Three-dimensional models of defective wave-soldered connections, shown 10 times their normal size to make them easier to see, are used in the standard.

The Westinghouse standard contains carefully chosen categories based on probable cause, not on superficial similarities or differences in appearance. Probable causes (never more than four) are listed for each defect.

Association of probable causes with a comprehensive list of defects can result in all defects being eliminated, not just identified and counted. All causes of defects can be identified; and, if eliminated, the result is zero-defect soldering.

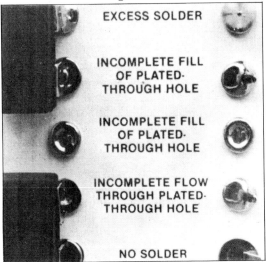

Wave-soldering defects are associated with their probable causes in the Westinghouse standards. This photo, showing some of the defects magnified to 10 times their normal size, is a section from a chart with the complete list of defects, cross-section drawings, descriptions, probable causes, and comments on the defects

An ideal solder connection is the reference point for all defects. The essential feature of an ideal connection is a fillet that is everywhere smooth and concave, joining the lead and land at a small angle. If near-eutectic 63/37 or 60/40 tin/lead solder is used, the fillet is shiny.

Excess solder

Leading off the defects list for wave soldering is excess solder, which occurs only on the solder side of the board (unless the board is submerged). This is a category that has often been regarded as subjective, but for purposes of accept/reject, it can be designated precisely. A connection has excess solder only if one of the following conditions applies:

- The fillet is convex to the edge of the land (preventing judgment of the quality of wetting of the land).
- The fillet is so large that no part of the lead is visible (preventing confirmation that the lead is present). A normal-sized connection in which the lead is not visible is not a solder defect but an insertion defect, namely 'lead too short'.
- Solder extends from one land, trace, or lead to another (bridging).
- Solder exceeds the height requirements.
- Solder is likely to break off (an icicle).

A connection with a greater than normal volume of solder, but that does not meet one of the previous conditions, may indicate a lack of process control. It is not, however, any less reliable, and should not be touched up. In this case, the process should still be corrected.

The probable causes of excess solder are:

- *Poor design.* Examples are leads of components that are spaced too close are clinched towards each other, or there is narrow spacing between traces without solder mask. These solder bridges should not be reported as solder defects, but as design or insertion defects, if that is the cause.
- *Conveyor too fast.* This can be seen from the fact that the size of the icicle on the end of a wire increases with the speed of its withdrawal from a solder pot.
- *Inadequate flux.* Either insufficient flux is applied, the flux used is of ineffective

Defect	Probable causes
Excess solder	• Poor design • Conveyor too fast • Inadequate flux
Incomplete fill of PTH	• Gas evolution blew solder from hole • Excess oil mixed with solder
Incomplete flow through PTH	• Insufficient heat • Crack or exposed intermetallic compound at rim of barrel
No solder	• Part of board did not contact solder wave
Poor wetting — leads	• Poor lead solderability • Inadequate flux
Exposed lead end	• Lead basis metal oxidized after cutting to length
Poor wetting — lands	• Poor board solderability • Inadequate flux (component side) • Insufficient heat (component side)
Hole in solder	• Gas evolution through crack or hole in barrel, PWB material defect • Oil or flux escaped as solder solidified
Inclusion	• Excess oil mixed with solder • Excess dross accumulation on wave solder pot • Inadequate process control before soldering
Dull	• Impurity metals • Solder low in tin
Disturbed	• Board was jarred as joint was solidifying
Solder on board	• Gas evolution produces solder balls

Westinghouse wave-solder standards for through-hole printed wiring assemblies

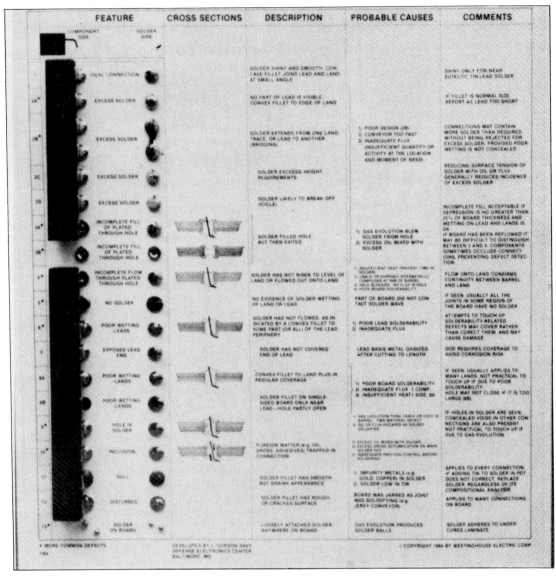

Standards for wave soldering for through-hole printed wiring assemblies (Courtesy of Westinghouse)

composition, or it decomposes before the board exits the wave.

Solder in PTHs

Incomplete fill of plated-through holes (PTHs), the next defect on the Westinghouse list, is a result of solder leaving a previously filled connection, either by draining out or by being blown out. The choice between these two causes should be apparent by examining the prevailing pattern. If the cause is gas evolution that blows the solder out, other connections should show holes in the solder, yet another defect.

The usual military accept/reject criterion for this defect is a depression of 25% of the board thickness, with adequate wetting of the lands and lead. The proposed DOD-STD-2000-1A sets a 10% limit for holes with leads. While the military documents do not make it very clear, the intent is that solder should flow out onto the land. If it then retreats some, that is acceptable.

For the next defect on the list, incomplete flow through a plated-through hole, in contrast to the similar-appearing incomplete fill of a PTH, solder has not risen to the level of the land or flowed out onto the land. The visual distinction between the two categories is based on the shape of the solder plug. Retreating solder leaves a concave fillet whereas advancing solder may be flatter. If the board has not been reflowed before wave soldering, the plating on the land may remain unmelted in the case of incomplete flow through a PTH.

Although military specifications require hole filling, many believe that as long as there is some solder in the hole, the connection is just as reliable as if it were filled. In fact, some believe a partially filled connection to be stronger, because it is less affected by differential expansion during thermal cycling.

Here are the arguments for rejecting every case of incomplete flow. First, one possible cause of solder not flowing onto the land is a crack in the rim of the barrel, where it joins to the land. Similarly, a crack around the circumference of the barrel may prevent solder from rising beyond it. Such a crack, if not covered with solder, may mean an intermittent open connection. It is not practical in each case to determine whether the cause of incomplete flow is a crack or something less serious.

Second, all the causes of incomplete flow can be prevented. The quality of PWB plating and its solderability can be verified by wave soldering and inspecting a coupon as a receiving inspection operation. Flux should be able to enter every hole, preheat (top as well as bottom) can be increased, and if there is insufficient thermal relief between a land and a ground plane, the design can be changed.

The next defect on the list, the case of no solder, is easiest to understand. If there is no evidence of wetting of land or lead, then that connection did not contact the solder wave. This could be due to a misshapen wave, pallet, conveyor, or board. Usually, if this defect is seen, all of the connections in some region of the board have no solder.

Lead wetting

Poor wetting of leads is a defect characterized by a convex fillet to some part or all of the lead periphery. While poor wetting may be caused by failure to provide adequate flux, the most common cause is lack of solderability.

A common way to reduce the incidence of poor wetting is to use a more active flux. There are three disadvantages to this approach:

- Not all cases of poor solderability are correctable with stronger flux.
- Any unremoved residues pose a corrosion threat.
- PWA manufacturers will have a reduced incentive to demand what they have paid for from the component manufacturer. The component supplier could, with sufficient motivation, deliver consistently good solderability with indefinite shelf life at virtually no extra cost.

Some cases of poor wetting of component leads can be overcome with a soldering iron, which runs hundreds of degrees hotter than the solder wave. Other cases cannot, and these, depending on the skill of the touch-up operator, end up:

- Having to be removed for stripping to the bare basis metal and solder dipping (the correct procedure, but expensive).
- Camouflaged (solder sculpture to simulate the appearance of wetting).
- Overheated (with damage to board or component).

The best way to prevent poor lead wetting is to ensure, by solderability testing, that the leads are solderable when the parts are received. Solderability testing, coupled with rejection of failed lots, motivates component manufacturers to send their best material — the remainder goes to those purchasers who do not test.

Exposed lead ends is another wave-soldering defect. Leads must be cut to length before soldering to avoid risk of cracking the solder connection. Unlike the rest of the lead, the exposed end is unprotected basis metal (copper or iron-nickel alloy).

Military requirements do not allow exposed basis metal because that indicates poor wetting and is regarded as a corrosion risk. However, the military also limits the activity of the flux to RMA, which is not completely effective against the oxides which begin to grow on these surfaces as soon as they are exposed. The best prevention is to minimize the time and heating between cutting and soldering. This is not always practical, though, and it seems appropriate to seek an exemption of lead ends from the requirement of solder coverage, particularly if the PWA is to be conformally coated.

Poor wetting of lands, the next class of soldering defects on the list, are characterized by a convex fillet to the land, coupled with irregular coverage of the land. In the case of a single-sided board, the defect is indicated by a failure of the solder to fully span the hole. However, if the solder fails to span because the hole is too large, the defect should be reported as 'hole too large' – a design error.

One way to determine whether the cause of poor wetting on a given board is due to poor solderability is to ensure the availability of flux and adequate preheating on the top of the board. A better way, which will also reveal other material-released defects, is to wave solder a coupon from the same plating lot when the boards are received. To provide a safety margin, the solder temperature might be set lower, and the conveyor speed faster, than for production PWA soldering.

If poor solderability of the board is discovered after it is populated, touch-up may be impractical since it is likely to be a problem all over the board.

Solder holes

A commonly offered reason for a hole in the solder connection, another type of defect, is solvent volatilization from the flux. In machines that mix oil with the solder, the hole may be attributable to oil escaping just as the solder freezes.

One sometimes hears that the hole is caused by lack of wetting of some part of the lead or barrel. However, it has been asserted that the only causes are water and air, and occasionally outgassing from organics in the barrel or component lead plating.

The amount of water in the board depends only on its history of, at most, the past few days, because water molecules diffuse fairly readily in both the polymer (epoxy or polyimide) and along the interface between polymer and fiber (usually glass). Water-originated outgassing can be eliminated by baking.

The case of air is different. Air is trapped during lamination at a pressure of about 10 atmospheres. The amount trapped appears to vary with the quality of adhesion between resin and fiber and the degree of cure before and after lamination. When holes are drilled, there are pockets of trapped air near the hole wall. Heating during wave soldering does two things: it roughly doubles the pressure and softens the polymer. Both effects help the air to escape. Then, one of three things may occur:

- If the copper plating is thin, less than about 0.0008in., there may be voids in it which provide an escape route.

- If the plating is good and less than about 0.0014in., the pressure may cause a rupture of the copper wall.

- With thicker plating, the escaping air causes disbonding between copper and laminate. Thus, air escapes around the land. Rupture or disbonding could cause a break between an internal trace and the barrel of a multilayer board. Such breaks have not been reported but if this mechanism can occur, then thick copper plating to prevent holes in solder is not a wise choice.

The risk posed by a hole in the solder is said to be that it may extend into the connection and terminate at exposed basis metal, thus posing a corrosion risk. While this is conceivable, it seems highly unlikely, and no study has ever reported such a finding.

Concealed voids in the solder that are not visible during inspection on a double-sided board pose no obvious threat unless the void is so large that the remaining solder, while continuous, is only a thin skin. Such a connection is unreliable because of the risk of rupture.

If it is determined that voids are a threat to the board, then touch-up of holes in the solder should be forbidden, because the two always occur together. Also, blow holes often prove difficult to touch up because further heating produces more bubbles.

Other defects

The final four wave-soldering defects on the comprehensive list are:

- *Inclusion*. The cause of this defect is determined by identifying the included matter. In the case of dross, for example, the cause is dross being sucked into the solder pump. If the inclusion is an adhesive or marking used during fabrication or assembly, this defect is materials-related, rather than solder-process-related.

- *Dull*. This defect is always process-related. It can only happen when samples of solder in the pot are not analyzed periodically for composition. This is the only defect caused by the solder. Since the cause is the solder, the defect appears on every connection of every assembly.

- *Disturbed*. This defect is sometimes combined with the dull defect, and is often referred to as a cold solder connection. Both are wrong. A cold solder connection may occur during hand soldering due to a lack of operator skill, but not during wave soldering. Its real cause is motion of the lead in the hole as the solder freezes. since many connections freeze at once, many disturbed connections should result if the board is jarred. Unlike dullness, the lack of smoothness is macroscopic; although the surface is rough, it is still shiny. An extreme case of 'disturbed' is 'cracked'.

- *Solder on board*. Solder balls are the result of gas evolution. If solder (balls or webbing) adheres to the laminate, this also indicates a material problem. Prevention is achieved by wave soldering a coupon at the first opportunity and rejecting unacceptable material.

WAVE SOLDERING

The most widely used automated soldering process is wave soldering. This established technique is undergoing re-evaluation and enhancement.

This re-examination is driven by two forces, one of increasing technical demands and secondly by the competition of other approaches such as vapor-phase soldering. This is discussed in the first article.

The hardware of wave soldering has long been established. Today great emphasis is placed on optimizing performance through study and greater process control. The other three articles in this section examine possible approaches to increasing reliability of wave soldering.

SOLDERING SYSTEM TRENDS

Quality of soldered assemblies, manufacturing productivity, and operating as well as initial cost of a soldering system provide the framework for examining automated soldering concepts and their implementation in machines. Wave-soldering equipment continues to be the key for most current PCB automated soldering applications with through-hole components. The anticipated growth of surface-mount component use, however, is forcing the industry to look at vapor phase and other reflow soldering systems.

Wave soldering

Although various assembly handling and control system components are crucial elements of a wave-soldering system, the centerpiece is certainly the wave itself and the dynamics that shape its properties. The symmetrical wave has been improved since its use in the first wave-soldering equipment. Opening the nozzle carrying the solder produces a wide symmetrical wave, and with the use of extender plates an adjustable symmetrical wave can be created. Asymmetrical waves logically follow, created with an extender plate on only one side of the nozzle.

Geometric characterization of the wave as its exit from the nozzle, and its relationship to the PCB traveling through the wave is only one aspect of the dynamics. Before reaching the nozzle, molten solder must be set in motion and its flow directed. For example, with its adjustable-dimensioned, uniformly distributed solder-wave apparatus, Technical Device Co. claims to have made a significant stride in solving the problem of redirecting a horizontal stream of solder to a controlled, uniform vertical flow direction.

In Technical Devices' innovation, a horizontal flow of molten solder from a centrifugal pump moves up an inclined distribution ramp where vanes form the vertical flow chambers. Each vane collects the same amount of solder, with equal kinetic energy, and redirects it to vertical flow. This assures equality of the solder emerging from each flow chamber, and thus uniformity across the entire width of the solder wave.

After the wave has been shaped and solder applied to the PCBs, the solder physics can still be exploited to improve the quality of the soldered assembly. As PCB circuit density increases, opportunities for bridging increases. Positioned immediately after the solder wave, Hollis Automation's debridging air knife is activated while solder is still molten on the PCB. The thin, heated airstream sweeps away excess solder, displacing any bridges or icicles that may have formed, explains Hollis.

Sound solder joints will not be distributed by the air-knife action, the company claims. The intermetallic between solderable copper of the substrate and the solder itself is much too strong to be distributed by the air at the knife. Poor solder joints, on the other hand, which lack a strong intermetallic, are stressed by the air knife.

The air knife, notes Hollis, thus serves as a fault detection system as well as a debridger. That is, it works to detect defects that cannot be corrected by the soldering machine alone, but must be corrected by better board design. Warren Abbott, technical services manager at

First published in *Electronic Packaging & Production*, December 1984, under the title 'Smaller PCB geometries shape soldering systems trends'.
Author: Ronald Pound, Associate Editor.
© 1984 Reed Publishing (USA) Inc.

With the solder wave next, a printed circuit board passes through a wave-soldering machine (Courtesy of Hollis Automation)

Jet-wave soldering uses a high speed, flexible, hollow solder wave (Courtesy of Kirsten)

Hollis, says about 75% of the questions he gets from users of soldering equipment relate to problems that are not machine controllable. These problems include, for example, the flux not being sufficiently aggressive to provide good solderability and component leads that are oxidized.

Jet wave soldering

Conventional wave-solder machines produce a fountain of molten solder. Departing from conventional wave dynamics, however, the Kirsten Jet equipment produces a hollow wave with solder, driven by an electrodynamic pump, moving at 3ft/s. This high speed, aerodynamic wave operates on the Bernoulli principle.

These jet wave-soldering systems have the capability to process a variety of assemblies, including conventional multilayer boards, boards with surface-mount components, and hybrids, says Scott Croce, director of marketing at Automated Production Concepts, the US distributor for the Swiss manufacturer.

Leaded through-hole components, on the board side opposite the wave, are pulled down on the substrate while being soldered instead of being floated as with conventional solder waves explains Croce. When soldering surface-mount devices, on the wave side of the board, the high-velocity hollow wave circulates solder on all sides of the component. This precludes a shadowing effect by the component body that can result in insufficient solder on some pads.

Fine lines down to 0.004in. can be processed with no bridging or solder balls, the

A dual-wave system for soldering surface-mount components has a turbulent wave and a laminar wave (Courtesy of Universal Instruments)

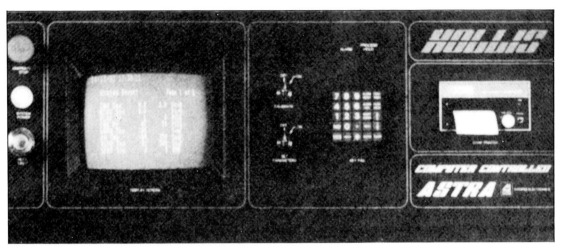

This microcomputer control panel for a wave-soldering machine has a display, key matrix, and strip printer (Courtesy of Hollis Automation)

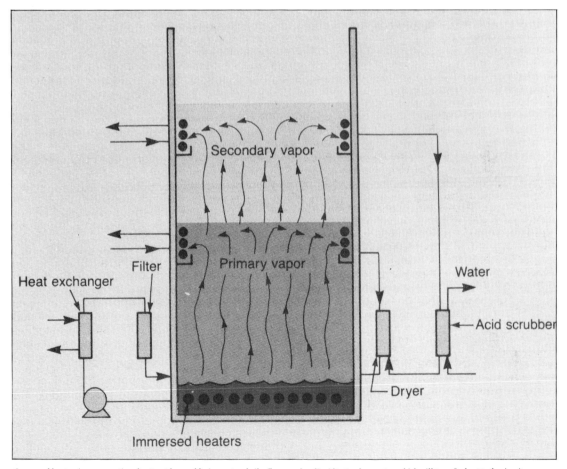

Immersed heaters in a conventional vapor-phase soldering system boil a fluorocarbon liquid to produce vapor which will transfer heat to the circuit assembly being soldered

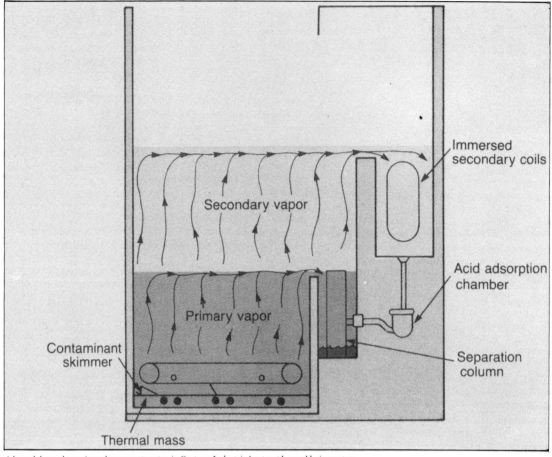

A heated thermal mass is used to generate vapor in Corpane Industries' vapor-phase soldering systems

company claims. Furthermore, the jet wave is flexible and adapts itself to the lower surface of the printed circuit board. Control of the wave height is not critical. It compensates automatically for differences in height when large boards bend over the soldering areas, thus eliminating unsoldered areas.

Operating on the basis of the force produced when a current flows across a magnetic field, the electrodynamic pump has no moving parts. A magnetic field is generated and passes vertically through the pumping zone and the solder. Transverse to the field, and also passing through the solder, a current is forced to flow. The resulting mechanical force is at a right angle to both the field and current, forcing the solder to move in that direction through a slot which forms the wave.

Discovered by Swiss mathematician Daniel Bernoulli, the Bernoulli principle says that, for the steady flow of an incompressible fluid, pressure is low where velocity is high. It is the principle that provides lift on an airplane wing. Since the solder within the hollow wave is moving with the same angular velocity at all points across its thickness, the linear velocity on an outside point, at a greater radius, is larger than that on the opposite inside point, at a smaller radius. A force is thus created that tends to lift the wave up to the board and around the components.

Surface-mount wave

Wave soldering of boards with surface-mount components held on the wave side of the board with adhesive involves, in most machines, the use of a dual-wave approach. As surface-mount technology becomes more prevalent, there is a need for specially designed turbulent-type waves, indicates Universal Instruments.

The first wave, turbulent rather than with laminar flow as in the conventional, smooth wave that follows, is intended to wet surfaces in tight spaces. The trailing smooth wave removes excess solder, assuring there are no shorts or solder bridges.

For instance, the DANCER wave, Technical Devices' concept, is independent of, and precedes, the standard laminar wave. Appearing to dance on the bottom side of a PCB, the wave is a series of solder columns or jets. With its own pump, solder pot heaters, and wave height and temperature controls, the DANCER, or chip wave, can be adjusted independently of the conventional wave.

Physics of the wave certainly impacts the quality of soldering, and hence manufacturing costs, through such aspects as soldered-assembly testing and rework required. In addition, wave characteristics affect dross formation and energy use, and thus operating costs as well.

Dross control

The most expensive part of a wave-soldering system is not the purchase price, it is the cost of the solder used to replace the dross, or solder oxide, says William Bokhoven, president of Cyclo-Tronics. A typical wave-soldering machine will generate five pounds of dross per day, the company claims.

Dross formation is accelerated in a wave-soldering machine due to the large air-to-solder interface area formed by the solder wave. Cyclo-Tronics' machines, with their anti-dross control option, seek to reduce exposure of solder to air by automatically lowering the wave height to just below the fountain overflow point between pallets when there are no boards to be soldered. The wave is restored to the height the operator had initially set when a pallet appears with a board to be soldered.

For a given conveyor speed, Cyclo-Tronics'

anti-dross control results in greater cost savings the larger the distance between pallets, or the fewer boards soldered per day. At higher production rates, and hence smaller distances between pallets, there is a shorter time period during which the solder and air interface area can be reduced.

Detailed surface characteristics of the wave, too, affects solder oxidation. Technical Devices Co., for instance, indicates that the solder wave in their machines is laminar, non-turbulent, and produces a minimum of solder oxidation. Still, a variable wave width adjustable to the board size being soldered and no more will produce less dross than a wave of fixed width. In addition, the smaller the wave width, the less energy lost to the surrounding air.

Claiming to have introduced the first wave-soldering system with an electronic processor and logic system for machine control, Technical Devices' president Douglas Winther explains that solder-wave width, as well as other parameters, automatically conforms to the width of the PCB being processed.

Computer control

High-quality soldering, with near zero defects, contributes to constraining the cost of manufacturing. The introduction of computer-controlled soldering systems has enabled many manufacturers to virtually eliminate soldering process errors by reducing operator intervention, according to Universal Instruments. Convenient generation of comprehensive reports for tracing the processing history of boards involved in reliability issues is another potential benefit of computerized systems.

With a custom-designed microcomputer, Electrovert's computerized wave-soldering system, for instance, monitors system parameters and adjusts them to meet product load requirements. Processing parameters for various boards are held in memory, once entered. Thus, the system can set up automatically for different jobs. A serial number entered on a key pad, or by a bar-code reader as the first production board enters the conveyor, actuates the automatic setup.

In Electrovert's standard software package are parameters such as conveyor speed, width, and angle; width and operation of the infrared preheater; top surface temperature of the printed circuit board; flux level and density; solder-wave height and temperature; solder-pot height; solder feed; and flux replenishment.

Functions on the five-by-five key matrix of Hollis' microcomputer control system are simplified for user operation, the company says. In the set parameters mode, the keys are used to enter information such as conveyor speed and solder temperature. The control mode enables an operator to initiate the soldering process with parameters already stored for a particular board.

Not limited to wave-type soldering equipment alone, programmable controls are also being employed in vapor-phase soldering equipment as well. There computerized functions include primary and secondary vapor control, elevator speed and dwell time in batch units, and conveyor speed for in-line systems.

Vapor-phase soldering

Emerging as a leading contender for soldering surface-mount components to their substrate, vapor-phase soldering offers several potential benefits in these applications. It provides uniform heating independent of substrate size and geometry. Vapor condensation, which provides the heat-transfer mechanism for the solder reflow, occurs almost simultaneously on all exposed surfaces of a circuit assembly. The soldering temperature is always the same, providing process repeatability. It is controlled by the boiling point of the fluid used to produce the vapor.

In a conventional vapor-phase system, the latent heat of vaporization is released as the hot vapor condenses on the cool assembly being soldered. The resulting cooler condensate falls to the boiling liquid below, lowering its heat density and reducing the vapor generation rate. An impractically large instantaneous heat input would be required in conventional systems to compensate for the tendency to reduce the liquid's heat density, and thus prevent vapor collapse.

Thermal mass

Conventional vapor-phase soldering equipment uses electric heaters immersed in the fluorocarbon liquid that is boiled to generate the system's working vapor. Corpane Industries' Thermal Mass innovation, on the other hand, utilizes a liquid metal which is heated, and is in direct contact with the vapor zone rather than a standing reservoir of the fluorocarbon liquid.

Vapor collapse is virtually eliminated, providing improved production and higher yield, says Corpane of their Thermal Mass innovation. Condensate is instantaneously reboiled, which is not possible in a conventional system. The thermal mass quickly gives up heat to the liquid.

Not only does this prevent vapor collapse, matching vapor production to fluctuating load demand, but it also results in elimination of a standing liquid reservoir. Fluid requirements are thus reduced to a minimal initial charge.

Single-vapor systems

Most batch-type systems in use are dual-vapor machines. A secondary vapor occupies a zone above the primary vapor used for heating in the reflow soldering process. Formed from less costly fluid, the secondary vapor prevents escape of the primary vapor, and thus reduces loss of its expensive fluid source.

Single-vapor machines have been developed, however, with the aim of simplifying vapor-loss control. The single-vapor concept was first applied to conveyorized, in-line reflow systems, indicates Dynapert-HTC. With the limits of the working vapor zone defined by a series of cool, mechanical surfaces, rather than a secondary vapor blanket, diffusing vapor is condensed and returned to the equipment's tank.

The in-line, single-vapor concept has been advanced one step by Dynapert-HTC with their integrated soldering system. Developed specifically for single-step soldering of assemblies utilizing both through-hole and surface-mount components, their integrated system combines vapor-phase reflow and a modified wave-soldering operation. It is designed to solder both the top and bottom of

PCBs using mixed technology with through-hole components and all types of surface-mount components, the company indicates. Mixed technology assemblies will be common as manufacturers make the transition from through-hole to surface-mount components with limited availability of some component types in surface-mount configurations.

WAVE SOLDERING: A STUDY IN PROCESS CONTROL

Automatic wave soldering has been a well-established manufacturing technique for about twenty years, and yet many basic facets of this process are not understood or controlled. This results in much higher rates of defects such as defective solder joints, pin holes/blow holes, incomplete fillets, bridging, and crinkling of resist. Effective control of the process requires an understanding of the basic characteristics of both the materials used and the production equipment involved. The goal is to achieve a consistent formation of ideal solder joints.

At Tektronix Inc., there are numerous solder-flow systems (e.g. vertical flow, drag, and dwell), and a serious attempt to understand, monitor and improve production soldering techniques for all systems was undertaken over a 15-month period. By fully understanding the heat transfer, chemistry, physics, metallurgy and mechanics of the materials and process equipment, a more effective program was implemented for each soldering system. Each soldering system required a separate solution because of the differences in the preheaters' method of heating (e.g. infrared or convection and length), thereby requiring different conveyor speeds for the same materials in order to achieve proper PCB preheat (top board temperature). This program resulted in an improvement from 21 errors per thousand connections before the program, to 1.5 errors/thousand today.

There have been many articles, books and lectures given about automatic soldering, but there is little or no process control informa-tion available on the topic. Thus, when Tektronix wanted to use process control methodology to predict solderability, it was forced to create a task force to examine the fundamentals.

Basics of process control

The automatic soldering system is run by an operator who is responsible for loading and unloading the system, checking the security of components, fixturing PCBs, and setting the operating speed as well as other sundry tasks. To maintain the system at defined manufacturing specifications, a periodic maintenance is performed and spare parts are stocked for minimum downtime. Both operating and maintenance procedures are written to assure new and old operators act under constant, known procedures.

Before attempting to implement any changes, the first step was to establish a

First published in *Electronic Packaging & Production*, February 1985.
Author: Lawrence Cox, Textronix Inc.

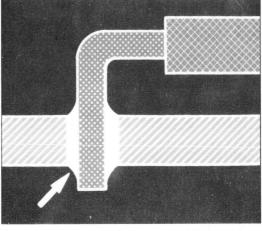

The ideal solder has proper fillet formation and is free of voids and pinholes

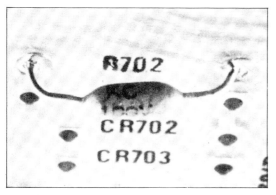

Poor solder fill on the component side prevents proper fillet formation and offers minimum support for component leads

standard of acceptable solderability. Tektronix's soldering standards can be summarized as follows:

- Good wicking and wetting.
- Fillets must be a minimum of 75% full.
- Pin-holes/blow-holes – variable, depending upon photographcd guidelines.
- No Bridging.
- Crinkling of resist – minor crinkling only.

Flux and metal considerations

The first element reviewed was the flux used in the present operation. By studying the flux contamination level and loss of activators in the flux during system operation, it was found that maintaining the proper specific gravities of the flux throughout system operation, and thinning the flux with the vendor's recommended thinner, the flux could easily be used for one week (40h), given the average number of boards through one system. (Tektronix manufactures approximately 10,000 boards per year throughout this operation).

In the soldering process involved, alloying or dissolving the metals, rather than melting or liquefying the metals, is the preferred method for bonding or making the connection. Whenever alloying takes place, the metals combine molecularly with the surface metals of the printed circuit board, and the metal on the printed circuit board gives up molecules into the bonding material (solder). When bonding or melting of the surface material occurs, contaminants may enter the bonding material (solder) causing its eutectic

point or melting point to change in temperature. Because of this, melting points of the surface metals had to be reviewed. Examples of printed circuit board materials used in the system, their metals and melting points are listed in the table.

With contaminants such as gold migrating into the solder pot or tin migrating out of the solder pot, the eutectic point or melting point of the solder can increase. Contaminates and allowable contamination levels were reviewed, and solder pot sampling criterion was documented for process control information.

Activation temperatures of flux had to be reviewed with the supplier. The flux contains amine hydrochlorides that start reacting immediately and very vigorously at higher temperatures. The supplier recommends at least 82°C for full activation.

PCB substrate materials

The next step in the analysis was to review the substrate material (the epoxy glass) of the PCB. The following thermal considerations were examined:

- Glass transition (T_g) temperature (the temperature point at which the epoxy glass starts to move or soften).
- Thickness.

The two types of substrate materials used in the Tektronix automatic soldering process are FR4 epoxy glass and CEM3 epoxy glass.

The glass transition temperature of both is

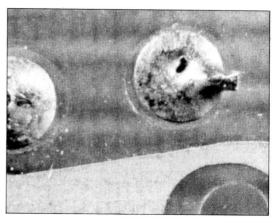

Pinholes indicate solder voids and unreliable solder connections, both electrical and mechanical

Shorts caused by solder bridging causes expensive rework and can be eliminated by strict adherence to process control guidelines

about 121–127°C. The average thickness of the substrate is 0.062in. Once the board material considerations were known, stability of variables during the preheating process and the soldering process were reviewed. These variables were conveyor speeds, solder pot temperatures and preheats.

Heat transfer data collection

Once the main variables in the system were known and eliminated through control of materials and operating procedures, PCB heat transfer was examined under the following criterion:

- Processed FR4 and CEM3 materials with standard 0.062in. thickness without metals or components.
- Processed FR4 and CEM3 materials without components and with large ground plane areas on the top surface.
- Processed FR4 and CEM3 materials without components and with large ground plane areas on the bottom (solder side) surface.
- Processed FR4 and CEM3 materials without components and with large ground plane areas on both top and bottom surfaces.
- Multilayer printed circuit boards (six layers or more) without components, with large ground plane areas in the middle layers (the two ground planes).
- Permanent (thermally cured) solder resist on the substrate (FR4 and CEM3) material

on the top and bottom (solder side) surfaces without components.

The thermal transfer of heat through the board was reviewed using a two-pen thermo-coupled chart recorder. The thermocouples were attached to the top of the board and the thermal bulb was immersed in a heat sink compound for positive heat transfer. All measurements were charted using the plain (void of metallic or plated through-holes) processed substrate (epoxy glass) as a constant reference with one thermocouple. Meanwhile, the second thermocouple was measuring the heat transfer of the different materials (e.g. ground planes).

When viewing the material heat transfer, the temperature of the top side of the board was the only concern. The automatic soldering systems were adjusted (either by slowing the conveyor speeds or increasing preheat temperatures) to achieve a top-of-board temperature of 121°C for the fastest heat transfer through epoxy glass material. The T_g (glass transistion temperature) where the

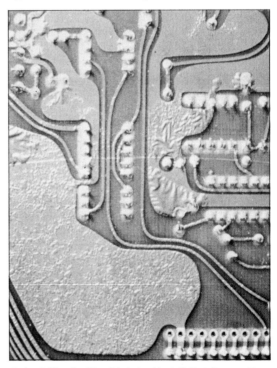

Resist crinkling should be minimized and controlled but often cannot be totally eliminated

Bright acid tin		
Base	Copper (Cu)	1.083 C
	Nickel (Ni)	1.453 C
Top surface	Tin (Sn)	232 C
	Acid tin	Approx. 230 C
Gold plated		
Base	Copper (Cu)	1.083 C
	Nickel (Ni)	1.453 C
Top surface	Gold (Au)	1.063 C
Hot-air leveled		
Base	Copper (Cu)	1.083 C
Top surface	Tin (Sn)	232 C
	Lead (Pb)	327 C
	63% Sn 37% Pb	183 C

Melting points of PCB surface materials

epoxy glass starts to soften is 121°C. Because printed circuit boards will warp and twist with high temperatures, it was felt necessary not to exceed 121°C on plain processed epoxy glass devoid of cladding.

When reviewing the charted data, the following variabilities are found:

- FR4 epoxy glass: ±6°C variations between the same date/lot batch material and different date/lot batches of materials.
- CEM3 epoxy glass: ±6°C variations between the same date/lot batch materials and different date/lot batches of materials. The heat transfer for the same preheat temperature and conveyor speed was found to be less than FR4 epoxy glass (18°C ± 3°C typically).

Heat transfer variables when viewing large ground planes on one thermocouple and using the second thermocouple viewing processed FR4 or CEM3 epoxy glass (void of metallics or plated through-holes) as a known constant were quite similar for either the FR4 or CEM3 epoxy glass. The results of these findings are summarized as follows:

- Large ground planes on the top of the PCB: typically 3–6°C higher in temperature than processed FR4 or CEM3 epoxy glass.
- Large ground planes on the bottom of the boards: typically 6–12°C lower in temperature than the plain boards.
- Large ground planes on the top of the board directly above large ground planes on the bottom of the board: typically there were 18°C lower temperature readings on these boards as compared to the plain FR4 epoxy glass boards and 9°C lower readings

than on the plain CEM3 epoxy glass.
- Large ground planes (two planes) in the middle of a six or eight layer multilayer board: typically a 3–6°C higher temperature than either plain FR4 or CEM3 boards.
- Thermally cured solder resist on FR4 or CEM3 epoxy glass: typically a 6°C higher temperature than plain boards.

The largest heat transfer variations were found between the epoxy glass substrate material FR4 and CEM3. The CEM3 epoxy glass material was found to be usually 18°C less in heat transfer than FR4 epoxy glass. Large ground planes on top, directly over large ground planes on the bottom of the PCB were found to be the greatest inhibitor of heat transfer when using the processed epoxy glass as a reference point. This heat transfer variation typically was 21°C less in temperature than FR4 epoxy glass and 9°C less in temperature than CEM3 epoxy glass.

The board in-solder-pot time was analyzed to see how much time was being allowed to solder the board under the constant set conveyor speed, and to see how high the top of board temperature became during the soldering process. The fastest in-solder-pot time recorded on the soldering system was 1.5s. The slowest time was about 4s. With the solder pot temperatures set between 232°C and 241°C, the top of board temperatures were between 176°C and 204°C.

Process control parameters

With the charted data on heat transfer variables of the materials of the board being known, (e.g. glass transition temperatures, time to solder, melting points of solder, board materials, and the activation temperatures of the amine hydrochlorides contained within the flux) process control could now be set up by asking the following questions:

- What is the PCB material used (CEM3 or FR4)?
- Are there many large ground planes on the top over large ground planes on the bottom of the printed circuit board?
- What conveyor speeds can be achieved to allow a minimum solder time of 1.5s and a

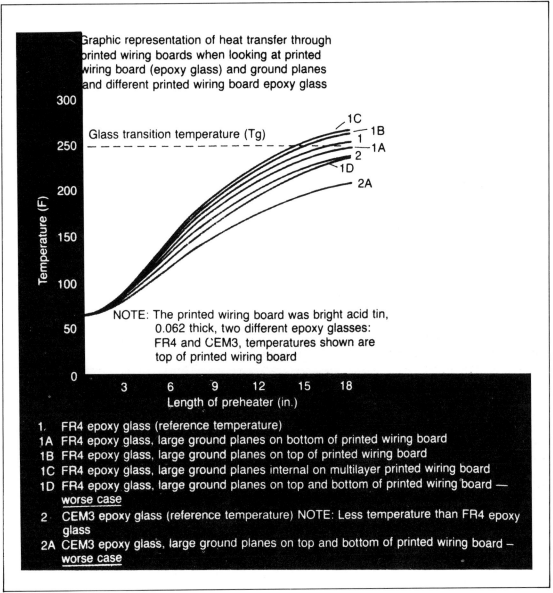

Graphic representation of heat transfer through printed wiring boards when looking at printed wiring board (epoxy glass) and ground planes and different printed wiring board epoxy glass

Glass transition temperature (Tg)

NOTE: The printed wiring board was bright acid tin, 0.062 thick, two different epoxy glasses: FR4 and CEM3, temperatures shown are top of printed wiring board

Temperature (F)

Length of preheater (in.)

1.   FR4 epoxy glass (reference temperature)
1A FR4 epoxy glass, large ground planes on bottom of printed wiring board
1B FR4 epoxy glass, large ground planes on top of printed wiring board
1C FR4 epoxy glass, large ground planes internal on multilayer printed wiring board
1D FR4 epoxy glass, large ground planes on top and bottom of printed wiring board — worse case
2.   CEM3 epoxy glass (reference temperature) NOTE: Less temperature than FR4 epoxy glass
2A CEM3 epoxy glass, large ground planes on top and bottom of printed wiring board — worse case

Heat transfer through PCBs of varied material and ground plane location presents a major process variable

maximum top of board temperature? Remembering that T_g (glass transition temperatures – e.g. 121°C) is best case (fastest heat transfer) on plain processed epoxy glass/ with large ground planes on the top of the printed circuit boards or multilayered boards with heavy ground planes inside.

• Can the conveyor speeds noted above still achieve reasonable (above 82°C) top board temperature for the worst case conditions? *Note:* 82°C is required to fully activate the amine hydrochlorides in the flux.

• What is the surface metal on the board to be soldered? (gold, tin, etc.)

• Are there any unwanted manufactured (soldered) conditions? (such as reflow of tin or crinkling of the permanent solder resist)

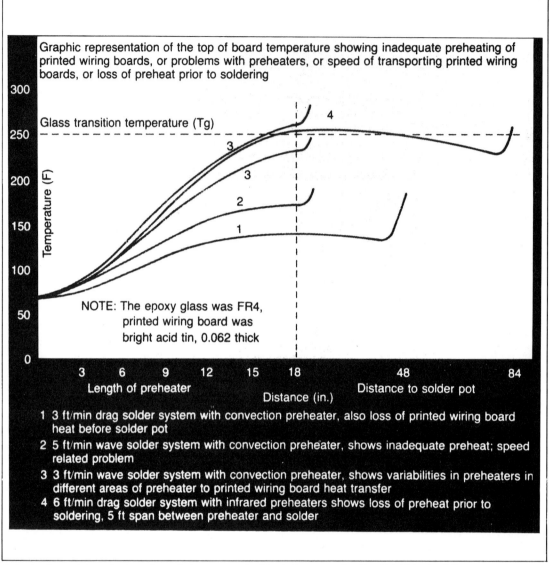

Graphic representation of the top of board temperature showing inadequate preheating of printed wiring boards, or problems with preheaters, or speed of transporting printed wiring boards, or loss of preheat prior to soldering

Glass transition temperature (Tg)

NOTE: The epoxy glass was FR4, printed wiring board was bright acid tin, 0.062 thick

Length of preheater

Distance (in.)

Distance to solder pot

1 3 ft/min drag solder system with convection preheater, also loss of printed wiring board heat before solder pot
2 5 ft/min wave solder system with convection preheater, shows inadequate preheat; speed related problem
3 3 ft/min wave solder system with convection preheater, shows variabilities in preheaters in different areas of preheater to printed wiring board heat transfer
4 6 ft/min drag solder system with infrared preheaters shows loss of preheat prior to soldering, 5 ft span between preheater and solder

Preheating characteristics as related to preheat method, conveyor speed and length of transport to the solder offer additional variables

- What conveyor speed is required to allow a minimum of 1.5s soldering time and to achieve a best case (FR4) board temperature of 121°C T_g?

Tektronix uses both FR4 and CEM3 epoxy glass for substrates and solders many different-sized circuit boards with a wide variety of large to small ground planes on both sides of the boards. Bright acid tin is the main process technology used with some hot-air leveled and gold-plated boards. The thermocouple data indicated that there was one conveyor speed per soldering system that could best handle both best- and worst-case board materials to achieve the maximum top of board preheat (FR4 or CEM3) temperature of 121°C (the glass transition temperature).

Resulting process parameters

With the necessary information about the materials and process having been collected and analyzed, the following control information was implemented for the soldering systems:

- Solder pot temperatures were set at between 227°C and 230°C to prevent crinkling of the permanent solder resist and reflow (re-liquefying) of the bright acid tin.
- Preheat temperature controls and conveyor speeds were set to assure top of board temperatures of a maximum of 121°C using FR4 (best-case materials) processed glass without metallics or through holes (this is the glass transition temperature). Conveyor speeds were set to allow at least 1.5s of soldering time. These control settings assure a preheat soldering temperature window of 82–121°C, while soldering for a minimum of 1.5s.

Automatic soldering system manuals were rewritten to include procedures for the following: system operation, system test, overall system process control, system maintenance, and safety. Thereafter the new procedures were implemented over a three-week period.

A review of solderability of PCBs with the defined settings showed minimum bridging, maximum fillets (100% full), excellent wicking and wetting of solder, little to no pin holes or blow holes, and little to no icicling.

Metallographic analysis was reviewed with no problems noted from lower solder pot temperatures (the original solder pot temperatures – derived from trial and error – were approximately 241°C versus the newly calculated maximum 230°C) or length of time to solder (a minimum of 1.5s of dip of board into the solder pot).

Keeping the systems within these defined parameters and known physical characteristics will assure predictable soldering results – regardless of board size, PCB material, or location of ground planes without having to resort to trial-and-error methods. In addition to providing reliable results with present materials and equipment, if new board substrates or plating materials, as well as new solder flux or bonding materials are introduced, or if new manufacturing equipment is purchased, a process is available for calculating the optimum process temperatures, time and procedures for the new process elements. This predictive ability is one of the more important advantages of implementing process control procedures to the manufacturing process.

In particular, for the successful high-volume use of automatic wave soldering, the following topics, as defined, must be addressed in developing a useful process control mechanism:

- *Heat transfer.* The transfer of heat through the PCB to the top of the board (where the soldering takes place).
- *Chemistry.* The study of which flux is to be utilized to properly prepare the board for soldering, the strength of the acids being used, the temperature activation points of the flux acids, the contamination of flux and finally the cleaning of the soldered board.
- *Physics.* The size, weight and thickness of the board and transfer of heat through to the top surface as the board is being conveyed through the automatic soldering system.
- *Metallurgy.* The metal substances used on the board, their melting points, the type of solder being used and allowable contamination levels.
- *Mechanics.* The maintenance and upkeep of the automatic soldering system to assure process control limits can be maintained.

Without investigating these concerns, process control is not possible and predictable soldering is then only guesswork. In high-volume production environments like Tektronix, where thousands of boards are manufactured each month, errors resulting from improper or ill-controlled automatic wave-soldering methodologies are unacceptable. Today's present rate of error, 1.5 errors per thousand connections, is below the minimum acceptable error rate; but it can be even further reduced through study of automatic insertion equipment analysis, incoming component solderability specifications, and PCBs designed for automatic soldering.

OPTIMIZING THE WAVE SOLDERING PROCESS

Wave soldering of circuit pack assemblies (CPAs) involves three main phases: fluxing, soldering, and cleaning. The function of each of these three phases is in turn dependent upon a number of factors, and variations in any of these factors can drastically affect the result of the soldering process and CPA reliability. Significant improvement in product quality can be achieved if this process is optimized and all soldered CPAs can go directly into automatic testing without prior inspection and touch-up steps. Moreover, a large amount of direct labor and process cost reductions, as well as appreciable savings, can be realized by reducing in-process inventory and floor-space requirements associated with various inspection, touch-up, and re-testing loops.

A properly chosen water-soluble flux (WSF) can effectively reduce the soldering defect rate and thereby help maximize the solder quality of circuit pack assemblies. In addition to the benefits in cost and quality, a WSF can have advantages in the following areas:

- Soldering of very high-density packages with which a mildly activated rosin flux would have difficulty.
- Eliminating from the mass soldering areas the use of chlorinated and fluorinated solvents by using aqueous detergent solution to clean the CPAs, thus helping to meet future EPA and OSHA requirements. Solvents are generally expensive to use, and many of them are suspected carcinogens.
- Reducing the electrostatic-discharge damages of very sensitive solidstate devices caused by the brush-cleaning operation with solvents.
- Soldering less solderable components and printed circuit boards is possible due to a higher flux activity.
- Reducing the potential for fire hazard in the plant since different types of flux vehicles are used.

First published in *Electronic Packaging & Production*, February 1986.
Authors: K.M. Lin and R.N. Kacker, AT & T Research Center.

However, due to the aggressive nature of the flux, the post-solder cleaning operation becomes critically important. Any excess amount of flux residue left on the assembly can cause serious problems in product performance and reliability. The cleaning process must, therefore, be closely monitored in order to meet the CPA cleanliness and reliability requirements. For this reason, the designs and materials used for the components and the PCBs must be compatible with the WSF soldering and the aqueous detergent cleaning process. When a rosin type of solderability preservative coating is used on the board, some detergent is added in the solution to enhance the cleaning efficiency. Both the components and the assemblies have to be made irrigable and immersible in the aqueous detergent solution to achieve thorough cleaning and rinsing.

Every factor involved in the entire soldering operation can affect the eventual outcome of the process. Hence, they must be jointly optimized to arrive at a 'robust process' from which high-quality products can be manufactured. In a robust process, the operating window of various parameters is made as wide as practical, while the variation in product quality is kept at a minimum. The goal is to achieve a high level of quality the very first

Factors	No. of levels	Associated columns in Table 4
Flux formulation		
A. Type of activator	2	3
B. Amount of activator	3	14
C. Type of surfactant	3	15
D. Amount of antioxidant	3	16
E. Type of solvent	2	5
F. Amount of solvent	3	17
Wave soldering		
H. Amount of flux	3	18
I. Preheat time	3	19
J. Solder temperature		
K. Conveyor speed	3	20
L. Conveyor angle	3	12
M. Wave height setting	2	7
Detergent cleaning		
N. Detergent concentration	2	1
O. Detergent temperature	2	2
P. Cleaning conveyor speed	3	22
Q. Rinse water temperature	2	6

Factors, levels and their association with OA-table

time the product is manufactured, without the need for inspection and touch-up operations.

The orthogonal array design method was employed in setting up a highly fractionated, designed experiment to optimize the entire fluxing-soldering-cleaning process. In the initial experiment, a total of 17 factors were studied in only 36 test runs. Results from this experiment were then used to set up some smaller scale follow-up experiments to arrive at a robust wave-soldering process which includes an excellent, new flux formulation as well as the proper soldering and cleaning procedures.

Soldering process

The study presented here illustrates a method for developing robust industrial processes. Further, it indicates that practical considerations in conducting the experiment play a more important role in an industrial environment than the conventional statistical assumptions and models that underlie the design and analysis of experiments.

The WSF soldering process is divided into three phases of operation: fluxing, wave soldering, and aqueous detergent cleaning. Within each phase, there are several factors which can affect the result. These factors and the number of levels of each to be used in this study are listed.

The functions of a soldering flux are to clean the joining surfaces, to protect the cleaned surfaces from re-oxidation, and to lower the surface tension for better solder wetting and solder-joint formation. The types and amounts of activator, antioxidant, surfactant, solvent and vehicle used in a flux mixture all have a definite impact on the soldering efficiency and the cleaning performance of the process. More importantly, these factors can influence the reliability of the product.

Soldering of the assembly is performed in a cascade wave-soldering machine. After flux application and preheating, the non-component side of the assembly is immersed into a solder wave for 1–2 seconds, and all solder joints are completed as the CPA exits

---

Insulation resistance test

Measured at initial room condition, 30 min, 1 and 4 days under:

- 35 C, 90 percent RH, no bias voltage
- 65 C, 90 percent RH, no bias voltage

Soldering efficiency

Visual inspection to record the number of no solder, insufficient solder, good solder, excess solder, and miscellaneous defects.

Cleaning characterization

Measure the amount of residues left on the board

Solder mask cracking

Visual inspection to record the number of cracked spots on the solder mask of the IR coupons

Responses

---

from the wave. The application method (e.g. foaming, spraying, etc.) and the amount of flux can determine how effectively the joining surfaces are prepared for solder wetting, protected from oxidation, and ultimately, the efficiency of soldering. Large amounts of volatile ingredients in the flux must be driven off by preheating the fluxed assembly to prevent microexplosions from occurring during rapid immersion of the assembly into a molten solder wave. At the same time, gradual preheating can greatly reduce the thermal shock on the assembly during soldering.

Cold solder joints are the result of insufficient heat and temperature during soldering, or premature cooling of the joint after soldering. Very high heat and temperature, on the other hand, can damage the CPA and heat-sensitive components. Conveyor speed can affect the level of preheating and the

Factors	Levels 1	2	3
Flux formulation			
A. Type of activator	a	b	—
B. Amount of activator—a	l	m	h
—b	l	m	h
C. Type of surfactant	c	d	—
D. Amount of surfactant—c	l	m	h
—d	l	m	h
E. Amount of antioxidant	l	m	h
F. Type of solvent	x	y	—
G. Amount of solvent	l	m	h
Wave soldering			
H. Amount of flux	l	m	h
I. Preheat time	l	m	h
J. Solder temperature (S.T.)	l	m	h
K. Conveyor speed—l (S.T.)	s	i	f
—m (S.T.)	s	i	f
—h (S.T.)	s	i	f
L. Conveyor angle	l	m	h
M. Wave height setting	l	h	—
Detergent cleaning			
N. Detergent concentration	l	h	—
O. Detergent temperature	l	h	—
P. Cleaning conveyor speed	l	m	h
Q. Rinse water temperature	l	h	—

Note: l: low, m: medium, h: high, s: slow, i: intermediate, f: fast

Factors and levels for the experiment

duration of solder-wave contact, while changes in wave height setting can change the contact length of the board with the solder wave. All these affect the temperature of the joint and the assembly.

The amount of flux and the wave dynamics at the peelback region, where a soldered assembly exits from the solder wave, play important roles in determining the number of solder defects on the wave-soldered CPAs.

The position and the shape of the peelback region can be changed by altering the conveyor angle.

In an aqueous detergent cleaning process, the assembly is first washed with detergent solution, then rinsed with water, and finally dried with hot air jets and heaters. Additional pre-wash, pre-rinse, or drying cycles may be added as required. As long as flux residues can be adequately cleaned, the lowest possible detergent concentration should be used so that less detergent is spent and less rinsing is required. The temperature of the detergent solutiion is raised to achieve more efficient cleaning and also to prevent excess foaming. Similarly, the rinse water is heated to obtain more effective rinsing. Any change in the cleaning conveyor speed changes the assembly dwell time in each cycle of the cleaning process, and that in turn affects the cleanli-

	N (1)	O (2)	A (3)	C (4)	F (5)	Q (6)	M (7)	— (8)	— (9)	— (10)	— (11)	L (12)	J (13)	B (14)	D (15)	E (16)	G (17)	H (18)	I (19)	K (20)	— (21)	P (22)	— (23)
(1)	1	1	1	1	1	1	1	1	1	1	1	1	1	1	2	2	1	1	2	1	3	3	1
(2)	1	1	1	1	1	2	2	2	2	2	2	1	1	1	1	3	1	3	1	3	1	1	2
(3)	1	1	2	2	2	1	1	1	2	2	2	1	1	2	1	1	3	2	2	1	1	2	1
(4)	1	2	1	2	2	1	2	2	1	1	2	1	1	3	3	1	2	1	1	2	2	1	1
(5)	1	2	1	2	2	1	2	1	2	1	1	1	2	3	3	1	1	2	2	3	1	3	3
(6)	1	2	2	2	1	2	2	1	2	1	1	1	2	3	2	3	2	3	3	3	3	2	1
(7)	2	1	2	2	1	1	2	2	1	2	1	1	2	1	1	3	3	1	3	2	2	3	3
(8)	2	1	2	1	2	2	2	1	1	1	2	1	2	2	3	2	3	3	1	1	3	1	3
(9)	2	1	1	2	2	2	1	2	2	1	1	1	3	2	3	2	1	1	3	3	2	2	2
(10)	2	2	2	1	1	1	1	2	2	1	1	1	3	2	1	1	2	3	2	2	3	3	2
(11)	2	2	1	2	1	2	1	2	1	2	2	1	3	3	2	3	3	2	2	1	2	1	2
(12)	2	2	1	2	1	2	1	1	2	2	1	1	3	1	2	2	2	1	2	1	2	3	3
(13)	1	1	1	1	2	1	1	1	1	1	1	2	2	3	3	2	3	2	3	2	1	1	2
(14)	1	1	1	1	1	2	2	2	2	2	2	2	2	2	2	1	2	1	2	1	2	2	3
(15)	1	1	2	1	1	1	1	2	2	2	2	2	2	3	2	1	3	1	2	3	2	3	2
(16)	1	2	1	2	1	2	1	2	2	1	2	2	2	1	1	2	3	2	3	3	1	2	2
(17)	1	2	2	1	1	2	1	2	1	2	1	2	3	1	1	2	2	3	3	1	2	1	1
(18)	1	2	2	2	1	2	2	1	2	1	1	2	3	1	3	1	3	1	1	1	1	3	2
(19)	2	1	2	2	2	1	2	2	1	2	1	2	3	2	2	1	1	2	1	3	3	1	1
(20)	2	1	2	2	2	2	1	1	1	1	2	2	3	3	1	3	1	1	2	2	1	2	1
(21)	2	2	1	1	1	1	1	2	2	1	1	2	2	3	2	3	2	1	1	3	3	3	3
(22)	2	2	1	1	1	1	1	2	2	1	1	2	2	3	2	2	3	1	3	3	1	1	3
(23)	2	2	1	2	1	2	1	1	1	2	2	2	2	1	3	1	1	3	3	2	3	2	3
(24)	2	2	1	2	1	2	1	2	2	1	1	2	2	3	3	3	3	2	3	2	3	1	1
(25)	1	1	1	1	2	1	1	1	1	1	1	3	3	3	1	1	3	3	1	3	2	2	3
(26)	1	1	1	1	1	2	2	2	2	2	2	3	3	3	3	2	3	2	3	2	3	3	1
(27)	1	1	2	1	1	1	1	2	2	2	2	3	3	1	3	3	2	1	2	3	1	3	3
(28)	1	2	1	2	2	1	2	2	1	1	2	3	3	2	2	3	1	3	3	1	1	3	3
(29)	1	2	1	2	1	2	1	2	1	2	1	3	1	1	2	2	3	3	1	1	2	3	2
(30)	1	2	2	2	1	2	2	1	2	1	1	3	1	2	1	2	1	2	2	2	2	1	3
(31)	2	1	2	2	2	1	2	2	1	2	1	3	1	3	3	2	2	3	2	1	1	2	2
(32)	2	1	2	1	2	2	2	1	1	1	2	3	1	1	2	3	2	3	3	2	2	3	2
(33)	2	1	1	2	2	2	1	2	2	1	1	3	2	1	2	1	3	3	2	1	2	1	1
(34)	2	2	2	1	1	1	1	2	2	1	1	3	2	3	1	3	1	2	1	1	2	2	1
(35)	2	2	1	2	1	2	1	1	1	2	2	3	2	2	1	2	2	1	1	3	1	3	1
(36)	2	2	1	1	2	1	2	1	2	2	1	3	2	3	1	1	1	3	1	3	1	1	2

Orthogonal array table OA (36, $2^{11} \times 3^{12}$, 2)

Flux formulation	Type of activator	Amount of activator	Type of surfactant	Amount of surfactant	Amount of antioxidant	Type of solvent	Amount of solvent
(1)	a	al	c	cm	m	x	l
(2)	a	al	c	cl	h	x	l
(3)	b	bm	d	dl	l	y	h
(4)	a	ah	d	dh	l	y	m
(5)	b	bh	c	ch	l	y	l
(6)	b	bh	d	dm	h	x	m
(7)	b	bl	d	dl	h	x	m
(8)	b	bm	c	ch	m	y	h
(9)	a	am	d	dh	m	y	l
(10)	b	bm	c	cl	l	x	m
(11)	a	ah	d	dm	h	x	h
(12)	a	al	c	cm	m	y	m
(13)	a	am	c	ch	h	x	m
(14)	a	am	c	cm	l	x	m
(15)	b	bh	d	dm	m	y	l
(16)	a	al	d	dl	m	y	h
(17)	b	bl	c	cl	m	y	m
(18)	b	bl	d	dh	l	x	h
(19)	b	bm	c	dm	l	x	l
(20)	b	bh	c	cl	h	y	l
(21)	a	ah	d	dl	h	y	m
(22)	b	bh	c	cm	m	x	h
(23)	a	al	d	dh	l	x	l
(24)	a	am	c	ch	h	y	h
(25)	a	ah	c	cl	l	x	h
(26)	a	ah	c	ch	m	x	h
(27)	b	bl	d	dh	h	y	m
(28)	a	am	d	dm	h	y	l
(29)	b	bm	c	cm	h	y	h
(30)	b	bm	d	dl	m	x	l
(31)	b	bh	d	dh	m	x	m
(32)	b	bl	c	cm	l	y	m
(33)	a	al	d	dm	l	y	h
(34)	b	bl	c	ch	h	x	l
(35)	a	am	d	dl	m	x	m
(36)	a	bh	c	cl	l	y	l

Runs for flux formation

ness of the assembly as well as the production rate of the line.

Experimental design

Based on the experience in the flux formulation and soldering/cleaning processes, a decision was made in the beginning to simultaneously study 17 factors in the initial screening experiment. Seven of them were at two levels each, and the rest were at three levels each. Test levels of the factors 'amount of activator' and 'amount of surfactant' depended on the 'type of activator' and the 'type of surfactant', respectively. Similarly, test levels of 'conveyor speed' depended on the associated 'solder temperature'. The levels for all the factors were deliberately set at extremely wide ranges to have large variations in the response. The trend in response variations thus obtained pointed out a path toward finding a 'robust' process.

A soldering flux is a mixture of five ingredients: activator, surfactant, antioxidant, solvent, and vehicle. If the amount of one ingredient is increased, the proportions of other ingredients are decreased accordingly in the mixture. Therefore, in dealing with mixtures, it is necessary to specify the relative amount of each ingredient. One convenient method for this specification (other than specifying percentages of each ingredient) is to write the amount of each ingredient as a fraction of the amount of one particular ingredient. Here, the relative levels were

Cleaning	Detergent concentration (percent)	Detergent temperature (F)	Cleaning conveyor speed (feet/min.)	Rinse water temperature (F)
(1)	l	l	h	l
(2)	l	l	l	h
(3)	l	l	m	l
(4)	l	h	l	l
(5)	l	h	h	h
(6)	l	h	m	h
(7)	h	l	h	l
(8)	h	l	l	l
(9)	h	l	m	l
(10)	h	h	h	l
(11)	h	h	l	h
(12)	h	h	m	l
(13)	l	l	l	l
(14)	l	l	m	h
(15)	l	l	h	l
(16)	l	h	m	l
(17)	l	h	l	h
(18)	l	h	h	h
(19)	h	l	l	l
(20)	h	l	m	h
(21)	h	l	h	h
(22)	h	h	l	l
(23)	h	h	m	h
(24)	h	h	h	l
(25)	l	l	m	l
(26)	l	l	h	h
(27)	l	l	l	l
(28)	l	h	h	l
(29)	l	h	m	h
(30)	l	h	l	h
(31)	h	l	m	l
(32)	h	l	h	h
(33)	h	l	l	h
(34)	h	h	m	l
(35)	h	h	h	h
(36)	h	h	l	l

Runs for detergent cleaning

expressed simply as low, medium or high to indicate the types of formulations considered.

For 17 factors at their respective 2 and 3 levels, there is a total of 7,558,272 possible combinations. Obviously, it is impractical to vary the experimental factors by the one-at-a-time approach, and some shortcut must be used to carry out such an optimization procedure. In this study, a highly fractionated experimental design was set up by using the orthogonal array design method. Considerations were also given to accommodate some difficult-to-change factors in order to simplify the running of the test.

An orthogonal array, OA $(36, 2^{11} \times 3^{12}, 2)$, employed in this design, can be used to study up to 11 factors at two levels each and up to 12 factors at three levels each in only 36 test runs. The factors are associated with the columns, and the test levels of the factors are indicated by the numbers in each column. The rows of the OA matrix provide the runs for the experiment.

Because some of the factors involved were more difficult to change, it was necessary in designing the whole experiment to have as

few changes as possible for these factors. 'Conveyor angle', 'detergent concentration', 'solder temperature' and 'detergent temperature' were such factors. The columns in the OA table were selected for various factors according to the degrees of difficulty in changing them. The levels of the factors associated with columns, 12, 1, 13, and 2 needed to be changed only 2, 5, 8 and 11 times, respectively, when the experiment was conducted according to the row numbers.

The arrangement with which the 17 factors were associated with 17 columns of the orthogonal array is shown in the table. Except for those factors whose test levels require infrequent changes (i.e. associating conveyor angle, detergent concentration, solder temperature and detergent temperature with columns 12, 1, 13, and 2, respectively), the manner in which the other factors were associated with the rest of the columns in OA was arbitrary. Other tables were thus obtained for each of the three phases of the experiment.

Results of the screening experiment were used to design a series of small experiments to confirm the initial findings and to better define the preferable levels for each factor. This series of small experiments consisted of the following:

- *Follow-up experiments* to confirm the findings of the screening experiment and to eliminate any ambiguity.
- *Set-up experiments* to better define the levels of significant factors in both the soldering and cleaning processes.
- *Confirmation experiment* to compare the performance of the newly developed flux with a selected commercial flux.
- *Formulation experiment* to fine-tune the composition of the newly developed flux.
- *Large scale trial* to find out the performance of the new flux on a large number of CPAs.

Responses

The aim of the experiment was to determine those factor levels which would result in the least solder defects, the smallest amount of flux residues, the highest insulation resistance (IR) for the IR test coupons, and the least

Soldering	Conveyor angle (degree)	Solder temperature (F)	Conveyor speed (feet/min)	Wave height (setting)	Preheat time (seconds)	Amount of flux (grams)
(1)	l	l	ls	l	m	l
(2)	l	l	lf	h	l	h
(3)	l	l	ls	l	h	m
(4)	l	l	li	h	l	l
(5)	l	m	mf	l	m	m
(6)	l	m	mf	h	h	h
(7)	l	m	mi	h	h	l
(8)	l	m	ms	h	l	h
(9)	l	h	hf	l	h	l
(10)	l	h	hi	l	m	h
(11)	l	h	hs	l	m	m
(12)	l	h	hi	h	l	m
(13)	m	m	mi	l	h	m
(14)	m	m	ms	h	m	l
(15)	m	m	mi	l	l	h
(16)	m	m	mf	h	m	m
(17)	m	h	hs	l	h	h
(18)	m	h	hs	h	l	l
(19)	m	h	hf	h	l	m
(20)	m	h	hi	h	m	l
(21)	m	l	ls	l	l	m
(22)	m	l	lf	l	h	l
(23)	m	l	li	l	h	h
(24)	m	l	lf	h	m	h
(25)	h	h	hf	l	l	h
(26)	h	h	hi	h	h	m
(27)	h	h	hf	l	m	l
(28)	h	h	hs	h	h	h
(29)	h	l	li	l	l	l
(30)	h	l	li	h	m	m
(31)	h	l	ls	h	m	h
(32)	h	l	lf	h	h	m
(33)	h	m	mi	l	m	h
(34)	h	m	ms	l	l	m
(35)	h	m	mf	l	l	l
(36)	h	m	ms	h	h	l

Runs for wave soldering

amount of solder-mask cracking. Normally, solder defects can be classified into four categories: no solder, insufficient solder, crosses, and icicles. However, the analysis of

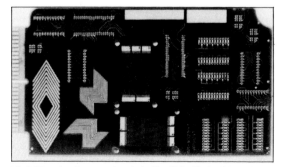

The PCB used as the test specimen is double-sided with plated through-holes, has special circuit patterns and is coated with solder mask in certain areas

initial soldering data indicated that almost all defects were either crosses or icicles. Therefore, only the crosses and icicles were investigated.

The amount of flux residue left on the assembly could affect the performance and reliability of the product. Therefore, chemical analyses were conducted to identify the quantities of residues on the assembly after it had been processed through the detergent cleaner.

The electrical performance of the processed test board was evaluated by the insulation resistance test of comb-patterned coupons in an environmental chamber. For each coupon, four IR values were taken: at initial ambient condition, after 30 minutes, 1 day, and 4 days

Flux	Mean joint crosses	95 percent confidence interval	Mean fine-line crosses	Total joint crosses	Soldering condition
Commercial	0.92	0.45	0.08	1.00	1
Flux B	0.54	0.22	0	0.54	
Commercial	0.65	0.28	0.23	0.88	2
Flux B	0.44	0.16	0.06	0.50	

* Normalized with the total number of defects for the commercial flux soldered under condition 1.

Normalised soldering defects per board

in the chamber. Two chamber conditions were used; one was at a normal stress level (35°C, 90% relative humidity, and no bias voltage), and the other was at a high stress level (65°C, 90% relative humidity, and no bias voltage). The higher the insulation resistance reading, the better the electrical performance.

If the solder-mask material on the circuit board were incompatible with either the flux or the process, damages or cracks would occur in the solder mask. This could result in the corrosion or staining of the circuit. After completing the IR test, all comb-patterned coupons were visually inspected, and the numbers of stained spots were recorded.

The responses examined for this experiment are summarized. The final performance of a particular fluxing-soldering-cleaning process was evaluated on the basis of both the soldering defect rate and the board reliability as measured by the IR test result.

Experiments and results

This experiment used a special test board fashioned after an actual product and very similar to the test board used in an earlier study. The board was a double-sided, rigid board with Sealbrite (a registered trademark of London Chemical Co.) copper circuitry and plated through-holes. The components selected were historically difficult to solder, maximizing the number of soldering defects on each board to produce optimal statistics with the smallest number of boards. Solder mask was used on both sides of the board except on the copper land and certain special circuit pattern areas.

Every board was clearly identified in six places so that various portions of the board, after sectioning for different tests and analysis, could later be traced for re-examination or verification.

Thirty-six different water-soluble fluxes were formulated according to the experimental design. Two boards in separate plastic bags at room conditions before they were placed into the chamber at the same time to facilitate the IR test. IR values of the coupons were recorded under the specified test conditions as mentioned before.

One of the 36 fluxes was used in the subsequent 'soldering follow-up experiment', and four new fluxes with slight composition variations were used in the 'flux formulation follow-up experiment'. From the available information up to that point, a new flux 'A' was formulated and tested in the set-up experiments to gain soldering performance information as well as to better define the factor levels in both the soldering and cleaning processes. A new flux 'B' was then developed for the confirmation experiment, and the soldering performance of the new flux B was evaluated against a selected commercial flux. Two soldering conditions with different solder temperature, top-side board temperature after preheat, and conveyor speed, were used for this experiment.

Condition 1 was commonly used in a

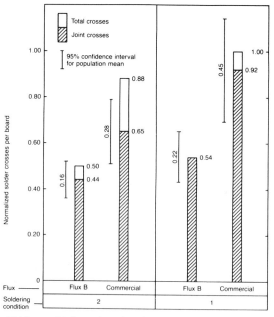

This bar graph indicates the total number of solder defects for each flux used under both soldering conditions

production line, while condition 2 was determined from the designed experiment for the newly developed flux B. Two major types of soldering defects were found – joint crosses and fine-line crosses.

Analysis of the result showed that there was a significant performance improvement by the use of flux B under either soldering condition. Furthermore, there was a smaller 95% confidence interval for flux B in both cases. It indicated that soldering with flux B gave a smaller variation and was a very robust process.

Results for both the insulation resistance test and the cleaning characterization were all very good for flux B; they passed all the required specifications. Visual inspections performed on the solder mask indicated that the smallest number of cracks occurred with the lowest soldering temperature.

The orthogonal array design method employed in designing the experiment is a very powerful technique by which a complex process with a large number of factors can be systematically and effectively studied with a series of small experiments, while practical considerations in industrial experimentation have been taken into account. In addition to being able to identify the significant factors, it can also provide information on the relative importance of various interactions. This method has been successfully applied in various industries in Japan to upgrade the product and lower the production cost.

The concept of such an off-line quality control method can be used both in the early stages of product designs and in the manufacturing processes to achieve a complete system's optimization and to identify a process which is relatively insensitive to the small variations in the process parameters. The high level of product quality is achieved by doing the job right the first time rather than trying to obtain it through corrective actions after the product is made. In most cases, both lower cost and high quality of a product can be realized at the same time.

The experiment discussed is an ambitious undertaking. A very lean design with only 36 test runs is used in the initial screening experiment. The process restrictions have also been taken into consideration so that a minimum number of changes are required of those difficult-to-change factors. The levels of each factor have been deliberately set at extreme conditions in order to obtain a wide range of responses were used in each run. Test boards were soldered in each run in a cascade wave-soldering machine and cleaned in an aqueous detergent spray cleaner. Then the boards were examined for soldering defects, sectioned, and analyzed for residues. Because of the capacity and the availability of the environmental testing chamber, comb-patterned coupons were accumulated which will in turn point out the path toward finding a 'robust' process. In the subsequent series of small experiments, more definite conditions and factor levels have been obtained, and a superior WSF, along with its appropriate soldering-cleaning processes, has been developed. With some minor adjustments in the composition for a flux formulation experiment, the newly developed WSF can be further fine-tuned to arrive at an optimum flux formulation for the circuit pack wave-soldering process.

COMPUTERIZED SOLDERING SYSTEMS

Wave-soldering systems are still at center stage, but other forms of automated soldering are becoming increasingly important, especially considering the rapid growth of surface-mount assembly. Advocates might even suggest that, in time, reflow soldering will become the new heavyweight.

"Unless you can achieve process control, putting a computer on a machine does nothing. It's a way of selling machines." So

First published in *Electronic Packaging & Production*, February 1986, under the title 'Computers power the soldering process'.
Author: S. Leonard Spitz, East Coast Editor.
© 1986 Reed Publishing (USA) Inc.

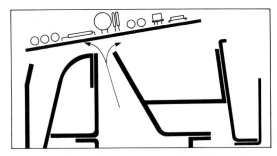

Cross section of the Lambda wave shows through-hole PCB assembly as it enters and is tangent to the wave (Courtesy of Electrovert)

says Michael T. Mittag, Electrovert's vice president for marketing, as a way of emphasizing his company's belief that machine and computer must be designed for each other from the bottom up in order to provide real synergism – not just the illusion of full automation.

Further confirmation of the substantive nature of computerization comes from Dick Brown, eastern regional sales manager of Hollis Automation Inc. Brown observes that "What people are after on computerized wave-soldering machines is to get hold of the process. The computer allows engineering to reclaim the process from the operator, thus permitting better control and better yields."

Computerization represents a giant leap beyond simple automation. Whereas automation constitutes the use of individual workstation sensors and closed-loop controls monitoring and modulating associated parameters, computerization affords integration and linkage of all workstations. Not only is each process parameter adjusted in relation to its own operation and needs, but also as it relates to all other system parameters for total process control. Moreover, process data collected in real-time permits, as no other system can, batch-to-batch and lot-to-lot comprehensive failure analysis. Finally, computerization provides both an equipment self-diagnostic tool and a thorough log for timely maintenance.

Fully automated or computer-controlled wave-soldering systems represent the latest technological wrinkle in the wave-soldering process.

Few concepts in the printed circuit assembly process have been as controversial as solder bridging, its avoidance or removal. One school of thought, led by Electrovert, holds that wave soldering, if done properly, will produce no bridges or other anomalies, and thus will yield literally zero defects. In support of this position, William H. Down, manager of R&D for the Canadian company, says, "Restricting our comments to leaded through-hole boards, we are satisfied that we have taken wave soldering, given good design and good parts, to the point where we can guarantee zero defects."

Not done with mirrors, this remarkable guarantee is the firm promise of the 'Lambda' wave, a fully engineered solder wave invented by L.V. Terdoskegyi and patented by Electrovert.

Wave soldering

Thermal properties of a solder wave depend on a number of characteristics, including solder temperature, area of contact between wave and workpiece, properties of the solder alloy, specific heat, thermal conductance, and surface tension. Flow characteristics of the wave are determined by the size and design of the nozzle aperture and by the pressure, speed and direction at which the solder is ejected.

The Lambda wave offers a mild and progressive preheating effect supplementing the heat transferred to the assembly while passing through the preheater. Subsequently, the wave gives an elongated planar active wave surface and a linear contact length between circuit board and wave surface. The back section of the wave provides a moderate

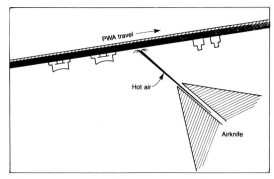

The angle at which the heated jet strikes the underside of the board is critical to the effectiveness of the airknife (Courtesy of Hollis Automation)

postheating effect composed of radiation and convection heat-transfer mechanisms. This exit heating retards assembly cooling and contributes to drainage by keeping the solder mobile.

The flow pattern of the Lambda wave is characterized by a wave surface area of limited extent that is essentially stagnating and slightly inclines with the horizontal. The major portion of the solder flows in the opposite direction of the PC assembly. The combined effect of the incline and the generated flow pattern creates an acute separation angle between board and solder, and establishes excellent drainage and peel-back conditions.

The position of Hollis Automation Inc. on the subject of bridging contrasts sharply with that of Electrovert. A highly placed spokesman for the New Hampshire organization, responding to the 'if done properly' tenet, retorts unequivocally, "If all the soldering parameters were absolutely perfect, which is hard to imagine, that is masking, large land areas, juxtaposition, spacing and widths of lines were all under absolute control, the chances of having solder bridges would be very small. But engineering design does not necessarily follow the desires of an optimum soldering layout any more than the choice of material and component leads can be controlled exactly. So the airknife is a corrective, a general corrective which, like aspirin, does no harm; and the user would be a fool to buy a machine that doesn't have the airknife."

The airknife directs a hot wind load to the underside of a printed circuit board assembly as it exits from the solder wave. The airknife's effectiveness in cleaning bridges, icicles and excess solder from assemblies with high interconnection density, special components, and close conductors and spaces depends upon the velocity, temperature, geometry, and stability of the air jet that issues from the knife's orifice. Other important determinants of effectiveness are the distance of the orifice from the still molten solder on the board and the angle at which the air jet strikes the board.

Specifically engineered for full computer control, Electrovert's Century 2000S uses a custom-designed microcomputer (8 bit/16 bit) to continuously monitor all wave-soldering parameters, adjusting them as product load dictates. Once the parameters for a particular board type are entered in memory, the system sets up automatically. When the first production board starts into the conveyor, either a serial number entered on a keypad or an

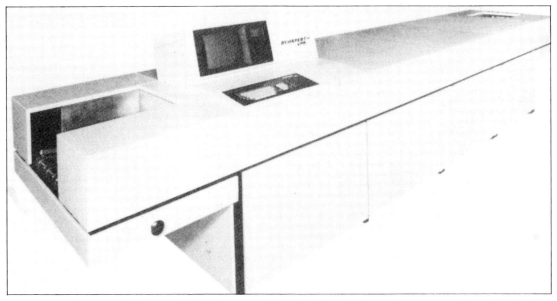

Designed for single- and double-sided attachment of SMDs, Model IL-18 computer-controlled vapor-phase soldering system handles boards 18in², transporting them by conveyor at speeds up to 96in./min (Courtesy of Dynapert-HTC)

automatic read-off bar-code activates the process. A scrolling display constantly confirms the accuracy of all parameters.

Operating dry or with an oil intermix, the GBS Mark II represents the Hollis entry into the fully automated wave-soldering systems marketplace. Complete with standard computer-controlled subsystems from conveyor speed set to variable solder-wave depth, the Mark II comes with the much vaunted, integrated heated airknife debridger. The Hollis mcirocomputer control system (HMCS) provides state-of-the-art process control and analysis by automating all significant control, monitoring, reporting, and recording functions. One 5×5in. key matrix and one 9in. CRT Supplants conventional instrumentation and controls. Three computer modes are available: control, set parameters, and calibrate.

Snapping at the heels of the front-runners, the John Treiber Co.'s Model 700-CCS fully automated system offers instant 'process recipe' recall and set up. After first determining process parameters and entering the associated data into the computer, the wave-soldering system is activated by keyboard entry or bar-code reading directly from the PCB itself. The 700-CCS management information system logs all parameters, program settings, and actual operating conditions.

Reflow soldering

Surface-mount technology has spurred the acceptance of both infrared (IR) and vapor-phase soldering. After acceptance, full automation cannot be far behind.

Not new, these soldering technologies are used primarily to reflow solder paste which has been previously applied to substrates and components by screening or dispensing techniques. Reflow soldering can be particularly advantageous as compared to wave soldering for surface-mount assemblies where extreme board density, high I/O counts, and very narrow line widths and spacings often produce shadow problems and solder bridges under the best of circumstances. Traditionally, however, these processes have been batch, off-line, are relatively slow and have the potential for creating production bottlenecks.

With the advent of both in-line systems in 1981 and today's computerization, reflow techniques are challenging the supremacy, if not the speed, of wave soldering – especially for surface-mount assemblies.

A pioneer in batch-type vapor-phase systems, formerly Hybrid Technology Corp., now Dynapert HTC, recently introduced their IL-18 computer-controlled, in-line vapor-phase reflow soldering system. Designed specifically for single- and double-sided attachment of surface-mount devices, the reflow system's microprocessor controls and monitors all single vapor soldering parameters. Providing computerized loop outputs, the menu-driven IL-18 reports management data malfunctions and maintenance requirements.

Reflow-soldering's proponents argue that such systems eliminate bridging, heat-induced tombstoning, solder balls, cold joints, board charring and delamination, and component failure, as well as solving problems related to mismatched thermal coefficients.

It is little wonder that vapor-phase soldering shares the reflow stage with an alternative technology, infrared soldering. Coupled with microprocessor control for impeccable regulation and repeatability of programmed profiles, infrared technology has reached a high level of maturity and sophistication.

Obsoleting point sources used in the past, non-focused, infrared area source emitter panels provide the IR energy in reflow solder machines made by both the Manix Division of Henry Mann Inc. and Screen Printing Technology (SPT) Corp. Non-focused IR permits processing at lower temperatures and assures uniformity of heat levels with even substrate heating. It diffuses infrared energy in wave lengths that match the absorption characteristics of the alloys and components being processed, and the emission is non-color selective in the substrate or the components. Non-focused IR converts 90% of all input energy into processing energy, minimizing power consumption.

SPT Corp. has developed an IR reflow-soldering system, Model 770 SMD, which creates and maintains an appropriate thermal profile as well as recognizes and indicates actual product temperature. The system corrects process profiles during operation based upon product temperature information. Interfacing with an IBM PC/XT, PC/AT or compatible computer, the menu-driven system, interacting with IR temperature sensors and an optical encoder, controls temperature profiling and conveyor-belt speed respectively. Outputs include all process temperatures, substrate temperature, and belt speed.

The last word

Perhaps putting solder-system computerization into perspective, John Treiber asks, "Does everyone need a computerized soldering machine?"

To which he responds, "Obviously not. A high-volume commercial electronic manufacturer may run 2000 boards per day of the same PCB assembly five days per week. Obviously there is no need to change set up. Assuming a high-quality soldering system is in use, albeit manually controlled, the process can run without interruption, and high-quality soldering will result."

Treiber adds a final footnote, "Accepting the fact the soldering system deserves a significant investment is a healthy trend in the industry. The sophisticated, computerized soldering system, and the price tag that goes with it, finally gives the (soldering) process the attention it deserves."

RE-EXAMINATION OF SOLDERING TECHNIQUES

The surface-mount technology revolution is altering the face of electronic manufacturing. Its tremendous impact is no better exemplified than in the changes in soldering techniques used. Soldering techniques heretofore foreign to most are now being installed at many electronic manufacturing operations.

Unfortunately, there is no universal panacea for surface-mount component soldering problems. Techniques which work well under one set of conditions may be totally inadequate for a different set of circumstances. For instance, soldering a surface-mount circuit with one of the common techniques such as IR or vapor-phase reflow may yield totally satisfactory results. On the other hand, if the same circuit in order to combat thermal dissipation problems requires the addition of a heat pipe in close proximity to a component being soldered, these techniques may not work. It might be necessary to use the highly concentrated beam of a laser to reflow the solder to form an acceptable solder joint.

A number of techniques exist for soldering SMCs to substrates. These techniques fall into two major categories, flow and reflow soldering. Flow soldering involves the application of a molten solder to the component attachment site. Under this category are processes such as wave soldering, drag soldering and dip soldering. The second category, reflow soldering, involves the melting of solder that has prviously been deposited on a component, its substrate attachment point or both. In this classification are found methods such as vapor phase or condensation, infrared (IR), laser, convection furnace, and conduction soldering systems.

Circuit construction

The selection of the appropriate soldering technique for the most part hinges on the type of surface assembly construction used and its complexity. For some surface-mount assembly constructions, the manufacturing scenario might even include the use of more than one soldering method.

The form and complexity of surface-mount component assemblies can vary widely. SMC assembly constructions have been subdivided

First published in *Electronic Packaging & Production*, February 1985, under the title 'SMT calls for a re-examination of soldering techniques'.
Author: Tom Dixon, Senior Editor.
© 1985 Reed Publishing (USA) Inc.

Differences in surface-mount component package styles and circuit constructions must be addressed when selecting a soldering process. No single soldering technique is optimal for all possibilities (Courtesy of Electrovert)

into three general categories or groups: types I, II and III. Type I assemblies use SMCs exclusively and include constructions having SMCs on one or both sides of the substrate's surface.

The second construction grouping, type II, is a mixed-technology structure that combines through hole and surface-mount components on the substrate's top surface. The bottom surface of the type II circuits may have SMCs, a mixture of SMCs and through-hole components or no components at all.

Type III circuits, the third grouping, are a special form of the mixed technology. Such circuits have only through-hole components mounted on the top surface and only simple surface-mount components, normally discrete components such as chip resistors and capacitors and SOTs, attached to the bottom.

Of all the possible soldering techniques available for building surface-mount component assemblies, wave soldering, vapor phase and IR reflow soldering are receiving a major portion of the industry's attention.

Flow soldering

The first significant use of surface-mount construction on traditional glass-epoxy substrates were type III circuits made by the

Wave-soldering system manufacturers have adopted a dual-wave construction such as the one pictured to overcome some of the problems encountered when soldering on a single wave. Behind the wave on the left is a hot-air knife which is used to help eliminate bridges on more complex components and circuit constructions (Courtesy of Hollis)

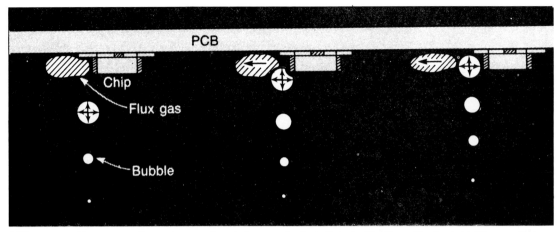

PCB

Chip

Flux gas

Bubble

While most wave-solder system manufacturers have turned to a dual-wave construction to eliminate solder skips resulting from trapped flux gases, Koki Limited of Japan has developed another approach which involved the injection of compressed nitrogen into the wave to dislodge the trapped flux gas

Japanese. The technique was adopted principally as a cost- and space-saving technique. This was accomplished by attaching chip resistors and capacitors to a substrate's bottom with an adhesive, then mounting leaded components to the top surface by inserting their leads into through-holes in the board and finally soldering the assembly.

The idea was that conventional single-wave soldering methods could be used to join the components to the substrate. Unfortunately, quality problems were encountered that resulted in a need to modify the wave-soldering process and circuit-design rules. Problems included solder skips and the formation of insufficient fillets as the result of shadowing effects from other devices and pockets of trapped flux gases which prevented solder from penetrating to the surfaces being joined. In addition, solder shorts and bridges were a problem.

There are two good reasons to use wave soldering to construct surface-mount component assemblies. First, the wave-soldering process is relatively familiar to a large segment of the electronics manufacturing industry. Secondly, a substantial capital investment already exists in wave-soldering equipment, and any change which can be made to systems which permits recouping of some of that original investment is desirable.

One scheme developed and patented by the Japanese to eliminate defects caused by entrapped flux gases was to provide gas escape holes into the board near the component mounting sites. A more common solution is the one that has been adopted by most wave-soldering system manufacturers to re-

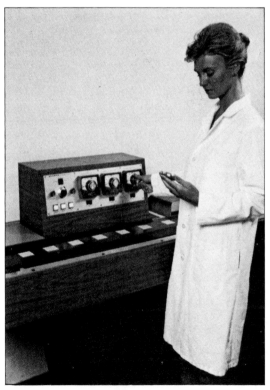

Reflow soldering by conduction heating has been an established process in the hybrid industry for a number of years. Pictured is a hot-plate style system which uses a conductive conveyor to transport assemblies across a series of temperature-controlled hot plates (Courtesy of MCT)

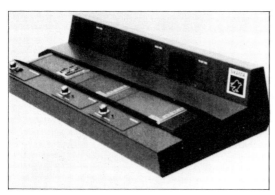

Hot-plate reflow systems such as the one pictured with chain-driven bars which push the assemblys' hot plates, and the conveyorized unit previously pictured, which work well on hybrids and simple surface-mount structures do not work on some complex surface-mount constructions and materials (Courtesy of Sikama)

solve the SMT soldering problems. That solution is the use of dual-wave soldering techniques.

Modified solder waves

The dual-wave concept for soldering surface-mount components uses a narrow, turbulent solder wave followed by a smooth, wider contact second wave. The first wave features a high vertical pressure, which permits better penetration of solder between tightly packed components while also providing some scubbing action action of the surfaces to be joined, thus enhancing solder wetting. The turbulent nature of the wave as well as its narrow contact width are intended to permit the escape of evolved gases.

The second wave is a more traditional smooth wave, like the single wave commonly

used to solder through-hole mounted components. This wave finishes the formation of the fillet by providing the correct zero velocity exit zone and helps drain off any excess solder.

Initial designs of the dual-wave solder concept used a single nozzle to generate two waves. While such systems did work, Electrovert's William Down notes that, because they used a single pump, they were difficult to set up and limited the height adjustments of the waves. Most of the dual-wave systems currently available have resorted to individual nozzles and pumps to resolve this problem. Some have carried this separation of controls even farther by offering two separately controllable solder pots.

Technical Devices' Douglas Winther notes that, based on favorable comments from sources at North American Phillips, they decided to provide separate solder pots for each wave to permit users to adjust the solder temperatures of each wave independently. This approach can also be found in at least one Japanese-manufactured wave-soldering system, explains Winther.

Wave soldering surface-mount component assemblies has some other notable differences over traditional through-hole component soldering. Fluxes with lower solids contents are used to get around some problems of the wave being unable to displace fluxes on the board. Since components run directly through the wave, new concerns have been raised about device integrity because of great

The programmable vapor-phase system from Centech Corp. controls preheating, soldering and cooling times and temperatures

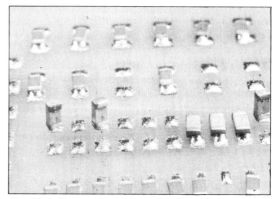

One of the problems sometimes encountered in reflow soldering operations is a phenomenon known as tombstoning, where chip resistors or capacitors raise up (Courtesy of Hollis)

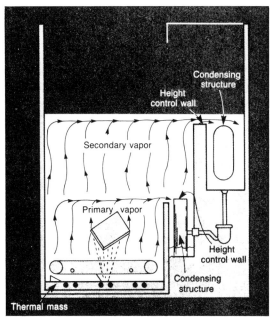

Unlike conventional dual-vapor systems which use two sets of cooling coils around the perimeter at different heights in the chamber to control vapor levels, the Coprane system uses cooling structures behind two walls in the chamber to draw vapors over the side by a partial pressure differential to maintain height control

thermal stress on the packages.

Wave soldering of surface-mount component assemblies typically involves longer and higher preheating than used on through-hole component assemblies. This change has a two-fold benefit. It drys the flux more thoroughly and it brings the temperature of the components up closer to the wave to minimize thermal shock.

However, the dual-wave concept still has some limitations. SOICs must be oriented so that the leads on each side enter the wave sequentially; that is, the longest dimension of the package body must be oriented so that it is in the direction of movement through the wave.

Soldering difficulties are particularly pronounced when it comes to large surface-mount components such as PLCCs and LCCCs, or tightly spaced ones, or both. One problem encountered with such devices is the formation of solder bridges and shorts. Many feel that present device packages have pushed wave-soldering technology to the limit of what it can consistently and reliably solder. Hollis'

engineers have been experimenting with hot-air knives after the waves as a means of removing bridges. Hollis' Warren Abbott claims that hot-air knives permit the soldering of larger and more tightly packed surface-mount components than solder waves can handle.

Approaching the trapped flux gas problem from a different angle is the bubble-soldering system from Koki. The system injects bubbles of nitrogen gas into the solder wave. These gas bubbles inflate in the hot molten solder and because of their high kinetic energy rise to the surface. Upon reaching the trapped flux gas pocket the high kinetic energy of the gas bubbles enables them to force the trapped flux gas pocket to move, allowing solder in.

Still another approach to the problems of soldering SMCs and solder skips resulting from trapped flux gas is the Kirsten Jet soldering technique. The system features a high-velocity hollow wave. This curved sheet of solder makes use of the Bernoulli effect, which also accounts for the lift in an airplane, to lower the surface tension of the solder. This enables the solder wave to quickly reform behind the trailing edge of the component as it passes through.

Reflow soldering

To many engineers, the use of reflow soldering as a component attachment technique is an unfamiliar procedure which must be learned. The typical process involves the deposition of a solder paste on the component mounting pads of the substrate. The leads or mounting pads of the component to be attached are then placed upon the solder paste.

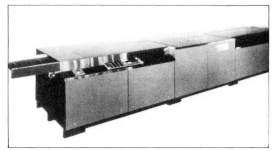

Corpane's new in-line vapor-phase soldering uses the dual-vapor approach to control loss of expensive soldering-fluid vapors

Type II surface-mount component assemblies often require the use of more than one soldering technique. The HTC type II soldering system incorporates a molten-solder application system with a vapor-phase reflow system in a single unit

In some cases, the paste is dried after placement to drive out volatiles which can be a problem during reflow. In other cases it is dried to more securely fix the components in place to prevent their movement during handling prior to reflow. Ultimately, the assembly is heated up to a temperature where the solder melts and forms a fillet between the pads on the substrate and the leads or pads on the component.

Reflow soldering is not a new technique. It has been used by the hybrid-circuit industry for a number of years. A process frequently used by the bybrid industry has been to reflow the solder paste by applying heat from a hot plate to the bottom side of the ceramic substrate which conducts it up to the top surface where the component and solder paste lie.

A commonly used system, based on this concept, is the MCT conveyorized reflow unit which passes assemblies over a series of hot plates with a conductive belt. Another variation of the technique is the Sikama system which uses a chain-driven pusher bar to move components across the surface of a series of hot plates.

These conveyorized hot-plate systems have limitations which make them unsuitable for some surface-mount applications. They rely on the conduction of heat through the substrate and thus require that the substrate be extremely flat and a good thermal conductor. In addition, some substrate materials cannot stand up under the heat and can warp, twist or delaminate. Another limitation of these systems is that they are difficult to use in processing double-sided circuit constructions.

It is possible to manufacture small surface-mount assemblies on glass-epoxy substrate materials with these systems, MCT's Jim

Hollomon notes, but he admits that they are not very practical for large glass-epoxy substrates. They do work well on ceramics and some other special substrate materials with good thermal conductivities, Hollomon points out.

Lasers

Laser reflow soldering is another technique with potential applications in attaching surface-mount components. YAG or CO_2 lasers are normally used for reflow soldering. Because of the laser's highly focused energy beam, it is a useful technique for special applications as mentioned earlier where delivery of a highly concentrated source of energy to solder-joint location is necessary. It is also ideal for applications involving high-temperature-sensitive components since it tends not to heat up the area around the solder junction. Earl Lish of Martin Marietta notes that one positive aspect of laser soldering is that the quick nature of this soldering process results in a minimal amount of intermetallic formation.

The laser is a quick means of forming solder joints on structures where only a few connections are being formed. Unfortunately, joints are formed sequentially one after another instead of simultaneously, making it relatively time-intensive on larger soldering jobs. The cost of additional lasers to permit

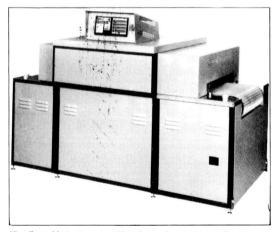

IR reflow soldering is one possible technique for attaching surface-mount components to substrates. Units such as the one shown from Radiant Energy tend to be much more enery efficient than either vapor phase or wave soldering

In response to the growing use of surface-mount components, hand-soldering tool manufacturers are designing new tips for surface-mount components (Courtesy of Hexacon)

simultaneous soldering tends to be prohibitive.

Vapor phase

Very few engineers involved in electronics manufacturing have not heard about condensation soldering by its more common name: vapor-phase soldering. It is a technique by which a vaporized fluid condenses on the surface of an assembly and gives up its latent heat. That energy transferal causes solder which has already been deposited on the components or substrate surface to melt (reflow). The melted solder then attached itself to the component and the substrate and forms a fillet. Upon cooling the solder fillet solidifies, forming both the electrical and mechanical connection between the component and the substrate.

Vapor-phase soldering tends to heat the entire assembly uniformly. When a part is put into a system's vapor it will never be heated above a fixed temperature. On the other hand, vapor-phase soldering does not allow components to remain at temperatures lower than needed for soldering. Thus it is not advisable to use this soldering technique for highly temperature-sensitive parts. Because of the rapid immersion of the work piece into the elevated temperature of the vapor, a prebaking of the solder paste is usually necessary. This eliminates rapid outgassing of solvents which can shift parts around and degrade the solder joints.

The Centech in-line vapor-phase system

has incorporated a preheat stage in its design. This preheater, notes Centech's Nile Plapp, can be used to drive out solder-paste solvent and bring component temperature closer to the vapor temperature, thus minimizing problems of rapid solvent boil off and thermal stresses on the component when the assembly enters the vapor.

Initial condensation soldering system designs were set up as batch processes. A simple system consisted of a stainless-steel chamber with immersion heaters on the bottom and condensing coils around the top. Unfortunately most of these systems tended to be somewhat inefficient at preventing vapor losses. Much has been made of the cost of the fluids that are used in these systems. Because they are relatively expensive, vapor losses can be costly.

Attempts to control the vapor losses led to the development of dual-vapor systems. These units use a blanket or secondary vapor layer of a low-cost freon material to prevent the expensive soldering fluid vapors from escaping.

Unfortunately the secondary vapor, while performing an excellent job of controlling vapor losses, is prone to degrade and form acids because of the thermal stress placed on it by super heating at the primary and secondary vapor-zone interface. ACid neutralization systems are thus required.

A closely associated issue is the potential degradation of the primary fluid, also from thermal stresses. In a poorly maintained system the primary fluid can break down to form HF and toxic PFIB (perfluoroisobutylene). The levels of PFIB and HF generated are virtually non-existant, according to 3M. In a properly maintained and ventilated system these materials should not be a problem, just as with potential breakdown products accompanying other processes involving organic materials such as vapor degreasing.

Recently 3M released FX-38, an improved version of FC-70, with better thermal-stability properties. Other materials with high thermal stabilities have also become available.

Aside from improvements in the thermal stability of fluids, changes to machine designs such as increased use of stainless steel are being implemented in response to corrosion problems, explains HTC's Jim Hall. Hall points out that even with improvements in system construction and fluids, it will still be necessary to incorporate neutralization units since acids will continue to be introduced from external sources such as the flux in the solder past.

The picture regarding vapor-phase soldering fluids is changing. At the inception of vapor-phase soldering technology, the only fluids available for use in vapor-phase soldering systems were the Fluoroinert fluids from 3M such as FC-70 and FC-71. While FC-70 and its improved version, FX-38, are the dominant fluids used in condensation soldering there are alternative fluids now on the market. Montedison, ISC and Air Products all are offering vapor-phase soldering fluids. ISC's product will be marketed in the USA by 3M as FC-5311. It has the same properties as FC-70 and FX-38. While prices of these fluids are not dropping appreciably, they do make claims of improved thermal stability.

In-line vapor-phase systems

While the first vapor-phase systems were batch units, the need and desire for in-line soldering units has grown. HTC introduced the first in-line vapor-phase soldering systems a few years back. Unlike their batch-processing forerunners with large openings, these units had relatively small openings which permitted the use of a single-vapor zone. The single-vapor zone eliminated the complexities encountered in maintaining dual-vapor systems to prevent acid formation.

Recently, Corpane unveiled an in-line system using the dual-vapor approach to controlling vapor losses. The system uses a weir or fixed height walls to control the width of the secondary vapor. The approach uses two internal walls of different heights to control the vapor zone heights.

Cooling coils just outside the higher of the two walls creates a pressure differential across the top which draws vapors over the wall to the coils where they are condensed. This establishes the top of the secondary vapor blanket. The primary vapor zone height is controlled by a similar flow over a shorter wall. Primary fluid vapors are condensed out by a separating column cooled by secondary zone fluids from the upper cooling coils.

As the primary vapor condenses it heats up the condensed secondary fluid until it vaporizes. The condensed primary fluid is vaporized by dropping it on a molten thermal mass such as a tin-bismuth instead of by immersion heaters in a liquid as is commonly used in other vapor-phase systems.

Recently HTC announced the development of a new soldering system for type II surface-mount component assemblies. The system combines single fluid, reflow vapor-phase soldering and a hollow solder wave into one in-line processing unit.

Tombstoning

One frequently mentioned problem with vapor-phase reflow soldering is what has come to be known in the industry as the tombstone effect of chip components such as resistors and capacitors. This phenomenon involves the standing up of the component on one of its end terminations so that it is joined to only one solder attachment pad. While this can potentially occur with any reflow soldering technique, it is reported as being more prominent with vapor-phase soldering.

Several explanations are circulating as to the possible causes of tombstoning. One theory is that it is the result of significant differences in the solderability of component termination or mounting-pad sites. Resulting differences in surface-tension force allows one site to exert more force on the other, resulting in a lifting of the component.

Another explanation for tombstoning is that differential heating of the bond sites allows the surface-tension forces on site to come into play earlier than for the other, allowing it to lift the part. Still a third explanation is that the condensed vapor-phase fluids lower the surface tension of the molten solder, weakening one bond site's hold on the part more than the other. Then there is the explanation that it is the result of the escape of entrapped volatiles from the solder paste.

The problem of tombstoning is not a problem with wave-soldering techniques for

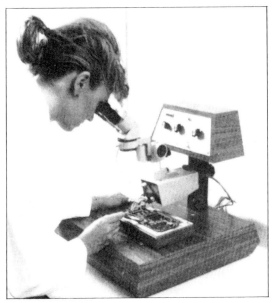

This PCB repair, touch up and microsoldering station from MCT features a pulse-heated tip at the end of an articulated arm. Foot-pedal actuation leaves both hands free

attaching SMDs since they are held in position with an adhesive. However, the use of an adhesive with wave-soldering techniques can create other problems. It does not allow surface-tension forces to realign the orientation of a poorly mounted device. If it holds the component too high from the substrate or spreads out over the bond site, it could possibly inhibit fillet formation.

Infrared reflow

Interest in reflow soldering with infrared (IR) radiation is growing. One reason is that the price tag of an infrared reflow furnace generally is lower than for a vapor-phase unit. Furthermore, the cost of operating an IR system is lower than for either vapor-phase or wave soldering because of their higher energy requirements.

The primary method of heating in most IR reflow systems is by radiation rather than by conduction or convection. The amount of radiant energy absorbed by the surface-mount component assemblies is related to the absorptivity of materials at the specific wavelength and the radiation flux intensity on the surface.

The transfer of heat by radiation has several major differences from transfer by convection, conduction or condensation, explains David Flattery of Radiant Technology. Radiant heat transfer has the potential to deliver energy beneath the surface of semi-transparent objects. The other heat-transfer mechanisms deliver energy exclusively to the surface of objects. Delivery of energy to the interior of objects, such as is possible with IR, can reduce or eliminate stresses induced during rapid heating, Flattery notes.

IR systems do not require the drying of the solder paste prior to reflow. The energy from the IR can penetrate into the paste and remove solvents without spatter or skinning, Flattery adds. Such systems can also direct the heat, thus it is possible to reflow only one side of an assembly at a time if it is so desired. Karl Seelig of Arconium claims that diffused IR tends to be better than focused for heat-sensitive components, but adds that there will be some sacrifice in system speed.

However, IR has its potential shortcomings. Because of differences in absorptivity properties of materials it is possible for hot spots to form as parts pick up energy at different rates. In order to combat this problem, it might be necessary to lower system conveyor

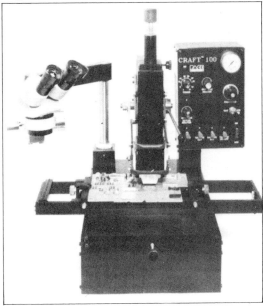

A number of special repair systems specifically designed for surface-mount technology, such as this one from Pace for removing and replacing chip carriers, are becoming available

speeds to allow time for the heat to be conducted away. Another potential problem is that difficulties could be encountered in soldering component leads and terminations shielded by parts which are not transparent to the IR radiation. Finally, setting up the proper reflow profile can be tricky and may have to be individualized for each assembly type.

Other techniques

The use of convection heating techniques which circulate hot air or gases is another method employed to reflow solder surface-mount component assemblies. These processes also allow the elimination of the drying stage since temperature profiles are established with a preheat stage which drives out the volatiles.

Furnaces can be found with multiple heater stages that allow profiling flexibility. As an example, BTU offers its three-zone TS-393 rapid reflow conveyor furnace which features a final heating zone that allows a sharp temperature peak to bring the solder to reflow temperature rapidly. This keeps the time that the solder is molten to a minimum, reducing intermetallics formation.

Vitronics offers another reflow soldering system which relies principally on the mechanisms of conduction and radiation to transfer heat. This unit is called an IR system by the manufacturer yet differs from other systems in that it does not use radiation as the primary means of energy transfer to the component assembly.

This system uses a panel emitter system instead of lamps or bulbs commonly found in IR systems. The radiation generated by the panel emitter system is of a much longer wavelength than that originating from more traditional IR sources, according to Vitronics. What differentiates this unit from other IR systems is that instead of heating the assembly directly, the radiation is absorbed by the atmosphere in the chamber which is then circulated around the assembly to raise its temperature.

This concept eliminates some of the differential heat problems associated with more traditional IR systems, according to the manufacturer. Vitronic's Steve Dow notes that other IR systems manufacturers claim that this system is really a convection reflow unit. However, Dow claims that they have been able to achieve reflow rate improvements ranging from 200 to 400% over conventional furnaces.

In response to the surging growth of surface-mount components, there is also a growing activity in the development of repair systems. New tip configurations for soldering-iron systems are becoming available. Small rework stations are available from a number of manufacturers. APE has announced a rework repair station specifically for surface-mount assemblies. MCT offers a foot-pedal actuated microsoldering station that allows the operator freedom to use both hands. Pace has designed a station for removing and replacing chip carriers that heats all the leads simultaneously.

ADHESIVES

As experience grows in adhesive use a great deal of research is being done to ensure reliability.

ADHESIVES FOR SMD WAVE SOLDERING

Much of what we term 'high tech' is really just an accretion of a lot of 'low tech'. In the hot new technology of surface-mount devices (SMDs), an important role is being played by adhesives in holding parts securely to the substrate during handling and soldering operations.

If they are not held fast they will fall off when the board is inverted so that leaded components, which are present on mixed technology boards, might be soldered. although the role of adhesives here might seem humble, it is nevertheless crucial, and a good example of how low tech is the infrastructure supporting high tech.

In practice, surface-mount devices are placed on beads of uncured adhesive which is applied by specially designed dispensing equipment. The adhesive, which is then cured by either heat or UV-based systems, must hold the parts securely during handling and subsequent soldering. A great deal of research has gone into the field of adhesives for surface-mount components, and the number of companies entering the area is growing.

Background

The preliminary attachment of surface-mount devices prior to their being soldered in place can be done in two ways. When the board contains SMDs exclusively, and vapor-phase soldering is being used, most likely the parts will be adequately held by the stickiness of the solder paste and no adhesive will be necessary. Bruce Murray feels that even reflow techniques will still demand the use of adhesives in some cases. He notes that larger surface-mount devices will need adhesive to keep them immobile during reflow, as well as to permit greater design flexibiltiy.

However, boards with a mixture of through-hole and surface-mount components will require the use of adhesive. The normal sequence of steps on such boards is to use adhesive to tack the SMDs in place, cure it, turn the board over, insert the leaded compo-

nents, solder them, and finally, clean the flux off the board. It is while such a mixed-technology board is being turned over that the adhesive serves to prevent the SMDs from falling off the board or twisting awry. After soldering, the adhesive no longer serves a purpose.

It is expected that the use of boards with a mix of SMDs and leaded devices (each technology on one side of the board) will continue for some time. Thus, adhesives for SMDs seem to be assured a role in the future.

Portrait of the ideal adhesive

The essential criterion for performance of an adhesive for SMDs is how well it holds the devices during the handling and soldering stages. This relates to the part-holding strength. There are other figures of merit for these materials, however.

For example, for simplicity of preparation, some have argued that the ideal adhesive should consist of only one component. It also should lend itself to dispensing by several automated methods and should provide sufficient bond strength to do the job during processing, yet be easy to remove if necessary. It should have a short cure time to ensure sufficient throughput, yet at the same time require a minimum of energy input to achieve the cure. In addition, the temperature of cure

Universal's Model 4713A dispenser module, which is part of their Onserter SMC placement system, works on the needle dispensing method. It is shown placing adhesive dots on a PCB

First published in *Electronic Packaging & Production*, June 1985, under the title 'For SMD wave soldering, adhesives still hold fast'.
Author: Robert Keeler, Associate Editor.
© 1985 Reed Publishing (USA) Inc.

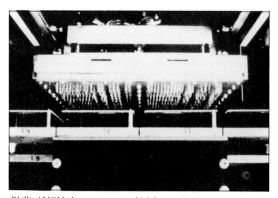

Phillips' MCM placement system, which features hardware control, works on the principal of pin transfer to mass apply adhesive to a PCB

must not be so high that it causes the board to warp. It should have sufficient body to fill in gaps and crevices while resisting sagging, yet be able to spread lean enough to make a thin-film bond. It should also be non-toxic, odorless, environmentally safe, non-flammable, non-corrosive, stable and non-conductive, and inexpensive.

Other desirable material and process-related attributes for SMD adhesives include a resistance to 'tailing' or stringing during dispensing. Resistance to detrimental change in properties due to aging, such as drift in resistivity, is another concern, as is the ease of repair of the PCB using the adhesive.

This last factor is underestimated in importance, claims Bruce Murray. Most adhesives are thermosets, meaning that they polymerize during cure, cannot be remelted and are highly resistant to the action of solvents. For the most part, SMDs are currently removed by touching a soldering iron tip to the part until the conducted heat remelts the solder and the adhesive. Any remaining adhesive residues are left on the site of the part, since by the time the board gets to the repair stage the adhesive no longer has a function.

Kinds of adhesive

The most prevalent adhesives used for SMDs fall primarily into two families: epoxies (either single- or two-component systems) and acrylic products. However, other kinds of adhesive not in these families include urethanes and cyanoacrylates.

In the area of epoxies, single-phase systems are the most common. They are known to cure rapidly and to provide a strong bond. Epoxies in general are cured thermally. They can be formulated into a wide variety of products, depending on the application, and can be applied by all of the popular dispensing methods.

They do, however, have a relatively short 'shelf life' – the usable life of the chemical as it sits in the container. This is a confusingly similar term (used in reference to two-part epoxy systems) to 'pot-life', which refers to the usable life of the epoxy after the two main components (catalyst and resin) have been mixed together and catalyzation begins. Ann Delmarsh of Epoxy Technology notes that two-part epoxies offer a faster cure rate than one-part systems, as well as a longer shelf life for the components. However, they also demand the use of a more intricate dispensing apparatus.

One way of delaying the premature catalytic reaction of multicomponent epoxy adhesives, which may contain accelerators, extenders, fillers and pigment in addition to the basic resin and catalyst, is to quick-freeze them. Says Hal Rocheleau of Poly Freeze, this can give the mixture a pot life ranging from several weeks to months. Bubbles, introduced by mixing or inherently generated by the chemicals themselves, must be removed by degassing before freezing.

Like epoxies, acrylics also come in one- or two-part systems, although they are unique from the point of view of chemistry. They can be dispensed through a needle or by a squeegee in screen printing and are processed for the most part by a combination of thermal and ultraviolet cure. Bruce Murray claims that, while acrylic cure times are known to be fast and they exhibit good resistance to the effects of moisture, solvent and temperature, the corrosion resistance and insulation stability to date remain unproven.

Cyanoacrylate adhesives (the family of chemicals that includes the commercial products known as 'Super Glue') see use with SMDs also. They give very rapid cures without the application of heat, although they

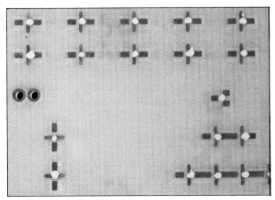

A pattern of adhesive dots applied to a PCB by the pin transfer method (Courtesy of North American Philips)

do not seem to be viable for use with pin-transfer or screen-printing techniques. They bond so aggressively that they can foul the pins or the squeegee. Also, they are not especially solvent-resistant and may not hold up to the heat of wave solder, since they are thermoplastics.

Other chemical formulations include urethane adhesives, which offer strong, flexible bonds (although they may require dry storage to prevent premature polymerization), and immide/esters, which offer high temperature stability.

A special class of adhesives are the anaerobics. The term 'anaerobic' refers to the fact that these substances can cure in the absence of air; the name thus relates to performance rather than to a particular chemical family. One formulation of anaerobics, for example, happens to be a modified urethane-acrylic.

Anaerobics can be cured by UV or heat, or as mentioned, even without any energy input. In the latter case curing begins in the center of the dot and literally proceeds from the inside out. Care must be taken with anaerobics to ensure that a complete cure has been achieved, or there may be lingering problems with surface tack and migration. Herb Krause of Ablestick observes that a solvent washing procedure is often necessary on anaerobics to remove the sticky outer layer to avoid contaminating the solder bath.

Methods of dispensing

There are three principal techniques for dispensing adhesives for SMDs: pin transfer,

needle deposition and screen printing. Pin transfer is a high-speed technique favored by the Japanese. It can be configured in two ways.

One approach uses a set of pins (approximately 0.015in. diameter or smaller) which sit submerged in a container of adhesive and are raised up to touch the PCB as it passes over the container. These pins, which correspond in position to the points on the board that must receive a dot of adhesive, pick up a fixed amount as they rise up out of the container. Gravity ensures that a relatively uniform amount is carried by the pins each time.

The other configuration for the pin-transfer method has the set of pins move downward to pick up the adhesive, which has been previously leveled off by squeegee into a film of uniform thickness in a tray. Once the pins have picked up the adhesive, they are moved over to the PCB, and come down to make the deposition.

Pin transfer does not permit much flexibility in the shape of the adhesive dot applied, but does, however, give more control than screening over the thickness of the deposition. Thus, says Bruce Murray, this may be the preferred method for mounting small-outline components where dot size must allow enough space for leads. Pin transfer is prone to tailing, and adhesive viscosity control is very important here.

Cliff Pukaite of North American Philips' SMD Technology Center points out that software-controlled pin-transfer equipment approaches the problem a little differently than the hardware-controlled machines which bring pins down en masse to the board. Here, the part is held in a set of jaws, a pin moves in to deposit the dot, and then the part is applied to the board.

Pukaite says software-controlled machines are more versatile in terms of where components can be placed, although he notes that hardware-controlled machines are much faster. Thus, firms with lower volumes and a great variety of product configurations might be better served with software control.

Another dispensing technique, screen

printing, uses a squeegee to push adhesive through a screen having holes in it corresponding to the sites for adhesive deposition. This method is fast and accurate, and lends itself to a very high throughput since, as with pin transfer, the entire board is done in one stroke.

However, it demands that boards be flat and free of obstacles (such as previously placed components) also, the size and cross-sectional profile of the deposited dots depend on the thickness of the screen. For very dense boards when the thickness of the mesh is very small, the size and volume of the drop may become too small to be effective. Screening does not seem to be very prevalent in industry at this time.

By far the most common adhesive dispensing method, though, involves the use of a hollow needle, usually mounted at the bottom of a large adhesive-filled syringe. The adhesive is forced into the needle from the syringe barrel by air pressure and the dispensing head can be programmed to index to successive dot sites.

Important parameters to control here are dot size, timing and tailing of adhesive. Not as fast as some of the other methods, needle dispensing has a typical rate of deposition of about two dots per second. A potential problem with the syringe-needle dispensing approach is possible entrapment of air bubbles in the adhesive, says Cliff Pukaite, who notes that the bubbles can cause intermittent drop application.

Regardless of the dispensing method used, once adhesive is deposited it must stay in place sufficiently long to hold the part without allowing movement prior to curing. The adhesive used must be 'thixotropic' and must resist the tendency to flow outward to cover up solder or a connection until after the part is placed and the adhesive cured. A thixotropic fluid is one that behaves as though it had low viscosity when it is being pumped or pushed but displays high viscosity when there is no energy added to it. The property of being thixotropic will also help ensure that the dot of adhesive will remain high enough to reach from the PCB surface to the underside of the mounted part.

With leadless devices as low as they are this may not be a problem, but with leaded devices such as small-outline transistors, the board-to-part spacing is greater and the drop must stand higher without sagging.

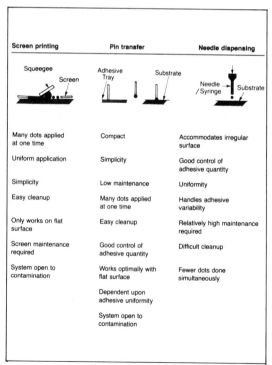

Screen printing	Pin transfer	Needle dispensing
Many dots applied at one time	Compact	Accommodates irregular surface
Uniform application	Simplicity	Good control of adhesive quantity
Simplicity	Low maintenance	Uniformity
Easy cleanup	Many dots applied at one time	Handles adhesive variability
Only works on flat surface	Easy cleanup	Relatively high maintenance required
Screen maintenance required	Good control of adhesive quantity	Difficult cleanup
System open to contamination	Works optimally with flat surface	Fewer dots done simultaneously
	Dependent upon adhesive uniformity	
	System open to contamination	

Characteristics of the three dispensing methods for SMD adhesives

Curing techniques

Adhesive has to be completely cured and non-sticky by the time it gets to the wave-solder operation, where it must withstand the ravages of the high solder temperature (up to 260°C), solder flux and cleaning solvents, as well as the mechanical stresses of board flexing and possible physical impact.

The two principal methods in use for curing SMD adhesives are heat curing and UV cure followed by infrared. Heat cure is the most prevalent technique, using either forced air or an IR oven. The cure cycle time can take up to 20 minutes, but care must be taken that the cycle temperature and time do not warp the board or damage components.

TRA-BOND 2116 is a two-part epoxy staking compound, available in bulk or in smaller packages for repair jobs as shown. It is suitable for surface-mount application

The use of thermally curing, two-part adhesives help to avoid this problem, since they cure faster than one-part systems – and at relatively low temperatures. However, they do have a relatively short pot life and might involve some waste in mixing.

UV curing, used in tandem with an IR curing step, is used primarily in lower-volume production. Adhesive tracks must be extended out from the sides of the part for a short distance so that the UV radiation can reach it for curing. Otherwise the adhesive actually under the part remains hidden and is shadowed from the energy of the UV. Thus, the impingement of UV actually just tacks the part on, and is followed by the infrared to further cure the shadowed areas of adhesive.

As SMD density increases, the space between parts shrinks and there is less room available to extend a track of adhesive beyond the edge of the part for UV tacking. This factor alone may hinder the future application of UV curing for adhesives. Bob Batson of Loctite notes that an exception to this is seen with anaerobics which can harden without the IR step. When used, the secondary IR cure stage also helps to avoid any problem with uncured adhesive contaminating the solder bath.

Both acrylic and anaerobic adhesives, in addition to many epoxies, can be cured with UV. Some of them get their final cure in the heat of the solder operation without going through IR. A completeness of cure is especially important with anaerobic adhesives which might be more prone to stickiness and migration problems.

Once the board has been soldered, the adhesive performs no useful function and is really just going along for the ride. At this stage it might even be a nuisance. Thus, a cured adhesive must be free of any tendency to absorb moisture or to decrease in resistivity after soldering is performed. Herb Krause notes further that an important but often overlooked consideration with moisture resistance is that the adhesive must not cause current leakage.

The high bond strength of the adhesive which was so desirable before soldering may become a liability after soldering. If a part must be removed and replaced in a repair operation, the strong adhesive joint may prove stubborn and difficult to remove.

Conductive adhesives

The use of metal-filled conductive adhesives is a promising technology. It holds the potential for reducing cost and eliminating the need for high-temperature processing associated with wave soldering, as well as improvement in design performance.

Such materials generally do not require the use of fluxes and post-assembly cleaning. They are for the most part resilient than solder, offering an improved resistance to thermal-stress cracking. Also, automating the dispensing operation for conductive adhesives is fairly easy.

On the other hand, there are technical problems here that still remain to be worked out. For example, conductive adhesives are still neither as conductive as solder nor as cheap. Under humid conditions, migration of metal (causing short circuits) can sometimes be a problem with conductive adhesives when the metal filler is silver. A conformal coating helps to entrap the silver in these cases.

ADHESIVE EVALUATION

One way of meeting the demand for low cost, high-density printed wiring board assemblies is by combining surface-mount techniques with traditional insertion of leaded devices. The leadless devices can be mounted on the solder side of the board, and the leaded devices on the normal component side. In this way, both sides of the board are used.

The critical factor in this method is the reliability of the adhesive in securing the SMDs during placement, assembly, and subsequent processing. Xerox's Electronics Division evaluated the properties of several adhesives, especially cure time, working life, adaptability to time-pressure dispensing, and cost. The objective was to select a thermal-cured adhesive which would hold in place discrete surface-mounted resistors, capacitors, and small outline transistors (SOT-23s) through the company's standard assembly, wave solder, and cleaning operations. The material would have to be compatible with the company's solder flux and withstand the mechanical and thermal stress of assembly and flow solder. It could not absorb flux or moisture which later might become a source of current-leakage paths.

After reviewing vendor data, Xerox chose three epoxy-based, commercially available adhesives for further study. Adhesive A comes as two separate components that can be stored at room temperature and must be mixed before using. Adhesive B is a one-part epoxy that is stored at 4.5°C. Adhesive C is a pre-mixed material, stored at −40°C. Each was tested in five categories:

- Adhesive cure time and flux compatibility.
- Adhesive drop size and working life.
- Adhesive/part stability.
- Wave soldering.
- Adhesive/part holding strength.

The first test

The printed wiring board assembly process requires an adhesive which can be cured

First published in *Electronic Packaging & Production*, August 1984, under the title 'Adhesives hold parts firmly for surface mounting'.
Authors: E. St. Peter, Fred C. Martin and Will A. Reyes, Xerox Corp.
© 1984 Reed Publishing (USA) Inc.

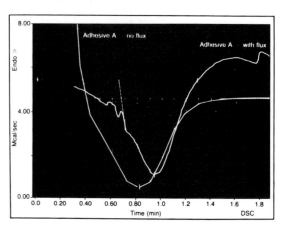

The differential scanning calorimetry curing plot for adhesive A. The curve is divided into intervals of 6s, and the percent cure for each interval is shown

rapidly in-line. In order to test the cure rate of the three adhesives, Xerox used differential scanning calorimetry (DSC), a thermal analytical technique which measures heat flow to the sample as a function of temperature. The sample is heated isothermally; 150°C was specified in this instance as being safe for extended exposure of surface-mounted components. The sensor monitors the differential current needed to maintain the sample at the pre-selected temperature relative to an inert reference. The DSC output plots heat capacity against time.

As the epoxy is cured (an exothermic reaction) the sensor sees a lower net rate of heat flow into the sample. The area of the exothermic curve is related to the degree of cure of the adhesive. The percent curve can be calculated by dividing the curve at regular intervals and forming a ratio of the area of one or more intervals to the whole area. That

Adhesive	Sample weight	Time for >95 percent cure at 150 C
A	4.48	1 min
B	2.55	1.5 min
C	4.83	3.5 min

Curing time for adhesives under test

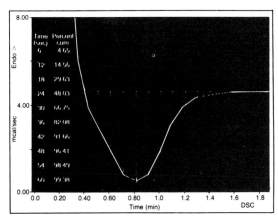

Superimposed on the curing curve for adhesive A is the curing curve for adhesive A after it was exposed to soldering flux

is, where H is enthalpy $(E + pv)$ and n is the number of intervals, ΔH of cure is equal to $\Sigma \Delta H n$ at each cure time interval.

The curing times achieved at Xerox are given in the table. These cure times are values obtained on laboratory samples under optimum heat transfer conditions. More realistic production times should include heat losses and thermal sinking effects of the printed wiring board assemblies. These would have to be determined by considering the heating rates, conveyor speed and the size and number of boards in the curing oven. At the Xerox facility, increasing these times by a factor of three proved satisfactory.

Flux compatibility

Problems ranging from adhesive failure, leading to loss or misorientation of parts, to current leakage can result from reaction of the adhesive with the soldering flux. Xerox tested the compatibility of the three adhesives with its production flux, once again using DSC. The adhesives were exposed to the flux after being 50% cured, 70% cured and >95% cured. Also shown are the curing curves for adhesive A; one is for the sample that was exposed to flux, and the other is the control sample that was not. There is little difference between the curing curves of the two samples at 70% and >95%, but at 50% the curves show a 20–30% reduction in cure. Xerox recommends curing an adhesive to at least 70% cure before using it with a solder flux.

To eliminate the need for time-consuming adjustments to the dispensing system, Xerox decided that a single drop size would be used for all applications whenever possible. The drop size required was the largest that could be used on resistors and capacitors without encroachment on the soldering pad area, and the smallest that could hold the SOT-23. The resistors and capacitors have a nominal 0.065 × 0.078in. area for adhesive application, while on the SOT-23 the adhesive has to fill a 0.007in. gap between the board and the component bottom.

Xerox first calibrated various drop sizes by laying down 100 drops on a pre-weighed strip, weighing the strip on an analytical balance accurate to ±0.01mg, and dividing the weight by 100. The different-sized drops were then used to place the SMDs on the boards, and all the boards were run through the entire processing sequence. The researchers found that a drop size of 0.2 ±0.05mg was optimum for Xerox's operation. This size ensures that all devices will remain in place

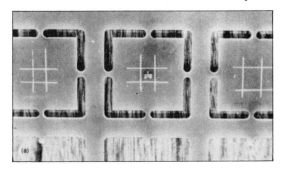

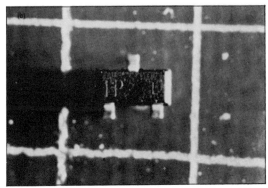

An SOT-23 is placed between four scribed reference points on the printed wiring board to test for part stability (top). After four passes through the wave solder and cleaning operation, the SOT-23 has not moved (bottom)

Wave solder passes	Adhesive A (lb)	Adhesive B (lb)	Adhesive C (lb)
Surface mounted resistors			
Initial no soldering	9.0 to 16.0	9.6 to 23.2	14.4 to 25.0
2	7.2 to 14.6	12.5 to 28.4	10.6 to 25.0
4	6.4 to 10.6	10.0 to 25.3	10.0 to 21.8
Surface mounted capacitors			
Initial	8.8 to 17.5	7.4 to 21.3	8.5 to 15.4
2	6.0 to 10.0	6.2 to 15.8	9.7 to 24.2
4	3.0 to 11.5	4.2 to 14.2	11.5 to 21.9
Surface mounted transistors			
Initial	6.2 to 14.4	9.0 to 17.7	3.4 to 8.0
2	1.5 to 7.2	6.0 to 14.4	6.2 to 22.9
4	3.5 to 8.2	1.3 to 12.4	7.7 to 21.3

Shear force range data

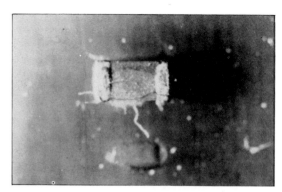

The surface finish on the particular capacitors used by Xerox was found to hinder adhesion between the capacitor and the board

with minimal adhesive spread onto the soldering areas.

The work life of the three adhesives was judged by how long the drops could be dispensed in reproducible sizes. The pre-mixed, frozen adhesive C proved to have a work life of only one hour. After an hour, the rheological properties had changed and constant readjustment of the dispenser was necessary. Both adhesives A and B had work lives that exceed the standard production work shift. Vendor data specified a work life of four days for adhesive A and two days for adhesive B.

Part stability

To test the ability of the adhesives to hold a part in place during curing and assembly, Xerox set components within four scribed reference points on the printed wiring board. The devices were photographed after initial placement, after the cure cycle, and after four passes through the solder-clean operation. There was no movement of the component with any of the three adhesives. A second test was conducted to look at the adhesives after wave soldering. Fourteen SMDs were mounted on printed wiring boards made up of 1in.^{2} 0.059 FR-4 material with solder mask applied on a push-out 4 × 6in. panel. The panel was flow soldered in a dual-wave machine for four consecutive passes at 6ft/min. After each pass the researchers inspected for any missing or moved component. There were none for any of the three adhesives under test.

In order to obtain a quantitative result of the holding strength of each adhesive, Xerox measured the force required to shear off the component bonded to the solder mask FR-4 board surface. An Instron Model 1122 Universal Tester measured this force prior to flow solder (after cure), after two and after four passes through the wave-solder operation. No loss of components occurred with any of the three adhesives. The average range recorded for all devices and adhesives was 10lb.

The SOTs showed the lowest values, as predicted; the gap between the component back and the surface of the PWB, coupled with the drop size selected, resulted in the average weakest bond. However, the holding force was sufficient to keep the SOT in place. Xerox's data showed a minimum of 1.5lb shear strength was required to hold the three kinds of components in place during SMC processing. After 17lb of shear force, the devices fractured, although in many instances the adhesive still held.

Xerox noted three types of shear failure in their experiments, one per component. With the resistors, the failure was in the adhesion of the solder mask to the board and had nothing to do with the adhesives being tested. With the SOTs, failure was a combination of solder mask pull-away and adhesive failure. When the capacitors moved or came unstuck, it was due to failure of the adhesive bond between the component and the board. Rather than being the fault of the adhesive, Xerox found this was attributable to the surface finish on the capacitors used in the test.

THE ROLE OF SOLDER PASTE

Wave soldering introduces the solder during the soldering process. Other processes such as reflow soldering pre-place the solder in paste form prior to entering the soldering equipment. The board preparation involved is described in the next two papers.

SOLDER COATING AND LEVELING

Solder coating of bare-board circuitry, followed by a leveling operation to clear the component holes and remove excess solder, is a PCB fabrication technique which offers major benefits to the user. These include such technical advantages as improved board solderability through the formation of a thick coating of eutectic solder on pads and plated holes, improvement in the product's appearance and an extended shelf life. Coating and leveling, as a process, has become closely associated in recent years with the application of solder mask over bare copper (MOBC) to prevent solder bridging under solder mask and wicking through PCB component holes.

Process background

Coating is achieved by dipping a circuit board into a molten bath of eutectic solder. Following this, a leveling process, using either an impinging spray of leveling fluid or a hot air blast, removes the excess solder and clears the component holes.

All solder-leveling systems, whether liquid or hot air, are faced with the problem of getting the highest possible clearance rate on component holes, while still leaving behind the proper amount of true eutectic (63% tin, 37% lead) solder coating on the surface of the holes and pads. If the force of the blast is too great, the board might be left with just the intermetallic compound of copper and tin, which is more subject to degradation than bare copper itself; bare copper only has to be cleaned to bring it back to a solderable condition. Extended shelf life comes about if the thickness of the coating is great enough; with a coat that is too thin, however, the intermetallic bond between the copper and solder tends to migrate outward, and this alloy is harder to solder than pure eutectic. If, on the other hand, the force is too weak, component leads might jam in their holes during insertion due to narrowed inner hole diameters. High aspect ratio holes (large ratio of length to diameter) can be a problem here.

The use of solder coating and leveling is growing fast, but the use of tin/lead as an etch resist is still a strong competitor. When circuit boards are fabricated by the pattern plating process, the pads and traces are plated up with copper after the circuit imaging resist (plating mask) is applied. A tin/lead plate is then applied over the copper to prevent the plated layer of copper on the circuit traces from being etched away again after the plating mask is stripped away to expose the broad areas of background copper for general etching. The tin/lead is usually fused before solder mask is applied.

This technique serves its purpose, but a problem arises when the board circuit resolution calls for fine lines and spaces. When such a board goes through wave soldering, tin/lead on the closely spaced circuitry remelts and circuit lines can bridge across to each other. Solder could actually wet out the entire bottom of a fine-line board were not solder mask present.

Even though solder mask helps here, the tin/lead can still remelt under the mask in the solder wave, causing the finish to crinkle and blister. In addition to ruining the esthetic appearance of the board, this can lead to particles of mask falling off into the wave solder. Flux can get into the cracks and cause corrosion.

This technical problem has motivated manufacturers to strip off the tin/lead and directly apply SMOBC. When SMOBC is adopted though, exposed portions of the circuit must be protected to prevent oxidation of the copper. Otherwise these sites, whether for leaded component or SMD boards, could present solderability problems. However, re-plating these sites with tin/lead again is not possible since the etching of the circuit pattern has destroyed continuity of the copper across the board.

Even if bare copper sites could be plated up again with tin/lead for protection, it is hard to control the plating thickness of tin/lead on the walls of the component holes. According to Cutech Inc., a firm which supplies tin plate

First published in *Electronic Packaging & Production*, February 1985, under the title 'Solder coating and leveling improves PC board solderability'.
Author: Robert Keeler, Associate Editor.
© 1985 Reed Publishing (USA) Inc.

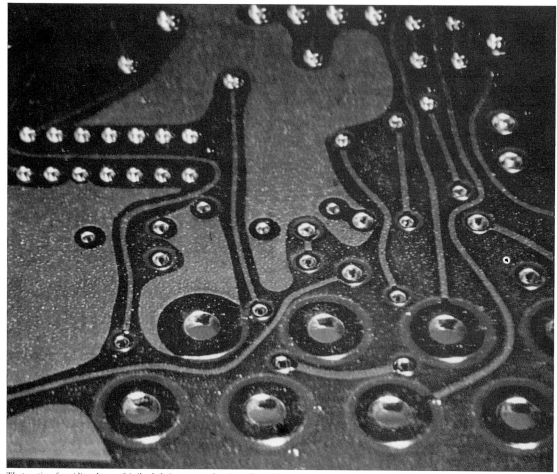

The practice of avoiding the use of tin/lead plating as an etch resist and applying solder mask over bare copper (SMOBC), as implemented on this board, is growing more popular and in turn is stimulating the adoption of solder coating and leveling (Courtesy of Hollis Automation)

stripping solutions as well as other soldering chemicals, some firms are looking at selectively plating tin on pads and holes with an immersion bath. They are solder coating and leveling over this, utilizing the fact that clean tin is more solderable than copper. Other manufacturers simply use tin as an etch resist and fuse it.

Stripping of plated-on tin/lead to yield bare copper again, however, presents problems. If it is not done properly, a film can be left behind on the copper that will inhibit solder wetting. Some manufacturers point out that boards are harder to clean at this point, too. In addition, with electroplated solder, the alloy composition can vary across the plated area.

Thus, PCB manufacturers are turning to the practice of coating a board to protect the copper, and using some sort of leveling technique both to clear solder out of the component holes again and to strip off excess solder. With coating and leveling, there is no solder bridging, no crinkled solder mask. Further, the solder thickness is relatively easy to control. Perhaps most important, a good eutectic coat is applied.

Although SMOBC seems to imply the use of solder coating and leveling, not all boards which are fabricated by the use of coating and leveling involve solder mask. MultiWire boards are a notable exception, although in a sense the embedment of the wires in substrate material in a MultiWire product constitutes a sort of solder mask.

One of the great disadvantages of applying solder mask over fused tin/lead electroplate is the tendency of the solder mask to wrinkle during wave soldering (Courtesy of Hollis Automation)

The use of solder coating and leveling to 'repair' a tin/lead plated board is fairly common in the industry. One solder leveling service admits that this sort of work constitutes up to 35% of its workload. When such repair is called for, it is usually due to a plated layer that is too heavy in lead or tin – either situation indicating that the layer will not fuse properly.

Other cases demanding such repair include incomplete converage of traces by the tin/lead plate and poor cosmetic appearance. When the badly plated board goes into the molten solder for coating, the plating tends to melt out in the bath, leaving a shiny layer of eutectic behind.

Process limitations

However, some industry experts caution that when coating and leveling is used to merely reflow a tin/lead plate, which is likely of non-eutectic composition, the metallurgical ratios in the eutectic solder bath can become imbalanced. Selective solder coating and leveling is not without other limitations and problems. As a process, it seems to be extremely intolerant of improper cleaning. Ralph Woodgate, a private consultant in the field of soldering, agrees that copper cleanliness prior to coating and leveling is vital, and points out that coating and leveling shows up every weakness in the fabrication sequence; in

this sense, he says, it is a fine quality-control tool.

One example of a fabrication problem that can make for a bad leveling job, and ultimately poor solderability, is solder mask 'bleed' due to poor registration; another is the fact that thermal cured screened-on soldermask inks have a bad habit of depositing organics on the surface of the copper. In addition, improper stripping of tin/lead presents problems. If it is left on too long, it oxidizes and is hard to strip off again.

Solder in the leveling-system tank can become ruined by the presence of copper, dross and other contaminants quite readily. Warren Abbott of Hollis Automation notes that in the ANSI/IPC-5-815A specification, over 0.3% by weight of copper contamination is called out as the limit. This amount of copper can result in a very sluggish solder flow, not to mention a gritty appearance and poor solderability. Abbott adds that most solder-leveling installations have a high turn-over of solder for this reason.

A major complaint about the solder coating and leveling process to date seems to be related to the high rate of maintenance required by leveling machines. This problem stems largely from two causes: the presence of flux, and the high heat levels developed in these systems. These factors can help to wear

The Caratsch solder coater and leveler, Type 1, is a manually operated, Swiss-made machine

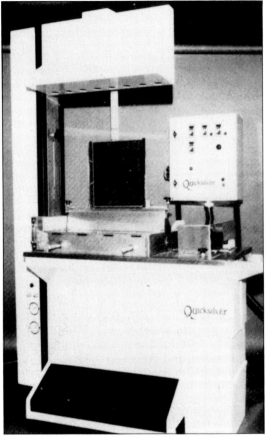

Machine Automation offers the Quicksilver hot-air leveler, which has a throughput capability of up to 360 boards per hour

out electromechanical parts. For example, pumps can break down in both kinds of leveler, liquid and hot air. Electric heaters in the solder pots of machines can fail. Solenoids and air valves, present in older hot-air levelers, are prone to malfunction.

The hot-air leveling process is inherently dirty, given the fact that air under pressure is blowing flux and dross around. Still another problem here is that there is occasionally a slight increase in the trailing edge of the solder on the pad as the part is drawn out through the leveler. This effect seems to be present to some degree on all boards, but its impact is greatest on SMD boards where a level surface is important for mounting a leadless part. The effect can run to thousandths of an inch when uncontrolled; adjustment of the parameters of solder

temperature, air temperature, air pressure, dwell time, part speed in and out, and flux composition can help to limit the effect.

Liquid levelers, on the other hand, at least those that still use some sort of hot oil for purposes of leveling, present the danger of contaminating the board with oil since it can actually seep into the fibers of the board. There is also the problem of disposing of spent leveling fluid with some of these machines.

Automatic vs. manual

Most coating and leveling machines currently on the market process the board in a vertical orientation. There are, however, a few hot-air leveling machines, fairly expensive and massive, for high production rates and full automatic operation requiring essentially no human intervention. They take the board horizontally and are more easily adaptable to loaders and stackers. With horizontal systems, however, it has been reported that solder tends to sag at the bottom of a component hole due to gravity. In vertical systems, a corresponding solder thickening is sometimes seen at the sides of the barrel of the hole. Strictly manual systems, which are all vertical machines, have no automatic control of dwell-time. They therefore suffer the likelihood of stripping off too much solder in the leveler, since human operators sometimes elect to put the board through the leveler several times.

When shopping for a solder-leveling system, easy maintenance should be a key consideration. Simplicity of design is important, although this must be balanced with consideration of individual production throughput requirements. For example, fully automated horizontal machines, which offer a high production rate, will be more complicated in design. They achieve their speed by using a conveyor or roller-feed system, as opposed to vertical machines on which the board usually has to be clamped by a human operator for dipping and leveling. Whichever design is chosen, a consistently leveled thickness of eutectic coat should be the most important consideration.

Liquid levelers

Historically, liquid levelers can be traced back to the first such machine, which was called the Hydrosqueegee. The inventor was Keith Oxford, now president of Oxford Engineering – a supplier of liquid leveling systems. This early machine is no longer being sold but is still being used in many shops. It sprays a hot, water-soluble fluid on the solder-coated board, under pressure developed by a motor-driven pump.

Oxford Engineering will introduce the Flash II liquid leveler, which coats and levels in one module, as opposed to their earlier Flash I, which merely leveled. The Flash II uses no motors or pumps, and has no moving parts other than two slide valves. The pressurized fluid used is based on polypropylene glycol. The firm claims the system leaves on between 0.0002 and 0.0004in. of solder. Although it is a manual machine, the Flash II will process two to four 12 × 18in. boards per minute. In the Flash II, the board moves through vertically, but no special fixturing is required. The firm claims the Flash II deposits a solder coat that is smooth enough for surface mount devices.

PC Products offers another liquid leveler: their Lightning Solder Leveling System. It features a patented laminar ribbon flow manifold design and a proprietary heat-transfer fluid. The use of a manifold, like the air knife of a hot-air leveler, avoids the problems of clogged nozzles, notes PC Products. The manifold pressure is only ¾lb to 1lb, so solder is not being removed by mechanical pressure, adds the firm. To minimize thermal shock, the board is lowered into a preheat tank filled with biodegradable water-soluble oil (basically a surfactant), at 120°C, using a modified pair of vise grips. After the solder pot dip, the board goes into the leveler, where two passes are recommended for opening the component holes. PC Products says the machine is designed to serve the market niche of the small to medium shops that cannot afford an air leveler. The equipment is available with an optional post-rinse.

Dynapert-HTC offers the SL-1824. It uses a heat-transfer fluid to effect the heating and leveling of the PCB. The leveling medium used is the Fluorinert vapor-phase fluid, which is used commonly used in vapor-phase systems. The heating vapor here is used to preheat the board prior to immersion in solder, a feature that allows the solder pot to run a little cooler than in some other systems, claims the firm. The system temperature is limited to the boiling point of the Fluorinert. Thus, it runs at the fixed temperature of 215°C, which in turn is the temperature of the saturated vapor of the system; it cannot get any hotter than this. Fluid pressure, as well as the spacing from the board and the angle of the nozzles (which are above the surface of the fluid), is adjustable. The board is processed at a 15° angle from horizontal plane. The system is designed for the smaller shop, although it can be adapted to higher production.

Monitrol Inc., a solder-leveling service, has developed the Solder-Clad system, a horizon-

The Gyrex Model 515, like the Model 510, is an automated vertical solder coating and levelling system. If differs from the 510 in that it features in-line fluxing

tal liquid coating and leveling system. They have recently sold the manufacturing rights for this machine to Phoenix, AS-based Magatronics. Neither the copper nor the solder are ever exposed to air in this machine. As a result, says the firm, there is no dross formation, and this helps to eliminate many maintenance problems. Solder consumption is reduced, too. As opposed to using a conveyor, the board is pulled through with a series of baffling rollers. The recirculating leveling-fluid system capacity is sufficient to offset leakage between the rollers as the board enters and exits. The firm claims that the machine will deposit a uniform solder thickness of 100 millionths of an inch. Both single- and double-sided boards can be processed, and acceptable panel thicknesses range from 0.020 to 0.125in.

Hot-air levelers

As with liquid-leveler manufacturers, there are only a handful of companies making hot-air levelers in the USA. Hollis Automation, which formerly offered the SCL machine, has dropped out of the market; not because of any lack of belief in the future of this technology, but rather for purposes of redefining their strategic market goals.

The most recent machines from Gyrex include the models 510 and 515. The difference between the two is basically that the 515 features automatic fluxing, while the 510 is for use with off-line fluxing. Both of these are semi-automatic (in the sense that a human operator is required) vertical-orientation machines. Both systems allow processing of boards of up to 0.187in. thick, and having holes as small as 0.021in. diameter. Maximum panel size accepted by these machines is 24 × 24in. With these designs, Gyrex has incorporated simplified electrical and mechanical features. The firm says it has made it easier to change the heaters on the solder pot, which used to be a fairly major job. They have also done away with heaters on the air knife, and instead draw heat from the solder pot for this purpose. The air knife has been made easier to change, too. Gyrex' GH-500 in-line automatic solder coating system moves PCBs along horizontally on a continuously moving conveyor. The need for clamps or fixtures has thus been eliminated. Production rates in excess of 2000ft^h (500 24 × 24in. panels) of either single- or double-sided boards are claimed for this system by Gyrex.

Electrovert USA Corp. offers its Levelair

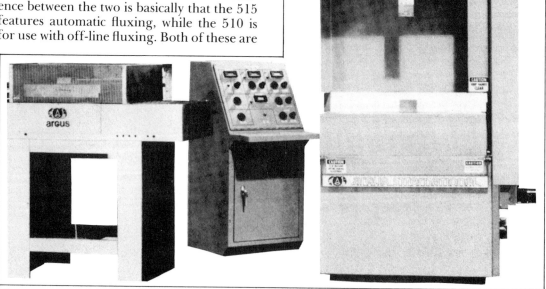

The Argus International HAL5000 hot-air leveler (on the right) is shown with the PC3215 flux-application station. Control console is in the middle

A panel is shown about to descend into the solder pot of Electrovert's Levelair system, which is available either in a manual or automatic configuration

solder coating and hot-air leveling system. The machine will work with additive, semi-subtractive or subtractive printed circuit boards. Electrovert claims that their machine is capable of accurately solder coating on circuit tracks less than 0.100in. wide and will clear plated holes as small as 0.020in. in boards up to 0.125in. thick. Minimum line widths and spacings of 0.008–0.010in. can be processed without solder bridges or solder balls. The automatic version of this machine includes a pneumatic conveyor device and ducted hood. The pneumatic clamping fixture permits rapid removal and attachment of panels for processing. The clamp is actuated by a foot switch, leaving the operator's hands free. Electrovert points out that, despite the fact that ordinary compressed air is used, oxidation of the applied solder will not occur during the process, because the air jets drive a fine ridge of flux across the board in advance of the air as the board is withdrawn. This, they clai, ensures a bright, shiny finish; a layer of flux on the surface of the solder in the pot inhibits dross formation.

Argus International offers the HAL5000 system, which boats a production rate of 200 panels per hour. The system offers variable conveyor speeds, self-metering of fluids and a continuous monitoring of fluid levels for added effectiveness. Requiring only one operator, this hot air leveler will accept panels either manually or automatically stacked. The complete Argus system consists of the load stage; a pre-clean module; the flux, coat and pre-heat stage; the hot-air level stage; a cooling, post-clean and dry stage; and the end stacker. The board actually remains in the horizontal position until it is coated and leveled, during which time it is oriented vertically. It then returns to the horizontal position. The HAL5000 can process additive, semiadditive, MultiWire, thin foil and flexible PWBs.

The Quicksilver hot-air solder leveler is being offered by Machine Automation Inc. Pre-treatment and post-treatment operations are carried on in a modular-style conveyorized system which has been specially developed for the Quicksilver; a throughput of 360 boards per hour is claimed. Pre-fluxed panels are located on the insertion arm by means of a two-pin fixturing system. The controls for cycle start, insertion and withdrawal speed, air-knife pressures, and emergency return are all positioned so the operator has convenient access to them. Insertion of the boards into the solder pot is effected by means of a rodless air cylinder; the length of travel can be adjusted to panel size in order to eliminate wasted motion and air consumption. A patented feature of the Quicksilver is a system which pumps a proportion of the solder through vertical ducts in the sides of the solder pot, thereby creating a liquid cushion between the PCB and the sides of the pot during immersion.

Several European machines are now available on the US market, and another is soon to make its appearance: Christopher Group, an importer of overseas technology, will be importing the Voss machine from Germany. European printed circuit fabrication operations tend to be captive shops and automatic systems are popular.

The Swiss firm Caratsch, whose products are being marketed in the USA by Solder

Station One Inc., offers a number of machines. Some are fully automatic models, some are semi-automatic. With their Type 1, a rate of 150 pieces per hour is possible with manual insertion and removal. This unit solders boards which have been already fluxed and activated. The Type 2 is a semi-automatic unit which can do 80 pieces per hour. Feed-in and take-off are again manual, but the sequence is automatic through the flux station, the drier, and the leveler. The Type 3 is a fully automatic, microprocessor-controlled conveyorized machine which does 160 pieces per hour.

A German machine marketed in the USA by Plus Methods Corp. is the Schmid. The board is moved horizontally, with no change in orientation; a pre-heat zone is present to prevent thermal shock. The four air knives can be adjusted individually for pressure, angle, distance and temperature. The company says special attention has been paid in their design to easy service and cleaning of the machine. For example, the air knives can be easily tilted or removed.

The future

Paced by the growing popularity of SMOBC, the use of solder coating and leveling should increase. Argus International seems to give credence to this prospectus in that they now offer both tin/lead fusing equipment and solder-leveling systems. A spokesman for the company sums it up by noting that Argus could not ignore the market demand for a technique to yield fine lines free of solder bridging.

Warren Abbott of Hollis Automation also feels that coating and leveling has a very high potential, ideal for surface-mount technology and basically much easier to control than tin/lead electroplate. Abbott, who is chairman of the Solderable Coatings Task Group for the IPC, is currently running a round robin test program comparing boards prepared with tin/lead electroplate (both fused and un-fused), solder coating and leveling, and other protective coatings based on flux/lacquer combinations. The specimens under study will be aged for periods of 6, 12 and 18 months with solderability tested at these intervals and correlated to coating thickness. Specimens will include SMD boards.

In one sense, the solder leveler has an easier job with SMD boards since there are few or no component holes to be cleared. On the other hand, the leveler must not apply too much pressure or the pads will be stripped down to the intermetallic, with a resultant loss of long-term solderability. The intersection of SMT, selective solder coating and leveling, and solder mask (either dry film or screened-on) is bound to make for some interesting problems. For example, some workers have found that if the solder mask is too thick, surface tension makes it difficult to coat an SMD board.

Solder coating and leveling also seems to have a promising future with the additive fabrication process, as it does with MultiWire. Neither of these operations require the use of tin/lead plating, so in these cases there is no need to abandon existing plating equipment. Hollis Automation's Warren Abbott speculates that boards made from these processes might even require a more immediate protective coating than do standard subtractive process boards because electroless copper is often more porous than electroplated copper, and more subject to oxidation.

Concerning the question of the ability of solder coating and leveling to apply sufficient solder thickness, Ralph Woodgate feels that the thickness issue has perhaps been over-emphasized. He claims it is more important to make sure that the bare copper is clean to start with, and that an 'adequate' thickness is deposited. Experiments performed by Woodgate seem to indicate that variations found in the coated thickness of solder from pad to pad or hole to hole on the same board are due not to the process but to variations in solderability at these sites. This, he says, strongly relates to copper cleanliness. Woodgate adds that merely using a more aggressive flux is not the answer.

The military specification covering the permissible coating thickness as applied by solder leveling is MIL-55-110C. This standard, which is currently under review, carries minimum requirements of 0.003in. of solder at the crest (the thickest point, as measured from the copper surface) of the conductor and 0.0001in. at the crest of the plated-through hole. It is interesting to note that one PC fabrication operation manufacturing avionics equipment as a captive shop for the US Navy is using solder leveling nearly exclusively in their product and meeting this specification.

More and more, solder coating and leveling will be the rule and not the exception, as manufacturers come to see it as part of the standard way of fabricating the board and not just a special requirement of certain customers.

SCREEN PRINTING

Screen printing, a process with its roots in antiquity, has become an enormous industry involved with such widely different activities as the decoration of kitchenware and the fabrication of state-of-the-art PCB hybrid electronic circuits. Despite the fact that the need for accuracy in the electronics world is vastly greater than what the graphics business demands, there has been an unfortunate tendency for some firms to treat their electronics screen-printing process as though they were turning out coffee mugs.

Screen printing has thus suffered market share attrition from the aggressive suppliers of dry film in the area of solder mask, and has been reduced to fabrication of simple boards in the area of primary imaging. Admittedly, the later situation has come about to a great extent because of the innate line and space limitations of the technology, but there is still a great need for a heightened sense of professionalism in the area of electronics screen printing.

Fortunately, the PCB and hybrid screening industries appear at last to be embracing the level of sophistication that the work demands. The power of the microprocessor is being applied to help control the many variables in the process. Always obsessed with production throughput, amnufacturers are now making use of automation and the fast-curing ink chemistries it demands. A concern for process cleanliness is just beginning to burgeon in the industry, too. The electronics screen printing industry is finally beginning to take its role seriously.

First published in *Electronic Packaging & Production*, December 1984, under the title 'Screen printing evolves from an art to a science'.
Author: Robert Keeler, Assocate Editor.
© 1984 Reed Publishing (USA) Inc.

The field in general

Screen printing is carried out with a relatively simple complement of equipment. A very basic system would consist of a mesh screen, some sort of stencil bearing the image of the circuit, a holder for the screen mesh (called a chase), a squeegee apparatus, some ink, and a drying stage. A fully automated system would do loading and unloading and require no operator intervention.

According to Bob Nersesian of Tetko, the state of the art of PCB screening is about 5 mils now, though not many are doing it. Ten mils is more typical for PCBs, while the thick-film hybrid industry is doing 0.002in. lines and 0.003in. spaces, benefiting from the control afforded by working on smaller substrates. Nersesian feels that the question is not what kind of lines and spaces can be achieved, but rather what kind of resolution can be held.

John Stewart, president of The Screen Room, a private consulting firm, claims that the frontier in screen printing lies in improving the screening environment instead of improving materials and machines. Stewart points out that job shops, which outnumber captive shops about 20 to 1, are particularly apt to lose control of their process because they do careless things when they hurry – for example, leaving doors open between the screen-making room and dirty areas like plating or routing. Temperature and humidity are other factors needing control that are often forgotten in the screen room. Advance Process Supply says that most problems come from inadequate screen preparation, rather than problems with the machine.

Universal Instrument Corp.'s Model 4114A Large Area machine for screen printing solder paste on a PCB

Screen printing is now battling with dry film for supremacy in the area of solder-mask application. The struggle is intensified by the strong marketing efforts of dry-film suppliers like Dynachem, with their Laminar line, and DuPont, with their Vacrel. Dynachem, which supplies both film and screened-on resist, is impartial in its evaluation of the pros and cons of these two contenders.

Says Dynachem's Kerry Grimes, dry film offers a high-quality solder mask – true reproduction of the circuit image where you want it. He notes that not only large companies use it; any company moving into the leading edge of technology will be forced to use it for fine-line resolution. On the negative side, he admits that a drawback to using dry film is that air bubbles sometimes get trapped under the surface of the mask.

Screening, on the other hand, is currently much less expensive than film, costing roughly five cents per square foot versus a dollar per square foot for dry film. The lower cost is due in part to the fact that screening can be applied with less waste, and also because there

is no throwaway cover sheet to pay for. However, smeared liquid mask sometimes drys on circuit pads, covering them (see sidebar).

An occasional problem with UV-curable screened solder mask is poor adhesion (a difficulty sometimes attributed to dry film, too). LeaRonal is offering their Rona Screen 1600 GFB Series to address this difficulty. They claim that they can even screen it over large ground areas of reflowed solder and still get good adhesion. It can furthermore be used, says LeaRonal's Dick Kessler, through the Gyrex process and hot-air leveling.

Inks and innovation

The most innovative development in the area of screen inks in the past several years has been the advent of UV curables, which involve the actual cross-linking of the ink molecules during drying, as opposed to simple evaporation of a solvent. UVs dry extremely fast, and therefore lend themselves naturally to use with automated systems. Even in conveyorized IR drying systems, the speeds

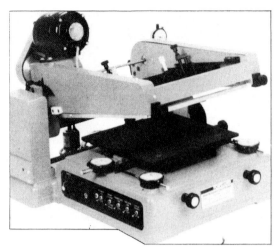

Crystal Mark's Semi-Automatic Circuit Printer

involved are low compared to UV systems. UV curables are premixed, in contrast to some two part epoxy-based inks which have a definite, and sometimes unforgiving, pot life.

UVs require less energy to cure than IR or convection-dryed inks and they permit the use of a finer mesh because they do not clog the screen as much. This ultimately means that less ink is consumed. They cut down on 'bleeding', (ink going where it is not wanted), and because they contain no solvents, there is no danger to the worker or the environment.

MacDermid Corp. is one example of a company marketing a fast cure UV-curable solder mask. MACu Mask is MacDermid's top-of-the-line UV-curable solder mask. They claim it has good adhesion to both copper and tin/lead, and a very fast cure rate.

Training 'n' Technology is a firm of technical consultants which offers both business plans and technical assistance to people getting into the PCB business. Company president Victoria Allies observes that many of her customers resort to thermal-cure inks when they cannot control surface cleanliness. They feel that UVs are not as forgiving in that department when good adhesion is at stake.

John Jandrey of the Hysol Division of Dexter Corp., which supplies both UV and thermal-cure inks, points out another drawback. According to him, although UV inks are a major breakthrough and have a great future, they still do not offer the end performance of currently available heat-

curable inks. Thus, people building high-quality screened boards for the military and for computer products often still rely on heat-curable ink.

To capitalize on this market, Hysol has formulated an epoxy-based ink with a faster cure time than was previously available with heat curables. Their SR 2020 is a two-component system that can be baked in an oven in as little as 15 minutes, versus 30 minutes to an hour for most existing epoxies. It can also be IR dried on a conveyor in a matter of seconds – thus putting it, as Jandrey points out, in the range of cure times for UVs.

Colonial Printing Ink Co. has a two part thermal-cure epoxy solder mask, their ER-7040, for which they claim outstanding electrical properties. It also offers high resistance to bleeding, a problem that can adversely affect later processing steps.

In the area of UV inks, Enthone Inc. claims to have made a useful innovation. Enthone's Gary Weidner notes that most companies offer one resist for acid etch and a separate one for alkaline etch. This is motivated by the relatively high pH level of the alkaline etchant (somewhere around eight or nine). If the resist is itself alkaline-strippable at a pH level of about 12, there is only a narrow margin of safety before the resist begins to be stripped by the etchant.

To solve this problem, Hysol is offering a UV curable primary-image resist called 570AD that is universal in the sense that it can be used in both plating and etching applications. This alkaline-strippable resist can be used successfully with the two most popular etchants, cupric chloride and ammoniacal. Furthermore, when stripped it merely dissolves, as opposed to breaking up into stringers as do most such resists.

A spokesman for Thick Film Systems, a division of Ferro Corp., remarks that the vehicle present in a paste is an important factor that is often overlooked. Their Partilok organic screening vehicle formula, present in most of their thick-film paste materials, prevents settling of particles so that paste stirring is not required before use. TFS's pastes are optionally available in prepackaged, ready-to-use syringes.

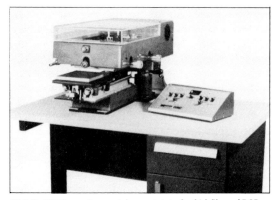

Weltek's 68MC console-mounted screen printer for thick film and PCB production

Though screening standard PCBs and hybrid circuits presents different problems, polymer thick film is one area related to screening inks where these two domains are coming closer together. Polymer thick-film circuitry on ceramic substrates, is now being applied to standard PCB substrates. Polymer thick films will help to usher in surface-mount technology, with its ensuing space savings, speed and lower cost.

Wayne Martin of Emca Inc. points out that like UV curables, polymer thick film involves curing rather than just drying. It can serve on flexible as well as rigid substrates, and sees applications in digital displays as well as membrane keyboards. Typical polymers include phenolics for resistors, urethanes for flexible substrate applications, and epoxies.

Compared to standard thick-film work, polymer thick films involve the use of a considerably larger printing screen. For example, entire TV circuits can be screened with this material, leading to economies of scale. However, because of the larger area to be screened, and because the printability of the polymer material tends to vary more widely than say, cermet, the user is faced with having to achieve much greater control of the screening process.

In addition, even though polymers are being used to deposit the same functions as standard thick films (resistors, conductors and insulators) they show different flow characteristics than standard thick films, and demand both a different squeegee design and speed of application.

Machines and screens

Perhaps the most exciting thing to happen to screen-printing machines in recent years has been the trend to add computer or microprocessor control. Advance Process Supply markets a microprocessor machine called the Viking. Tetko offers a machine called the H-41.

Says Tetko's Bob Nersesian, the use of a microprocessor means that screen printing is finally growing up. The whole key to a successful future in screening, he adds, is to automate. He points to the fact that in screen printing the process parameters are tied together in a complex relationship, with screen tension affecting off-contact, which can affect squeegee pressure, which in turn relates to elongation, which impacts on registration. The microprocessor, he says, can help to pull it all together.

AMI Corp. believes in the idea of computer control of the screen printer. Their machine, the Model 1505, which has three microprocessors offering closed-loop control over various parameters like squeegee pressure,

Chemcut's Model 704 Screen Printer

squeegee speed, and programmable setup, can also act as the host of a network, for controlling loading, unloading and drying. It can be programmed on a tape cassette, with the cassette being reusable later. It has both position and speed sensing capability, so that any perturbations having to do with screen tension or unevenness of material on the screen will be smoothed out.

An interesting machine for small volume and prototype work is Micrajust's manually operated 2000UA. On this machine, it is possible to raise the chase as much as 8.5in. and thus print cabinets and thick panels. Micrajust claims that they are unique in this capability, with other machines being able to open up to 2in. at most. The 2000UA also offers the availability of dial indicators to facilitate X-Y positioning. For ergonomics, a counterbalance mechanism helps to lessen worker fatigue.

Crystal Mark is another manufacturer of small screen printers. This firm contends that electromechanical printers are substantially cleaner machines than hydraulic or pneumatic models. They point to the tendency of any actuator-driven machine to leak oil onto the floor, and in some cases to vent it into the air.

Crystal Mark offers five sizes of electromechanical machine under the Forslund trade name, none of which are automatic or

Aremco's Accu-Coat 3230 wide area screen printer

intended for high-volume production. Nevertheless, these machines, says Crystal's Jawn Swan, are ideal for people who want to do their homework in preparation for getting into SMT. Prototype and low-volume production work is a natural for these machines, too, says Swan.

Autoroll offers their A5100 high-speed printer, again an electromechanical machine. A cylinder printer, the drive, and the screen and the squeegee are all totally mechanically synchronized.

Universal Instruments also agrees to the premise that electromechanical is cleaner, and they have based their machines on that technology. Universal's Alan Cundy points out that oil vapor put into the atmosphere can ultimately cause a problem in the firing furnace. Electromechanical screeners are also quieter, he says, and since they are simpler in design, they are less apt to break down. Yet another advantage cited by Cundy is that it is possible to do more with electronic controls and software on these machines than with pneumatic or hydraulic models, especially apropos of microprocessor software. Repeatability is improved too, he claims, because the electric motor can supply enough torque to smooth any fluctuations in the power supply.

Weltek is a company that has chosen for its niche the desk top and console screen printer market, although they also offer an automated screen printing and parts handling system. Gary Petersen, president of Weltek, sees an extremely strong trend now toward SMD, driven by the demand for greater packaging density and finer lines, as well as an interest in multiple image printing to achieve economies of scale. The firm is offering their Model 68 MC screen printer with microprocessor control. It addresses both the hybrid thick-film device market and the emerging surface-mount technology area. The squeegee here is driven by a 'threadless' lead screw that delivers constant speed and precise repeatability.

The Electron C-240 screen printer from Cugher Inc. offers the option of being convertible to fully automatic operation. It possesses a unique new design feature in its expandable registration pins. Rather than force panels down over pins that are slightly different than the diameter of the pilot holes, Cugher uses expanding pins, working much like collets. The panel drops comfortably over the pins while they are relaxed, and the pins then expand as much as 0.5mm to fill the hole. This eliminates the usual sloppy or overtight condition that exists when a solid pin is used in conjunction with a drilled or punched pilot hole. Cugher says their expanding pin concept improves both speed of operation and positive repeatable registration.

In the area of screens, IPM/ARGON Corp., jointly owned by Argon Industries in Milan, Italy, and the Dutch conglomerate, Stork, will soon be introducing what promises to be a very significant new concept. This firm, which currently markets a fully automated screen press called the Astromat and a machine for upscale SMD format boards which is known as the Hydra, has developed a non-woven pure nickel mesh, called Himesh, for which they claim a resolution of 0.004in. lines and spaces. The screen, which is not an etched mask, is

made by a proprietary additive plating process. It features hexagonal-shaped holes, both for strength and for higher resolution.

John Moore of IPM notes that a conventional mesh has to be precisely tensioned and each wire or thread (some screeners use polyester) has to be tensioned individually. One wire out of tension will distort the screen. Moore explains that since Himesh is a metal plate, it is quite difficult to distort, and does not have to be stretched.

A key feature of Himesh seems to be that the thickness of the mesh plate and the percentage of open area can be varied independently of each other. By contrast, in a conventional screen, the wire diameter controls the thickness of deposit. With Himesh, the thickness of the deposit ultimately depends on the amount that the screen was plated up.

In surface-mount technology, it is necessary to lay down 0.006–0.008in. of solder paste with a coarse mesh screen – up to a No. 80. This represents the largest wire diameter available in stainless steel, notes Moore. Yet,

AMI's Model 1505 intelligent screen printer

for the percentage of open area this affords, there is also a tendency to print a sawtooth effect when emulsion is applied. Moore says the Himesh will avoid this because its surface is smooth, while still being able to retain sufficient amounts of emulsion.

Advance Process Supply, meanwhile, has recently introduced a direct screen emulsion called Zero Plus, which they claim can offer the user a circuit image accuracy of .0060.00375in.

SMT

Screen printing plays an important role in surface-mount technology in the area of solder-paste deposition. According to Cesar Frustraci of Chemcut, the first goal of a solder-paste printing machine is to be able to print on-target repeatedly on surfaces of up to 18 × 24in., while holding accuracies of ±0.001in. Next, it should print the whole surface witha consistently equal thickness of ink deposit, adjustable in thickness up to 250m.

Solder paste, being thick, viscous and abrasive, tends to wear down the emulsion on a screen. Any screen through which paste is deposited must have sufficient open area to let it through, and for this reason there has been a trend to use a stencil instead of a screen for solder-paste work.

When a stencil is being made, as the acid starts to etch downward into the metal plate, it also etches outward. On a fine-pitch stencil, such as one for a product with leadless chip carriers, the acid might actually break through the thin bridge separating stencil holes.

Universal Instruments has addressed this problem with a photographic compensation technique that ultimately results in smaller pads. Universal's Alan Cundy hastens to point out, however, that further work will be necessary to meet the anticipated resolution required in the future.

Aremco is now in the process of developing a machine that automates the solder-paste deposition process. In addition, on their Model 3245, they have devised a way of controlling the planarity of the screen surface.

To do this, they have made the upper head of their machine slightly adjustable by letting it rest on 3m stops. This allows the user working with larger screens to tune the snap-off between the bottom of the screen surface and the workpiece. Thus, it becomes irrelevant if the screen is a little warped.

According to Aremco's Dick Weir, this approach will reduce the role of the squeegee to that of merely wiping off the top of the stencil; each new setup can be conveniently altered while preserving a uniform gap from corner to corner of the screen.

Chemcut Corp. offers its Model 704 screen printer for high-volume printing of solder paste on SMT boards, although it can apply resist, solder mask and legend ink as well. For solder paste, it features 0.006in. line and space reproduction. Though the machine is not fully automatic, Chemcut's Frustaci points out that for this kind of precision, no one has yet introduced a fully automatic machine.

The 704 is described by Chemcut as a pin-registered 'three-quarters automatic' machine. The operator loads the board on the printer top, but from that point on, the entire process is automated, including the takeoff and evacuation of the board from the printing area into the next production step. Chemcut observes that all existing fully automated machines today use edge registration, as opposed to using tooling pins. Tooling pins provide a much higher level of repeatability, says Chemcut. In addition, the firm claims large panel surface area (18 × 24in.) printing capability of 0.008–0.010in. accuracy.

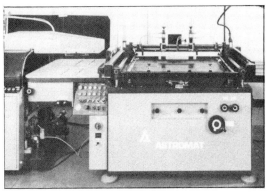

IPM/Argon's Astronaut automatic screen printer

Hybrid thick film

Hybrids require superior three-dimensional control, since a variation in the thickness of the resistors, conductors and dielectrics laid down can alter the values drastically. Another problem is proper registration of thick-film layers, especially for multilayer hybrids.

Thick-film printers are usually smaller and more precise than PCB printers. In many cases, the manufacturer wants to deposit components in layers (up to 100 in some cases) so both accuracy and registration are important here.

AMI is offering a new thick-film printing machine which they claim is large while still being very precise. They have added a machine-vision system to their microprocessor controlled thick-film printer so it can monitor itself and ensure that the position of the print stays where it is supposed to be. The system also features a thickness-sensing capability – in this case a non-contact system since the emulsion is still wet when it leaves the screen-on stage. If the system senses a problem, it signals a change in the snap-off, squeegee pressure, or other parameter.

Stencils are not much used now for standard inks in the manufacture of hybrids, says Universal's Bob Jillings, since there are enough fine wire meshes on the market to easily get down to 0.005in. lines and spaces. Jillings notes that some new meshes coming on the market will probably allow a resolution of 0.003in. lines and spaces.

Universal is also looking at an automatic thickness sensing capability for their thick-film printer, the Model 2000. In addition, they have just developed a system to solve another specific problem associated with thick-film production. Currently, on large area substrates ($>16in^2$), it is not possible to fire a thick-film circuit at the same production speeds at which screen printing can be performed. This step thus becomes a bottleneck. Universal has designed an IR thick-film firing furnace, the Model 4311, that offers a higher capacity of throughput – one which matches the output of a printer processing substrates larger than $16in.^2$ It should be of interest to surface-mount component suppliers.

CLEANING SOLDERED ASSEMBLIES

Fluxes which facilitate the soldering process must be removed after soldering to ensure their corrosive action does not damage the assembled circuits. Present day density of insertion makes such flux removal difficult. Environmental concerns about atmospheric contamination and waste discharge must be considered.

SOLVENT FLUSHING

Proper post-solder cleaning of printed circuit boards is a very important step in the assembly process because it affects the ultimate reliability of the board, says Greg Unruh, director of research at Tech Spray, a supplier of electronics production chemicals. It is important to weigh the differences in cost of various cleaning systems and future electrical reliability of the printed circuit board.

Reliability is why soldered circuit assemblies are cleaned. Solder flux removal describes the cleaning process, but it is reliability of the printed circuit board or hybrid that is the reason for cleaning.

Contaminants

A common problem with solvent cleaning of printed circuit boards is the assemblers' lack of familiarity with the type of solder flux contaminants to be removed, notes Unruh. Contaminants or residues will be polar, non-polar, or somewhere in between with respect to polarity. The rule of thumb, 'like dissolves like', will hold true in every case, says Unruh. If the contamination were predominately non-polar, the user would want to use a non-polar solvent for removal, such as a chlorocarbon or fluorocarbon. On the other hand, if the contaminants were very polar in nature, the user might consider deionized water or deionized water and low molecular weight alcohol mixtures. Since contamination composition is always mixtures of polar and non-polar materials, solvent blends or azeotropes are the method of choice to assure complete removal of contaminants on PCBs, concludes Unruh.

Ionic contaminants from flux activators, unless removed, can lead to electrical leakage and dielectric breakdown in the presence of humidity. The importance of removing non-polar rosin is sometimes understressed, says Susan Dallessandro, research engineer in Dow Chemical's Inorganic Chemicals Department. Ionic contaminants may be encapsulated in rosin left on the board, and leak out

First published in *Electronic Packaging & Production*, August 1984, under the title 'Solvents flush solder flux from tight spaces'.
Author: Ronald Pound, Associate Editor.
© 1984 Reed Publishing (USA) Inc.

over a period of time. Rosin left on a PCB can also cause problems with adhesion of conformal coatings.

Choosing solvent cleaning

Myriad, interdependent issues enter into assessment of cleaning system alternatives. The influence of solder flux activity, environmental and occupational safety and health regulations, and the ability of the cleaning agent to penetrate a small space between component and board are important considerations in assessing one of the most important alternatives – solvent versus aqueous cleaning.

Highly active, organic acid (OA) water-soluble flux consists of low molecular weight organic acids with activators. This flux can be removed with water or water and alcohol mixtures. Rosin fluxes, consisting of rosin acids and activators, may be cleaned with chlorinated or fluorinated solvents. Designated by the amount of activator present, these fluxes are: rosin (R, rosin mildly activated (RMA), rosin activated (RA), and rosin superactivated (RSA).

Synthetic activated (SA) flux, introduced two years ago by DuPont, is a factor that may influence the choice between solvent and aqueous cleaning. Consisting of alkyl acid phosphates and activators, SA flux provides the same fluxing action and oxide removal that water-soluble OA flux provides, says DuPont. The SA flux can, however, be removed with fluorinated or chlorinated solvent cleaning systems. This means that if aqueous cleaning is being considered to get better solderability from stronger OA flux, there is now an alternative available with SA flux and solvent cleaning.

The industry started with rosin flux; then it began to require more active flux because of loss of component solderability, says Dow Chemical's Dallessandro. Highly activated water-soluble flux was developed, and recently SA flux appeared, with the activity of OA flux. For military applications, the view is to control solderability of components, rather than using a corrosive activated flux in the OA

Contaminant		Solvent	
Category	Material	Category	Material
Ionic (polar)	Activators	Polar	Alcohols Water
Non-ionic (non-polar)	Rosin	Non-polar	1,1,1-trichloroethane Fluorocarbon 113

Contaminants and corresponding solvents for cleaning soldered circuit assemblies

category. The trend to more highly activated fluxes for commercial applications, however, and water-soluble fluxes in particular, has resulted in increased use of aqueous defluxing, says John Neidhart, sales vice president at Crest Ultrasonics.

Military tests

Solvent cleaning of solder flux, and its effectiveness weighed in relation to that of aqueous cleaning of rosin fluxes, has been studied at the Naval Weapons Center, China Lake, CA. The Weapons Center concludes in its report of the study findings that the best methods for cleaning rosin fluxes are vapor degreasing using solvents containing a high percentage of alcohol, and combination cleaning using vapor degreasing followed by a deionized water rinse. Under the conditions and procedures of their study, it was also revealed that water alone cannot adequately remove flux, and that the working life of an added detergent must be closely monitored to assure flux removal. Even with detergent cleaning, RA fluxes could not be cleaned to an acceptable level. It is recommended that non-rosin, highly active, water-soluble flux not be used for military applications.

Testing of four fluorinated solvents in the China Lake study showed DuPont's Freon TMS cleaned the best for this category of solvents. Containing methanol, a more polar constituent than the ethanol in the other solvents, the Freon TMS removed more of the ionic contamination than did other fluorinated solvents tested.

Dow Chemical's Prelete, containing mixed alcohols, was the only chlorinated solvent among the three tested that removed sufficient ionic contamination from RA fluxes. It removed ionic contamination as well as Freon tMS, the report said.

The other fluorinated and chlorinated solvents tested cleaned RMA fluxes well, but did not remove sufficient ionic contamination from RA fluxes.

DuPont attributes the effectiveness of their fluorocarbon solvents to low surface tension, low viscosity, and high density which allows all parts of an assembly to be thoroughly wetted by the solvent, and to the bifunctional action of the fluorocarbon and alcohol additive which removes both non-ionic and ionic contaminants. Fluorinated solvents, having a low boiling point and low latent heat of vaporization, require low energy input for use in a vapor defluxer and, during vapor rinsing, large volumes of condensate are formed.

The base chemical for Prelete, 1,1,1-trichloroethane, is an excellent solvent for rosin, and the alcohol additive has been optimized for effective ionic removal, says Dow Chemical. The condensation rate on the PCB, determined by the temperature differential between the solvent's boiling point and the board, provides an effective flushing action. Because of the increased solvency of a chlorinated solvent, Dow recommends tests to determine compatibility with plastics.

Other solvents for solder-flux cleaning include multicore PC81, GAF M-Pyrol, and Allied Chemical Genesolv. Multicore's PC81 has a higher boiling point than fluorinated solvents, but lower than chlorinated solvents. As a result, says Multicore, this solvent will perform its vapor-cleaning function longer, and the solvent evaporation rate at room temperature is lower than that of fluorinated solvents.

Regulatory issues

In recent years the trend has been to aqueous

Contaminant		
Category	Properties	Effect on printed circuit board
Ionic	Conductive Corrosive	Degrades electrical properties with electrical leakage and dielectric breakdown. Attacks circuit and component materials.
Non-ionic	Visible Insulating	Degrades cosmetics. Encapsulates ionics which escape over time. Prevents electrical contact; for instance connectors and bed-of-nails test probes. Prevents adhesion of conformal coatings.

Contaminants' possible effects on PCBs

cleaning for reasons of economy and OSHA and EPA regulations, says Hollis Engineering. Now, companies who adopted, or are investigating, aqueous cleaning systems are finding that they are not simple, cheap, nor problem-free, remarks DuPont.

While detergents improve contaminant removal initially in an aqueous system, they are expensive to monitor and require frequent additions to keep the proper concentration level. Waste disposal of contaminants removed in cleaning is also becoming a problem where the waste water cannot be simply discharged into a municipal sewer system, notes DuPont. Solvent cleaning, on the other hand, lessens the severity of some waste disposal problems, claims Neidhart of Crest Ultrasonics. It also eliminates the need for complex rinsing and drying, and in many cases is more energy efficient than aqueous cleaning.

Solvents, of course, also pose questions regarding environmental regulations as well as worker safety. All used solvents are classified as hazardous waste by the EPA and can be shipped only when in compliance with Department of Transportation shipping regulations. Individual state regulations may apply to both fluorinated and chlorinated solvents.

Fluorosolvents have relatively high allow-

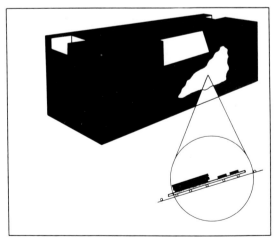

Cleaning equipment configurations that produce enhanced mechanical fluid forces such as Detrex Chemical Industries' inclined positioning of PCBs in the spray zone of an in-line defluxer, combine with properties of solvents, such as low-surface tension, to remove solder flux contaminants from dense circuitry

able vapor concentrations in the workplace, says DuPont. The threshold limit value (TLV) for fluorosolvents is sufficiently high that normal room ventilation is all that is required to keep vapor concentrations within allowable working limits. The odor threshold is at a relatively high level so that employees working in the vicinity of a vapor defluxer are not annoyed or affected by solvent odor.

People are concerned with toxicity, and ask about the TLVs, says Dow environmental specialist Stan Dombrowski. The TLV, however is not a measure of toxicity. It is a guideline for use in control of exposure to vapors, explains Dow. For example, one material may have a higher TLV, and thus may be incorrectly assumed to be less hazardous than another. If the higher-TLV material evaporates faster, however, exposure to a potentially hazardous vapor level could be greater than that of a less volatile, lower-TLV material.

Surface-mount defluxing

With pad spacing on closer centers, now around 0.050in. and likely to get closer, and spacing of components off the board around 0.003in., surface-mount printed circuit assemblies present tough cleaning problems.

Hybrids, which employ surface-mount components, are being successfully cleaned. They do not, however, present the same type of cleaning problems as surface-mounted devices on PCBs, says Don Gerard, manager of application engineering at Detrex Chemical Industries. On a ceramic substrate and without large plastic chip packages, aggressive solvents can be used on hybrids without materials compatibility problems. With surface-mount PCBs, chip carriers 1in.^{2} or larger were seen for the first time right down on the board, observes Gerard. Manufacturers coming to Detrex had selected less active RMA solder flux because they thought flux could not be adequately removed from surface-mount PCBs.

Aqueous cleaning, even with detergent to lower surface tension, may be limited in reaching contaminants trapped in the tight space between board and component. Detergents lower the surface tension and allow

Surface-mounted assemblies are cleaned in an ultrasonic vapor degreaser. Staodynamics, a manufacturer of medical electronics, exposes its assemblies to chlorinated solvent in the vapor zone for 15 seconds; then immerses them in an ultrasonic bath for 15 seconds (Courtesy of Dow Chemical)

penetration to areas that pure water cannot reach. Rinsing contamination and detergent out of entrapment areas is particularly difficult since rinse water does not have the benefit of detergent to lower the surface tension, says DuPont.

Research in the industry indicates that solvent can penetrate well below 0.006in. spacing, while the lower limit for detergent systems is 0.010in., says Dallessandro of Dow. It is difficult to measure cleanliness with surface-mount devices, and this creates a problem in determining the limit of solvent wetting. A solution of isopropyl alcohol and water, intended to flush out contaminants and measure their presence by the change in conductivity of the solution, will not flow where solvent cannot reach.

Not only is the tight space a problem, but the rosin left on a surface-mount assembly is more stable and less soluble than that on a through hole, leaded-component PCB. Exposure to a more direct heat source for a longer

period, whether vapor phase, IR reflow, or wave solder, causes this difference, according to Dallessandro. The ionic contaminants left on a surface-mount assembly are similar to those on a leaded-component board, with only the rosin residue differing.

Lower surface tension may not be the entire key for a solvent to penetrate and clean in small spaces. The heat associated with higher boiling point, chlorinated solvents may give them some help in contaminant removal from tight 0.003in. spaces, compared to lower surface tension, and lower boiling point, fluorinated solvents, notes Detrex's Gerard.

Wave soldering is a viable method of soldering surface-mount components to the board, says Hollis Engineering. Adhesive holds components in place as the board goes through wave soldering with the component side down. This adhesive also prevents flux residues from penetrating beneath the component body, making removal less of a problem. Hollis' preliminary tests, using leaded chip carriers, SOICs, and chip components assembled on a test board with adhesive and then wave soldered, have shown that there is no significant difference in cleaning whether aqueous or solvent methods are used.

Ultrasonics

Just immersion in vapor or liquid is not enough; ultrasonics enhance the ability of solvent to get in and clean surface-mount assemblies, says Don Vidoli, assistant product manager at Branson Cleaning Equipment. Ultrasonic cleaning poses concern in the industry, and military specifications prohibit it for PCB cleaning. There are ways, however, to use ultrasonics without damaging components, remarks Vidoli.

The temperature of the cleaning liquid can be adjusted to affect the intensity of cavitation. There is a bell-shaped curve when one plots cumulative intensity as a function of temperature, explains Vidoli. At the cooler end of the temperature scale, it takes a lot of ultrasonic power to create cavitation because much of the power goes into heating the fluid. Thus, the cumulative intensity is low because little cavitation activity is generated. As the temperature is increased, however, the intensity increases until a maximum is reached. Further increases in temperature decreases the intensity, although there is a lot of cavitation activity. When the boiling point is reached, intensity drops sharply as cavitation ceases.

It is on the downward sloping portion of the curve that Vidoli talks about operating ultrasonic cleaning equipment, rather than at the maximum intensity. Operating at a higher temperature exposes the parts to less intense but active, cavitation and the chemical cleaning action is enhanced, says Vidoli.

Trends in packaging, such as narrowing line widths as well as surface-mounted components, are establishing the requirement for more elaborate and effective cleaning techniques, says Neidhart of Crest Ultrasonics. Simultaneous multifrequency ultrasonic energy leads to more effective solvent cleaning, he claims.

Low-frequency transducers generate large, but relatively few, cavitation bubbles with high cleaning power in the cleaning fluid. High frequencies generate a large number of

Ultrasonic vapor degreaser from Branson Cleaning Equipment is portable for convenient use anywhere in a manufacturing facility

smaller cavitation bubbles, with less intensity, but with the ability to penetrate small spaces. The use of a simultaneous multifrequency approach provides penetrating capability and high cleaning power at the same time, and eliminates the problem of selecting the correct frequency, says Crest. In their approach the dominant frequency is 40kHz, but resonances are produced in the 50–90kHz range.

Cleaning process

Because of closer spacing, a little contamination can do more damage than it did formerly when line spacing and clearances were larger, says DuPont. Circuits with smaller dimensions cannot stand many particles, and there is more emphasis on thorough cleaning, says Branson's Vidoli. As degreasing tolerances get tighter, there will be more emphasis on monitoring and controlling the cleaning machine with microprocessors. Detrex is now offering an RS-232C interface module which enables an in-line defluxer to be monitored by a host computer.

Cleaning of dense surface-mount boards is aided by positioning them on an angle in the last spray zone of an in-line defluxer, notes Detrex's Gerard. An incline of 10°–20° from the horizontal will result in hydraulic action from gravitational forces, helping to force solvent under the components.

Detrex has built a few special machines for cleaning surface-mount assemblies that are vertically oriented in the defluxer. However, handling is generally a problem with a vertically-racked, in-line degreaser following a wave-soldering machine where the boards come out horizontal. The preference is to stay with a belt and keep the assemblies flat, says Gerard.

Mixed-technology boards with surface-mounted components on one side and leaded components on the other side have been satisfactorily cleaned in an all-spray design, says Wayne Mouser of Detrex. These boards are normally wave soldered in a dual-wave machine with RMA or RA flux. Where specifications permit less than complete flux removal, sprays without immersion are normally adequate, notes Mouser. Spraying from both sides of the board, Detrex increases the pressure on the surface-mount side. High-pressure spraying, which takes advantage of impingement force, is more effective than just immersion alone for surface-mount assemblies, says Detrex.

Some design engineers are looking at the possibility of putting conventional leaded DIPs over tiny surface-mounted components. Once the DIP is inserted, it will shield the surface-mounted devices from spray. Immersion is likely to be required to get to the surface-mount components, says Mouser.

When surface-mount assemblies require immersion in the cleaning cycle, a spray-over-immersion process creates a flushing action which aids cleaning. Assemblies are immersed about 1in. deep in the solvent, and a spray impinges on top of the solvent to create mechanical agitation. Although not quite as effective as ultrasonics, this technique can alleviate users' fears of component damage.

Equipment operation

Frequently errors are made in adapting existing cleaning equipment that is improper-

Microprocessor-controlled vapor degreaser hoist from Unique Industries is programmed with a keyboard and thumb-wheel switches to set the parameters of each dwell cycle

ly designed or sized for the intended use, or in purchasing an inadequate machine in the first place, says DuPont. High solvent consumption can result from inadequate free-board, insufficient cooling, or excessive leaks. Overloading of the machine with contamination can occur with inadequate capacity or poor maintenance. An improper cleaning cycle, where important steps are omitted or altered by cleaning personnel, may lead to cleaning performance below what should be attainable with the equipment and solvent used.

Cyclo-Tronics claims to offer in their vapor degreaser the deepest freeboard in the industry at 125% of the working width of the machine. The degreaser is designed with the immersion cleaning tank suspended in an enclosed boiling tank which is the vapor generator. Solvent losses are reduced 50%, and the cleaning area is enlarged, says Cyclo-Tronics.

A microprocessor-controlled vapor degreaser hoist is also designed to reduce solvent use up to 50%, says Unique Industries. Their AutoArm unit moves baskets of assemblies into and from tanks of a batch degreaser at the OSHA-approved rate of 11ft/min, thus reducing loss of solvents through dragout. Speeds closer to 100ft/min were observed with manual operation, says Art Gillman, president of Unique Industries. The resulting solvent dragout causes uncomfortable conditions for the operator as well as excessive solvent use.

Solvent cleaning in high-volume applications is more oriented toward the use of in-line defluxers which attach directly to the output of a wave-soldering machine. This provides immediate cleaning of the freshly soldered boards, and they are cool, dry and ready to be tested only minutes after being soldered, remarks DuPont.

Baron-Blakeslee's CBC-18 in-line defluxer is designed for cleaning boards up to 18 × 18in. with operating speeds up to 10ft/min. The unit uses a five-stage cleaning cycle with the initial vapor stage followed by three spray stages in individually filtered and isolated spray zones, and a final vapor stage.

Using a closed-loop system for heating and cooling results in up to a 50% reduction in

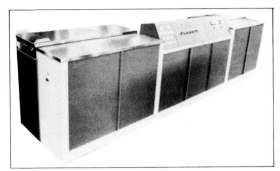

In-line solvent defluxer from Baron-Blakeslee uses a five-stage cleaning process

energy consumption of the model PCBD in-line defluxer compared to standard electric-heated designs, claims Detrex Chemical Industries. This system utilizes its refrigeration compressor's hot refrigerant gas to boil and vaporize the defluxer's fluorinated solvent. The refrigerant then cools the solvent vapor that is not cooled by the assemblies being cleaned, until it condenses back to a liquid.

Solvent and aqueous cleaning can be addressed in a complete process, and not as adversaries, says Unique's Gillman. Unique supplies a combination solvent and aqueous in-line cleaning system for rosin or SA flux users. Used by military contractors who must meet the ionic cleanliness specifications of MIL-P-28809, multistage solvent cleaning is followed by closed-loop, deionized aqueous-rinse stages.

ROSIN SOLDER FLUX RESIDUES

More and more, the ultimate reliability of a printed circuit board is considered to depend on its cleanliness level. The three main reasons for this are the increased use of faster soldering speeds which demand fluxes with higher activity levels, developments in PCB technology which permit higher density conductor lines and spacings needed for higher frequencies and opeating levels, and the assembly of surface-mounted components, introducing extra problems with regard to flux entrapment between the board surface and components.

The assumption that board reliability is solely a function of a detected level of cleanliness excludes other relevant parameters. One of these parameters for solder flux is the potential of resins to encapsulate the activators in their molecular matrix, resulting in an inert residue once the resin is hard. Rosins can even be modified with that particular purpose in mind. This is why some fluxes with a relatively high level of halide activators are able to pass corrosion tests designed to evaluate fluxes with a significantly lower halide content.

It is wrong to assume that the previously mentioned ability of rosin is only relevant for fluxes used in cases where cleaning is omitted. Obtaining complete cleanliness is virtually impossible in the normal day-to-day handling and assembly of PCBs, in particular with the fast-growing SMT miniaturization of circuits.

These factors all lead to the conclusion that desired level of cleanliness can be defined by its reciprocal: the acceptable level of dirtiness or the amount of remaining residues or contaminants. Henceforth, the word 'residue' will be used when discussing the level of dirtiness.

The permissible amount and categories of residue are given in documents from government agencies, such as military specifications. The value of any of these specifications hinges on the reliability of the tools and procedures used for process evaluation and quality control.

Flux residues

In order to understand the true contribution of solder fluxes to the residues left on the circuit board after post-solder cleaning operations, it is necessary to look for a relevant

First published in *Electronic Packaging & Production*, February 1985, under the title 'Rosin solder flux residues shape solvent cleaning requirements'.
Author: Eli Westerlaken, Cobar Europe BV.
© 1985 Reed Publishing (USA) Inc.

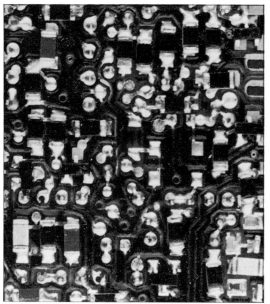

Surface-mount components on PCBs generally introduce extra problems with regard to flux entrapment between the board surface and the components. Fluxes active enough to adequately wave solder such components require sophisticated technology in order to achieve both thorough cleaning and high dielectric properties of their residues

systematic approach to classify all residues. A classification based on the stages in the production process which results in the discrimination between solder fluxes and other residues is inadequate and confusing. The most relevant contribution to the desired level of cleanliness is not caused by the solder flux residues themselves but by their reaction products which are formed under exposure to heat and by the presence of residues from other stages in the process. Some of the typical solder flux residues that might exist on a PCB are: rosin, polymerized rosin, unattered acti- vator, decomposed activator, and metal salts. These residues only represent a small portion of the residues that may be found on a board.

The term 'white residue' is sometimes used to identify a certain group of residues. The only thing they have in common, however, is their white appearance. Their probable causes, as well as their exact chemical natures, are as varied as the chemicals used in the PCB fabrication process. Solder flux residues and white residues will be discussed here as individual categories.

A logical approach to classifying residues is

to follow their basic chemical nature, since detection and dissolution is a task exclusively within the spectrum of chemistry. Even within this approach several criteria are possible. Residues may be basically soluble or insoluble. They may be organic or inorganic. Most important, however, is to know the chemical characteristics of the residues once they are left on the board. The latter classification is the one dealing with the discrimination between ionic or polar, and non-ionic or non-polar residues.

This differentiation and the one dealing with the residue's organic or inorganic nature encompass the basic questions concerning the solvent system required for the removal of these residues. The classifications non-ionic and organic, ionic and inorganic tend to correlate. Non-ionic residues principally are organic. This correlation, however, is not total.

One of the crucial distinctions is in the solvent systems: water is an ionic fluid but it is also inorganic. Alcohols too are ionic, yet they are organic. This is why water is an excellent solvent for most activators, but it is not capable of removing organic residues such as rosin unless it has been modified for that particular purpose.

Most alcohols are able to dissolve both ionic residues as well as organic residues. However, pure alcohols create safety problems and should not be used alone as a cleaning solvent. Since, in most cases, both ionics and non-ionic residues are left on a PCB, a solvent system

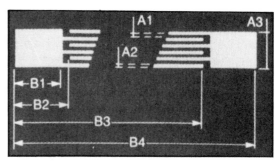

Example of typical IR test pattern providing the surface resistance in megaohms per square surface. The shape and dimensions of test patterns may vary to some extent. Each test pattern has its own multiplication factor for the calculation of the sheet resistance, with the effective length of the electrodes and effective width between them as the most distinctive parameters

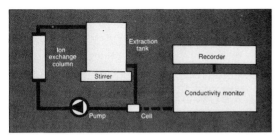

A typical configuration of a so-called ppm-meter, featuring quantification of ionic residues through solvent extraction. The development of a correlation factor between IR test data and the results found through the ppm-meter will upgrade the validity of this in-process QC tool significantly

with both an ionic and a non-ionic/organic constituent is generally required.

Ionic residues

Whether soluble or insoluble, ionic residues are the most harmful of all residues to be left on a PCB. Species of ionic residues include salts, acids and alkalines left on the PCB from operations like etching, plating, cleaning, reflowing, manual handling (finger salts), fluxing and soldering. The majority of these residues are soluble in water with the exception of some organometallic and inorganic salts. For the most part, the strong hydrogen-bonded alcohols too will adequately remove these residues. However, a salt like lead-bromide is virtually insoluble, even in hot water.

Residue particles embedded in or below the surface of the PCB, such as those coming from processing operations like pumice scrubbing, can often become ionic and hygroscopic. These residues can have considerable effect on the surface resistivity of the end product. It is virtually impossible to remove these residues in a post-solder cleaning operation.

Ionic residues are hygroscopic. This means that in a humid environment the water in the air is attracted by the residues and dissolves them. When this occurs, ions can carry electrical current and enter into chemical reactions like corrosion. Consequently, once ionic residues are left on a PCB, humidity can initiate as chain reaction that will result in high concentrations of conductive ions.

This process becomes quite obvious when the boards are under electrical load. It causes lower surface resistivity between the conductors on the PCB, ultimately resulting in

electrical leakage and shorts. This phenomenon is known as electromigration or dendritic growth.

Detecting ionic residues

A number of tests exist for detecting the presence of ionic residues. One of the most common is the insulation-resistance test. This test is normally carried out on PCBs with a special insulation-resistance test pattern.

The test boards are exposed to predetermined relative humidity and temperature conditions while biased at a specified voltage level for some fixed period, usually at least 20 days. While under this condition, the surface resistivity between the conductors is monitored and measured in ohms or megaohms. In this way the effect of any ionic residue can be measured, be it soluble, partly soluble or insoluble, encapsulated in the molecular matrix of the rosin or not, or embedded in or below the surface of the board itself. The measurement of the effect of any ionic residue makes this type of testing one of the most accurate quality control tools available in regard to cleanliness and reliability.

When decisions on new process parameters have to be made, the humidity test is the most reliable indicator. Variations in the duration of the test, the temperature, the relative humidity and the voltage used should not be ignored when comparisons are being made.

One means of quantification of ionic residues left on a PCB is the use of solvent extraction techniques. Ionic residues can be

A printed wiring assembly covered with white residues left after cleaning in a bipolar solvent

quantified in terms of micrograms per square surface area of NACl equivalents. This is related to the concentration of ions in a predetermined volume of an extract. The latter is to be made with a deionized solution of water and alcohol over a certain period from a PCB with a known surface area. This is a widely used method most often executed with a so-called ppm-meter. It is a relatively quick in-process quality control tool to assure that the standards of a process are maintained and that the levels of ionic cleanliness meet the company's specification or the specification of the contracting authority, such as MIL-P-28809 requiring a maximum of 8.35gr NACL-equivalents per square inch.

For the basic evaluation of any change in relevant process parameters, or the implementation of an entirely new process, this method is questionable. It only detects the concentration of ions going into solution, not the concentration and effect of other contaminants still left on the board.

The preceding remarks may indicate a somewhat reluctant attitude towards a method so widely used. That is because of the fundamental discrepancy between the simplicity of the tool itself and the myriad of misdirections a printout from such a ppm-meter may give to an unexperienced engineer who is not completely accustomed to the many nuances of specific cleaning processes.

Following is an illustration of this viewpoint. Flux A is easy to remove in a certain

Extrusion of sap from pine tree, providing rosin. Rosins differ from place to place and time to time of extrusion from the tree sap

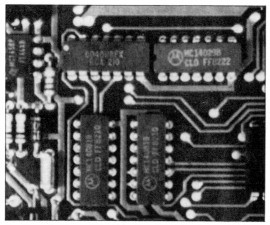

A well-cured solder mask provides a thoroughly polymerized, non-porous, smooth, hard surface that results in perfect cleaning without white residues

solvent. Flux B is relatively difficult to remove using the same process. Suppose both fluxes are to be cleaned in the same fluorocarbon solvent, but with the handicap that either the dwell time in the cleaning system will be too short, the solvent blend not effective enough or the solvent will be practically saturated before the start of the test. The result of the insufficient cleaning operation will be that both boards still have a balance of residues. Consequently, Flux A most probably will provide the highest readings of ionic residues. This is because the balance of the residues of this flux easily come off in the virgin water/alcohol solution in the ppm-meter. Unfortunately, Flux A may be erroneously rejected because of a suspected higher rate of ionic contamination.

Ion-chromatography is a method by which a selective quantification of ionic residues may

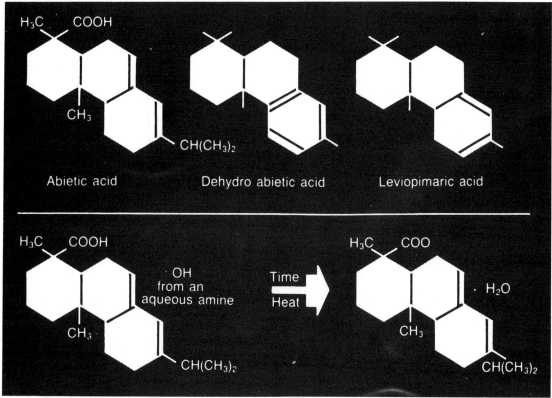

Abietic acid forms the major constituent in natural rosins. Dehydroabietic acid and leviopimaric acid may form a minor part. This latter acid is detected in rosins from some of Europe. Abietic acid is not water soluble, so conventional rosin fluxes have to be saponified prior to removal with water. The saponification reaction is also illustrated

be made. This relatively new method is a carefully controlled laboratory procedure. The advantage of this procedure is the potential to separate some probable causes of the ionic residues.

UV spectral analysis is yet another procedure which can be used to provide accurate selective quantification of both ionic and non-ionic residues. Making measurements in the parts per million range, it can separate multiple unknowns in muddled spectra of residues.

Ionic residues from fluxes

In order to evaluate the specific contribution of a certain flux to the ionic residues on a board, the most reliable procedure is to measure the effect with an insulation resistance test. By comparing boards that were fluxed, soldered and cleaned with different fluxes, and also to a control group of boards

with no flux at all, it is possible to get a good idea as to which way to decide. UV spectral analysis is also a reliable, all-around technique for determining the contribution of flux technology towards ionic residues as well as non-ionic residues. Compared to the IR test, UV spectral analysis is a relatively quick procedure, carried out with real PCBs. The disadvantage may be that this is a destructive test and therefore should be used for process evaluation only.

Non-ionic residues

Non-ionic residues are not as harmful as ionic residues. This type of residue, however, may form insulating films on plug-in contact surfaces resulting in intermittent open circuits which are difficult to trouble-shoot. Non-ionic residues may interfere in the making of electrical contacts with pogo-pin probes during testing operation, or they may flake off into open components, accumulate airborne dust and dirt, etc.

Non-ionic residues typically come from machine oils, greases and fluids, skin-care products used by operators containing ingredients like silicone and lanolin, soldering oils, organic cleaning solvents, mold release, masking materials, plasticizers in plastics and plastic masking products, and photoresists. One of the main species of non-ionic residues, however, is the rosin from the solder flux.

Generally non-ionic residues can be removed adequately by an effective bipolar solvent system, unless the mechanics of the cleaning process have been improperly specified or where undesired interactions occur between the chemistry used in sequential processes. Metal-oxides such as PbO and SnO are non-ionic but can slowly hydrolize, degrading the insulation resistance. They are formed as a reaction product of the base metals with air and are insoluble.

Typical tests for the detection of non-ionic residues vary from the cumbersome tests exclusively for rosin, like the Lieberman-Storch and HalphenHicks tests (Ref. 5), to the more advanced and universal detection method like the UV spectral analysis.

White residues

White residue is basically not a third category of residues. It is either ionic or non-ionic. The reason why this defect is not always that easy to eliminate is that the common symptom of the white color covers a variety of different functional defects related to a number of possible causes, some of which can be traced back to the PC fabrication operation. White residues can be identified as etch or plating residues, residues from reflow chemicals, wax from components or even impurities in the solder alloy. However these causes for white residues are relatively rare.

Activators added to the rosin in a flux formulation help speed up the breakdown of oxides on metal surfaces. Unfortunately, they also can create significant problems if not completely removed. The chain reaction between residues of an amine hydrochloride activator and a partially oxidized copper surface with a practically infinite reformation

of HCl from secondary reactions is depicted as follows:

$$NH_4Cl \leftrightarrows NH_3 + \mathbf{HCl}$$
$$Cu_2O + 2CH1 = CuC1_2 + Cu + H_2O$$
$$CuCl_2 + H_2CO_3 \leftrightarrows CuCO_3 + \mathbf{2HCl}$$
$$Cu_2O + 2HCl = CuC1_2 + Cu + H_2O$$
$$CuCl_2 + H_2S = CuS + \mathbf{2HCl}$$
$$Cu + 2HCl = CuCl_2 + H_2$$
$$CuCl_2 + H_2CO_3 \leftrightarrows CuCCO_3 + \mathbf{2HCl}$$

A white dusty residue is most often formed by remaining activator residues when the control over the solvent composition is lost. Principally this defect is caused by solvent cleaners equipped with a simple water separator operating in a room with a relatively high humidity. The need for relatively high humidity may be required by a program to protect sensitive circuitry from electrostatic discharge.

What happens is that the water condenses out of the ambient air around the vapor condensing coils and mixes thoroughly with the solvent in the holding reservoir. Alcohol, being the ionic constituent in the solvent has greater affinity for the introduced water, which is also ionic, than it does for the non-ionic constituent in the solvent. Thus the alcohols are scavenged by the water and lost through the water separator.

The end result is a depletion of the polar constituent from the solvent formulation. Because a polar solvent is lacking, the flux activator, being also ionic, will not go into solution. Ionic residues are thus left on the PCB. Most often a dip in alcohol will remove such residues. Saturation of the cleaning solvent, as well as short dwell time in the cleaning system, may also result in white dusty residues from flux activators.

Lead salts

Another type of white residue formed by reaction with the flux activators on PCB surfaces are lead salts. Included in this category of white residues are lead chloride and lead bromide.

Unaltered halogenated activators such as amine hydrochloride are easily removed by the regular bipolar cleaning solvents. However, much of the activator will decompose

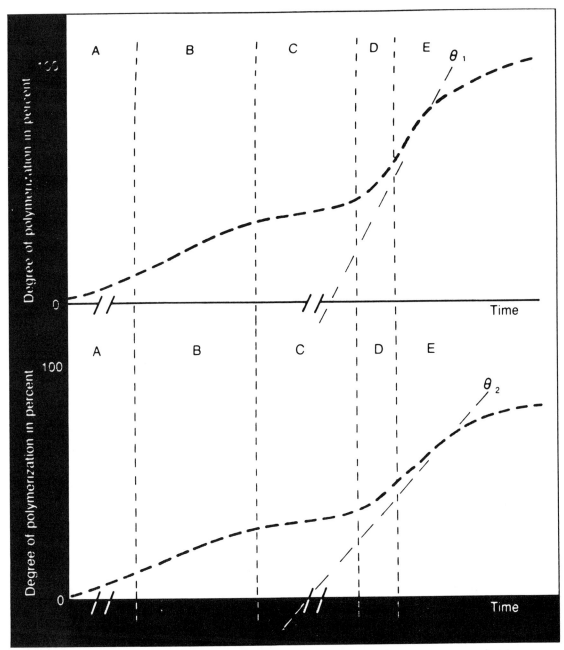

The typical polymerization profiles of commercial fluxes (a) and fluxes with modified rosins (b) are compared. Note the delayed and limited polymerization of the modified rosin (b) illustrated in phase E

because of the heat exposure on the solder wave releases hydrogen chloride and free amine. These decomposition products are lost as gases. Special consideration needs to be given to a possible reaction of the chloride with metal-oxides such as Cu_O, SnO and PbO. Lead chloride, however, is a salt of poor solubility.

Increasing safety requirements demand the use of flame-retardant PCB laminates implying the use of brominated resins. On the other hand, some of the higher activated fluxes used in high speed soldering operations contain activator systems yielding lead containing salts. These salts may react with the bromine from the laminate resin resulting in an insoluble salt known as lead bromide. The symptom is also a white dusty, ionic residue.

Mealing is another type of ionic residue with the symptom of white granular spots formed between the surface of the PCB and the conformal coating. A change to another type of conformal coating sometimes eliminates the symptom but not the cause, which is a balance of ionic residues due to improper cleaning.

Polymerization problems

The most common cause of white residues is that one of the polymerization processes is out of control. Polymerization is the chemical reaction in which two or more monomers or polymers of the same kind are united to form a molecule of a higher weight. It is a reaction process influenced by parameters like time, temperature exposure and UV exposure. Problems can arise when there is either insufficient or excessive polymerization.

Thorough polymerization is required for the consistency of the base laminate, to avoid the possibility of releasing epoxy smear and to obtain a smooth, non-porous, hard surface of solder resist that cannot be attacked by the solvent. Insufficient polymerization in the above-mentioned applications will cause white residues.

The causes of improper polymerization of solder resists are numerous, varying from improper mixing of the two-part screenable mask to improper design of the thermal-curing oven or UV tunnel. Often, performing a post-curing procedure on the boards before assembly will improve the level of polymerization and thus the level of cleanliness.

Excessive polymerization is more frequently observed than insufficient polymerization. This defect is a concern with water dip lacquers, but is even more so for the rosin in the solder flux. The rosin in both products tends to polymerize during the heat exposure in the manufacturing process of the rosin itself and during the soldering operation.

Repeated heat exposure, like that in solder-cut-solder operations, has catalyzing effects on the polymerization process. The polymerizing rosin becomes a very long chain molecule which cannot be dissolved in the commonly used solvents. Standard cleaning solvents are designed to dissolve the short chain rosin segments.

The tenaciously adhering white residues are the polymerized part of the rosin. Once the polymerized rosin has been formed, not even the most effective fluorinated or chlorinated solvent is capable of dissolving it and cleaning it off. One way to correct this problem is by dipping the boards in a rosin solder flux. The rosin in the flux will dissolve the polymerized rosin, after which the board can be re-cleaned. This double cleaning operation, however, is not cost-effective. That is why modified rosins ahve been developed in order to settle this major white residue problem.

Rosins are made from pine tree sap. Like most natural, organic substances they are subject to polymerization. Polymerization starts at the moment of extrusiion from the pine tree, and continues as time passes. This progressive polymerization is dramatically influenced by parameters such as UV and thermal exposure. When held at low temperatures and kept away from the daylight, progressive polymerization is practically zero (phases A and C).

Thermal exposure of the rosin, however, is required in order to prepare the product as raw material. During preparation, the rosin goes through processes like steam and vacuum distillation which purifies it and separates out the turpentine (phase B).

A second progression in the polymerization process, although not as significant as during the manufacturing process, starts when the solder flux is brought into the fluxing unit of the soldering machine. In this case, the UV and thermal influences on the polymerization of the rosin can be noticed when cleaning difficulties arise. This happens when the flux is kept too long in the fluxing machine without changing it. (phase D).

The crucial phase in the polymerization process starts when the fluxed circuit boards are brought up to soldering temperature (phase E). For rosins that are not particularly modified for this purpose, the polymerization slope of progression becomes quite steep. The discernible point of the onset of this change is at approximately 145°C for most solder fluxes.

In solder-cut-solder operations where the fluxed assemblies are soldered twice before they are cleaned, another progression on top of the one just discussed may cause serious cleaning problems with regard to white residues. At temperatures above 250°C, there is also some catalyzing effect of the activators on the progress of the polymerization process.

Rosins, however, can be modified so that the onset point of rapid polymerization is situated at a higher temperature, >200°C and the slope or rate of polymerization is dramatically reduced. This modification results in delayed and limited polymerization, yielding higher levels of cleanliness in particular with regard to white residues.

Halide-free activators

The assumption that the selection of a halide-free flux followed by a cleaning operation in an effective solvent system would yield the highest levels of cleanliness is a misunderstanding. Some non-halide acids, like adipic acid, have a considerably lower potential to ionize than a halogenated activator like an amine-hydrochloride. This means that most of the milder acids do not dissolve easily in the cleaning process. A small amount of halogenated acid in the activator system generally promotes the ease of cleaning, resulting in lower levels of ionic residues. Decomposition products of some organic acids may be very complex and insoluble while encapsulating ionizable residues. For the sake of completeness, however, it must be emphasized that non-halide acids principally are less corrosive than halogenated acids when left as residues on the board after cleaning.

AQUEOUS CLEANING

For many years controversy has raged over which method of removing flux from a soldered printed circuit board is best — aqueous or solvent cleaning. Through the middle 70s most printed circuit board cleaning was done with solvent, recalls Jeff Hilgert, worldwide product manager for Branson Cleaning Equipment Co., a manufacturer of ultrasonic aqueous cleaning machines. The pendulum later swung away from this solvent orientation to the point where the market breakdown seemed to favor aqueous cleaning in the late 70s and early 80s. Although Hilgert contends the trend is now back to solvent cleaning, aqueous cleaning appears to hold a strong lead in cleaning high-volume production, commercial PCBs.

There are some good reasons for this,

First published in *Electronic Packaging & Production*, August 1984, under the title 'Aqueous cleaning challenges solvent in solder flux removal'.
Author: Robert Keeler, Associate Editor.
© 1984 Reed Publishing (USA) Inc.

including the fact that the highly active organic acid (OA) fluxes, which are especially well-suited to high-throughput wave soldering, are formulated strictly for water removal. Ideal for removal of ionic contamination because of the polar nature of its molecule, water is also recyclable and non-toxic.

Armand Karolian, product specialist for Universal Instruments, points out several other reasons for the enduring popularity of aqueous cleaning, including the development in recent years of components capable of tolerating water processing, the increased availability of specialized in-line cleaning equipment, and the adoption of stringent pollution standards in some communities which make the use of solvent cleaning unrealistic.

There are drawbacks to aqueous of course, among them an apparent problem with aqueous cleaning of surface-mounted devices

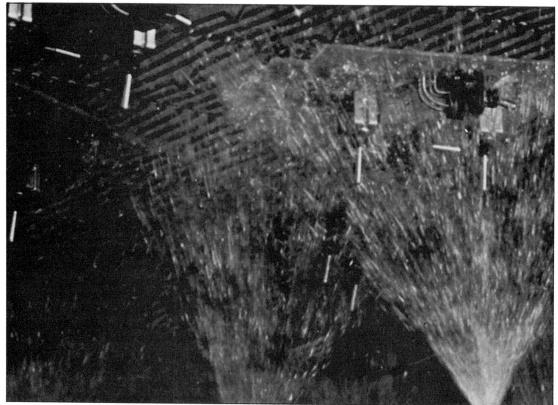

A PCB is shown being spray-cleaned by a Hollis Polyclean II aqueous cleaner. Of polypropylene construction, the machine uses the proprietary Trivent airknife nozzle design to achieve flash drying of the board. The board shown contains water-washable L series toggle switches from Arrow Hart Division of Cooper Industries

and a lack of military approval. Nevertheless, aqueous cleaning remains a well-established flux cleaning technique and will be around for some time to come.

Aqueous flux removal

There are several reasons for cleaning flux from the soldered PCB. Pre-eminent among these is the need to remove residues which might later cause current leakage. Cleaning is also done to prevent the active element of flux from corroding circuit components – a problem which is of special concern with organic acid fluxes. In addition, notes Warren Abbott of Hollis Engineering, cleaning is done after soldering to insure that bed-of-nails probes can access test points properly, to facilitate solder joint inspection and repair, to permit better adherence of conformal coatings and to generally improve the appearance of the board.

There are three main families of fluxes – rosin, organic acid and synthetic-activated resin. Of these, aqueous cleaning is used both with the rosin-based fluxes and the organic acid fluxes. The latter are formulated strictly for water cleaning. Synthetic-activated fluxes, on the other hand, which were developed by DuPont in 1982 to have roughly the same activity level as the water soluble fluxes, are exclusively for solvent cleaning. DuPont claims that by using SA fluxes, touch-up costs arising from poor solderability can be reduced while all the other benefits of solvent cleaning can be retained. However, Joel Camarda, sales and product manager for John Treiber Co., states that he hasn't seen any real impact of SA fluxes on the commercial market. He feels SA fluxes will only prolong the life of solvent processes in the military market.

With aqueous cleaning of rosin fluxes, an

organic alkaline cleaning solution which is based on amines and called a saponifier is used both to react with the rosin acid to form a water-soluble soap and to lower the surface tension. That content of the rosin which cannot be saponified must either be removed by means of a water-soluble solvent, such as glycol ether, or be physically scrubbed away. Because of foaming associated with the saponified rosin soap, an anti-foam agent is also often put into the wash water or the detergent. In addition, sequestering agents are sometimes added to prevent metal ions from entering the effluent.

Over the past 10 years, the manufacturing trend has been toward more and more use of water cleaning. This has been strongly driven by consumer products manufacturing concerns, including computer products. The more active water soluble, organic acid-type fluxes are said to give better soldering results than the rosin groups. When processing high-density (i.e. thousands of solder joints per typical PCB) circuits in large volumes, the first-pass quality of the soldering process is a critical factor in maintaining manufacturing efficiencies in highly competitive industries. The water-soluble fluxes contribute significantly to a more forgiving solder process, and aqueous removal methods for these fluxes yield clean, reliable circuit assemblies, says Treiber's Camarda. However, the residue of directly water soluble OA flux is very

The dryer module shown, which is part of the Treiber TRL-HSE aqueous cleaner, uses a combination of turbine supplied, heated-air blow-offs and infrared energy to dry the PCB. The seven-station cleaning system makes use of several high pressure airknife stations to eliminate water dragout and cross-contamination between consecutive stations

corrosive and cleaning must be performed in a thorough way. At their best though, the advantage of OA fluxes is that their high level of activity results in a very good soldering job.

Bill Haines, vice-president of Westek, says OA fluxes are easier to clean because they break down into solution more readily than the heavier rosin fluxes. Water soluble residue flux can frequently be cleaned with common tap water, although in cases where water is more than seven grains hard, water softeners are recommended to reduce the water's surface tension, which, in turn, helps break down the flux residue, notes Haines. Concentration of 1–2% detergent in the wash module might be considered if the PCB has picked up excessive amounts of finger oils, dirty components or associated contamination during the assembly procedure. For rosin-flux cleaning higher concentrations of saponifier, usually 5–10%, in hot tap water are required. There are also available water soluble rosin fluxes that do not need to be saponified.

Cleaning assemblies

The aqueous cleaning system itself can have process-monitoring instruments, such as those to measure pH, conductivity and resistivity, each reflecting the quality of the incoming water or the water being used in the various rinse and wash sumps. A degradation in water quality in the machine would be indicative of possible degradation in the cleaning process. The aqueous user thus will set process limits for water quality throughout the system.

Treiber's flagship cleaning system, the Model TRL-HSE, monitors water quality and automatically regulates the flow of fresh incoming water to preserve constant process conditions in the machine. Although such machine monitors are useful, they are not as accurate as an actual ionic test.

Aqueous cleaning provides a safety advantage over solvent by the elimination of solvent fumes on the production line, the environmental aspects of which can range from annoying to toxic. Since all of the cleaning solvents have low boiling points, there is always some measure of vapor involved in handling operations, as well as in filling and

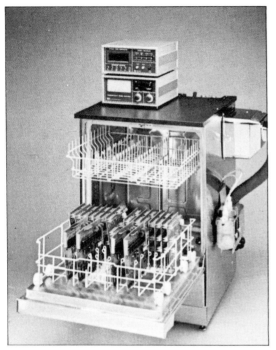

The new 630uP PCB batch aqueous washer from Electronic Controls Design is shown. The topmost instrument is a programmable controller, and just beneath it is a conductivity rinse control. The attachment on the right is a convection dryer

draining solvent cleaning system sumps, loading and unloading PCBs, etc. Treiber's Camarda maintains that the general trend in enforcing higher levels of safety in the manufacturing environment is a strong factor driving the popularity of aqueous cleaning.

The high costs of the cleaning solvents are another factor in favor of water cleaning. Water is cheaper than solvents. Even large volumes of pre-treated water are less expensive than moderate volumes of the various cleaning solvents. The higher energy consumption involved in running a small-to-moderate size aqueous cleaning system vs. a batch vapor degreaser can sometimes be offset by eliminating the solvent cost, with some users paying back their water cleaner purchase cost in only a few months via chemical cost savings and large productivity improvements.

The design of aqueous cleaning machines constructed of polypropylene, as an alternative to stainless steel, has lessened the concern about energy consumption in the cleaning

stage to some degree since polypropylene is a better thermal insulator than steel. Another side-benefit of the use of plastic construction material for the machine is that repairs are simplified. A hole punched in the side of a machine by a forklift truck may be repaired relatively easily and inexpensively by plastic-welding a patch of the plastic material over the puncture.

Art Gillman, president of Unique Industries, points out that polypropylene is more subject to heat stress and tank sidewall bowing than stainless steel, a material which Unique uses in the construction of their Hydro-Kleen family of in-line and batch cleaners. In general, he claims, polypropylene is popular mostly because it involves a lower capital cost to the user. Yet, although water is much cheaper than solvent, water cleaning machines are still not often lower in overall costs, contends Urban Chemmical's Allen Sluizer, who claims the marked cost advantage of water over solvent is frequently offset by the larger system-wide expenses for energy, equipment and overhead.

According to Warren Abbott, manager of technical services at Hollis Engineering, it is important to clean assemblies as soon as possible after soldering. For this reason, in-line aqueous cleaners usually accept PCBs directly from the exit conveyor of soldering systems. PCBs enter the warm pre-wash or wash chamber while they are still warm. Many users who have standard low-to-medium density boards operate aqueous cleaners without detergent, especially if water-soluble flux and oil intermix is used. However, high-density PCBs and those with surface-mounted components benefit from detergents because it reduces the surface tension of the wash water.

The method of removing bulk water from an assembly after final rinse and prior to radiant drying is of vital importance, says Hollis' Abbott. If bulk water is merely evaporated, possible residual contaminants from flux, saponifier and even water itself will remain. A turbine-generated airknife is used to squeegee rinse water off under high pressure in most machines. For example, Hollis offers their Trivent airknife with a

The Hydro-Kleen Model HK55 aqueous cleaning machine from Unique Industries features stainless steel construction. The duct heater and blower system allow drying temperature selection to be independent of conveyor speed

special tapered design to achieve flash drying action.

Cautions

There are a number of issues to consider when looking at the purchase of an aqueous cleaning system. Problems can arise when using water soluble, organic acid fluxes if there is any tendency for flux residues to get trapped under circuit components. Since such residues are very corrosive – the price one must pay for their efficiency – they must be cleaned off the board thoroughly to prevent component damage. Residue-laden water can get trapped in a component such as an open-winding transformer and will not readily evaporate.

Bill Bokhoven, president of Cyclotronics Inc., notes that his firm builds in-line aqueous cleaners that rack the PCBs vertically to help the wash and rinse water flush out any flux that is trapped under components. He claims vertical racking not only speeds up the drying time, since most of the water tends to run off the PCBs before they reach the drying section, but also results in a smaller machine, thus saving on floor space. In addition, he claims, it reduces the capital cost of the machine by 50% or more. Treiber's Joel Camarda maintains, however, that the force of gravity to aid in draining water is insignificant compared to the forces delivered by the turbine blower airknife in an in-line machine.

Horizontal processing is much more amenable to fitting in with other in-line processes which are also horizontally oriented, he says, such as soldering, test and component insertion.

As in solvent, cleaning, reminds Westek's Haines, some components are simply not compatible with the aqueous process – for example in the area of inks and markings. Since the electronic industry originally evolved around solvent cleaning, the number of solvent-compatible components available currently outweigh the number of water-compatible parts. Haines emphasizes, though, that component manufacturers are addressing the incompatibility problems for both aqueous and solvent processes, and have already made great strides in reducing the quantiy of post-cleaning operations.

It takes less heat to boil solvent than to heat water to the appropriate temperature for adequate cleaning, so the power consumption of the aqueous machine will be correspondingly higher. Also, because solvents have a higher volatility it will take less energy to dry a solvent-cleaned board than one cleaned with water.

Environmental considerations

There are other concerns for the potential buyer, involving water. It is important to consider one's geographical location and draw some conclusions about the future availability of a sufficient supply of fresh water – something that may be a problem in the Sunbelt states, for example. Branson's Jeff Hilgert points out that even though the aqueous process is often claimed to be ecologically safe, disposal of contaminated water might still be a problem. Water from the cleaning cycle is laden with detergent, flux residues, oils and other chemical additives and it is important to know what local and federal pollution regulations have to say about the discharge of the effluent down the sewer in the user's community.

Final pH and lead content are two of the most serious effluent chemistry concerns. Urban Chemical's Allen Sluizer reminds that there are some areas in this country where fresh water is in such short supply that local regulations either forbid new water cleaning systems outright, or ban those that cannot recycle the water. Furthermore, the mineral content of water, varying as it does on a geographical basis, can result in the need for special treatment like water softening. For high-reliability applications where a final rinse in deionized (DI) water is necessary, this can become expensive.

On the other hand, it has been pointed out by partisans of aqueous cleaning that solvent disposal poses an even tougher problem than water disposal, in the sense that it cannot be put in the drain at all. Typically, a solvent-using company pays the same company who supplied the solvent to haul it away again,

The Aquapak Model 526 from Electrovert USA offers a choice of 12 standard modules intregrated into one in-line sequence, each available with variations and options. Depending on the soldering operation and flux used, as well as the level of cleanliness desired, any combination of modules can be put together to form a system

either for treatment or distillation at the supplier's factory.

Aqueous cleaning has not won military approval, so it is currently not suitable for any manufacturer contemplating mil-spec work. A study at the Naval Weapons Center at China Lake, CA, has been carried out on solvent and aqueous cleaning. Among other things, it concludes that the two best methods for removal of rosin fluxes are vapor degreasing with solvents having a high percentage of alcohol, and 'combination' cleaning consisting of vapor degreasing followed by a rinse with DI water. The report maintained that water cleaning by itself, whether tap or DI, was not able to properly deflux a printed circuit board with RMA and RA fluxes to an acceptable level of cleanliness.

The study concluded that the working life of a detergent must be closely monitored in order to ensure removal of flux. Furthermore, RA fluxes cannot be cleaned to an acceptable level of ionic cleanliness even by detergent cleaning declares the Navy report, stating that in general, non-rosin fluxes degrade the resistance characteristics of the PCB. The study group recommended that these fluxes should not be used on military hardware.

Future issues

Perhaps the most ominous drawback for aqueous cleaning is the challenge posed by surface-mounted devices. Says Don Elliott,

Universal Instruments Corp.'s Soltec aqueous cleaning system features a modularity that permits easy add-ons for future expansion of the system when called for by changes in production speeds, PCB types or fluxes. All modules are prepared for expansion with a return conveyor system underneath

manager of technical support services with Electrovert's Consulting Service, the biggest question today for aqueous is how people are going to properly clean aggressive flux residues from printed circuit boards when the PCB has tiny surface-mounted components on board, with almost no space between the top of the board and the bottom of the components. The problem stems from the fact that the relatively higher surface tension of water, as compared to solvents, prevents it from easily penetrating under SMDs to loosen flux residues and carry them away.

It has been pointed out, in a partial defense of aqueous as a viable process for SMDs, that if the surface-mounted device is not too large, water should have no trouble getting under it. Preliminary results of a study done at Hollis Engineering have shown that the proper amount and formulation of adhesive used for holding SOICs in place, prior to wave soldering, might act as a barrier to entry of flux contamination, at least in that case vindicating aqueous cleaning as a satisfactory cleaning medium for SMDs. Ultrasonic cleaning systems, such as those offered by Branson Equipment Co. and Crest Ultrasonics, might help the viability of aqueous for SMDs too, although these are likely to be batch-mode applications.

Dr Jack Brous, Senior Research Scientist at Alpha Metals, contends that for SMDs, the route to go is rosin fluxes – in particular RMA type flux, because of its lower degree of corrosiveness compared to either OA or SA types. Brous, whose firm supplies both aqueous and solvent cleanable fluxes, feels that if the board design raises any doubt about the feasibility of getting residue out from under an SMD, (and therefore about future circuit reliability) water cleaning should be avoided in favor of solvent.

However, Westek claims that their extensive testing on PCBs with conventional, SMD and planar-mounted devices has shown that the proper combination of nozzle configuration and angle, as well as volume and force of water impingement, can overcome the alleged surface mount problem for the aqueous process. They say that higher pressure airknife generators, positioned between wash and rinse modules, are required for SMDs to

There are four designs in the Formula series of Westkleen aqueous cleaners from Westlek. They all feature variable speed in-line filters. The multidirectional cascade system for process control allows the user to isolate wash and rinse sections. The machine will clean either rosin fluxes or water soluble organic acid fluxes

flush water in and out of hard-to-get-at areas. They also say that larger and more efficient dryers to evaporate the residual water are called for. They are currently working on a machine that will accommodate a large spectrum of cleaning requirements, including both SMDs and planar mounts.

A spokesman for Asahi, the Japanese firm which supplies soldering fluxes to about 45% of the Japanese market, claims the use of water soluble flux is almost nil in Japan. There are two reasons offered for this fact: the Japanese don't trust the ability of aqueous cleaning to totally remove ionic contamination from the PCB, and they wish to avoid the expense of water treatment, which is apparently necessary in that country. The Asahi spokesman feels strongly that water-soluble flux will never be accepted by the Japanese PCB industry, and says alkline cleaning (saponification) of rosin-based or phenolic-based flux may only be accepted in Japan for hybrids.

Treiber's Joel Camarda, however, feels that despite the fact that Japanese manufacturing is pre-eminent in the area of automation, they are several years behind US technology in flux chemistries. He says he already sees aqueous cleaning beginning to make inroads in European PCB manufacturing, which has traditionally been solvent-oriented. He speculates that European divisions of Hewlett-Packard and IBM, which are currently making a strong commitment to aqueous, may convince native European firms by the example of their success.

CLEANING SURFACE-MOUNTED ASSEMBLIES

Cleaning is normally one of the final assembly processes, therefore it is one of the last processes which can affect the quality and reliability of the product. Overlooking the complexities of solvent cleaning surface-mounted assemblies (SMAs) can lead to inconsistent process control and product field failures. Fluxes, solvents, machines and PCB design all can produce detrimental effects if not properly used, maintained and chosen.

The cleanliness of the assembly, both around and underneath the components, is important. Conventional assemblies having DIPs and cylindrical devices with axial and radial leads are not as difficult to clean as SMAs. SMAs which have leadless chip carriers (LCCs), small-outline integrated circuits (SOICs), small-outline transistors/diodes (SOT 23s), quad packs, etc., provide a minimum spacing (off-contact distance) between the component and surface of the PCB. The small spacing and the large area underneath SMDs is what makes effective or acceptable cleanliness levels difficult to obtain.

Pros and cons

Batch solvent cleaners or degreasers have been used to clean SMAs that have been soldered using mildly activated rosin (RMA) fluxes for high reliability applications such as those found in military and aerospace electronics. Batch cleaners are effective for low-volume production and when RMA fluxes are used.

However, for large-volume production or when using activated fluxes, the batch system may no longer be suitable or as efficient when compared to conveyorized (in-line) solvent cleaners. The cleanliness uniformity of SMA, when cleaning with batch systems, can vary since this process can be very operator-dependent. When strict and effective operator procedures are enforced, batch systems can provide acceptable results. Most importantly, when flux types change from mild to more activated, operator instructions should

First published in *Electronic Packaging & Production*, August 1984, under the title 'Surface mounted assemblies create new cleaning challenges'.
Author: Carmen A. Capillo, ECR Corp.
© 1984 Reed Publishing (USA) Inc.

SMA type	Typical component varieties	Component types	Assembly cleanliness difficulty
Type I	Passive-discrete Active-discrete	Chip resistors & capacitors, SOT-23, SOT-89	Easiest
Type II	Integrated circuit packages (leaded)	Flat-packs SOIC Quad-Packs Leaded chip carriers	Difficult
Type III	Integrated circuit packages (leadless)	Leadless chip carriers	Very difficult
Type IV	Passive-discrete Active-discrete Integrated circuit packages Conventional devices	All types	Very difficult

Cleaning difficulty of some SMAs

be revised to allow for additional solvent vapor, spray and immersion exposure or a combination of these in order to remove the greater amount of flux residues present on the assembly after the soldering process.

Worldwide, both commercial and industrial electronics are now converting to surface mount technologies. High-volume cleaning systems should, in turn, convert from in-line aqueous to in-line solvent cleaners. Aqueous cleaners are used to remove water-soluble organic fluxes which usually have 10 times the activator concentration, therefore increasing the difficulty of removing flux residues to acceptable cleanliness levels on the assemblies as well as underneath the components.

If water-soluble organic fluxes are produced with lower concentrations of ionic halide and organic acid activators, removal of sufficient quantities of these activators and other flux residues still remains a difficult process. The higher surface tension of water, as compared to solvents, makes water (aqueous cleaning solutions) difficult to penetrate under large SMDs with small off-contact distances.

For those who have no choice but to use an aqueous cleaner, it would be advisable to use the mildest of water-soluble fluxes and perform IPA/water extract resistivities of the assemblies rgularly. If extract resistivities are within acceptable limits, the amount of contamination underneath the devices should be determined before the process is accepted. To

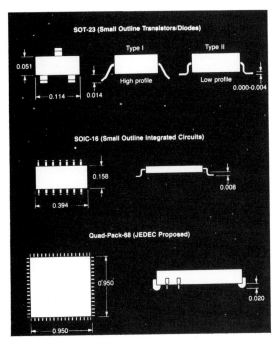

Some typical leaded SMDs showing off-contact distances and maximum surface area covering the substrate (Dimensions are in inches)

do this, remove all devices from the substrate to which they are soldered, without the use of flux, and then place the parts and the circuit board in the same test cell for determination of ionic levels. As an extra precaution, functional/environmental tests of the assembly or product should be performed.

Contamination

Cleaning is a process by which contaminants or soils are removed. In cleaning SMAs, the prime cleansing task is basically the removal of fluxing agents and flux carriers after the reflow process.

Contamination that produces electrical and mechanical failures on printed circuit boards and assemblies can be classified by type and origin. Contaminants may be organic compounds, inorganic insoluble compounds, organometallic, soluble inorganic, and particulate matter. They may originate in fluxes, solder mask, photoresists, PCB processing chemicals, and airborne matter.

The several different types of contaminants can be characterized as polar or non-polar. Polar contaminants are compounds that have

eccentric electron distribution and are sometimes called ionic contamination. Non-polar contaminants are compounds which do not have an eccentric electron distribution and will not disassociate into ions or carry an electrical current and are sometimes called organic (or non-ionic) contamination.

Effects on cleaning difficulties

The difficulty of removing flux residues from SMAs increases dramatically as component off-contact distance (air-gap underneath the device) decreases and as component density, component surface area and flux activation increase. The difficulty of cleaning SMAs is also related to the component off-contact distance, or how far the bottom of the component is from the surface of the substrate material, and also the surface area of the substrate which is covered by component. The larger the component surface area and the smaller the off-contact distance, the more difficult it is for solvent to penetrate, dissolve and carry away the flux residues from underneath the component. Therefore, one should expect cleaning to be more difficult for assemblies which have components on both sides, and components such as SOICs, quad-packs and chip carriers. Leadless devices such as ceramic chip carriers, chip resistors, and chip capacitors obviously have no off-contact distances other than the solder and pad thickness, which adds to the off-contact distance normally by 0.002–0.005in. For this reason, it is recommended that leadless chip

Azeotropic solvent (major ingredient)	Boiling point (F)	Latent heat of vaporization (BTU/16)	Flash point	KB value
Alpha 1003 (tetrachloro-difluoroethane)	180.0	100	None None	71 71
Freon TMS (1, 2, 2-trichlo-ro-2, 2, 1 tri-fluoroethane)	103.5 117.6	90.7 63.12	None None	45 31
Freon TMS (1, 2, 2-trichlo-ro-2, 2, 1 tri-fluoroethane)	111.9 1217.6	76.7 63.12	None None	37 31
Genesolve DMS-A 1, 2, 2-trichloro 2, 2, 1-trifluoroethane	108.0 117.6	88 63.12	None None	48 31

Physical properties of azeotropic solvents

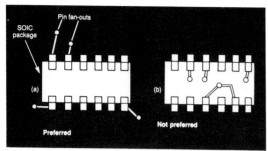

Plated through holes can be placed to prevent flux from travelling to the top side of the board during wave soldering: (a) preferred plated though hole fan-out patern; (b) non-preferred, fan in pattern

carriers should only be soldered with mild activated rosin fluxes, mounted on spacers, used with socket-leads, or replaced with leaded plastic carriers such as quad-packs if hermeticity is not required.

On the other hand, even though chip resistors and capacitors have no off-contact distance, as described, contamination removal under these devices can be easy. These components have relatively small surface areas as compared to other components. Furthermore, when wave soldered, these components are always bonded to the substrate with adhesives which act as a barrier to flux contamination.

Solvents

Solvents currently being used in solvent cleaners for vapor/spray degreasing of SMAs are normally organic-based solvents – solvents whose main chemical skeleton is composed of carbon. These solvents can be classified under three groups: hydrophobic, hydrophilic, and azeotropes of hydrophobic and hydrophobic/hydrophilic solvent blends.

Hydrophobic solvents are not miscible with water at concentrations normally exceeding 2700 ppm and consequently exert little, if any

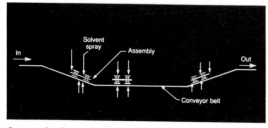

Downward and upward slopes in variable heat recovery solvent cleaners

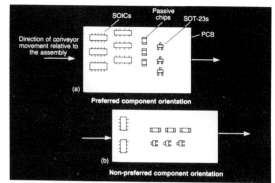

Lead extensions can be oriented on the board to take advantage of the cleaning system's slope and encourage the solvent to flow underneath the component

dissolving action with ionic contaminants. Due to this non-polar characteristic, most hydrophobic solvents will dissolve non-polar or non-ionic contaminants such as rosin, oils and greases.

Hydrophilic solvents are miscible with water and can exert dissolving action with non-polar and polar contaminants. Normally hydrophobic solvents have greater dissolving action for non-polar contaminants than hydrophilic solvents. To overcome these various degrees of dissolving action, azeotropes of the different types of solvents can be formulated to maximize the degree of dissolving action.

An azeotropic solvent blend is a mixture of two or more different liquids (solvents) which behaves as a single liquid, in that the vapor produced by its partial evaporation has the same composition as each liquid which makes up the azeotropic solvent blend. The azeotropic solvent blend has a constant boiling temperature which is lower than the highest boiling liquid and higher than the lowest boiling liquid composed in the azeotrope.

The basic ingredient of azeotropic solvent blends is either 1,2,2-trichloro2, 2,1-trifluorethane (freon 113) or tetrachlordifluorethane. These are then combined with alcohols and stabilizers. Stabilizers, such as nitromethane, are used to prevent a corrosive reaction which can occur with metal parts on the assembly and the basic ingredient in the azeotropic solvent.

A comparison of the physical properties of common azeotropic solvents which are used for flux removal is given in one of the tables.

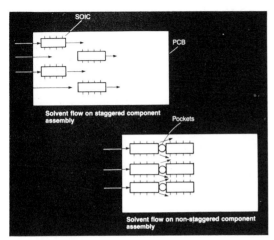

By staggering components (a), pockets of reduced solvent coverage (b) can be avoided

The physical properties of the basic ingredients in these azeotropes are also compared.

One of the important physical characteristics is the kauri-butanol (KB) value. The KB value is a measure of solvent power and expresses the number of milliliters of solvent that is required to produce a cloudiness when added to a solution of 20g of kauri gum in 100ml of 1-butanol. The higher the KB value for a given solvent, the higher its relative solvency power, or ability to dissolve flux.

PCB design and component layout

Circuit board design and component layout can complicate the task of cleaning SMAs. Cleaning difficulties and field failures can result if circuit board design does not adequately address these potential problems. The following design considerations should be investigated when designing a board for ease of flux or contamination removal:

- Plated through-holes should not be placed underneath component bodies, to prevent flux from traveling to the topside of the PCB (if used) during wave soldering. Fan-out patterns are preferred.
- Substrate cutouts should be avoided or kept to a minimum.
- Proper board thicknesses should be used in relation to board width so that stiffeners will not be required for wave soldering. Flux will become entrapped between the stiffener and substrate, leaving flux re-

sidues on the substrate after cleaning.
- Maximum conductor spacing should be used when substrate surface area permits.
- Solder masks that can maintain good adhesion after several reflow soldering processes should be used.

Component layout, still another factor which can affect assembly cleanliness, is one of the most important considerations which must be taken into account during the preliminary stages of design.

Two major parameters relating to component layout, which can contribute to the cleanability of the assembly, are lead extension direction and component body staggering. These parameters have a great effect on solvent flow uniformity and turbulence, both underneath the components and over the assembly, while the assembly passes through a conveyorized solvent cleaner.

Typically the SMA travels on a flat conveyor belt into the solvent cleaner's throat or entrance and downward on a slope of approximately 8 to 12°. During the last 18in. on the downward slope of the conveyor, solvent spray hits the SMA on an angle from both sides.

At this point the bulk of the flux residues

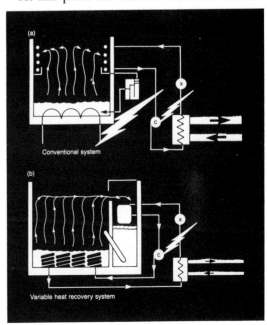

Batch solvent cleaners

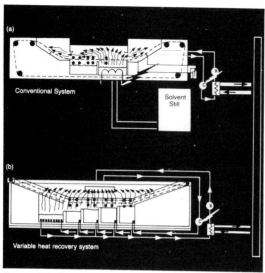

Conveyorized solvent cleaners

are removed from the assembly and from underneath the components, since at this angle the solvent runs underneath the devices more easily. The same effect takes place when the assembly travels through the last sprays on the upward exiting part of the conveyor.

To take advantage of solvent cleaners designed in this manner, the lead extensions on components such as SOICs, SOT-23s, and chip resistors or capacitors should be positioned perpendicular to the downward travel of the PCB. In this orientation, the downward flow of solvent across the board is not blocked or deflected and easily passes underneath the component bodies.

Furthermore, component bodies should be staggered, so that 'shadowing' of surrounding components does not create pockets where solvent impact and flow is reduced.

This design parameter should be considered if room on the substrate permits. Due to current advancements in solvents, fluxes, and cleaners, this parameter does not play as critical a role as lead extension direction.

Solvent cleaners

As mentioned previously, there are two types of solvent cleaners: batch and conveyorized. Of these two types there are conventional electric/refrigeration systems and variable heat recovery refrigeration systems.

The normal heating and cooling cycle

encountered in the conventional batch and conveyorized systems includes boiling and vaporization of the solvent in the sump(s) by means of electrical immersion heaters. Cooling of the solvent vapor is accomplished by helical pipe coils located at the desired solvent vapor level. The solvent vapors condense back to liquid, after which the condensed liquid is returned to the boiling sump for revaporization. The cooling of conventional systems is accomplished by circulating water or refrigerant through the cooling coils. For high-volume production, a separate solvent-recovery still is required for continuous distillation (purification) of solvent. The solvent-recovery still, which is only required for conventional systems, is used in conjunction with the solvent cleaning system for reducing contamination build-up in the solvent. Conventional batch systems usually have from one to three tanks of solvent from which vapor is generated. When spraying is required, a spray wand is used to spray cold solvent from the second or third tank on the work piece.

Variable heat recovery refrigeration systems operate on the same solvent cycle as described above, but without the use of a separate heater to boil and vaporize the solvent. A refrigerant is passed through a close-loop plumbing system to boil and cool the solvent. In this case, electrical immersion heaters are not used for boiling the solvent for vaporization or circulating water in coils for vapor condensation.

A standard close-loop variable heat recovery refrigeration solvent-cleaning system operates on the same principle as standard

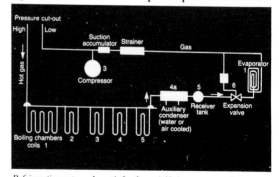

Refrigeration system schematic for the variable heat recovery solvent cleaners

refrigeration cooling systems, with the exception that it uses a compressed hot-gas refrigerant to boil the solvent in the cleaning system.

The heating (boiling of solvent) and cooling (condensation of solvent) cycle of the variable recovery refrigeration cleaning systems is shown in the figure. In the condenser (2), refrigerant which enters as liquid evaporates and absorbs heat, changing to a low pressure, heat-laden gas. This part of the cycle is where the solvent condenses on the cooling coils.

The compressor (3) pumps the low pressure heat-laden gas (refrigerant) out of the evaporator into the compressor cylinder. Here the gas is compressed by the action of the pistons and is delivered through the discharge tube as a gas under pressure to the helical coil(s) (4).

Here the compressed gas gives up its latent heat to the surrounding liquid solvent, causing it to boil. The gas turns to a liquid while continuing on to an auxiliary condenser (4A), sized to remove any remaining heat introduced to the gas by mechanical and electrical inefficiencies of the compressor.

When the gas loses heat, it reverts back to a liquid. This change of state is complete at the last pass of the auxiliary condenser. The liquid then passes from the auxiliary condenser to the receiver tank (5). The tank stores liquid for use by the evaporator (1).

Between the condenser and the evaporator is an expansion valve (6) which controls the flow of liquid according to the demand of the evaporator. From there, the liquid returns to the evaporator to repeat the cycle of heating and cooling.

Utility cost comparisons

When employing the heat recovery system for just cooling or condensing the solvent and using electrical immersion heaters for boiling purposes, it is normally expected that these types of conventional systems can save as much as 30% in utility costs.

Furthermore, cleaning systems which employ the variable heat recovery system that does not use electrical immersion heaters for boiling solvent and water for solvent condensation have been proven to save as much as 90% of utility costs in comparison to conventional cleaning systems.

As mentioned before, for high-volume applications a still for additional solvent purification is usually required in conventional systems. In variable heat recovery systems, a still is not required since there are five boiling tanks of solvent for vapor generation versus one tank of boiling solvent in conventional systems.

Measuring ionic contamination

The functional life of an assembly may be severely limited by ionic contaminants remaining on the assembly after cleaning. Therefore, it is important to measure the amount of ionic contamination after the solvent cleaning process.

A simple measuring process, available as a production line inspection test, is specified in MIL-P-28809. This test involves hand-spraying a test solution of 75% isopropanol/ 25% deionized water solution (by volume) from a wash bottle onto the test board surface, collecting the washings, and measuring their conductivity/resistivity. A wash solution of resistivity greater than 2×10^6 -cm^2 is acceptable.

The development of more elaborate measuring systems has made the process less operator-dependent. Currently available systems include the Ionograph, Omega Meter and Ion Chaser. These systems typically employ several test tanks in which the assembly is immersed in test solution. As the test solution dissolves ionic contamination from the assembly, the solution is pumped to a conductivity cell. An output voltage is produced over the test time and then converted to a representative contamination level, normally expressed in gNaCl/cm^2 to equivalent readings in -cm^2 as related to type of machine used.

REPAIRING SOLDERED ASSEMBLIES

Testing may indicate circuit defects. Some may require soldering changes. Others involve replacement of defective components. This must be accomplished without further damage to the assembly.

DESOLDERING COMPONENTS

Although desoldering is an operation to be avoided, because it represents rework and is therefore unproductive, there are many valid reasons to remove a component from a finished PCB. The primary reason for desoldering is to remove a faulty component, and to do so without incurring damage to the board. However, there are also instances requiring that the component itself not be damaged during desoldering, such as removal of a component for failure evaluation, removal to correct a component incorrectly keyed to its footprint, removal of an expensive component for future reuse, and component removal for memory expansion or programmed component changes.

Desoldering methods are generally based upon convective or conductive modes of heat transfer. The former consists of hot gas (air, nitrogen, argon), while the latter encompasses all forms of soldering-iron tools configured for desoldering, and it also includes reflow via molten solder dip and the use of solder braid for wicking. The most popular method for lead-by-lead desoldering is a combination of conductive heating accompanied by vacuum, both integrated into a single hand tool.

In recent years, the importance of proper desoldering techniques has been more widely publicized. Coverage in trade magazines and special seminars at trade shows have brought the subject of desoldering to a much higher level of awareness. One of the more informative seminars is that given by Pace Inc. at most NEPCON shows. Pace, among others, supplies a complete line of desoldering products, typically in the form of workstations. The importance of desoldering is also emphasized by the growing number of companies offering diversified desoldering products, both for leaded and surface-mount devices.

The goal, just as for soldering operations, is to attain zero-defect desoldering.

Solder braid

One of the simplest methods of removing

First published in *Electronic Packaging & Production*, December 1984, under the title 'Desoldering strives for zero-defect status'.
Author: Howard W. Markstein, Western Editor.
© 1984 Reed Publishing (USA) Inc.

A solder extraction tool is shown performing a lead-by-lead desoldering operation removing DIPs. As the solder melts, an integral vacuum draws the solder from the joint (Courtesy of Pace)

solder without the need for special tools is by use of braided fine copper wire. The braid, prefluxed, absorbs the molten solder via a wicking action (capillary principle). The proper desoldering technique using braid is to place the braid on the solder joint and then press gently on the top of the braid with a soldering iron. An advantage of braid is that the solder is absorbed and removed from the joint at the instant it becomes molten, thereby minimizing the potential for pad damage due to overheating.

According to braid suppliers, the proper

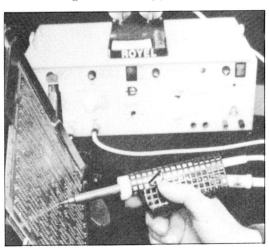

Modern desoldering tools feature a remote control unit, variable temperature control and electrostatic discharge (ESD) protection (Courtesy of Royel)

size braid to use for a specific joint size should be as wide or slightly wider than the joint diameter. Too narrow a braid may not be able to efficiently absorb the required amount of solder. Too wide a braid will restrict heat transfer to the solder because of the excessive mass. For best results the soldering-iron tip should be at least as wide as the braid. The absorbed solder portion of the braid is clipped off after each joint is desoldered.

Simultaneous reflow

The reflow method of desoldering is most avantageous when removing multilead components such as DIPs. The technique involves bringing the soldered leads on the bottom of the board into contact with a molten solder bath configured for the lead pattern, while at the same time extracting the component by use of a clip or tweezer mechanism. A built-in air module then provides a blast of air from the top side of the board to clear the holes of solder.

Reflow using heat guns or hot gas systems are also ideal for simultaneously reflowing a group of solder joints. These systems are applicable for surface-mount devices (SMDs) as well as for DIPs. A hot-air system capable of removing large SMD gate arrays, either surface mount or pin grid, is shown in one figure. Hot air is discharged from the top outlet and impinges on the component and exposed leads (if any). Hot air is also directed from under the board. When the solder becomes molten, the top air outlet is retracted and the component removed by tweezers or tongs.

Another common method of melting SMD solder joints simultaneously is by use of a special soldering-iron tip configured to contact all joints. Even unconventional tools are finding application in SMD desoldering, as indicated by an alternate use of the hot filament wire strippers.

As mentioned previously, one of the more popular methods of lead-by-lead desoldering is by use of a special desoldering tool having an integral vacuum pickup for solder removal.

Desoldering tools are also available employing a hot gas jet (also used for soldering)

Desoldering and board repair are integrated into a single workstation. This unit, in addition to desoldering tools, features a conformal coating remover, thermal wire stripper solder tweezer and abrading tools (Courtesy of Pace)

to melt the solder while a separate vacuum tool is used to remove the molten solder. Common electrical soldering irons are also often used in conjunction with these separate vacuum tools. However, today's technology favors the integral vacuum solder extraction unit.

The solder extraction unit is essentially a soldering iron with a hollow cylindrical tip for accepting the end of the component lead. This hollow portion is an extension of the vacuum system such that air is drawn through when the vacuum is actuated. Molten solder extracted from the joint is deposited in a removable collection chamber. Most units feature variable temperature control of the heated tip. Some units generate the vacuum in the control box; others generate the vacuum directly within the tool.

Solder can be removed from a joint by capillary action. A soldering iron in contact with a braid made of prefluxed copper wire forces the solder to wick up through the braid (Courtesy of Solder Removal)

Reflow is a viable technique for removing multilead components at a single pass. A small molten solder wave contacts all of the component solder joints simultaneouly, allowing the component to be withdrawn. An air blast attachment then clears the holes of solder (Courtesy of Electrovert)

The desoldering operation

When using a solder extraction tool, the heated tip is placed over the lead to be soldered and in contact with the solder. If the lead is clinched, the desoldering tip can be used to straighten the lead when the solder becomes molten. The vacuum is actuated as the solder melts, and the component lead is oscillated within the hole while the vacuum is engaged. This draws cool air past the lead and prevents a sweat joint from forming between the tinned lead and hole wall.

A preferred rule to follow for most two-sided boards is to apply heat at a tip temperature of about 302–316°C for about two seconds after initial solder fillet melting on the lead side. Then apply vacuum with the lead kept in motion for at least one or two seconds more to complete the desoldering.

When desoldering DIPs it is recommended that the corner leads be desoldered; the rest in a non-adjacent sequence to avoid heat build-up.

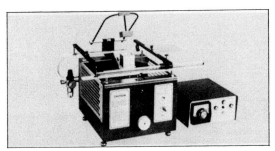

In reflow desoldering, the reflow solder wave size can be configured for a specific component lead pattern by interchangeable nozzles. The unit can be preset for solder wave temperature, height and duration of contact (Courtesy of Air-Vac Engineering)

Thermal considerations

When using the vacuum desoldering tool, the amount of heat required to melt a soldered joint is dependent upon the joint's thermal mass. More heat will be required for larger diameter component leads and pads, and if the pad is connected to multiple traces or an internal or external power plane. How this variable joint mass compares with the mass of the heat source (desoldering tip) determines the rate of temperature rise and the length of heating time required. In general, the desoldering time for a single joint should not exceed three seconds.

To facilitate heat flow to the joint, the surface condition at the area of contact between the desoldering tip and joint should be free of oxides, coatings or contaminants. The surface should, if necessary, be cleaned of any obvious impedance to the transfer of heat.

The main transfer of heat between the

Hot-air systems have proven to be efficient for desoldering surface-mount devices as well as leaded. The heated air is directed at a specific area on the top and bottom of the board, each under separate heat control (Courtesy of Nu-Concept Computer Systems)

The SMD desoldering tool contacts and melts the solder simultaneously on all pads (Courtesy of Nu-Concept Computer Systems)

desoldering tip and the joint should be via the solder mass, and secondarily through the component lead. This ensures heat flow to the high-mass elements. Direct application of heat to a pad or trace should be avoided.

Solder joints should also be inspected to ensure that there is a sufficient solder fillet at the joint surface. If not, then solder must be added to the joint before the joint is desoldered. This is necessary to provide fast, efficient heat transfer and minimum dwell time during desoldering.

Avoiding board damage

Correct desoldering procedures are required to prevent damage to pads and plated through holes. Excessive heat and pressure can lift pads and traces and cause measling of the base material. At temperatures necessary to melt solder, the bond strength of copper cladding is reduced to 20% of its normal value. Excessive pressure from a desoldering tool can easily result in delamination and pad lifting.

For leaded components, only a gentle pressure should be applied to the desoldering tool, allowing the tool to lightly descend and touch the pad as the solder fillet melts. For SMD joints, a hot gas or air jet is preferred rather than allowing a hard tool to touch the surface. This avoids potential damage due to combined heat and pressure.

Surface-mount chip components are easily desoldered by a notched desoldering tool (Courtesy of Hexacon Electric)

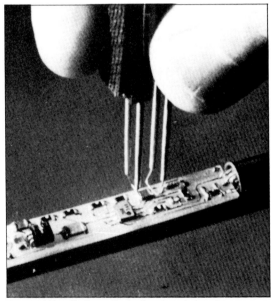

An unusual but effective application for a thermal wire stripper: desoldering and extracting chip components (Courtesy of American Electric Heater)

A plated through-hole can be damaged by excessive heat if it is used as the primary heat-transfer medium. Damage to the hole also occurs from sweat joints. Therefore, solder must be added to the joint if there is an insufficient solder fillet or when reworking sweat joints. A lead sweated to the hole wall will damage the hole plating when the component is pulled away.

Proper desoldering tip size is also a factor in preventing PC damage. A tip inside diameter of 1½ times the lead diameter is considered optimum. To avoid burning the laminate material, the tip outside diameter should not exceed the pad diameter. The tip wall thickness provides the heat path to the solder joint, therefore thicker tip walls are preferred when desoldering large solder joints and high thermal mass connections.

Tool maintenance

Some simple rules will ensure that the vacuum desoldering tool will perform as designed. Suppliers recommend the following procedure:

- Maintain a clean open hole in the desoldering tip and solder conduit. Use the proper diameter cleaning tool to clear the tip hole while the tip is hot.
- Wipe the tip on a clean wet sponge at frequent intervals, and tin the end of the

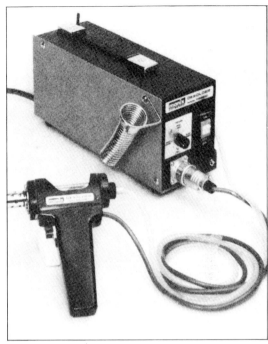

The Manix desolder unit is a typical design featuring remote vacuum generator and tip temperature controller (Courtesy of Henry Mann)

tip with rosin-core solder.
- Empty accumulated solder from the solder collector frequently. Excessive buildup will degrade performance.
- Periodically remove, clean or replace any air filters in the system.

Desoldering and ESD

Desoldering tools can cause damage to components due to electrostatic discharge (ESD) and voltage spikes. It is advisable that the surface the board is placed upon for rework is

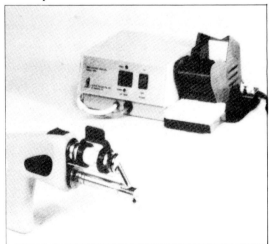

Plato's Vac-Kit desoldering unit uses a pulse vacuum rather than the conventional continuous vacuum. Pulse duration is 250ms each time the vacuum is actuated (Courtesy of Plato Products)

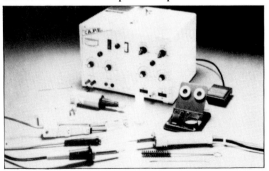

Complete repair workstations are popular for supporting PCB rework operations. These systems are also portable for field use (Courtesy of Automated Production Equipment)

grounded, and the operator is also grounded with a wrist strap or similar device. The desoldering tool must also be properly designed to minimize the magnitude of voltage spikes.

Bjorn Westly, chief engineer at Plato Products, says that electromechanical desoldering tools involving vacuum pumping systems can cause electrical damage to components during the desoldering operation. The damage is caused by transient voltage spikes transmitted to the device, destroying the layer between the gate and channel on MOS devices, even on gate-protected devices if the rise time is fast enough. Bipolar devices, because of reduced geometries, are also susceptible.

"RF voltage bursts in excess of 100V have been measured on some desoldering tools" says Westly. "These voltage spikes are usually caused by the electric circuit that controls the air or vacuum solenoid valve or vacuum pump motor. Special design and construction precautions must be taken to eliminate these sources of transient voltages."

According to Westly, several commercial and military electronics manufacturers have recognized the potential device damage caused by these voltages and have established internal standards on the allowable magnitude of transient voltages and available current that can appear on the desoldering tool tip. The standards call out maximum voltages in the range of 75–300mV peak-to-peak, with maximum current of 10–50mA.

Factors in board design

Desoldering operations can either be simple or difficult depending upon how the PCB is designed. In summary, the following board design considerations will assist any required rework involving desoldering:

- To minimize the potential for heat damage, pads should be the maximum allowable size compatible with board density requirements.
- For reduced thermal mass, power plane connections should be thermally isolated by etching an area surrounding the hole and leaving only a single or double trace connection.
- For efficient solder removal, component lead-to-hole clearance should be the maximum tolerable.
- For heat tolerance, polyimide/glass is preferred over epoxy/glass.
- For easy access, space the components uniformly and avoid unnecessary density of layout.

CHAPTER 7
TEST AND INSPECTION

The question of test, inspection and verification grows increasingly important in electronics assembly. Electronic failures in space launches have cost literally billions of dollars of waste and untold dollars in the loss of potential benefits.

In an era of consumer rights, even the least expensive electronic toy or appliance can be costly in terms of warranty costs and the loss of customer confidence in such products. Increasing use of electronics in automobiles has produced a new form of frustration when these electronics fail.

From a manufacturing standpoint improper verification of each incremental step in the assembly process can cause one to continue adding value to an already defective product.

Test and inspection must encompass each incremental step in the assembly process from verification of the quality, value or tolerance of each component all the way through to final functional testing. Increasingly, documentation of each test procedure for every unit of production is being demanded.

For each assembly process, therefore, different tests must be made. Human vision and mechanical probes prove increasingly ineffective in a world of surface-mounted devices and hybrid circuits.

New tools, machine vision, lasers and x-rays are increasingly being employed, while the nature, sequence and mode of testing are being controlled by computers.

Three basic areas of test and inspection are examined: verification of component quality, verification of insertion and joining, and verification of product functionality.

VERIFICATION OF INCOMING COMPONENTS AND CIRCUIT BOARDS

Since the circuit board is in itself the foundation on which all component insertion is based, it must be examined to see that the manufacturing processes succeeded.

Traditionally, simple circuit boards were tested by contacting the current-carrying lines on the board with hand-held probes or by gauged probes (commonly called bed-of-nails) to determine shorts, opens or other faults. Today these are being displaced by a variety of automatic probes and vision-type systems.

BARE-BOARD TESTING

In response to its internal need for bare-board testing, Kollmorgen's Electronic Equipment Division, based in Melville, NY, has developed a system that the company believes will bridge the gap between hand-held probes and the bed-of-nails tester. The system uses two computer-controlled moving probes that access any point on any bare board, then test the board for shorts, opens and high-resistance leakage. The system, packaged and sold under the trademarked name of Integri-Test 4500, permits sequential testing of any PCB up to 32,000 points over a 21 × 24in. circuit area, is suitable for testing boards designed for surface-mount devices, and because it is essentially gridless, the system will test at any location on the board with test point spacing as close as 0.020in. The moving probes eliminate the need for expensive dedicated fixtures, but the system's sequential nature limits the test speed to a rate of 350 points per minute.

Test early

It has been shown that up to 50% of all circuit board faults can be found at the bare-board level. Thus when considering the development of a moving-probe tester for inspecting bare circuit boards, the testing concept that one selects is as important a consideration as the probe positioning system. The most obvious method of testing is with two moving probes using continuity to detect open circuits. On close examination of this method, however, it will become apparent that if shorts testing is necessary, the number of tests and hence the testing time can become prohibitively long. This number can be easily calculated using the relationship $(n^2 - n)/2 =$ number of tests, where n is the number of nets to be tested. This relation shows that as the number of nets increases, the number of tests required to identify shorts will become very large.

To overcome this problem, the testing method selected combines both capacitance and resistance measurements. The two probes are directed to holes on the circuit board with one probe traversing the board in an 'S' pattern for efficiency, measuring the capacitance of every active point in its path. The other probe will move simultaneously to the end points of each net and the two probes together will perform resistance testing, identifying opens and resistive segments. With this testing concept, both shorts and opens can easily be uncovered by using capacitance for detection and resistance for verification. This method eliminates the need for redundant net testing when looking for shorts and ensures the accuracy of all test results.

Testing concept

Fundamentally, testing is performed in the following manner. A board is placed on a dielectric surface on the bottom of which a conductive plate is mounted. The capacitance between this plate and each net is measured using a precision high-speed capacitance meter. At the start of a test, a small quantity of boards are 'learned' by measuring the capaci-

The positioning system for the moving probe tester uses four stepping motors under computer control. Notice the structural stiffness, yet light weight, achieved through the use of castings and light-weight materials to obtain the necessary high speeds and accuracy. The system is capable of measuring about 350 points per minute (Courtesy of Kollmorgen Corp.)

First published in *Electronic Packaging & Production*, June 1985, under the title 'Bare board testing takes a new form'.
Author: Nikita Andreiev, Editor.

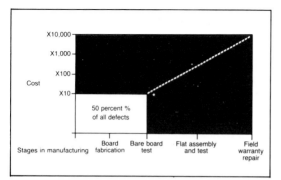

The figure shows that 50% of all board assembly defects can be eliminated if detected at the bare-board stage

tance of every point in each net with respect to the conductive plate. The computer then statistically derives the expected capacitance of each net on a perfect board, creating a test standard.

With this concept, testing any point in a good net should give the same capacitance reading as the test standard value. If a net has a fault, such as a short or open, it will be flagged because of its variation from the expected value. For example, if a good net should measure 15pf, but actually measures 5pf, an error record is written as a suspected open. If that same net measures 40pf, an error record is written as a suspected short. Since a short will have created increased and matching capacitance values in two independent nets, one has two nets that will be flagged as probably being shorted. When all testing is completed, these faults will be verified, by reprobing, using the resistance meter.

Testing of resistive segments and opens in a net are performed during the first measurement taken of each net. At this time both probes are directed to the end points of each net to test for continuity. A current of 300mA is forced through the net via the two probes and the resulting voltage drop is compared to a corresponding voltage that is converted from a programmable resistance threshold of 1–27 set at the beginning of the test. This test will quickly identify a resistance segment in the net as well as an open.

High resistance leakage between nets is detected during capacitance testing and confirmed during the verification phase of testing. After all the testing has been completed,

the probes will be directed back to the nets that were identified as suspected short. A 40V pulse will be placed across these nets via the two probes and the net resistance will be compared to one of five user-selected resistance values (usually set at 10 M), to verify the suspected shorts and leakage.

During the test phase, if the probes should measure a point and obtain an incorrect reading, the system is programmed to tap the point again to be sure that the probe has not made a bad connection. At the conclusion of this test phase, a printer will produce an error report for the board under test. The report will identify the nets by number that are shorted or opened and the X-Y coordinates of any holes involved.

System hardware

The tester is controlled by a DEC PDP 11/23 minicomputer with 256 kRAM running the DEC RT-11 operating system. The terminal used is a DEC VT 103, which provides a 'Q bus' backplane to support the 11/23 computer system. Mass storage is provided by a DSD 880 5 Mbyte Winchester with a 1 Mbyte floppy disk drive. Data input is accomplished via an

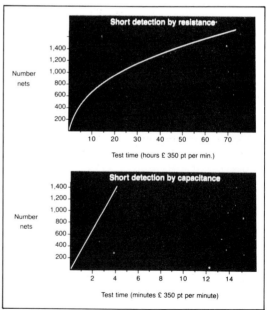

A comparison of short detection capabilities using resistive and capacitive measurement methods clearly shows the advantage of the latter in terms of time saved

8in. floppy disk. The system operating program directs the probe position, data collection, fault diagnosis and verification as well as error reporting. Probe positioning motion is controlled by four independent microstepping motors, with communication between the motors and computer over separate RS-232C serial lines. Communication between the capacitance and resistance meters is over separate DRV11-J parallel interfaces. Error reports, which are generated at the conclusion of each test cycle, are printed by an IDS 480 printer, with communication over an RS-232C serial line. Communication for other machine functions, such as table control, is over a separate DRV11-J parallel interface.

The capacitance meter used for these measurements was designed specifically for this system. It is a prcision high-speed meter capable of taking measurements in under 10ms. Measurement ranging is performed automatically in ranges of 0–80, 0–800, or 0–8000 pf. The testing frequency is 10kH and the accuracy is 0.25% of the reading, with resolution of 0.01pf.

Positioning system and probes

When developing a moving probe tester, the selection of a positioning system can have a major impact on the size and configuration of the overall machine. Since our testing concept requires two moving probes and we wanted to keep our system as small and simple as possible, we selected a dual cable driven beam and carriage configuration. This decision, arrived at after reviewing many designs, enabled us to develop a compact, quiet running, lightweight machine that would not require any special utility outlets, or floor loading considerations. The drive system we designed for our machine uses four computer-controlled microstepping motors which have 25,000 steps per revolution. They move the beams and carriages, which carry the measuring probes in X or Y coordinates with a resolutiion of 0.001in. They can accelerate the beams and carriages to 100in.s^2 and attain a terminal velocity of 10in./s. with a positional accuracy of ±0.005in. using software compensation.

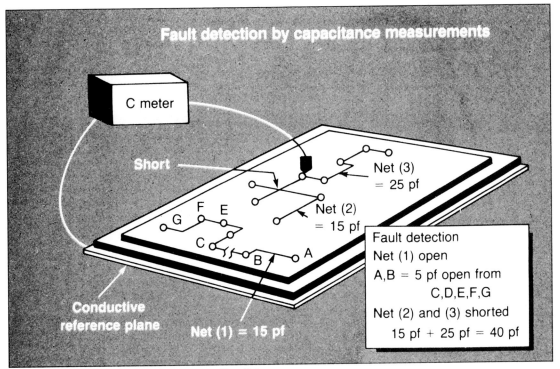

Fault detection by capacitance measurements

C meter

Short

Net (3) = 25 pf

Net (2) = 15 pf

F E
G
C
B
A

Conductive reference plane

Net (1) = 15 pf

Fault detection
Net (1) open
A,B = 5 pf open from
C,D,E,F,G
Net (2) and (3) shorted
15 pf + 25 pf = 40 pf

Measuring probe movement patterns have been programmed for efficiency

Integrated into the moving carriages are the probes used for measuring circuit integrity. They are entirely different than the 'pogo' pin type used in a bed-of-nails tester. The probe method used in the tester consists of a solid, hardened beryllium bar that is connected directly to its holder and thereby avoids uncertain electrical contact. The tip of the bar forms a pyramid that enters the hole such that it forms a three-faceted point, and is driven into place by a solenoid activated linkage and buckling spring.

This method ensures good electrical contact and provides the ability to change contact force easily by changing the buckling spring. Another advantage of this probe configuration is the ability to probe center spaces as close as 0.020in. with no modification to the system. This system can test boards of various thicknesses up to 0.18in. To ensure the proper relationship between the probe tips and board features, the testing surface must remain constant. To achieve this, a method of automatically gauging and adjusting the board mounting surface has been incorporated.

AUTOMATIC OPTICAL INSPECTION SYSTEMS

Deregulation of the US telecommunications market has triggered an increasing influx of small, feature-packed telecommunications equipment to the printed circuit board market. Rapid evolution of office and engineering computerized systems is driven by enhanced computing power, packaged in small units. In addition, military and aerospace equipment, as well as consumer electronics, compete in a market demanding 'bells and whistles' provided in the minimum possible space.

The move to fine lines

Electronic equipment manufacturers achieve their high-density computing power through extensive use of large and very large scale (LSI/VLSI) integrated circuits. These components require increasingly higher density of interconnections per square inch of PCB, forcing PCB designers to use finer line technologies and to increase the number of layers per finished board.

Yesterday's 0.020in., 0.015in. and 0.012in. width line technologies are being rapidly replaced with 0.008in., 0.005in. and even 0.003in. lines, with the projection that 0.0015in. and 0.002in. lines will be commonly used by 1990. Moreover, double-sided and four-layer boards are losing market share in favor of the 6- and 10-layer, and up to 30-layer, boards.

The rapid change in product requirements, coupled with highly competitive market conditions, forces the PCB industry to venture into the manufacture of such fine-line multilayers (MLBs) in a faster than expected pace. However, the manufacture of such boards is realized to be costly, with lower than expected yields. Yesterday's 'don't care' process inaccuracies and circuit flaws become harmful under a new set of requirements, and should be identified and dealt with in order to gain a high quality, cost-effective PCB production.

Circuit defects as viewed on the operator console of the Orbot PC AOI system

First published in *Electronic Packaging & Production*, September 1985, under the title 'AOI assists process control of fine line PC board production'.
Author: Hanan Gilutz, Orbot Systems Ltd
© 1985 Reed Publishing (USA) Inc.

AOI and process control

The drive to eliminate faulty layers from being incorporated in a finished board follows two paths. First, minimize the number of faulty inner layers through tight refinement and control of the manufacturing processes; second, identify and either repair or scrap the faulty layers prior to lamination. Process refinement may require upgrading of the production equipment, and even more important, much tighter handling and operation procedures.

Automatic optical inspection (AOI) has emerged as a tool that maximizes the benefits of visual inspection, while eliminating the shortcomings associated with human inspec-

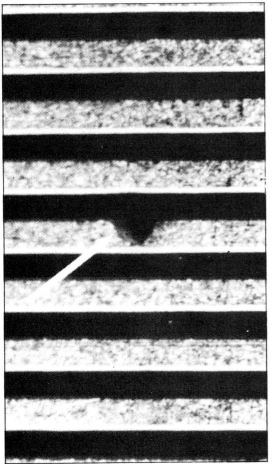

A deep dish-down cuts the circuit trace's cross section to a thickness that is below specification. This functional defect will not be caught through electrical test or by visual measurement of the line width at the line footprint

tion. Pinholes (0.003in.) or copper specks that could be harmless on a 0.012in. line/0.012in. spacing circuit are unacceptable on a 0.005in. line/0.005in. spacing circuit. This is because they cause 'near shorts' or 'near opens'.

Eliminating these flaws requires a more prudent handling of artwork and thorough rinsing of the panel after photoresist development. Harsh scrubbing of the copper-plated panel prior to resist-lamination may cause deep scratches that could be harmless when 2oz copper is used, yet cause 'near breaks' on half-ounce copper circuits.

These and other processes should be monitored and controlled in the area of etchant Ph, temperature, developer-liquid cleanliness, scrubbing-brush wear, etc. However, the ultimate control is achieved through inspection and quality assurance of the product emerging from the process. Early detection of improper artwork, leftover resist flakes on imaged panels or over-etched or underetched copper lines can support prompt identification and retuning of a malfunctioning process and prevention of repeat production of faulty layers.

Further, early flaw identification makes it easier and less costly to repair the flawed panel or to scrap it and to prevent the cost associated with fruitless further processing and handling. Meeting this goal requires an inspection technique that will handle the different mid-manufacturing products, support efficient flaw data collection and analysis, and have a cost-effective operation at production rates.

Traditional visual inspection is becoming increasingly inadequate to support this task; proper inspection of 0.008in., 0.005in. or 0.003in. wide lines and spacing requires magnifying glasses, microscropes or projection viewers to verify the thousands of pads and lines. A proper visual inspection of such an 18 × 24in. panel could take 30–60min, and the inspection accuracy would be heavily influenced by operator experience and fatigue.

The pressure to keep up with previously set production rates frequently causes operators to 'race' through the panel, resulting in high flaw escape rates that may reach up to

40–50% when large panels with fine lines are verified. The results of such high escape rates become catastrophic when those layers are integrated into multilayer boards.

Further, human visual inspection does not lend itself to systematic, on-line collection of flaw data for process control as this approach would be time consuming and would double the already high investment cost.

Electrical testing has become a common practice in the industry in verifying the circuit continuity on finished boards; it sees frequent use for inspection of the etched signal inner layers. However, electrical test does not provide for identification of line- or space-width violation, nicks, pinholes, spurious copper or similar circuit flaws. It is geared primarily to inspect for shorts or opens.

These tests cannot be performed on interim

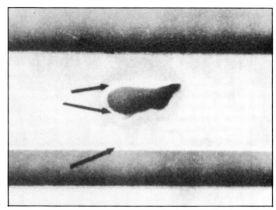

A large pinhole reduces the effective line width below specification and may cause eventual line break-out. An AOI system should distinguish between such critical pinholes on signal layers versus don't care pinholes on power or ground planes

products such as phototools or imaged panels prior to copper etch, and do not uncover the type of flaws required to support process malfunctions analysis. Thus, electrical testing remains primarily as a final-test tool that verifies line continuity but does not replace visual inspection as a tool for quality assurance and process control.

AOI systems are geared to provide an accurate and consistent circuit inspection, cutting down the high flaw escape rate experienced by visual inspectors. They are further capable of collecting the flaw data at the different production states, thus allowing process engineers to promptly analyze, identify and retune malfunctioning processes.

Their contribution is therefore two-fold: by cutting down the flaw escape rate on inner layers from 10% or more down to 0.5%, AOI supports yield enhancement of an eight-layer board from below 50% up to 96%. Further, through early identification of flawed panels and process malfunctions, AOI supports lower cost rework and minimization of scrap.

Artificial intelligence in AOI

The task of automatic visual inspection requires the AOI system to be capable of repeating the steps performed by a human inspector. The circuit image should be clearly viewed, and the circuit features should be identified and verified based on their quality requirements. The flaws found should be

---

Artwork master: silver-halide film or glass
• Systematic line (or space) width violation; may indicate over (or under) exposure, improper development or improper performance of the silver-halide coating emulsion.
• Local scratches, pinholes, or surface contaminations; usually the result of improper manual or mechanical handling during the exposure and development stages.

Phototools: diazo film, silver-halide film or glass
• Systematic line (or space) width violations; may indicate under (or over) exposure, improper development or improper performance of the light-sensitive coating emulsion.
• Regional line (or space) width violations; may indicate uneven contact between the phototool and the artwork master during exposure.
• Local scratches, pinholes, or surface contamination; may result from repeat handling when used for panel exposures.

Imaged layers after photoresist development
• Systematic line (or space) width violations; may indicate a faulty phototool or improper exposure and development.
• Regional line (or space) width violations; may indicate improper contact with the phototool during exposure.
• Cuts, pinholes, nicks; may indicate faulty phototool.
• Shorts, local spacing reduction, resist flakes; may indicate improper rinsing after development, or foreign particles (dirt, hair, etc.) that were deposited on or under the phototool during exposure.
• Minimum annular ring width violations (on outer layers); misregistration of the phototool with the drilled panel, caused due to misalignment or dimensional distortion of the phototool, or mispositioned hole drilling.

Etched bare copper layers
• Systematic line (or space) width reduction; may indicate over (or under) etch.
• Regional line (or space) width violations; may indicate inconsistent etching.
• Very rough copper edges; may indicate improper rinsing of the photoresist after development prior to copper etch.
• Nicks and dish-downs along the copper edges; may indicate poor adhesion of the resist to the copper surface, allowing the etchant to penetrate under the resist and to "bite" into the copper. Such flaws may critically reduce the cross-section area of conductor lines and may cause the line to break or to be etched open during microetch, prior to final oxidation before finished board lamination.
• Fine straight cuts; may indicate too harsh copper scrubbing prior to resist film lamination. The deep grooves allow the etchant to penetrate under the resist coating and to cut open the line.
• Shorts, copper specks or resist flakes; may indicate improper rinsing of the resist prior to etch or incomplete resist stripping after etch.

Reflowed solder outer layers
• Opens, substantial line-width reductions; may indicate too excessive copper etch prior to tin-lead fusion.
• Very fine shorts; may indicate too excessive tin-lead fusion, leading to excessive overflow that creates such shorts.
• *Note:* Actual line and annular ring width measurement are performed at the dry-film photoresist stage.

Analysis of flaws identified by AOI at different production stages

recorded for further verification or repair.

Those human tasks are translated in an AOI system to an analogous series of functions. In image acquisition, the system employs an electrooptical head to sense a light reflected from the inspected surface and transform it into an electronic signal. The system then has to differentiate between high light intensities, typical of reflective materials such as copper, and lower light intensities that represent less reflective materials, such as the base laminate.

Copper discolorations or regional surface contaminations vary the intensity of the light reflected back to the optical head. The system has to analyze the distribution of the light intensities in order to make an optimal decision as to what intensity range represents copper, and what intensity range represents the substrate. Following this determination, the system proceeds to divide the acquired image into a grid of picture elements (pixels), each one assigned a value of '1' for copper of '0' for substrate.

A further refinement of this digitization process can be achieved by reviewing questionable pixels based on their neighboring pixels. A marginal light intensity in a pixel whose neighbors are clearly identified as copper is likely to also be copper, and will be identified as such. Automatic calibration and judgment-based digitization reduces the system sensitivity to surface impurities or color inconsistencies. The same procedure is used to calibrate the system for the inspection of films, or dry film resist.

Line height reduction due to photoresist lift during etching can lead to a line break when microetched prior to being oxidized. This critical defect may be overlooked by an electrical test or by a visual inspection that relates strictly to the line footprint

The orientation of the optical head determines its capability to 'see' three-dimensional defects: a vertical head acquires a two-dimensional image of the circuit, ignoring minor surface scratches as well as deep cut or dish-downs as long as they do not affect the footprint image of the line. A slightly angled optical head allows for the identification of dish-downs and deep grooves that cut into the copper.

This approach requires good sensitivity control in order to allow the operator to adjust the system to identify flaws that are considered critical for the inspected product. For example, a given dish-down can be considered as 'don't care' on a 2oz copper, 0.010in. line, while being critical on a one ½oz copper, 0.004in. line.

When inspecting outer-layer photoresist, the width of the horizontal copper lines as viewed by the optical head may be reduced by 0.0003in. to 0.0005in., depending on the thickness of the resist layer. The system should take this effect, however minor, into consideration when performing line-width measurement.

No such compensation is required when inspecting inner-layer photoresist, or etched-copper inner or outer layers. Determination of the image pixel size (also referred to as 'inspection resolution') is equivalent to the selection of an optical magnifier by the human visual inspector. Use of a small pixel size (0.0003in.) is analogous to the use of a

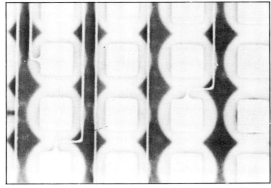

Severely overetched signal lines are detected by an AOI system, resulting in prompt identification and adjustment of the etching process

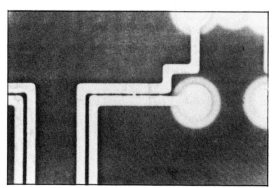

Underetched circuitry is identified through minimum spacing design rule violation by the AOI system

higher-power microscope, while a large pixel size (0.001in.) resembles the use of a regular magnifying glass.

A small pixel size leading to high magnification is required to inspect very fine features or to verify very tight width tolerances. However, when high-power magnification is used, the system becomes more sensitive to small surface dents, or foreign particles that may be deposited on the inspected area.

Further, the speed and throughput of the system are significantly reduced as many more picture elements should be processed and analyzed. For example, when the size of the pixel is reduced from 0.001in. to 0.0005in.^{2}, the system will have to handle four times the number of pixels, thus reducing the inspection speed to one-fourth of its previous rate! Large pixels will enable a high-speed inspection with less sensitivity to small nicks or rough edges. However, the system may be limited in its capability to inspect fine cuts, shorts or width reductions of less than 0.0015in. to 0.002in. size.

For optimal operation, the AOI system should allow for different pixel size (or resolution) settings. A continuously variable resolution will enable a system to operate with the optimal resolution needed to assure the required inspection accuracy without compromising the inspection speed and throughput.

Image analysis

The acquired circuit image should be analyzed by AOI in order to identify and record

functional defects. Human visual inspectors use several approaches for circuit correctness verification. In one, PCBs stuffed with components are often inspected with the use of an optical comparator that allows the operator to verify that the proper components are mounted in the right place and in the required orientation, by comparing a magnified image of the inspected board with that of a known good board.

However, this pure comparison approach is ineffective when two bare boards are compared to each other. Allowed panel dimensional distortions or feature width variations will cause a miscorrelation between the two images that could be erroneously interpreted as a flaw. Therefore, in the other approach, the conventional visual inspection of inner and outer layers and phototools is done based on a set of design rules. Here, the inspector looks for improper shorts, opens, ring breaks or unacceptable feature width variations without verifying that a given line is required to connect two given pads.

Early attempts to design AOI systems that would use straightforward comparison techniques suffered from the same problem: lack of perfect correlation between a stored reference circuit image and the image of the inspected panel resulted in improper identification of actual circuit flaws.

Today, commercially available systems primarily use the design rule concept. The system analyzes the circuit image and recog-

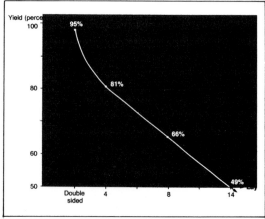

With an increasing number of inner layers, the probability of buried flaws increases exponentially, thus reducing yields

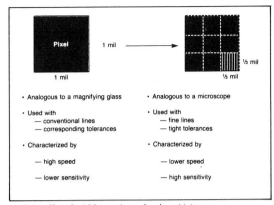

Pixel size affects the AOI system's speed and sensitivity

nizes the different features such as lines, spacings, pads, holes, or annular rings, and each feature is inspected based on its engineering specifications. The systems identify shorts, opens, mouse bites, line- or space-width violations and similar flaws.

Data collecting and reporting

The AOI system identifies the position of the flaws found, allowing the operator to determine the type of defect and to repair or scrap the inspected panel or phototool. This information, including the position and type of the flaw, can be analyzed to indicate the process that is causing these defects. Systematic copper line-width reductions may indicate an overetch, or successive photoresist panels with a cut or pinhole in the same given location may lead to a defective phototool.

Other than process-related analysis, the inspection data also supports on-line monitoring of panel production throughout and midprocess yields, allowing the user to control his work-in-process inventories and to cut down the need for systematic production overruns.

Most PCB production equipment has a fixed processing time per panel, regardless of the panel size or circuit complexity. AOI system load and unload operation is also a fairly repetitive process, requiring a total of about 10 seconds. However, the actual panel inspection time is heavily influenced by the size of the area to be inspected; even more so by the inspection resolution and the resulting speed.

While large panels (18 × 24in.) with 0.010in. line circuitry can be inspected in 10 seconds, the same panel with 0.003in. line circuitry may require one or more minutes of inspection time. In assessing the required AOI capacity, the user has to analyze his product mix based on the number of panels to be inspected, and the panel area to be inspected of the different circuit types.

The expected throughput of the AOI system will depend on the mix of those circuit types to be inspected, and will be affected when the portion of fine-line panels increases, thus requiring more use of finer resolution inspection. Flawed-panel verification and repair are done off-line and do not affect the throughput of the automatic inspection unit.

Panel inspection programs are stored in the AOI system's database and can be recalled instantaneously when a lot consisting of the same type panel is to be inspected. This short setup time allows efficient inspection of small production batches with minimal reduction in the system's overall throughput.

The Orbot PC-20 was introduced at NEPCON East '84, and was followed by the higher speed Orbot PC-30 that was introduced at NEPCON West '85. The systems are designed to be field upgradable to accept product enhancement, thus allowing users of the PC-20 system to upgrade their equipment to PC-30 when additional speed and AOI capacity are required.

Present users of the Orbot Systems, includ-

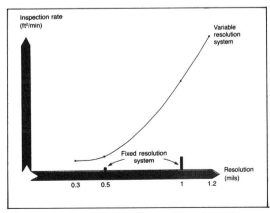

Since pixel size affects speed and sensitivity, continuously variable resolution provides the highest speed and optimal sensitivity for a given set of inspection requirements

ing Parlex Corp., Printed Circuit Corp. or Benchmark Technology, have reached inspection rates of a thousand panel sides per shift, and were able to eliminate electrical tests of inner layers.

The models, developed and manufactured by Orbot Systems Ltd of Yavne, Israel, are marketed in the USA by Orbot Inc. These systems are also marketed in Japan and in Europe.

Automatic optical inspection (AOI) systems play a major role in assisting the PCB industry in gearing up to meet the increasing demand for very fine-line MLBs. In-process optical inspection and early flaw identification supports tight process control and quality assurance of phototools, inner and outer layers. This results in cost-effective, high-yield manufacturing of advanced-technology boards.

MACHINE VISION

Machine vision (MV) is a process that involves the use of an optics system and computer to automatically acquire an image, store it in the form of digital electronic signals and perform processing on it, with the ultimate goal of making a decision about part acceptability, process adjustment or control of a machine's movement.

In parts inspection for electronics manufacturing, MV systems are being used to read part nomenclature, inspect PCBs for size and location of NC-drilled holes, look at the underside of boards to assure that leads are protruding through holes, and make sure board components are present and are oriented with correct polarity.

In addition, MV systems from various manufacturers can inspect bare boards, PCBs loaded with conventional components which come in a multiplicity of colors and shapes, and boards laden with miniscule surface mount devices. Machine vision checks out products before parts are soldered in place, inspects hybrid thick films before over-firing hardens them and checks inner layers before the stack is pressed. In all of these inspection applications, machine vision's great strength is that it catches defects before a great deal of value has been added to the product, and before mistakes become irreparable.

AOI systems

Machine vision systems all begin with a process called image acquisition, in which a camera scans a scene and transforms its optical image into an array of picture elements, or pixels. Each pixel, which has associated with it some analog value based on its brightness, can then be processed in one of several ways. In the case of automatic optical inspection, this is usually a binary digitization process where the intensity level of the pixel is compared to some reference threshold value stored in the computer. If the pixel has less intensity than the threshold, it is assigned a value of binary zero, meaning that it is converted to black. If the pixel has more intensity than the threshold, however, it is assigned a value of binary one and is converted to white. Varying the threshold settings will influence the accuracy of feature measurement and the probability of detection of different kinds of defects.

Thus, a manufacturer can shift the weight toward fewer or more false alarms, depending on how much human reinspection time is to be invested. Binary digitization will yield a rougher approximation to the true image than will some other imaging techniques, although this might be quite acceptable for applications like AOI.

The term 'automatical optical inspection' (AOI) as it is used in the industry today seems to imply inspection being performed on boards before they are populated, although

First published in *Electronic Packaging & Production*, January 1985, under the title 'Machine vision helps to acquire the image of quality'.
Author: Robert Keeler, Associate Editor.

there is an iron-clad semantic understanding here. With this meaning in mind, various designs of AOI systems are being used to inspect inner layers, bare boards, artwork film, glass masters, substrates, flexible circuits and phototools.

In these roles, AOI is specifically being used to identify, quantify and analyze such defects as shorts, opens, nicks, pinholes, spurious copper, mousebits, missing features (in some cases), traces and spaces being out of dimensional tolerance, and over- or under-sized pads, as well as incorrect hole size.

AOI can also help to close the loop for process control. Says Orbot Inc., a supplier of AOI systems, a system might identify the presence of excess copper somewhere on a board and feed this information back to the imaging stage (in the case of artwork) or the etching process (in the case of inner layer) to bring about the needed process adjustment.

Optrotech points out that up to 30% of the time and cost of manufacturing PCBs arises from visual inspection at various stages of the manufacturing process, with makers of these products being strongly motivated by the need to catch product defects early in the game. DIT-MCO International concurs that human visual inspection represents the largest labor content left in the manufacture of PCBs today. They point out that multilayer

The Electronic Assembly Verification System from Automatix is shown inspecting a surface-mount board. The artificial vision system can be retrofitted to a wide variety of chip placement machines. Blue colored fiber optic cables deliver strobe flashes to illuminate parts for the camera

boards are so expensive today primarily because of their high failure rate; manufacturers often do not inspect inner layers before lamination, even though MLBs are increasing in layer-count and each additional layer added should have an inspection.

A major difficulty for AOI systems when inspecting inner layers and bare boards is a lack of contrast in reflected light between copper and substrate. Generally, AOI systems are designed to either detect diffuse reflection from the insualtor or specular reflection from the (relatively) shiny metal. Copper will reflect light as does a mirror; substrate material, however, will typically reflect it in a scattered fashion, with the angle of incidence no longer equal to the angle of reflection.

Without proper positioning of the light source and the detector relative both to each other and to an imaginary vertical axis placed between them at the point being scanned, a patch of oxidation or other blemish on the copper can be read as part of the insulator. Still another potential problem is that missing copper can be overlooked.

Typically, AOI system hardware consists of an X-Y table and an illuminator scanner with a sensor such as a CCD camera. Most AOI systems work on the principle of binary digitization, and AOI defect detection schemes generally fall into one of two categories: comparison of the board to be inspected with a known good board as a reference, and inspection by spotting deviations of individual circuit features from design rules.

Some firms have been pursuing an approach which is a combination of both techniques. In one possible approach to a combination scheme, software calls for comparison to be done at lower resolutions, while the principle attack is still design rules checking.

The correlation method (also called the database method, 'golden board' method, comparison or pattern verification technique) makes use of the data from a known good board stored in memory. Each point of a board under inspection is compared to this standard, pixel by pixel.

The design rules method (also called the feature-checking method) looks at each cir-

Control Automation's Interscan 1500 is designed to inspect the underside of PCBs for presence of clinched and straight leads. Both illumination and sensor are contained in a module below the board. The monitor shows a live camera view of the underside of the board

cuit feature individually, checking it out against some preset circuit design rules programmed into the system. An example of a rule universal for all boards might be that there must be a pad at the end of each conductor. Other rules, peculiar to a specific design, might address line and space width.

Each type of inspection technique has its pros and cons. The known good board method is able to detect missing components, something that the design rules method is unable to do. It has also been pointed out that there is a potential speed limitation with design rules due to the need of the system to follow sometimes elaborate, decision trees for each flaw considered; to accommodate them, sophisticated pattern recognition algorithms must work at very high speeds. In addition, design rules which are too general may cause some flaws to be overlooked.

In response, proponents of the design rules method point out that missing components are becoming rare, especially with the trend to see more and more artwork being imaged on a laser plotter. Design rules advocates observe that the correlation method requires a large memory capacity and demands the rapid movement of large amounts of data. They also point to the need of the correlation method for precise edge registration (a

constraint which does not apply to design rules) and question the sometimes shaky concept of a known good board. In partial rebuttal, suppliers of correlation type systems note that the price of memory has been dropping sharply in recent years, so that this is no longer a major disadvantage.

DIT-MCO, an ATE firm which markets a version of a correlation system under the name P-SEE, contends that the throughput on both the pattern verification system and the design rules system should be about the same, if it is assumed that the speed of a system is a function of the resolution, which is ultimately related to the system lensing – a technology that in principle is available to suppliers of both kinds of system.

A spokesman for Machine Vision International feels that the bare board inspection problem is more difficult than that of the loaded board, because the resolution required is much higher. For inner layer detail on a board with 0.003in. traces, they estimate that a resolution of a 0.00025in. is required – resolution in MV is a measure of the smallest

The checkpoint 5500 Automatic Visual Tester from Cognex finds defects on both sides of each PCB simultaneously

feature which maintains a certain minimal contrast in the image; it is related to both pixel size and the inherent contrast of the object against its background.

In either kind of approach, correlation or design rules, when defects are found there are several ways an AOI system can report them to the user of the equipment. Among these are printing out the coordinates of the flaws, downloading data to other devices (such as a host computer), marking the board with ink or a sticker, displaying the results on a screen or simply storing it on disk or tape for future reference.

Optrotech USA was the first company to deliver a commercially available AOI system for inner layers, artwork and phototools. Their Vision 104 system works on design rules checking. In 1984, Orbot Inc. joined Optrotech as a member of the AOI club with the introduction of their PC-20 system, also based on the design rules method.

Other companies now courting the AOI market include Everett-Charles with their Kryterion 250, and Mania with their MOP 5000 (offered through Testerion). Automated Systems Inc. offers a system for checking inner layers incorporating Optrotech's Vision 104 system. Automation Engineering Inc. offers an AOI system for artwork and inner layers and finished bare boards, called the Circuit Board Inspection Machine (CBIM). It works by an enhanced design rules method.

Membership in the AOI system supplier club is expected to swell even more this year with Hughes (the OPTI-III system), DIT-MCO International (the P-SEE system), and Itek (the PWB Inspection System) planning to move out of the beta-site stage and deliver production systems to the marketplace. Machine Vision International will soon be introducing a system that makes use of both design rules checking and pattern verification.

Loaded board inspection

Loaded board MV systems are concerned with tasks like making sure component holes have been drilled properly, leads protrude through holes, and parts are present and oriented properly. One user of MV for a loaded board inspection task is NCR Corp. in Cambridge, OH. The company is using the interscan 1500 system from Control Automation for checking the presence of leads through holes, producing about 1/3 million PCBs per year, with about 70% of these multilayer boards.

NCR's Don Sargent, a principle engineer in the firm's Industrial Engineering Group, states that without proper inspection, it is

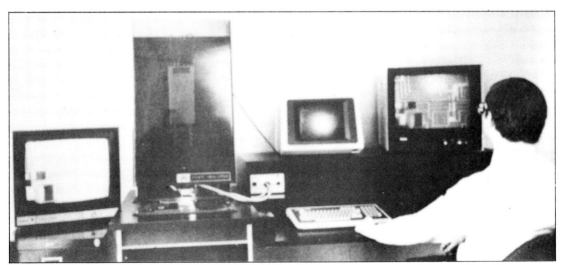

The SVS Image Processing Research System is shown analyzing a thick film electronic circuit. Left to right is shown the data acquisition unit (camera, lighting, etc.), keyboard with Cytoprocessor and 6800 controller, and the CRT display

possible to see up to 35% of his boards fail due to leads not going through holes. He points out that it costs his company about 50 cents to replace a part before it's been soldered in. After soldering, however, the cost is five dollars, and if the problem board gets to the field, it is more like 50 dollars.

Of all of the most typical kinds of loaded board problem – leads not through hole, missing part, wrong part and reversed polarity – 55–60% turn out to be pin not making it through the hole. Yet human inspectors only catch about 80% of them at most, says Sargent, who is just completing a human inspector versus machine inspection correlation study.

However, Sargent admits that even though NCR's use of MV for lead checking has been successful, he is not sure the technology of machine vision is completely ready for checking presence/absence and polarity. To illustrate, he points out that no two IC manufacturers mark their parts the same way. A system must be able to recognize many different part appearances or 'aliases', and this entails a tremendous amount of processing that ultimately means a loss of speed. He feels that humans are still faster here. Still another problem is variation in color from one resistor band to another of the same value on another resistor. There is, furthermore, even a lack of standardization in the background color of a resistor barrel. Sargent suggests that parallel processors might be the key to licking the speed problem, allowing a much quicker reduction of data.

Synthetic Vision Systems seems to concur that the biggest unsolved problem in MV today is lack of speed. An image is a very complex collection of data, involving millions of bits of data, requiring a great deal of processing. The MV firm notes one aspect of the problem concisely: once a board is put under magnification, the area that can be inspected at one time is reduced. This impacts speed, because now it becomes necessary to look at things ten times faster to keep the job within the same time frame as previously.

Besides their Interscan 1500 for checking leads for insertion through holes and proper clinching, Control Automation offers two other systems oriented toward the loaded board. The Interscan 1400 is an MV system designed to inspect for the presence, location, and completeness of holes in bare single layer and inner layer boards after fabrication operations. Interscan 1000XT, which at the time of this writing is still in prototype, is a system offering optical inspection of component presence/absence and polarity on the top side of PCBs following PCB assembly. Control Automation also markets two machine vision camera systems – the Intervision 2000 and the CAV 1000 – and they claim to have provided for proper fixturing of the board for MV inspection with their 'Easyloader' system, which uses indexing pins to match up with tooling holes in the PCB.

Cognex introduced their Checkpoint 5500 system at the ITC show in late 1983. It is a complete visual tester for loaded PCBs, detecting assembly errors on the top and bottom of each board simultaneously and integrating automatic board location, character reading, inspection, verification and datalogging technologies. After Checkpoint 5500 locates a board, it reads the board's identification code and calls up the appropriate test routine. It then captures and analyzes images of the board top and bottom to find assembly errors such as missing, misoriented and damaged components.

Object Recognition Systems Inc. is developing a system for inspecting loaded boards which will be modular in design. It will check leads through holes and direction of lead clinch, as well as looking at the top side of the PCB for presence and absence of components. The system will further have capability for inspecting SMD boards.

Itran offers the Itran 8000, an MV system for looking at presence and absence of discrete parts on the loaded board. This product distinguishes the part's edges and boundaries, and can still recognize these features despite changes in the reflections of light. As many as eight solid state or Vidicon TV cameras can be directed by one controller to do different jobs at the same time.

IRI builds the IRI PCB256 Vision System which provides real-time machine vision for applications ranging from image sensing for

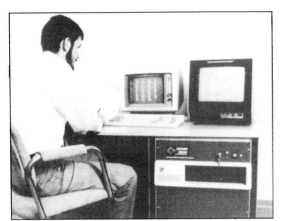

The EV 5000 from Machine Vision International can inspect both SMD and standard printed circuit boards, and uses a combination of the design rules method and correlation technique to find defects. The menu driven system also has application in process control

robots to quality control inspection of PCBs. The P256 system boasts three processors to speed up the processing and allow it to take advantage of gray scale image processing. It can inspect 100 lead insertions per sec, cover four square in. per sec per image, inspect DIPs, resistors, capacitors and other parts. The system can also inspect hole registration and part orientation.

Penn Video's system is built for PCB inspection, determining discrete component presence/absence, polarity, value, circuitry correctness and solder integrity in boards ranging up to 24 × 24in. in size and having several hundred discrete components attached. VIDOMET II and FORESIGHT are MV systems from Penn Video that are application engineered for 100% inspection of parts at production line speeds in the factory environment.

CAMSYSTEMS' MV product, called the AQC-10, is another presence and absence checker. It can accommodate unlimited numbers of components and board sizes within its 14 × 18in. scan area. Able to store user programs on 5¼in. floppy disks and print errors (missing components) on 20 column paper, the AQC-10 does not require any programming knowledge in use.

Automation Tooling Systems Inc. in Kitchener, Ontario, is offering a family of MV products for electronic assembly and robot guidance. Although their products only work on either binary processing or for gray levels, a company spokesman notes that simplicity makes high speed possible at lower cost than with some other, more sophisticated, systems. The company claims it can still handle 80–90% of vision applications with this power.

Anorad offers the ANOSCAN X-Y-Z positioning system for checking drilled holes, as well as for assisting in tuning radial components like coils before product is shipped.

Micro-Vu offers a general purpose optical inspection equipment system. Their Visual Robotics CNC Inspection System features software that is compatible with CNC software; its solid state camera and processor work with 128 gray scales, in a thresholding technique. An IBM PC works with the operational software. Two models which are capable of three dimensional measurement are offered. In particular, the Model 90-144, will inspect drilled holes on PCBs, among other things.

The Punched Hole Inspection System (PHIS) from Octek can inspect a PCB for unpunched or mispunched holes. All system operations are menu selected. It will accept a board 9 × 12in. maximum, and the operator defined inspection window, can be set as large as these dimensions. The IPCB printed circuit board inspection system, also from Octek, will inspect a board for presence of leaded components.

IBM has a system for checking the presence and location of component leads through circuit boards. This system, showed potential, says the company, for applications in raw circuit board inspection for hole presence, location and size, inspection of SMD devices and standard components (ICs, resistors, relays, etc.), for proper parts presence, location, orientation, and, in some cases, polarity.

Surface-mount and thick-film layers

Hybrid thick-film and surface-mount device inspection present special problems to machine vision. For one example, surface-mounted capacitors can be an especially difficult problem for MV systems (depending on the source of manufacture, color and substrate they are mounted on) since these

The IRI PCB256 Inspector can detect errors in fully loaded printed circuit boards

devices often present very little contrast against the ceramic substrate; their image can be confused with circuit traces and screened resistors.

Furthermore, while inspection of solder paste is easier than looking at components, after reflow it becomes a tough job. In fact, solder represents the most difficult gray level distinction problem, principally because of the very broad dynamic range of gray levels associated with a solder pad, given the specular nature of the light it reflects. Solder fillets compound the difficulty, presenting a shape analysis problem.

There are various algorithmic approaches available on the market today for inspection – some generic and some proprietary. One of the earliest on the market was a family of algorithms, binary in nature, called the SRI algorithms (they were developed at the Stanford Research Institute) and these are still fairly widely used.

While binary systems, such as are typically used in AOI for inner layers, etc., can theoretically work acceptably for the inspection problems like those presented by SMD and hybrid thick-film layers, the detection problem must be structured simply for these systems. This might result in a sacrifice in utility to the user.

As a representative of Synthetic Vision Systems points out, the ideal would be to provide a single solution for all inspection situations. Even gray-scale processing, however, a technique which allows the computer to analyze complex scenes, is not enough to totally solve the general case – no technique yet available can do this. Thus, if the need is to look at a populated substrate and make sure that components are oriented properly, a gray-scale processor will be advantageous, but it will not be enough.

With true gray scale processing, as opposed to simple thresholding, a range of gray shade values are stored in memory, and a given pixel can be compared in intensity to each of these shades. When the closest match is made, a byte of information (8 bits is enough to distinguish 256 gray shades) is assigned to describe the pixel's value. Thresholding, on the other hand, would ultimately choose between just two available values.

Today, highly sophisticated MV systems perform both pattern recognition and image processing. Pattern recognition (PR) is concerned not with the picture itself, but rather with descriptions of the objects in the picture. Thus, it deals with numbers that measure features like area, width and length – i.e. spatial relationships between objects in the picture. Pattern recognition can be divided into statistical PR, which deals with measurements, and syntactic PR, which deals with the arrangement of things in a picture.

Image processing, on the other hand, is the transformation of the image or picture in order to enhance those details of the image which are important, and to suppress those which are unimportant. Image processing can be broken down into two domains: signal processing to reduce extraneous detail (examples of two-dimensional signal processing include the Fourier Transform, certain edge detectors, and certain noise averaging techniques), and a technique called Mathematical Morphology.

The most general way of processing gray scale images is by using the method of neighborhood analysis, which technically can be subdivided into the specific techniques of two-dimensional signal processing and mathematical Morphology. It is also useful in handling binary images. With neighborhood analysis, pixels are not assigned values individually, but rather with reference to their surrounding eight neighbors (some systems look just at neighborhoods of 3 × 3 pixels, while other neighborhood arrays are considerably larger). Neighborhood analysis allows the user to do edge detection and to suppress noise spikes, among other things.

Mathematical Morphology, a specific case of neighborhood analysis, is a style of image processing developed originally at the Environmental Research Institute of Michigan (ERIM). It is basically a mathematical logic for dealing with images on a conceptual basis. Akin to Boolean algebra, this method is used in Synthetic Vision Systems' approach to the problem of inspecting hybrid thick films, and in modified form in Machine Vision International's attack on the SMD machine vision problem.

In practice, Mathematical Morphology uses a probe called a structuring element to examine the spatial characteristics of the traces which are being inspected, moving it around the trace, as SVS says, like an 'electronic marble' in operations called erosion and dilation which are carried out to search for both breaks in the traces and dimensional rule violations. Mathematical Morphology, however, is highly computation-intensive, so that systems using this technology demand the use of parallel processors.

Machine Vision International notes that proper inspection for each of the cases, of bare board, loaded board, SMD and hybrid thick film, requires both statistical and syntactical pattrn recognition, as well as two-dimensional signl processing and Mathematical Morphology.

Machine Vision International is soon to introduce an MV system using their Image Flow Computer, which evolved from the original machine vision technology of ERIM, and is adapted for use with SMD boards. The company's EV-5000 will have an IBM PC as the system controller. All database and configuration data will reside in the bulk storage device associated with the IBM/PC. The EV-5000's software comprises application/

The Circuit Board Inspection Machine (CBIM) from Automation Engineering is an automated optical inspection system which combines a proprietary optical system with CCD array sensors, a multiprocessor controller, and pattern recognition software to solve circuit board inspection problems

inspection packages, interfaces for operator and computer, database management and system control. Application/inspection software is developed on MVI's GENESIS Development System, using MVI's BLIX language, which directs morphological operations to extract specific features even from complex low-contrast images. Typical application tolerances for SMD component placement are plus or minus 0.0003in. for x and y, and 3° for theta. System resolution can be as tight as 0.001in.

The Teknispec 1000 from Integrated Automation is a machine vision system intended for high speed inspection of surface-mount boards after component placement, capable of inspecting up to 75 component placements per second. This system, which is knowledge-based and requires a stored description of each board, uses pattern recognition to find the exact location of each SMD and its solder pads.

The Autovision 3 from Automatix is a SMD inspection system that works either as a standalone system or integrated into existing chip placement equipment. A high speed and gradient pre-processing capability here enable edge detection for verification of chip presence, location and rotation.

The Pixie 1000 system from Applied

intelligent Systems Inc. is a gray level vision system which uses neighborhood processing rather than the SRI approach to machine vision. It has a parallel processor with a pipeline architecture which sees application both in hybrid component presence checking, and in verifying the presence of leads through holes.

Photo Research Vision Systems of Kollmorgen Corp. is offering the MIDAS (Modular Inspection Data Analysis System) PR-830 Hybrid Circuit Inspection System series. It is a fully automatic system that can inspect and evaluate thick film and thin film hybrid circuit substrates.

Synthetic Vision Systems has had a hybrid thick film inspection system installed at Ford Motor Co. for approximately a year now. The Cytocomputer-based system, known as the Thick Film Inspection System (TFI), is designed to provide fully automated process control capability for the inspection of complex high-density hybrid circuits. It offers full gray-scale processing to provide the detailed segmentation required in thick-film inspection – segmentation meaning the differentiation between the multiple thick film layers visible to the camea. This system is designed to detect a wide variety of flaws, holes and misalignments, in thick film circuitry, checking for minimum path width, path spacing, voids, edge alignment, internal layer alignment, and resistor/conductor overlap. It will also check the proximity of void edges.

Micro-Vu also offers a system for inspection of hybrid thick-film layers called the Visual Robotics CNC Inspection Machine. This system can also inspect loaded PCBs for presence and absence of parts, location of drilled holes and part polarity.

The V-100S inspection system from Intelledex, using the SRI and other algorithms, can inspect PCBs for presence of parts. It also finds application in checking that the right amount of adhesive has been applied to PCBs for surface-mounted components.

VERIFYING COMPONENT QUALITY

From the very first development of automatic axial lead insertion machinery for printed circuit board assembly, emphasis was placed on verification of the nature, value and/or tolerance of the inserted item. Since these devices had metal lead wires it was relatively easy to contact these leads with probes and measure their value at the moment of insertion.

The nature of today's assembly, the density of insertion, and other factors, mandate other approaches to this problem.

COMPONENT TESTERS AND HANDLERS

In a recent interview, a consultant in the field of ATE remarked that Incoming Inspection is often the lost orphan among the departments of a manufacturing company. For years it has been caught in the double bind situation of being given little budget allotment while having to test or inspect an enormous variety of parts. Typically, a PCB manufacturer's mix will consist of such components as discrete devices (both passive and active), linear and TTL devices, and possibly some microprocessors, peripheral chips and memory.

This problem has been recently compounded by an important new trend to make chips with both analog and digital technology on board – a fact that is presenting Incoming Inspection with one of its toughest challenges. The availability of new testing systems, ranging in price from several thousand dollars to almost $1 million, is finally beginning to pull Incoming Inspection off the horns of its dilemma.

Solutions

Traditionally, Incoming Inspection departments attempt to solve their quandary in a number of ways. One method is to employ a testing service – which means farming out to someone else the headache of dealing with the tremendous variety of components and the burden of purchasing expensive equipment. A variation of this approach is to set up a central test service within one's own company. This system, which because of its cost is really only feasible for large companies, involves having components for the entire corporation move through one large facility, even though divisions are geographically far-flung. Economically, this has its advantages, but it can lead to some obvious scheduling and logistics problems. Yet another method is to carry out vendor auditing, with the goal of ultimately certifying suppliers as being reliable. Hopefully, it then becomes necessary to do only occasional sampling of parts.

First published in *Electronic Packaging & Production*, May 1984, under the title 'Incoming inspection needs component testers and handlers that are versatile'.
Author: Robert Keeler, Associate Editor.

This solution for Incoming's problem may be the wave of the future and is the path chosen by many Japanese companies. It is not without limitations, however. It often locks a company in to only a few suppliers – those that can pass muster – and in some cases forces them to be single sourced. The applicability of this approach has a lot to do with how well a given part has been characterized, how mature its technology has become and what sort of failure rate it has. In fact, these factors are usually more important than the actual level of integration present on the part. Still another solution available to Incoming Inspection has been to purchase inexpensive yet versatile test equipment, something that happily is becoming more available all the time.

Mixed-signal testing

There is a major trend now to put digital capabilities on what were once strictly analog chips. This affects a number of applications, including consumer electronics such as digital hi-fi and digital television. Having both analog and digital circuitry present is called mixed-signal, also known as dual domain or analog LSI. In this kind of technology a tester with the flexibility to test a wide range of parts is needed. Typical analog-digital devices include digital-to-analog and analog-to-digital converters, CODECs, tone generators, filters, analog microprocessors, solid-state image sensors, read/write head amplifiers, flash converters and digital-to-analog converters with a non-linear transfer function (COM DACs).

A good mixed-signal tester should be able to do digital-signal processing, generate complex bit patterns at megahertz rates, and supply programmable waveforms like sine waves, triangular waves, etc. It should also be able to force and measure very accurate signals with a resolution of microvolt and picoamp resolution levels, and convert analog signals to digital form, analyzing the results. Most linear testers do not have a lot of digital test capability and most digital testers fall short when it comes to linear capability. In the

The test electronics of the MCT 2000 test system are shown exposed in the test head, which will support up to 64 of the pin cards shown with no retrofit

past, therefore, the incoming inspection user had to make some compromises on one side or the other of the test. It is only quite recently that we see testers appearing that can do justice to both needs.

Component tester performance

The first order of business when considering the purchase of ATE for Incoming Inspection is to ask what sort of components will be tested, including the families of logic that will be seen. Any good bench-top tester for Incoming Inspection of PCB components and devices should meet certain performance criteria. Cost and versatility will probably be the most important factors for the smaller company. A unit of ATE should deal with as many parts as possible, at an affordable price. Tester manufacturers that can offer large test program libraries and quick-change family boards are most likely to satisfy this need. Some systems offer easy menu selection of tests, but a good system will allow the user to alter the kind of tests offered on the menu by using a common programming language, such as BASIC. This can be important when the part to be tested simply does not fit existing menu categories.

Another capability which helps is being able to correlate data taken at Incoming Inspection with data gathered by the outgoing test facility of the parts suppliers. Ideally, Incoming will use the same kind of test as their supplier, calibrated to the same standard. One obvious way to achieve this is to purchase test

equipment made by the same manufacturer as is used by the supplier. This allows test personnel to better judge the manufacturing capability of their supplier. Other important considerations are simplicity of operation, as well as proper accuracy and resolution for the kinds of parts that will be seen. Furthermore, a good system will do some sort of bin sorting of tested parts to separate bad parts from good.

Finally, note must be taken of the trend toward linking testers in a networking scheme. Those ATE suppliers that offer some sort of networking capability should be given a second look. Networking allows the experience gained in testing incoming parts to be put into a database – this can pay off later when selecting those vendors with the most reliable parts and in choosing the most appropriate tests to perform. It can also help to determine if burn-in is economical, by keeping track of pre-burn-in and post-burn-in data.

Some testers offered

Siemens-Allis builds an MPU-based bench-top digital IC tester called the Model 725. This 24 pin, high-speed IC tester offers functional and parametric test capability, along with other features such as lot reporting and data logging. For the low- to medium-volume end-user such as Incoming Inspection, a standard configuration with a single manual test site is available. Siemens offers a library of magnetic tape cassettes on which device programs are recorded, with each library program tape containing approximately 50 programs. This format is intended to facilitate the rapid program setup and changeover necessary for Incoming Inspection applications. Access to data logging and reporting features, and the English language-prompted program development capability, is obtained in this system by connecting optional terminals to the standard RS232C or 20mA loop I/O ports.

GenRad offers an entire line of bench-top component testers that are designed primarily for Incoming Inspection. Their 1735 Component Test System is for digital, linear and discrete components, while the 1731M is primarily a linear IC test system and the 1732 is purely a digital IC test system. The 1734 is a memory IC test system and the 1689M is an RLC bridge – part of a family of four bridges from GenRad. The 1735 tests linear, digital and discrete devices – it is for those testing low to moderate volumes needing a broad coverage. The 1731M, 1732 and 1734 are for higher volumes. GenRad systems program with menus using 'fill-in-the-blanks' programming, where the user answers the prompts with appropriate limits and specifications, and are backed up by a large program library. GenRad's networking system, the Semiconductor Analysis Network (SCAN), is a work-center automation product. They advertise it as being able to turn testers into workstations. It permits off-line program generation and test data management, including data logging of test results and report generation. This last feature helps with vendor assessment and decisions regarding burn-in and test optimization.

Eagle Test Systems provides their LSI-3 and LSI-4 systems for testing digital and linear ICs, with AC and DC parametric capability. The LSI-3 will test microprocessors with up to 64 pins; the LSI-4 will handle up to 128 pins. The pin electronics in both systems are identical. However, in the case of the LSI-3, what drives the pins is a single-user, single-tasking 8-bit computer, while in the case of the LSI-4, it is being done by a multitasking, 16-bit computer with a full 16-bit operating system. The LSI-4 lends itself to networking with either Ethernet or DR Softnet. With the LSI-3, programming modifications can be done in BASIC or PASCAL, while with the LSI-4 these can be done with BASIC, PASCAL, C, FORTRAN and other languages.

Micro Component Technology Inc. (MCT), markets a digital IC test system called the MCT 2000, which is capable of AC and DC parametric testing as well as parallel dynamic functional testing. For DC parametric testing, the firm has implemented three voltage ranges and six current ranges, using D/A and A/D converters. For AC parametric testing, four measurement ranges are available with a clock rate capability of up to 16MHz. Highly

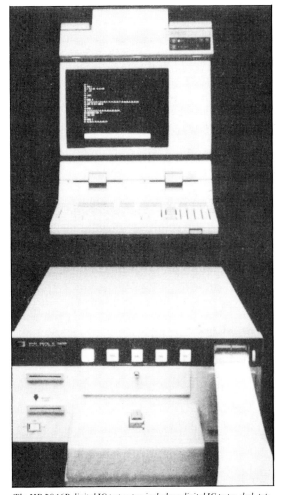

The HP 5046B digital IC test system includes a digital IC tester, desk-top controller and thermal graphics printer

accurate measurement is claimed here for rise and fall times, edge transitions, and delay times. Regarding dynamic functional testing, the system will test up to 16MHz. Each pin card has its own programmable data formatter, high-speed pin driver circuit and high-speed comparator circuits.

The 5045A digital IC tester from Hewlett-Packard can be used by an unskilled operator. Testing a device involves inserting a pre-programmed magnetic card into the front panel slot. The tester is then ready to test any of the devices listed in the program catalog, which has over 2300 device programs currently available. The universal pin electronics in the 5045A allows each pin to act as either driver, receiver, clock, power supply, input or output. HP notes that this provides the flexibility to test circuits from basic gates to ALUs, ROMs and RAMs without hardware changes. Each device's program package comes with both pass/fail and diagnostic program cards, plus backup cards for each. HP's 5046S digital IC test system is built around the 5045 tester, the 9826 or optional 9836 desktop controller, and 2671G printer. This system offers the ability to change parameters in minutes, and write a completely new program in days. Also, the user can edit and program on-line with the company's 9826 or 9836 controller. Networking is through HP's Q-STAR system.

Analog Devices Component Test Systems offers several lower cost bench-top tester models. Both their LTS-2010 and LTS-2000 are flexible linear test systems that include software controlled automatic handler interface, IEEE 488 interface, integral printer and integral statistical analysis package. The LTS-2010 can be programmed in BASIC; the LTS-2000 can not. However, the LTS-2000 is suitable for basic go/no-go testing for compo-

nent selection and grading, for qualification testing or as a diagnostic tool for component evaluation. The LTS-2010 provides an overall system measurement accuracy to 16 bits, as does the LTS-2000. Both can be programmed in minutes using 'fill-in-the-blanks' programs. Test results can be stored on floppy disk for later processing, or they can be communicated directly to a host computer via the bi-directional RS232C port. Devices interface to the LTS-2000 by means of family boards and socket assemblies. The LTS Series allows the user to test digital, linear, passive and discrete devices with a single bench-top test system.

From Tektronix comes the TM 5000 line of programmable instruments, which include among their capabilities the ability to be used during incoming inspection testing of digital and analog devices, RAMs and ROMs. The TM 5000 includes two counters, a digital multimeter, two power supplies, a function generator, an audio distortion analyzer and a sine wave generator. There is also a 16-channel scanner which is a switcher for handling rf signals. This can be configured as four 4 to 1 switches, or one 16 to 1 switch or

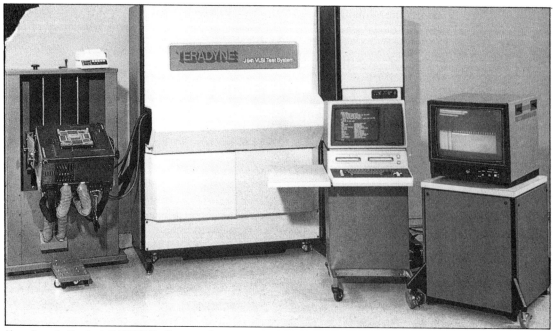

New software for Teradyne's J941 VLSI Test Systems speeds program generation, debugging, and system calibration

two 8 to 1 switches. The particular instrument that will carry the brunt of the load for the Incoming Inspection station here is a multifunction interface system called the MI 5010, which with its extender unit will accept up to six function cards capable of controlling a variety of functions within a test station. These cards include A to D and D to A converters. There are also two different relay scanner cards. One is a 16-channel relay scanner card for single-ended switching. The other is a differential low-level scanner card with less than 1V of differential offset in any signal path. Additionally, there is a digital input/output card which will put 16 channels of TTL logic out and will accept 16 channels of logic in. Thus, it can be used as a controlling device for many non-GPIB programmable instruments, or as a sensing mechanism which can be used to sense the condition of up to 16 lines of various sorts. Alternatively, it can be used strictly for driving TTL devices and determining what the output is. Thus it can act as a logic tester.

Pragmatic Designs Inc. offers a tester with 22 pins of digital testing capability, and 28 pins of linear test capability. Their tester, called the TM7B, is suitable for testing mixed-signal devices, and it has the capability to do parametric measurement on digital pins. Standard configuration cards are available for testing op-amps, comparators, and D/A converters. Users can design their own configuration cards, or the company will design configuration cards for special applications. The test head here can be interfaced directly to the firm's Inspector 200 Industrial Microcomputer System, which communicates with it over a general-purpose S100 bus. The 200 features an 8-bit 8085 microprocessor operating at 6MHz, and a 16-bit 8088, operating at 8MHz.

For discrete component testing at Incoming Inspection, CIS Testronics offers their model 201A programmable Curve Analyzer. The firm advertises this system as being able to test a wide variety of discretes, including bipolars, Darlingtons, SCRs, triacs, optocouplers, LEDs, IREDs, zeners, diodes, FETs, regulators, alphanumeric displays, and multiple transistors in DIP packages. The 201A will perform six types of component test: leakage current, breakdown voltage, component gain, gate triggering parameters, and holding parameters. Testing here may be either go/no-go testing to preset limits or can involve measurement of actual limits. Data logging of absolute values is available, and tallies of test passes by bin category is also offered. Networking of up to eight of these units on a single HP-85 computer is possible.

Megatest Corp. offers a memory tester called the Q2/52 for Incoming Inspection. The user simply loads in a new program when parts change. It is particularly suitable for testing non-volatile memories, although it is also used for testing dynamic RAMs – up to four in parallel. Networking here is accomplished through MegaNet, the proprietary networking system developed by Megatest Corp. The system will also test EPROMs and EEPROMs. Two independent options are available: a data buffer memory to test ROMs, and a high-voltage module to electrically repair redundant memories. Off-line analysis of the test data is possible here, with output in tabular or histogram format.

Megatest also manufactures several testers that are aimed at the user with larger volumes of parts. One is the Q2/62 Logic Test System, which was created basically as a production machine, although it can serve in Incoming Inspection with the addition of a Signal Routing Card (SRC). This will take the signals from the instrument and route them to the pins. The high-volume user would design his own routing card. The same company also offers the MegaOne – a VLSI test system which has what the firm refers to as 'tester per pin architecture' – putting complete and independent tester resources behind every pin of the test socket. The user basically describes what he wants to happen behind each pin, using the PASCAL language, and the tester takes over. The system can test many kinds of digital device, including memories, serial devices and microprocessors, among others.

LTX Corp.'s LTX77 Integrated Test System finds application in the testing of pure digital and pure linear devices and in mixed-signal circuits where higher volumes exist.

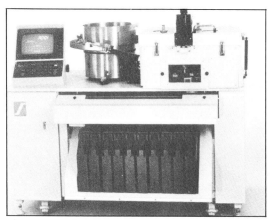

The system 600 from Sym-Tek is a discrete component environmental handler. Input is a bowl feed and the system is CRT programmable. Temperature range is −55°C to +155°C

LTX has integrated their instrumentation and software for testing pure linear and pure digital parts so as to make the system effective for testing devices that include both linear and digital circuitry. For linear testing the system includes a precision DC source and measurement modules, rf test electronics, time measurement circuitry, and simple and complex analog signal generation; for digital device testing, it offers a versatile pattern-generation capability (both stored and algorithmic), formatting, independently programmable high-speed drivers and comparators, both window strobing and edge timing and array processing. The LTX77 includes a special 4k × 1 memory behind each driver/detector pair, and high-speed burst interface.

Production Automation's CTS-2030 discrete component test system enables users to perform full functional testing of many different components. Active and passive discrete components testable with this system include resistors, capacitors, inductors, potentiometers, transformers, passive networks, bipolar transistors, diodes and SCRs. In addition, the company says it will customize the CTS-2030 to enable the user to add or delete component types tested. As well as measuring parameters critical to each part, the CTS-2030 provides complete test data upon demand. These might include a summary of all measurements and parameters requested for any test run, test-run identifica-

tion information, and test results by measurement for each part tested. The CTS-2030 system is composed of a variety of Hewlett-Packard instruments controlled by an HP 200 series computer.

Fairchild Test Systems Group offers several systems of use to the Incoming Inspection department. For digital testing, the Series 10 and the Series 20 are modular systems in which pin count can expand to 120, if desired. The Series 20 offers up to 40MHz data rates, and is available in three configurations – a 60 pin LSI tester and both a 90 pin and a 120 pin VLSI tester. The Series 20 can handle MOS and bipolar memories, microprocessors and peripheral chips and others. Fairchild's Sentry 21 VLSI tester is a high accuracy automatically deskewerable system which can also be used in Incoming Inspection. The Fairchild Series 80 analog tester is specifically targetted to meet the needs of inspectors doing linear and mixed signal testing. The company's Models 5588 and 5582 are memory testers suitable for Incoming. The 5588 can directly address 16 Mbits of data. The 5582 is a lower-priced version of the 5588. Fairchild calls its networking system the Test Area Manager (TAM).

The A360 test system from Teradyne is a low-cost system that can be customized for different applications since it is modular. It features compact packaging and will handle a wide variety of mixed-signal devices. Modules added-on allow users to test different classes of devices. A new universal test head will allow users to test devices either manually, with automatic probers or an automatic handler. The A360 system can be equipped with up to four of these test heads. Teradyne features a networking system called Teranet, and several Teradyne systems can be configured as part of a Teranet system. Two host computers, which Teradyne calls Test System Directors, are offered as options: one based on the DEC PDP/11-44 and the other on the PDP/11-24.

The J386A Memory Test System and the J941 VLSI Test System from Teradyne are both for use in Incoming Inspection. The J386A will test all common RAMs and ROMs including memories with up to 256K addresses. It interfaces to standard handlers with a single cable interface set, and uses the same PASCAL-T language common to all Teradyne memory test systems. Teradyne's JN41 VLSI Test System can test devices up to 96 pins. The pattern generator here holds up to 264K bits of pattern for each pin to allow long device patterns to be stored intact in memory. Each device is tested completely in one pass.

The T320/44 test system from Takeda Riken has been developed for use in production testing as well as Incoming Inspection of multipin, multifunction LSI devices. It can accommodate up to 128 pins with test rates of up to 20MHz. A 16K word-pattern generator, 8 Mbyte super-buffer and high-quality waveforms are available at the test head. The T320/44 uses a data processor to enhance data analysis and evaluation, making it useful in both R&D and Incoming Inspection. Systems can be interfaced to a local area network through the TESTARIUM data-management system, which permits centralized management of test programs.

Handler performance

Perhaps the two most important things to consider when choosing an Incoming Inspection component handler is whether or not the system can do thermal conditioning of parts prior to test, and whether or not it is easy to operate. Many companies feel that testing at room temperature is a waste of time, claiming that it merely uncovers catastrophic failures – like opens and shorts – problems that IC manufacturers probably already do a good job of finding. Thus, it is claimed that the real objective is to find those failures that the IC manufacturer does not test for on a routine basis. Thermal conditioning can comprise both cooling a part and heating it up prior to test, although usually only military and aerospace or automotive applications require cold testing. The basic idea behind hot testing is that ICs that are parametrically marginal will show up unquestionably bad when the parts are tested at elevated temperature. Most manufacturers refer to this environmental test capability as a 'hot rail'. The part slides along a temperature-controlled rail in an enclosed environment and absorbs heat from the rails until it is at the temperature specified

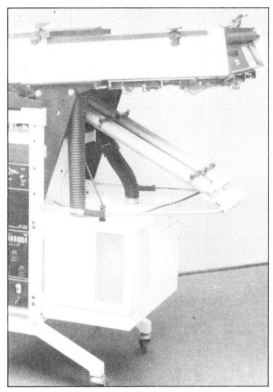

The Model 1108-HSS is an elevated temperature, dual-line IC handler for Incoming Inspection from Siemens-Allis, Inc. It features an option for direct test-head docking and air-track feed and sort

for a given test. Another way to do hot testing is to use hot turbulent air to heat the device leads, which conduct heat to the die. Some companies use a combination of the two approaches for maximum efficiency.

An Incoming Inspection manager will probably want a handler that does not require a high level of technical skill to use. Thus, ease of setup and operation are factors to be considered. For example, some handlers require that new rails be installed or that the equipment setup be modified if there is a part changeover. Small companies often deal with small lot sizes, so it is not uncommon here to have equipment changes a dozen times a day. Thus, it behoves the end user to insist on seeing the operating manual for the equipment he is considering, to get some idea of its complexity.

In small companies, the component tester is usually a bench-top model with a relatively slow speed, working even slower when testing complex devices. Thus, the throughput capability of a small company's handler is not as big an issue as it would be with a large company. Similarly, adaptability between most commercially available testers and handlers is not of concern these days except with really inexpensive testers.

It is wise to check out a handler for the ability to process ceramic parts without damaging them. Most handlers work on simple gravity feed, and while plastic parts present no problem in this area, ceramic parts must be slowed down in their descent down the handler rail to avoid the danger of inducing microcracks or chipping. The handler's design should take this into account.

Some handlers offered

Micro Component Technology Inc. offers a line of Incoming Inspection handlers, as well as its line of testers. Their 3608AE can handle SSI/MSI devices, LSI and memory devices at either ambient or elevated temperatures. The 3608E is designed to provide high speed, elevated temperature-handling capability with tight temperature stability, repeatability and linearity for SSI/MSI and memory devices. The 3608C handles the same devices at temperature ranges from ambient to −60°C within the tight tolerances demanded by automotive and military applications. In addition, MCT's Model 3616DTS is available for elevated temperature handling of memory devices. MCT is also offering something called the MATRIX evaluation system to optimize the handler buying decision. It is advertised as helping the purchaser take into account all criteria related to the device mix to be tested and the test environment before the final selection is made.

Exatron, which also makes digital IC testers, offers their Model 2900 Large Package Handler, as well as their model 2500, five sort DIP handler. They also offer their 2000 Series of handler options. The Model 2900 handles devices 0.600in. wide, to as large as 64 pins, 0.900in. wide Exatron claims that conversion from one size DIP to the next only takes seconds with very few adjustments. The Model 2500 can handle devices as small as

eight pins, 0.300in. wide, to as large as 40 pins, 0.600in. wide. On both models, universal guide rails and input/output tube holders require no operator adjustments. On all Exatron handlers, sample devices serve as gauge blocks to eliminate package thickness adjustments. The company's 2000 Series of Handler Options include a test station cart to provide a dedicated portable test/programming location, the Model 2510 Hot Rail for elevated temperature handling to heat 0.300/0.600/0.900in. devices from ambient to 125°C in less than five minutes. There is also a spare parts kit, and a test head stand diagnostic box which allows users to check all the functions of the handler, including solenoid, sensors, switch and I/O ports. Any of the eight sort signals can be sent to the handler by pushing one of the buttons on the diagnostic box. The handler will then respond to the signal allowing semiautomatic operation.

Trigon offers its Model T-2080 Twin-Tilt environmental handler for Incoming departments that are doing fairly high-volume testing. At the other end of the volume spectrum, Trigon offers its Model T-2045, also with environmental conditioning, and suitable for the very smallest users. Both units feature size adjustability, and are able to process devices with lead counts ranging from six to 40, in 0.300in. to 0.600in. sizes. Changing from one size to another is accomplished by using digital dials and calibrated scales for the measurement chamber, and two quick-change self-aligning parts for the soak chamber. In the case of the T-2080, in addition to the sealed chamber over the contacts there is a large soak chamber to preheat components before they reach the point of measurement. Trigon uses its Loop Chamber to heat the part by enveloping all surfaces of the package in a turbulent flow of heated air, in addition to the use of a heated rail. This, they claim, increases system throughput.

Daymarc's 757 QuadSite test handler is designed for use with 0.300in. DIPs, 14 to 20 pins standard, with 24-pin capability optional. The QuadSite will accept plastic, ceramic or side-brazed ICs. It will self-load, bring to temperature, transport to the test head, sort and reload into tubes up to 6000 of the 16-pin DIPs per hour. Proper soak time for each device is assured here by a unique Ferris wheel arrangement of holding tracks that prevents any shortcuts through the heating chamber. Daymarc claims this technique assures set point temperature uniformity throughout the heating chamber and contact areas to guarantee thermal accuracy. The device to be tested is always visible through transparent panels, enabling the operator to spot an abnormal condition. RS232 and 488 interfaces are available to facilitate networking, and device sorting is closed-loop to eliminate part intermixing. Besides their 757, Daymarc offers the Type 149 Contactor for axial/radial taped components test handling, and a variety of handlers for ICs and discrete components.

Sym-Tek Systems Inc. makes several tester handlers for Incoming Inspection. Their System 300 handler is suitable for 0.300in. DIPs, while the System 340 is designed for handling 0.300in. memory devices. In addition, their System 500 handles small outline surface-mount devices which are now replacing 0.300in. DIPs in many applications. All of these handlers can operate over a temperature range of −55°C to +150°C and all have automatic loading from part sleeves. All are CRT programmable, and the test site is totally enclosed to maintain temperature integrity (some handlers just heat the device 'on the rail', but have no provision for keeping the part in a thermally conditioned environment during the actual test). The System 500 specifically can test both 0.300in. and 0.150in. devices with only a change of the test-site mechanism being required. The Systems 300 and 340 are virtually the same machine, although the 340 can test four devices at one time, with each one of the test sites being able to operate independently.

Several DIP handlers from Siemens include one for ambient, and one for elevated temperatures. The ambient model is the 1108-AS, which is a dual-in-line IC handler that has an eight-bin output sort. Features here include a direct test-head docking option and a low velocity air-track feed and sort system. Parts are always exposed in this ambient handler so that feed interruptions can always be cleared with ease. Siemens' Model 1108-HSS elevated-temperature DIP IC handler can heat parts to 125°C, and it has a large-capacity hot chamber input (16 sticks). Temperature-controller accuracy at the test site is ±1°C and acceptable device body widths include 0.3in., 0.4in. and 0.6in. Audible alarms and a visual display are provided to identify feed interruptions.

HANDLERS

If you had asked the typical packaging engineer a decade or so ago about handlers, he very well might have responded with remarks about the airport porter. At that time, the automatic test equipment industry was in its infancy and handlers, for the most part, were simply mechanical devices used to pick and place components onto printed circuit boards.

The precise geneology of handling technology is unclear but, like ancient Gaul, having evolved into specialty segments, is divided into three parts. In no hierarchical order, one segment deals with 'pick-and-place' machines. That is, machine systems load components onto printed circuit boards or assist an operator, in the case of low to moderate volume production, in assembling boards efficiently. Contact systems of Danbury, CT, and Universal Instruments Corp. of Binghamton, NY, are examples of two leading firms who make and market sophisticated pick-and-place machines.

Universal's Model 4712 Onserter reflects the newest advances in PCB assembly: surface-mount technology. This component

First published in *Electronic Packaging & Production*, April 1985, under the title 'How to handle a million chips'.
Author: S. Leonard Spitz, East Coast Editor.

Universal's 4712 Onserter features a turret-style component application head which extracts each chip from 81mm tape, tests, verifies, rejects faulty chips, orients and places good components in programmed locations on the PCB. Operating as a parallel function of the component application head, the adhesive application head deposits a controlled dot of adhesive at each placement location

placement machine is a semiautomatic assembly system developed to place and secure surface-mount components directly and reliably on PCBs or substrates. The system cycles components from 8mm tape at rates up to 6000 pieces per hour with a placement accuracy of 0.008in.

Robotic devices represent a second segmentation within the overall context of handling systems. As distinct from other picking, moving, placing and locating devices, robots offer flexibility of motion over as many as seven axes, often include visual systems, and frequently have intelligent control capabilities.

Model 605 robot from Intelledex Inc., Corvallis, OR, represents the state-of-the-art approach to flexible automation in the assembly and test of printed circuit boards. A typical routine with respect to insertion of odd-shaped components in a PCB has the robot transferring the board from a rack to a fixture and then loading the components, using its automatic tool change capability. Next, the robot removes the board from the fixture and moves it to an appropriate carrier. The robot also performs go/no-tests on each component prior to insertion, eliminating bad components before loading them.

The third segment is that which is most closely allied to the automatic testing of components such as ICs and discrete components. This is the segment, estimated to be worth up to $200 million in 1985, which has staked a proprietary, virtually exclusive claim to the nom de guerre, handler.

Leading manufacturers

In probable descending order of sales volume, the top three leading manufacturers of handling equipment are Micro Component Technology Inc. (St. Paul, MN), Sym-Tek Systems Inc. (San Diego, CA) and Daymarc Corp. (Waltham, MA). No Fortune 500s yet, they tend to be mid-size (although MCT boasts over 700 employees), closely if not privately held, and fiercely protective of their positions with respect to business affairs, sales, market share, and technology. On the other hand, they all display remarkably versatile and innovative machines which are capable of feeding all manner of components rapidly and reliably to testers, then assigning each piece to an appropriate bin depending upon test results. And they can provide a controlled high and low temperature environment for the DUT during the test cycle, as required.

As described by Nicholas J. Cedrone, president of Daymarc, which employs about 100, the company designs, makes and vends automatic test handling equipment for taped components, transistors and integrated circuits.

Daymarc's 757 has a footprint of only 23 × 25in., yet can cycle 3000 40-pin DIPs per hour. Microprocessor-controlled, it can pro-

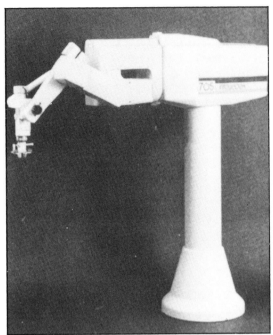

Designed specifically for light electronics manufacturing, Intelledex's 605 robot features an integrated vision system and a six-axis anthropomorphic arm

Component Technology, described MCT's philosophy this way: "MCT's concept of handlers has been dedicated handlers, dedicated towards a particular package size without adjustability. The reason we've done that is to ensure high package handling reliability. The more adjustments in a machine, the more potential there is for jams." "Typically," Vosika continued, "if machines allow for variations in package sizes (that is, adjustable) you're going to have increases in jam rates. For example, if you open up a track width to allow for bend leads, you are going to cause jams."

Industry sources suggest that MCT enjoys the number-one position in market share with total revenues for fiscal '84 of about $75 million.

vide as an option stable temperature environments from −10°C to +135°C. When banked with up to 36 tubes, the 757 can run unattended for long periods of time. To facilitate networking, industry standard RS232 and 488 digital interfaces are available for bin assignment, bin verification, temperature set, temperature monitor and other essential links. The handler is designed for use with 0.300in., 0.400in. and 0.600in. DIPs having 8 to 48 pins.

Because the JEDEC-standard device outline permits wide variations in dimensions, Cedrone believes that, "A test handler should be easily adjustable and relatively insensitive to variations within a given package type. Yet, the adjustments should not require critical mechanical skills." He adds, "For incoming inspection, test handlers should be convertible to one or more alternative packages."

Dedicated vs. adjustable

Voicing a contradictory point of view, Robert E. Vosika, vice president and general manager of Component Handling Products, Micro

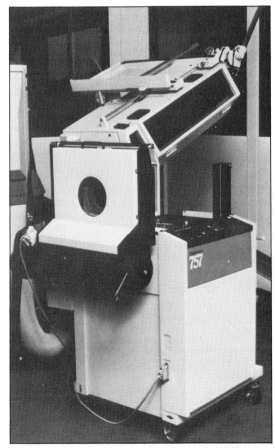

Daymarc's Type 757 handler nominally cycles 6,000 16-pin DIPs per hour as it sorts into eight bins

The microprocessor-based electronics of MCT's 3616E regulate environmental temperature, sort verification, handler throughput, self-diagnostics and status indicators

Representative of their broad product line, MCT has designed and built three different LSI handlers which gently accommodate these large and complex devices, usually 0.600in. with 24 to 48 leads respectively. The 3616E-6, 3608E-6 and 3608C-6 incorporate gently sloped input and output tracks to help control device descent velocity and velocity limiters located at intervals along the input and output tracks. Devices travel in the 'dead bug' position: leads pointed up to protect them from being bent or misaligned.

The 3616E-6 and 3608E-6 can provide a device environment of +155°C with 3608C-6 can achieve −60°C with a ±1°C stability in 35 to 40 minutes. All three units have optical sort verification for accurate device sorting, self-diagnostics to reduce maintenance time and large capacity input reservoirs.

More specifically, the Model 3616E-6 provides throughput for 48-pin DIPs of 3800 devices per hour. Weighing only 250lb, its footprint measures 30 × 34in. LEDs on the viewing panel indicate status and alert oper-

ators to machine delays as well as the reason for the delays. After ICs have been loaded onto the handler, the LEDs blink while the devices soak to the desired temperature. The 3616E-6 has two operating modes: normal, in which the handler operates in conjunction with a tester; and calibrate, in which the handler cycles without a tester.

With reference to environmental control and its effect on DIPs, a study undertaken by James W. Ramsey, Department of Mechanical Engineering, University of Minnesota, John A. Wedel and John D. Menken, both of MCT, originally published in the Proceedings of the January 1985 ATE West Conference, came up with the following conclusions:

- For any realistic handler design, negligible temperature variations occur within a DIP.
- Regardless of the external heat transfer method used to heat a DIP, the temperature of the DIP is exponentially related to

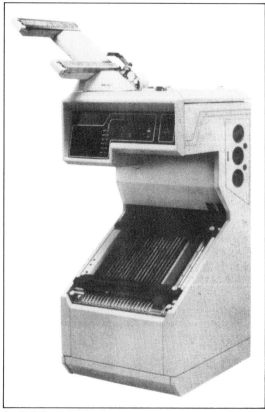

Sym-Tek's System 380 Octal feeds eight test sites simultaneously

Sym-Tek's System 340 Quad, shown with Fairchild's 5588 tester, provides true parallel testing

time and it is convenient to express this relationship in terms of a time constant.

- The time constant is a function of the thermal properties of the DIP and the method of heating.
- Measured time constant for a 24-pin CERDIP varied from a minimum of 9s when squeezed between a pair of heating surfaces to a maximum of 55s when heated by a hot air stream.

Although a miniscule sample and certainly not statistically significant, two out of the three leading handler manufacturers opt for adjustability over dedication as a virtue rather than a vice.

Particularly at incoming inspection, observes Ken Major, advertising and publications manager for Sym-Tek, "handlers must be adjustable. You don't want to force a user to limit himself to a particular IC."

Sym-Tek Systems Inc., founded in 1975 by Ray Twigg, now president and CEO, employs some 325 people and analysts suggest, enjoys the number-two position in terms of handler market share. The West Coast company designs and builds 11 different models of

ambient and environmental handlers for the semiconductor automatic-testing industry.

All Sym-Tek Systems' environmental handlers operate over a temperature range of −55°C to +150°C. The new System 300 Series handlers are equipped with IEEE 488 GPIB interface to allow networking the handlers with the host ATE system controller. The handlers can also be interfaced using RS232 or P849 logic.

Providing true parallel testing capability, System 340 is an example of advanced environmentally controlled handling technology available in the industry. Not only can this quad model load devices into four separate test sites for greater throughput, it also addresses the 'lights out' automated factory of the future with a continuous input belt that accommodates 25 device sleeves. After parametric testing, failed devices are sorted and replaced with untested devices. Functional testing takes place simultaneously on all four devices, after which devices are sorted in rapid sequence. The sorter has 25 (24 plus one reset) independent sort categories.

The most recent handler innovation and upgrading is represented by SymTek's System 380 Octal. This environmental handler directs memory devices to eight independent parallel test sites from a bank of four test sites on the right and four on the left. Each bank can operate independently from the other.

A look at the crystal ball

Glancing at the future, Sym-Tek's Ken Major notes that it is here. For he claims, "The multi-test site handler is becoming state-of-the-art, increasing throughput by number of test sites to which it delivers components."

In his vision of future handling technology, MCT's Robert Vosika has this to say, "Most devices have been packaged in DIPs. However, they are being packaged in a much broader proliferation of package types. Handler design concepts have to be different than what we are doing for DIPs to ensure high reliability package handling. That's the challenge for the handler companies. We must be able to provide products that will reliably handle all those package types."

LEAD INSPECTION

With the increasing use of automatic insertion techniques for all types of components, extremely costly assemblies can be ruined because a simple lead was bent.

LEAD SCANNING

Five years ago, the concept of automated inspection of electronic device leads was just that – a concept. Even a few years ago, scanning for bent or missing leads on dual in-line packages (DIPs) was considered a luxury in the electronics assembly process. Today however, with the explosive growth in markets for electronics products and the aggressive quest for perfect automation equipment to produce these products, lead scanning has become a necessity.

Successful assembly automation (either flexible or fixed) depends heavily on the speed and accuracy of increasingly complex and 'smarter' component-handling machinery. Greater production capacity per manhour, reduced manufacturing costs, and improved product uniformity are among the touted benefits. While efforts in this direction have been largely successful, assembly automation's ultimate profitability can be substantially reduced by low yields of finished assemblies due to component defects or assembly errors. Electronic component fabricators wee quick to recognize this potential problem and instituted elaborated process controls and in-process quality checks to maximize yield and limit the percentage of undetected defects in each lot of finished components.

Electrical component manufacturing long ago embraced the need for 100% electrical checks and tooled their factories accordingly. With respect to electrical properties, the need to cull out totally failed components is obvious. The desire to group devices by degrees of quality may even have been the motivating force for 100% inspection of electrical properties. Other errors, such as product intermixing and device misorientation within the packing tube, are also discovered during 100% electrical verification, providing additional incentive to invest in this process.

First published in *Electronic Packaging & Production*, September 1985, under the title 'Lead scanning – Luxury or necessity?'
Author: Eric J. Penn, Trigon Industries Inc.
© 1985 Reed Publishing (USA) Inc.

The DIP in the upper left-hand corner sports a bent lead, which prevents proper insertion in the board

Device geometry inspection

The recent introduction of automated circuit-board loading has focused new attention on the importance of 100% inspection for another production characteristic: correct device geometries, in particular lead straightness, especially on multilead devices such as DIPs. The chief problem caused by bent and missing DIP leads is misinsertion, causing either open or short circuit failures of assembled printed circuit boards. For example, the device shown in the figure will not insert and will lie across the top of the PCB, possibly causing both an open and short circuit failure.

The principle of 'garbage in garbage out' (GIGO) is a well-known to users of computers. This principle is no less valid for automated assembly equipment. Defective components yield defective assemblies. The more numerous and interactive the components per assembly, the more vulnerable the assembly to malfunction resulting from individual component defects, even though the defects occur infrequently. Nowhere is this more true than in the automated assembly of electronic circuits, but the principle also holds for other multicomponent assemblies.

A simple example may help quantify the magnitude of this problem. To achieve an average yield of 95% functional assemblies, a circuit assembly containing 100 components must be made from components having fewer than five defects per 10,000 parts or 0.05%. This quality level must be achieved for the combination of all defects that affect circuit performance. If there are five such categories of defects, then the average defect level per category cannot exceed 0.01%. These categories can include bent or missing leads, electrical defects, incorrect devices, misoriented devices, and solderability problems. Defect levels of 0.01% are both difficult to achieve and can only be verified by 100% inspection. Statistically based sampling plans such as military standard 105D do not provide sufficient assurance of defect levels unless the sample size is extremely large compared to the lot size of devices checked.

Some solutions

Up until the past few years, solutions to the problems caused by bent or missing leads were crude and labor intensive, employing armies of workers to manually inspect and straighten millions of device leads. As the volume of multileaded semiconductor devices grew, the dire need for a more cost-effective solution to this problem became apparent.

QA departments began asking the engineers to invent 'some form of lead scanner' to detect and segregate damaged devices. Obviously, such a scanner should also be fast, require minimal operating labor, and be very accurate.

Acceptance standards for degree of DIP lead bending and frequency of occurrence were quickly introduced by all major semiconductor manufacturers. Standards varied but, in general, limits were set at 0.010–0.015in. of deflection at the lead tip. However, compliance with these standards has been difficult to attain for the following reasons: connecting leads on DIPs are frail, being necessarily slender and made from soft solderable metal alloys such as copper; protruding at 90° to the DIP body, the leads are easily snagged during either manual or mechanical movements required during the manufacturing inspection, branding and packaging processes.

Several approaches have been adopted,

singly or in combination. Implementation of automated DIP production processes, eliminating many handling steps where damage to leads can occur, has reduced the frequency of these defects to a few percent. In addition, inclusion of 100% lead straightening in the process flow has also helped to reduce the frequency of bent leads in finished product. Several companies supply equipment for this purpose, to reduce the labor burden, but 100% lead straightening is still expensive due to the high cost and modest throughput of this equipment. Lead straightening involves deliberate flexing of all leads beyond the elastic limit to achieve final straightness. Consequently, weakening of device leads and damage to hermetic seals discourage many manufacturers from using this process.

Despite the enormous labor cost, visual inspection of all device leads is common practice. Only damaged devices are then subject to the stress of lead straightening. The burden of this process upon major manufacturers provided the incentive to automate inspection and many 'lead scanners' have been launched into this ready market. Many of these early scanners were unsuccessful but did establish a base for the development of today's very effective scanning equipment.

Early DIP manufacturing processes were not automated and provided numerous opportunities to bend device leads. Extremely high fall-out was commonplace, 40–60% reject rates were not unknown, leading many manufacturers to invest heavily in programs to correct this problem.

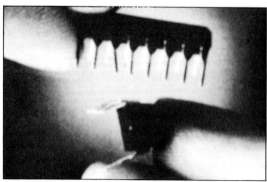

Bent DIP leads can bring a $500,000 autoinsertion machine to a grinding halt

Criteria to consider

Opinions vary regarding what to measure and how to measure pin straightness, but certain issues are essentially generic:

- All pins of every device must be inspected.
- Longitudinal (forward/aft) and lateral (inboard/outboard) bends must be detected.
- Both individually bent pins and combined or cumulative effects must be detected.
- Tolerances on degree of bending must be such that all the leads on a DIP will enter either the circuit board or socket holes without additional deformation or induce jamming of loader machinery.
- Reference datum should be consistent with those used by automated board-loading machinery.

How to sense and measure bent leads quickly and accurately has engaged several companies in extensive engineering and product development efforts. Two basic approaches have been followed: camera-based, machine vision techniques were among the first systems to be investigated. Usually several cameras are required to inspect alignment of each row of leads in two directions. To achieve required resolution the pixel count is fairly high, resulting in large volumes of data that must then be encoded and processed. This large data burden in turn requires substantial computing power and speed to achieve prompt sorting decisions. While these systems are extremely versatile, they are usually fairly expensive and relatively slow. These limitations led naturally to investigation of custom optical arrays in an attempt to reduce the data necessary for sensing misaligned leads.

Several major manufacturers of DIPs have already implemented lead scanning with notable success. End-user installations are also becoming commonplace. The points of application vary but in all cases substantial benefits are derived from lead scanning.

Lead scanning prior to test or other mid-process steps significantly reduces jamming and enhances the throughput of these close-tolerance handling machines. Signetics has adapted this approach in two major offshore facilities and derives the dual benefits of improved productivity and superior finished product quality.

Lead scanning immediately prior to final pack obviously enhances the quality of finished product shipped to customers, but other uses have proved beneficial. For example, Motorola not only scans finished product but also audits the output from all upstream process steps. In this manner, the sources of lead bending are traced and corrected in a timely manner. Timely adjustment to a malfunctioning marker or test handler can contribute to additional cost savings attributable to lead scanning.

Lead scanning at the end-user receiving point or immediately prior to use in PCB assemblies improves loading efficiency and directly upgrades the yield of finished quality assemblies. In a control test, one major manufacturer of communications circuit boards achieved a zero jam rate using 20,000 autoinserted DIPs following lead scanning, compared to an unscanned control group that jammed at a rate of 50 times per hour in the same loading machine. The reject data for finished assemblies were not published but were stated to be dramatically better using scanned devices.

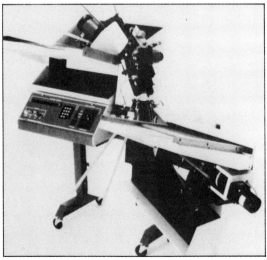

MCT Synerception's LS 2300-2 lead scanner inspects 8- to 20-lead DIPs at a rate of 14,800 units per hour

Scanning SMDs

Some of the problems that bedeviled the manufacture and use of DIPs are also present with the newer surface-mount multipin devices such as SOIC and PLCC packages. While great efforts are underway to fully automate production and use of these devices there will inevitably be some incidence of bent leads. The higher pin counts per device, and device count per circuit assembly, coupled with greater sensitivity to slight bends, all serve to drive a need for lead scanning from the outset. The integrated structure of the multistep encapsulation process for these devices, as well as the high-speed automated circuit-assembly processes, dictate that lead scanning be in-process. As might be expected, several companies are attempting to develop this capability.

MCT Synerception demonstrated a successful lead-scanning technique for surface-mount devices in May of this year and is currently developing this technology into scanning equipment for major industrial installations. The method meets the challenge of detecting lead deflections as low as 0.0001 with extreme precision and represents an advance in in-process inspection capability for the electronics industry.

When questioned on the need for bent-lead inspection, a number of semiconductor manufacturers tout plans for the eventual elimination of bent leads by the automation of their production lines. A key concept in the factory automation business is to eliminate the need for finished quality inspection. Including finished quality inspection can be translated to mean manufacturing will make a bad product. "The production machines of tomorrow will not be permitted to make a bad product," says a key Asian semiconductor plant manager.

"This objective in theory is admirable," says MCT Synerception's president, Thomas Brekka. "However, in practice, optical vision inspection technologies are crucial to achieving perfect product manufacturing. Inspection is necessary for process monitoring and control of all critical steps of component mechanical fabrication and handling equipment."

The equipment that is in use today bridges the 'real world' gap between 'perfect product' and the product that is shipped today with less than perfect quality.

As the need for quality is crucial to our international competitiveness, the final products that are produced in the USA must be perfect.

"For the electronics portion of these products, automatic optical visual process monitoring and control is crucial at every step in the industry. The component manufacturers must inspect at each point in the fabrication to assure perfect parts. The printed circuit board fabricators must inspect as the boards are assembled. true automation occurs only when the equipment itself knows when an error has been made and can react appropriately. To achieve this, the machines must have automatic optical visual inspection capabilities," Brekka concludes.

SPECIAL PROBLEMS OF HYBRID AND SMT INSPECTION

The very density of surface-mounted devices produces real challenges to probe design for tactile sensing.

PROBE DESIGN AND TEST FIXTURING

Several of the concerns for those involved in surface-mount technology include soldering, cost, component availability, component standards and product reliability through testing. The current methods of attaching surface-mount devices to pads require special attention because of the high temperatures associated with the soldering process. The cost of individual surface-mount components is higher than their conventional component counterparts. And there is the cost of new SMD board-assembly equipment, too. Furthermore, only 10–15% of the leaded-type components are available in the SMD format, and if that is not enough, there are currently different dimensional characteristics between similar US, Japanese and European manufactured parts.

Each of these issues is integral to maintaining the reliability of the product and has gained much attention over the last one to two years. Yet, until recently, product reliability through testing was generally of little concern, or at the most an afterthought. Today, it is considered by many to be one of the most important issues concerning SMT. The question used to be, "Now that we have it, how do we test it?" Today, testability is being considered during the design stage. There are two categories of reliability testing: hybrid circuit/component testing and the testing of SMD board assemblies. Stated in conventional PCB terminology – bare board and loaded board testing.

Hybrid circuit/component testing

Currently, there is a limited number of techniques and devices capable of accessing and testing hybrid (bare board) circuits and special components with test points located on centers of less than 0.039in. In addition, the scope or variety of hybrid circuits and special components with lead spacing down to 0.010in. is wide. Therefore, this discussion will cover general guidelines for testing

First published in *Electronic Packaging & Production*, February 1986, under the title 'Probes ease testing of hybrids and SMT assemblies'.
Author: Charles M. Tygard, Everett/Charles Contact Products Inc.
© 1986 Reed Publishing (USA) Inc.

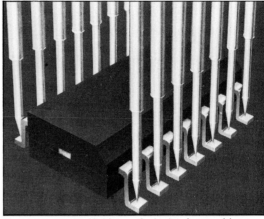

Test probes are a critical link between the circuit under test and the test system. As such, tip design, probe force, support fixturing and mechanical tolerances must all be integrated for optimum performance

products with center spacing of 0.010in. only.

Two well-known approaches to probing pad sites down to 0.010in. are the cantilever beam probe and the Euler-column probe. The Euler-column probe has the most versatility in regard to high-density circuitry within a given area, and as a result will be discussed in greater depth. The cantilever beam probe is mentioned for reference only.

A Euler-column probe consists of a long, narrow wire affixed to a terminal and supported by a terminal header with probe wires feeding through a registration or guide plate. These parts, when combined, are referred to as a test module. As a test module is lowered in contact with a test circuit, the upward force against the probe wire causes the wire to buckle at its midpoint. This action exerts a force of approximately 0.88oz at the probe wire tip. Although this appears to be a small load or force, it is sufficient for accessing most test points found on components or hybrid circuits. If the circuit traces can be damaged easily with forces between 2×10^3 lb/in.^{2} to 8×10^3 lb/in.^{2}, then the user should contact the supplier for information on alternate contacting methods.

However, in response to this concern, several manufacturing techniques are used to polish and buff the probe wire tip to produce a shape similar to that found on conventional

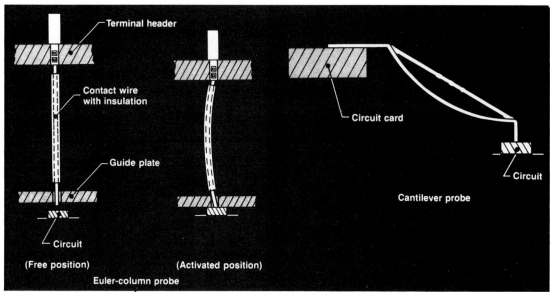

Two methods for probing pads make use of the column and cantelever principles

probes and one that will not deform the circuit under test. It is also possible to limit the wire size 0.005in. and to minimize the amount of deflection, thereby lowering the contact force. (*Note:* The current state of the art is the Euler-column probe for 0.008in. test centers, utilizing a 0.0035in. diameter wire.)

Clearance problems

When accessing components on a strip, such as semiconductors attached to a carrier tape, it may be necessary to provide additional clearance for the IC chip. Since most test-module assemblies are custom designed, providing a

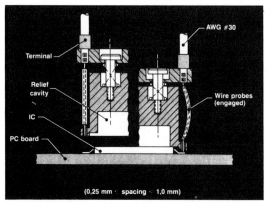

Probe module with Euler-column probes has a relief cavity to accommodate different component body sizes

relief cavity for various chips does not present any problems in the manufacture of the test-module assembly.

Normally, close-center test modules are designed for operation in room ambient and temperature conditions from −50°C to +105°C. Some test and burn-in fixturing for ICs mounted on TAB (tape automated bonding) tape require a much higher upper test limit of approximately 150°C to 175°C. Some manufacturers subsequently expose these TAB ICs to high and low oven environments from −100°C to +250°C. The choice of materials and manufacturing assembly techniques is very critical when designing to meet these extremes. Special wire, insulation and module-frame material must be used to provide proper conductivity, good oxidation resistance, high dielectric strength, and excellent structural stability at these temperatures. Currently, the flexibility of the wire insulating material at temperatures below −50°C is in question. However, there is promise that material will soon be available with the flexibility properties of the insulation for temperatures down to −100°C.

Testing boards with SMDs

The greatest challenge in SMT is testing the SMD board assembly. Long-travel spring

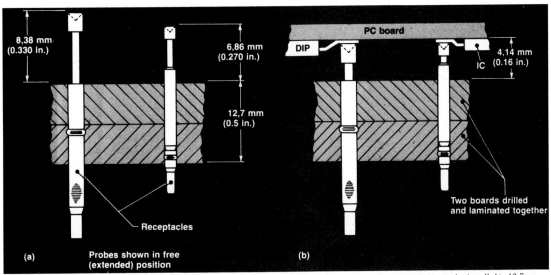

Several types of long-travel probes can be used side by side when properly mounted and positioned: (a) 0.100in. and 0.050in. probes installed in 12.7mm (0.50in.) test head, (b) probes shown in normal working position (²/₃ deflection)

probes for 0.050in. test centers are available in the ATE industry for accessing various types of SMDs. One type of probe is mounted into receptacles which are pressed into the wire side of a probe plate. Several types of long-travel probes, with a stroke of 0.16in. or more, can be used side by side with the more conventional 0.100in. test center probe.

The maximum component height that can be accommodated with this arrangement is 0.16in. Note that the probe plate is shown in two pieces laminated together, either mechanically or chemically. When attempting to drill the mounting holes through the 0.5in. board, the desirable length-to-diameter ratio limit of 10 maximum is far exceeded, making it virtually impossible to hold accurate tolerances. If the component heights exceed 0.16in., two basic options are feasible: to router cut the bottom side of the laminated probe plate; or to use a thinner one-piece, 0.375in. probe plate. The maximum allowable component height with this arrangement is slightly in excess of 0.288in. Other probes are also available with longer travel and can be mounted in a single probe plate.

Tip style and pad size

Selecting the proper tip configuration for testing various components of bare-board pads is important.

Tip applications that, at present, are the most commonly used for accessing SMD board assemblies are shown. The spring probe is shown accessing the solder pad which is extended beyond the component lead. This is the preferred technique if enough space is available on the PCB. Accessing a soldered component lead probably accounts for 75% of the actual applications. A badly soldered joint may test 'good' if the fixture is designed so

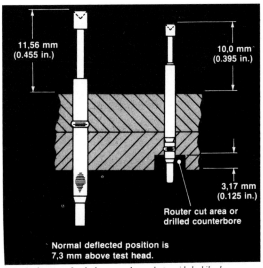

Required support for the larger probe can be provided while also accommodating the smaller probe by reducing the supporting depth

that the spring probes access the component leads rather than the solder joint. This is due to probe spring pressure forcing a weak joint to close under test. Most users find this problem to be almost non-existent, but it does require consideration.

Some test fixture manufacturers have experienced better probe performance using a radius tip instead of a more conventional pointed tip when the SMD board was cleaned well after soldering. This improvement (lower average probe resistance) is caused by increasing the surface contact area of the probe tip with the soldered component lead. Probing miniature chip capacitors and chip resistors requires the use of a pointed or radius tip probe due to the short width of the end terminations. These termination bands can run as short as 0.010in. When accessing round components, such as miniature diodes, the serrated-style tip has proven to be very effective.

Target size determination becomes a very critical issue when dealing with close center testing. Therefore, it is essential for the circuit design engineer to define a minimum size target area for probing. Another figure shows the relationship of tooling locating pins used to align the unit under test (UUT) and the test pad pattern with respect to the aligned edges of the PCB. The expected error encountered in drilling the fixture test head generated by inaccuracies in the CNC drill machine, drill bit flexibility and material properties is also shown. The fixturing error shown represents the absolute minimum ideal condition and does not indicate any mechanical errors in the probe itself.

The table shows the summation of the UUT and the fixture head manufacturing tolerance. The recommended target pad size of 0.035in. may be larger than the available board space will allow. More typical dimensions, like 0.20 × 0.020in., generally work very well because, statistically, the chances of all the tolerances being 'out' in the same direction are minimal. Based on experience, an average size of 0.025 × 0.025in. has proven to be adequate; but if space permits, an 0.035in. target may preclude any probe target misses.

Test probe evolution

SMD spring probes have been available in the ATE market for approximately one and one-half to two years. The many problems associated with testability, such as fixture and tooling pin tolerances, circuit pad and line closeness, probe tolerances and electrical specifications are just now becoming fully understood.

As a result, the next generation of the SMD spring probe will have the following characteristics: higher spring force, reduced mechanical radial (plunger side play) runout, longer plunger travel and less movement within its receptacle. These probe improvements, coupled with tighter fixture tolerances, should result in a much-improved system for testing SMDs.

Design for probeability

Improve your product reliability by designing it, when possible, for easy access by test

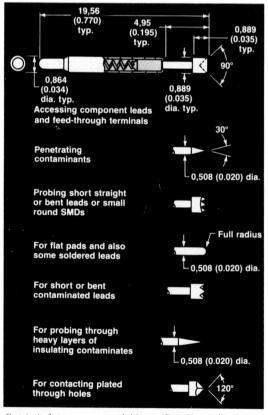

Certain tip shapes are recommended for specific probing applications

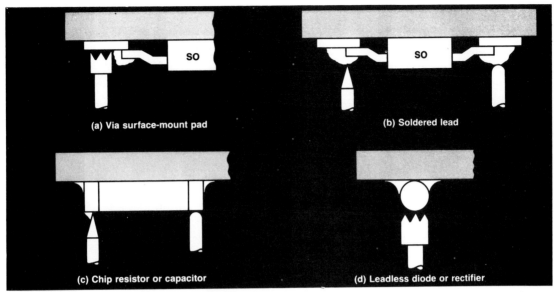

(a) Via surface-mount pad

(b) Soldered lead

(c) Chip resistor or capacitor

(d) Leadless diode or rectifier

Typical probe tips used for SMDs

probes. A few general guidelines are:

- Incorporate test pads into the artwork at the design stage. Provide one pad for each electrical node. The pad should be a minimum 0.25in., but if space permits, use 0.035in.
- Keep all test pads at least 0.380in. from tall components when possible.
- Group all components together when possible in case relief cavities are needed in the test head.
- Coat the test pads with solder, either wave solder or solder screen methods. This ensures a clean test pad area.
- Keep to a minimum the number of traces that run between the test pads.

Accessing leadless SMDs

Testing leadless SMDs is the most perplexing problem to the users of this type of component. If the PCB solder pads could be extended from each component to allow the use of test probes, then a relatively easy test solution would exist. Unfortunately, this is almost never the case because, if enough room existed for these test points, it would minimize their advantage, i.e. its space-saving characteristic. The answer to this dilemma is a unique test module: the access module.

An access module is designed to engage both leadless ceramic chip carriers (CCCs) and plastic chip carriers (PCCs). However, the use of such devices for in-circuit testing

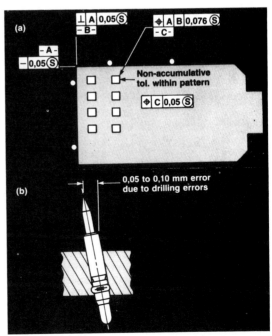

Tolerances in fixture design are critical. Given here is the relationship of pad pattern to locating pins and the edges of the board. Drilling errors can also induce misalignment

	mm	in.
Straightness of datum edge	±0.05	0.002
Squareness of datum edge	±0.05	0.002
True position of solder pad to datum edges	±0.076	0.003
Pad to pad tolerance	±0.05	0.002
Probe manufacturing drilling tolerance	±0.10	0.004
Probe radial play factor	±0.10	0.004
Total	±0.426	0.017

Note: Typical test targets must be two times larger than total tolerances (2 x 0.426 = 0.85 mm diameter target).

Summation of tolerances

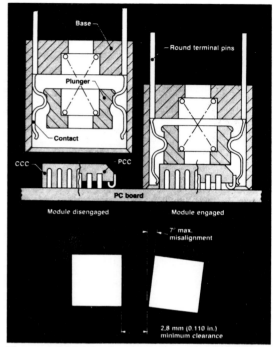

The access module is designed to engage leadless and leaded chip carriers

assumes that a number of practical constraints be present, namely:

- Adequate spacing of 0.110in. minimum should exist between the closest points.
- No more than 7° maximum chip misalignment should exist.
- The components should be centered on their respective solder pads, within a tolerance of 0.003in.

The access module is capable of testing either square or rectangular packages and is designed to meet the JEDEC standard of 0.050in. pin spacing. The modules can also be mixed with conventional probes for testing boards with a variety of components.

In summary, the current state of the art in ATE fixturing provides the capability to test both hybrid circuits with spacing down to 0.010in. and to test PCBs loaded with SMDs with 0.039in. spacing. As the SMD industry matures and begins to standardize on package sizes, test modules as described herein will be readily available in the marketplace. To do the job efficiently and effectively, a close liaison between PCB designers, probe and fixture manufacturers must be established. Once this rapport has been developed, the problems of test accessing will be minimized at all design levels.

INSPECTING SOLDERED JOINTS

No matter what type of component used, with or without leads, these components must be soldered to join them with the conductive paths on the circuit board.

The verification of the quality of these soldered joints has been made more difficult by the decreasing size and the increasing density of boards and components. In the end, the quality of the soldering operation will be determined by product function, but it may be critical to look more closely at the soldered connection itself.

INSPECTION EQUIPMENT

The unaided human eye and the eye aided by a magnifying glass – these are the two basic inspection approaches to determining the quality of soldered joints on printed circuit boards. For detailed study, a stereo microscope may be used with perhaps 10 or 20 times magnification.

Now, however, there are techniques which expand the options for solder-joint inspection. Single-lens systems that form three-dimensional images eliminate disadvantages of stereo microscopes. The technique, though, is still manual, and defects not in the line of sight remain hidden.

Automation of the inspection process is a response to the demands to increase manufacturing productivity and to improve the quality and consistency of products. An automated inspection system using a laser-induced infrared signature approach and systems using x-rays to form images of soldered joints are attempts to fill these needs.

New packaging technology utilizing high density surface-mount components, too, places tough requirements on the solder-joint inspection process. The number of solder joints in an area can be vastly increased as well as hidden beneath the component body.

3-D viewing

Stereo microscopes provide a magnified image with a large depth of field and field of view for macro inspection. The magnification range of these instruments may be from 2X to 250X.

In a stereo microscope based on two divergent optical systems, eye fatigue may be caused after a period of use. In addition, inspectors may not focus both optical paths and effectively use only one eye for viewing. Microscopes, too, require that the inspector's head be held foward, which tends to cause fatigue.

Depth of focus can be provided with a single-lens system, according to Ron Grant of

Hunter Equipment Sales. This UK-based company sells the Vista VX inspection system in Europe. Made in England by Plessey Controls, this system provides depth of focus without the eye strain caused by stereo microscopes.

Magnification by this screen projection system is continuously variable between 5X and 10X. Independent top and underboard lighting is provided by three halogen lamps (two 75W lamps for incident lighting and one 50W lamp for substage lighting).

The image is projected on a 229 × 152mm screen. When adjusted for 5X magnification, the field of view is 38 × 25mm. The work carrier for a soldered printed circuit assembly is a spring-loaded frame on PTFE feet.

A 3-D image can be seen through the single eyepiece of Metron Optics' scanner for solder-joint inspection, claimed to have a depth of focus twice that of a stereo microscope. The 3-D image does not come from fixed position

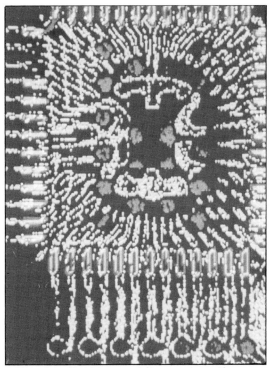

A chip carrier's solder joints are revealed in an x-ray image with pseudo-color digital processing (Courtesy of IRT Corp.)

First published in *Electronic Packaging & Production*, February 1986, under the title 'Inspection equipment exposes quality of soldered joints'.
Author: Ronald Pound, Editor.

Three-dimensional images of soldered joints are seen with one eye through Metron Optics' scanner

stereopsis (the two eyes of a person seeing two views of a scene).

Viewing is not confined to a single exit pupil when using this optical scheme in a 3-D inspection system, explained Russell Jones, Metron's vice president of production. Accommodating interchangeable lenses for 6X to 135X magnification, this system permits the operator to view the image with one eye. Movement of the eye enables the inspector to see a solder joint from slightly different angles. The inspector peers into the system with his head held in a natural, vertical position in relation to the rest of his body – not thrust forward as when using a stereo microscope.

System lighting can surround an object with light from six directions. Underlighting of a translucent PCB is particularly useful to spot solder balls on surface-mount assemblies.

Implementing improved techniques in manual inspection equipment is but one thrust. Eliminating the operator altogether with automatic equipment is the other.

Automatic inspection strikes at the problem of inspecting an increasing number of solder joints in a given area of high-density circuitry. It also offers more consistent inspection than the manual approach with a human. Looking at joints that are hidden from the sight of a human inspector, too, is one of the thrusts of new, automated inspection systems.

Automated inspection

The focus of automated solder-joint inspection systems today is laser-induced, infrared signature analysis and x-ray imaging. The IR signature technique is employed by Vanzetti Systems in their equipment. Real-time automated inspection features, though, are most highly developed in IRT's integrated radiographic inspection system.

The system sold by Vanzetti is based on heating a solder joint with a pulsed laser, then observing its cooling characteristics via detection of emission of infrared energy. The properties of a solder joint determine its cooling characteristics. Specifically, a solder joint with a defect cools differently than a good solder joint does – providing the basis for distinguishing between good and bad solder joints.

Surface-mount technology is an issue that has influenced the design of the second-generation Vanzetti laser inspection system. An adjustable optics head tilts to scan solder joints in cases where looking straight down onto a board will not produce a thermal signature that can be observed. Surface-mount devices which require tilted optics include leaded and leadless chip carriers and chip components.

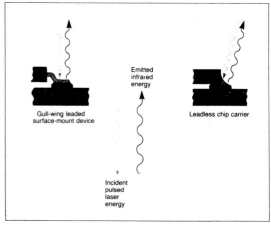

The infrared signature of a solder joint is observed as it cools after heating by a pulsed laser beam in Vanzetti Systems' solder-joint inspection system

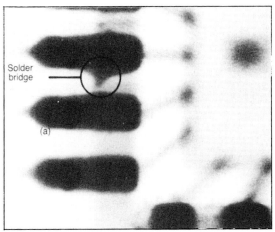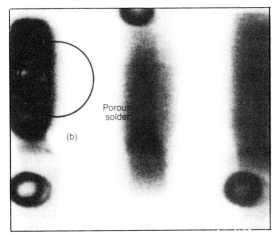

Solder-joint defects are shown in x-ray images: (a) solder bridge, (b) porous solder (Courtesy of IRT Corp.)

In x-ray inspection, the probing radiation passes through component packages and the PCB. Lead and tin in the solder are, however, more opaque to x-rays than the epoxy and copper in PCBs and ceramics in chip carriers. This results in the contrast between the solder joint being observed and the surrounding parts of a surface-mount assembly. Solder joints can be inspected even when obscured by component packages in the spectrum associated with human vision.

In IRT's circuit board inspection system, automatic recognition of good solder joints and bad solder joints is based on applying machine vision concepts to the x-ray image obtained. The x-ray image of surface-mount solder joints is digitally processed to obtain certain characteristic features of the soldered connection. These parameters are compared with preprogrammed solder quality rules to determine the acceptability of the solder joint under inspection.

IMAGE PROCESSING AND NON-DESTRUCTIVE TESTING

Many defects in electronic products can escape detection by some of the most sophisticated electrical testing and vision inspection equipment available for manufacturers' use.

Some defects are hidden from the human eye. Misregistration of the inner layers of complex multilayer boards is an example. Other hidden mechanical defects may not be revealed by electrical testing, but under thermal and mechanical stresses in operation can create electrical circuit problems. Solder

voids and poor component die-attach bonds are examples.

On the other hand, there are defects in component placement and soldering that are not masked by a visual barrier. With increasing circuit density, however, these defects may not be readily apparent to the human eye at production line rates. Substituting machine vision inspection for manual effort is an approach to solve this problem. Still, there are problems inspecting features of solder joints and spotting solder balls under a pin grid array or a leadless chip carrier.

The penetrating character of x-rays and ultrasonic signals suit them for the job of

First published in *Electronic Packaging & Production*, June 1985, under the title 'Image processing boosts the power of non-destructive testing'.
Author: Ronald Pound, Senior Editor.
© 1985 Reed Publishing (USA) Inc.

searching out hidden defects. X-ray transmission forms images of the interior of PCBs, solder joints, and components being inspected. Ultrasonic transmission is applied to inspection of hybrids, IC packages, and chip components.

The penetrating power of x-rays in the electromagnetic spectrum has been utilized for some time to inspect multilayer boards. New technologies in electronics, however, bring new inspection requirements while at the same time offering more sophisticated techniques to apply in meeting these requirements.

Now, the transmission properties of x-rays are being harnessed to digital image proces-

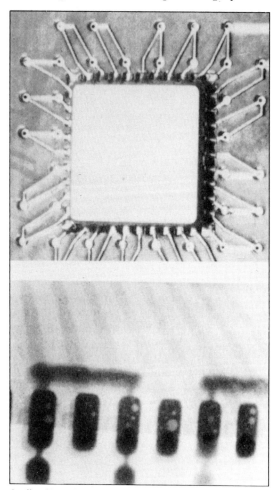

Leadless chip carrier on a PCB has voids in its solder connections revealed by Nicolet's Mikrox PC-1 x-ray system

sing techniques such as filtering and time integration, gray-sale histograms, pseudo-color, and contour synthesizers. These techniques can take one to higher levels of image enhancement and analysis than that provided by a simple revelation of the amount of probing signal that has passed through the object under scrutiny.

X-ray image

Fundamental elements of an x-ray system are the source and a means for creating an image of the interior of the object being inspected. X-rays from the source pass through the object, attenuated and scattered in accordance with the object's internal structure that is the subject of interest. The variations in the x-ray flux emerging over the area of the object can then be used to produce an image of the object's internal structure.

Sources used in x-ray inspection of electronic products are radioactive, iodine isotopes and high voltage x-ray tubes. Images are formed on film or displayed on a fluoroscope or video monitor.

In terms of resolution, an x-ray tube may reveal printed circuit lines and spaces less than 0.001in., depending on focal spot size and the method of image formation. Film provides better resolution than a fluoroscope. A radioactive isotope, operating with a fluoroscope display, will resolve lines and spaces down to about 0.005in.

With an isotope, x-rays are emitted in the radioactive decay of the element. Iodine-125 emits x-rays with optimum energy in the trade-off between penetration of multilayer boards and image contrast. Its half-life is sufficiently long to provide an x-ray flux level adequate for inspection of multilayer boards over a reasonable period of time. Isotope units are available which are portable and provide some inspection capability at relatively low cost.

The predominant source in systems suited to inspect electronic products is an x-ray tube. X-rays are emitted as the velocities of charged particles change in the high-voltage tube.

Voltage on the tube can be varied to shape the energy spectrum of x-rays emitted. The higher the voltage, the more the emitted x-ray spectrum is shifted toward higher energies. X-ray tube anode voltage for electronic inspection systems is normally variable up to around 90–160kV.

The tube voltage is adjusted to provide optimum image contrast for the object being inspected. X-rays of low energy that do not penetrate many areas of the object result in low contrast. On the other hand, high-energy x-rays that readily penetrate most areas also produce a low-contrast image. There is a voltage range that produces optimum contrast.

Viewing an image formed by passing x-rays through an object can be done directly, in real time, with a fluoroscope. Phosphor converts the x-rays to light emission, forming an optical image. This image, amplified by an image intensifier, can be viewed directly by a human or displayed on a video monitor via closed-circuit TV.

An x-ray sensitive videcon camera can be used to directly convert x-rays for display of the object's image on a monitor. For instance, TFI Corp.'s real-time PCB inspection system uses an x-ray sensitive videcon capable of resolving 10 shades of gray.

With a fluoroscopic system, however, there may be adequate light coming off the image intensifier to drive a standard, optical videcon system. The cost is one-tenth that of an x-ray videcon, says Gil Zweig, president of The Glenbrook Co.

In the analysis of an image, tonal range is as important as the resolution. If one is dealing

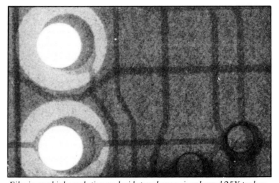

Film image high resolution and wide tonal range is enlarged 25X to show the plated-through walls of buried vias (approximately 0.0015 in.) a six-layer board. The image (photographically reversed here) was obtained on Glenbrook's High Detail Film System

with an 18-layer board, for example, many shades of gray have been taken and compressed into the limited display capabilities of a video monitor. Very limited tonal information is conveyed in a fluoroscopic or video image. There are not that many shades of gray on a video monitor, Zweig says.

When one is performing a detailed analysis of a multilayer board, film radiography is required, continues Zweig. Requirements are a film that has a density range at least as high as two, a resolution capability better than 0.001in., and at least 25–40 times magnification.

Image processing

Video peak store, or time integration of the image, can be used to reduce noise and thus reveal more of the image's detail on a monitor.

The video peak store is like a time exposure and frame memory at the same time, Glenbrook's Zweig explains. With their unit, the user controls the length of the integration period. One starts the integration, then lets it proceed, finally terminating the integration period when the desired image detail appears on the monitor. The image is frozen on the screen, and the actual object could be removed from the x-ray machine after the integration period is complete.

Integration, for example, is useful in conjunction with a real-time, fluoroscope system using a radioactive isotope x-ray source. A fluoroscopic visual image scintillates because of the high amplification of each x-ray photon into light photons. Bursts of light photons are emitted as each single x-ray photon strikes the screen. This results in the scintillating, time-dependent, noisy image.

An iodine isotope used as an x-ray source has a limited life. As the isotope activity gets weaker, scintillation becomes more pronounced. The video peak store can effectively extend the life of the isotope. For a strong isotope, the integration period may be only one second. If the isotope is weak, one may want to integrate for four or five seconds.

Normally the fluoroscopic image is adequate without the video peak store if the operator is interested in geometry no smaller

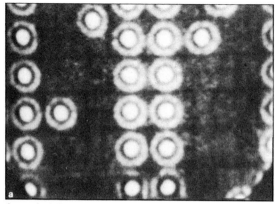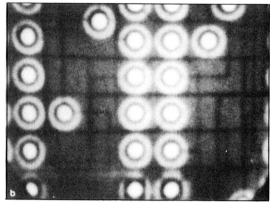

Image of a six-layer board as it appears (a) on Glenbrook's Lixiscope (isotope source with fluoroscope), and (b) using the firm's 493G video peak store to enhance the fluoroscope image

than 0.005in. lines and spaces. For applications requiring viewing of 0.001in. or smaller features, the peak store is essential. In component work, for example, the peak store is essential, and also for boards with very fine lines and spaces, observes Zweig.

Other video tools and image processing techniques available in systems on the market include video micrometers, contour synthesizers, and pseudocolor synthesizers. Video micrometers generate calibrated spaces between lines that can be moved over features to be measured on the displayed image. A contour synthesizer creates a three-dimensional effect on the video image.

Pseudocolor synthesizers assign a specified color to a given gray-scale level. A pseudocolor image is generated from the array of gray-scale picture elements in the image from a monochrome TV camera. The pseudocolor image is displayed on a color monitor. It should be noted that a color is assigned only to a gray-scale value, and not to a given layer in a multilayer board under inspection, nor specifically to any other features one desires to observe.

Pseudocolor does not provide any additional information not contained in a gray-scale image. An operator viewing an image can use it as an aid to make certain areas stand out very sharply against the rest of the image. For automated inspection, however, pseudocolor does not provide any benefit, indicates Bruce Baker, product manager for automated inspection systems at IRT Corp.

This firm has taken the path of creating an automated inspection system by combining x-ray penetration with the digital processing techniques used in conventional, optical machine vision. In their system, the CXI-5100, the videon image of a fluorescent screen is divided into 512×480 (or 245,760) picture elements, or pixels. The ADR^2 image evaluation processor in the system operates on this information to automatically provide an inspection result.

Inspection of soldered assemblies is the target application for IRT's system. Its x-ray capability permits inspection of solder joints and the area under pin grid arrays and leadless chip carriers. Sensing components to determine their presence and orientation, though a task which can be performed by optical machine vision, is also done with an x-ray source.

There are, however, some types of inspection that cannot be done with an x-ray source, notes Baker. For example, color codes or markings on a device will not be revealed. If it is required to detect such features, inspection in the visual spectrum can be combined with that in the x-ray spectrum.

A camera for inspection directly in the visual spectrum can be added to the ADR^2 image analysis system, thus simultaneously supporting x-ray and optical inspection. The processor is not fully loaded, notes Baker.

During inspection the system views a specific, small area of the circuit assembly for about ¼s to average out the noise from the x-ray signal. Then it moves to the next neighboring area (or field of view) for inspection. While collecting the image data from that region, the system is also analyzing the image from the previous field of view.

Image data can be analyzed very quickly, and is thus not a limiting factor, Baker says. What sometimes turns out to be the pacing factor as inspection proceeds is motion of the X-Y table moving the object to provide different fields of view to the camera.

Windowing is the most basic and most used processing technique for circuit assembly inspection with the CXI-5100 according to Baker. The system takes a rectangular area and constructs a histogram of gray-level values within that window. That reveals characteristic parameters of a feature in the window.

For instance, a window may include a solder connection. Looking at the darkness or filled area of the histogram provides a measure of how much solder is there.

The resolution of an image processing system like IRT's CXI-5100 is influenced by the field of view. The smaller the area viewed at any time, the less the area that has to be represented by a single pixel or gray-scale value. Hence, the smaller the field of view, the better the camera resolves the image on the fluorescent screen. For circuit board applications, says Baker, a field of view from ½in. to 2in. on a side gives the best results.

Multilayer inspection

Principal applications of this non-destructive, penetrating x-ray energy and image processing power include inspection to spot inner-layer misregistration and drill hole and pad misalignment in multilayer boards. Increasing use of surface-mount technology has an impact on multilayer inspection techniques.

In addition, along with pin-grid-array technology, surface mount is forcing industry to take a fresh look at inspection techniques for the assembly steps following multilayer fabrication. Inspection of solder-joint quality using x-ray systems is a result.

Inspection of multilayer boards intended for use in surface-mount assemblies can introduce a radiographic problem. Invar, says

Gray-scale of the image area is displayed on one monitor, along with the touch-screen menu for programming IRT Corp.'s CXI-5100 automated inspection system. The x-ray image of chip carriers being viewed is shown on the other monitor

Glenbrook's Zweig, is a tremendous x-ray scatterer. Use of copper-Invar-copper power and ground planes in the core of a multilayer board is one approach to solve the mismatch in thermal coefficient of expansion between a leadless, ceramic chip carrier and its FR-4 substrate.

Invar does inhibit the image somewhat, says nicolet's marketing manager, Adrian Sabater. The image is not as sharp, and somewhat lighter. Scattering from Invar, however, does not reduce the ability to inspect boards, he states.

Copper power and ground planes also have the effect of scattering x-rays. There is progressive degradation of the image as the number of signal layers and power and ground planes increases. Because of this, users inspecting surface-mount multilayer must anticipate using coupon areas where they can eliminate as many scattering components as possible, like copper ground planes and invar cores. Otherwise they will have to get image information through material that tends to scatter the x-rays.

Glenbrook proposes now to their customers with anything more than about 16-layer boards, where they have numbers of internal signal layers and power and ground planes, that coupons be used. The firm is becoming active in standard-setting activities for x-ray inspection coupons which recognizes that scattering occurs. It is desirable to encourage

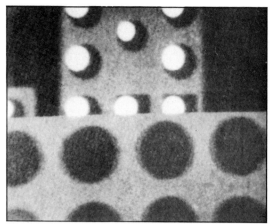

Multilayer Inspection reveals hole – pad alignment and lamination registration simultaneously using the split-screen feature of Glenbrook's video peak store

PCB designers to anticipate that x-ray inspection will be used. They can then incorporate lamination and drilling x-ray targets in the initial design.

The intent of coupons is to make them recognizable in the scattered x-ray background, and to clear away material around the coupon areas to reduce the scattering. Coupons may be just a small area where all the inner-layer material has been etched away and there is some diagnostic target.

Coupons are usually on peripheral sections of the board that will be removed during routing. They are desirable on all four corners, says Zweig. Drill misalignment usually occurs as a result of rotation and not translation. One can check the corner around which the rotation is taking place and that will look perfectly aligned. But when one checks the farthest corner from that, the rotation will be noticed.

Assembly inspection

With acceptable multilayer boards from the fabrication process, aided by x-ray inspection, component placement and soldering quality is the next manufacturing concern.

Real-time, x-ray imaging of solder connections in surface-mount assemblies is provided with resolution of details as small as 0.0004in. by Nicolet's Mikrox PC-1 system. This system

is intended for multilayer inspection applications as well as for soldered assemblies. For an inspector's viewing, the unit presents a color-synthesized image on a 13in. color monitor with magnification from 20X to 50X. The unit, using a 90kV x-ray tube and sensitive videcon camera, is capable of imaging through eight or more copper-clad, inner layers in a PCB.

If an x-ray system is to automatically provide an inspection result, it must be provided with software and data from the inspection process to make pass and fail decisions without human intervention. One approach is to enter into the processor's memory the image of a perfect or 'golden' board, to which all subsequent production boards are automatically compared. Another approach is to automate a set of rules defining what is a good feature and what is a bad feature for parameters automatically measured by the inspection system.

Inspection is guided by a solder-quality rules set in IRT's CXI-5100, not by a golden board. The rules define what is an acceptable solder connection, and what constitutes a connection that is to be rejected. To automate such rules, IRT defined a set of parameters or thresholds for various solder-joint characteristics measured during x-ray inspection. These parameters can be adjusted by the user.

Thickness, density, and shape of acceptable solder connections, as determined by x-ray inspection, fall within a fairly narrow range for different families of devices, such as leadless chip carriers or DIPs. IRT's system, looking at up to 60 solder connections per second, measures these parameters using an x-ray source and applies the solder-quality rules to the measurements.

A library of inspection routines for standard electronic component packages and configurations is created. When IRT works with a customer on the initial board, algorithms are devised to inspect the solder connections for each of the various component types used in production. With a new board design, using the same components, the computer will thus already know how to inspect components' solder connections.

For a new component, the main part of

creating a library entry may already be done once fundamental solder-inspection algorithms and parameters are established. Simply changing from a 44-pin quad package to a 68-pin quad package, for example, is not difficult, claims IRT's Baker. One uses the same fundamental algorithm over again on the solder connection for each pin. The basic algorithm provided is for a pin or solder pad.

A board design can be loaded into the IRT system from a CAD system to set up the inspection routine. For boards not in a CAD system, an operator can guide the system through a first-part inspection using menus on a touch screen.

When using data from a printed circuit board CAD system, component type, location, and orientation are specified for the inspection system. Then, the x-ray inspection system calls from its library the appropriate algorithms for inspecting those specific components, including solder-joint quality.

X-ray tube voltage is adjusted from the control panel on Nicolet's Mikrox PC-1 x-ray system to obtain the sharpest image for the particular assembly being viewed

Inspection process

Assemblies with soldered, leadless chip carriers on both sides of a PCB are being inspected with x-ray systems. The components on the top and the bottom of a board can be seen at the same time.

Images are also formed of the interior structures of individual components on a board. This is rather fascinating, and customers are finding that they are kind of doing inspection within the semiconductor packages at the same time, notes Nicolet's/Sabater. The radiation absorbed by a semiconductor is far below the level that would cause damage.

The x-ray energy is tuned and filtered to get the highest image contrast within the solder joint itself, says IRT's Baker. Normally other features are made practically transparent, such as the copper traces. Solder, containing x-ray absorbing lead, is relatively dense compared to anything else on the board.

In most surface-mount applications, the board's plane is oriented perpendicular to the axis of the x-ray beam, indicates Baker. For inspecting solder joints of through-hole com-

ponents, a board is normally tilted slightly so that the view is not directly along the axis of the solder column, he notes.

With chip carriers on both sides of a board, some solder connections on opposite sides may be superimposed in the x-ray image. In these cases, the board can be tilted so that they are not superimposed, then each can be seen individually.

Board layout designs that do not place components of similar form in direct opposition on each side of a board can also be used to avoid this problem. This approach enables boards to flow through the inspection process in a flat, horizontal position without having to be tipped. Nicolet's/Sabater indicates they recommend to their customers that boards be designed with x-ray inspection in mind to avoid the overlap of solder-joint images.

The x-ray tube voltage for the best image contrast tends to be different for surface mount solder-connection inspection than for multilayer work. An operator, though, can simply adjust the voltage to obtain the sharpest image. The number of layers in a board is of greater impact in determining the tube's voltage than the lead in the solder joint, Sabater says.

SPECIAL PROBLEMS OF SURFACE-MOUNTED AND HYBRID CIRCUIT CONSTRUCTION

The very nature of the construction of hybrid or full surface-mounted construction makes any type of manual inspection extremely difficult.

It would be useful to examine these problems before looking at functional and automated testing procedures.

DESIGN OF SMA FOR TESTABILITY

With the advantages of SMT, one might be tempted to wonder why the SMT assembly has not seen wider use by now and why everyone has not rushed out to acquire SMT assembly capabilities and equipment. One of the reasons for the reticence is that few people have developed and implemented any kind of a workable, comprehensive strategy for testing SMT assemblies!

The main difference between the SMT assembly and the traditional printed circuit board assembly (PCBA) is that the majority of the components used on an SMT assembly are installed onto, rather than into, the assembly. While this may seem quite a simple difference, there are many not-so-simple ramifications that stem from that fact. One affected area is testing.

The designs for testability and testing disciplines are beginning to take into account the significant differences between standard through-hole printed circuit board technology and the soon-to-be ubiquitous surface mount technology (SMT). To date, many test people are using the tried and true approaches of using larger pads for electrical access, fanning out closely spaced leads to wider spacings for access using regular fixtures, and spreading out the components, which of course negates many of the benefits available through surface-mount technology, so that traditional fixtures can be used. For those who do not wish to spread things quite so far, some automatic test equipment vendors, most notably GenRad, Hewlett-Packard, and Zehntel, are bringing out so-called 'clam shell' fixtures that can access both sides of an SMT technology PCB. For the long term, however, approaches such as more functional testing and more built-in design for testability features will be required.

Testing strategy is a must

With an SMT assembly, the component packages are only as large as necessary to

First published in *Electronic Packaging & Production*, September 1984, under the title 'Design of SMA for testability requires new rules'.
Author: Jon Turino, Logical Solutions Technology Inc.

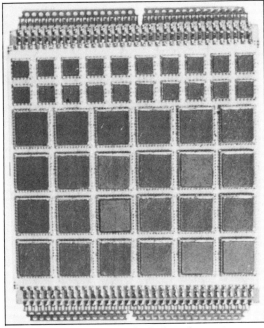

Typical PCB for surface-mount application has surface mount components on both sides of the board (Courtesy of RCA)

protect the delicate silicon chip and the circuit connections are only as far apart (0.010in. centers are typical) as electrical performance requires. And the susbtrate or PCBA upon which the components are mounted needs to have far fewer holes, and is thus perhaps less expensive to produce, than the traditional PCBA. While a through-hole PCBA might have an average of 2000–3000 holes, an SMT board with the ability to perform the same functions might have only one-tenth as many holes.

Smaller components, closer mounting, and shorter circuit paths mean that SMT assemblies can perform more functions, and perform them at higher speeds, than traditional PCBAs and do so at lower device and assembly fabrication costs. An example of the size reduction possible with surface mount devices is illustrated by looking at the various package styles that a vendor like Motorola offers for its MC68451 Memory Management Unit. The standard dual in-line package occupies about 1780mm^2. By contrast, the pin

The calm shell fixture has recently been introduced by Zehntel for use with ATE to test PCBs using surface-mount components. Note that both sides of the board are being accessed by the bed-of-nails fixtures

grid array package occupies less than 50% of that space, or 730mm², and the even smaller leadless chip carrier provides another 50% reduction, down to approximately 360mm².

The very advantages of SMT assemblies from the fabrication, assembly, and performance standpoints present large challenges to traditional testing approaches. Smaller packages mean more function per assembly. The translation of 'more function per assembly' into testing language is 'increased complexity'. Increased complexity leads directly to higher test programming costs, longer test times, and increased difficulty of fault isolation. However, the complexity can be broken down to manageable size by building more smaller boards into the system.

The tighter spacings (currently holding at around 0.050in. but moving rapidly toward 0.025in. and even 0.010in. centers) and sometimes totally inaccessible circuit leads (e.g. the pin grid array package) on SMT assemblies, preclude the use of one of the most popular PCBA-to-tester interconnection strategy: the bed-of-nails fixture.

In the face of these testing problems, SMT assembly proponents point out that the increased use of SMT assembly automation equipment will reduce the number of assembly errors and defects to a point where fault isolation and rework costs will be at acceptable levels, and that the test programming costs increases brought about by increased complexity are someone else's problems. But with testing costs approaching, in many instances, 35% or 45% (or more) of total product costs, can anyone afford to take such statements at face value? Only at their peril.

Because the increased proliferation of SMT assemblies seems inevitable, professionals involved with testing, or those impacted by testing as part of their overall responsibilities, must begin to formulate effective strategies for testing these new assemblies.

The traditional approaches of in-circuit testing and functional testing must be fundamentally altered for SMT assemblies. In-circuit testing, which tests each component on a component by component basis, requires physical access to all unit under test (UUT) nodes via a mechanical fixture.

In-circuit testing approach

While work is currently under way at several automatic test equipment (ATE) companies, and at several independent and in-house fixture groups, to develop reliable 0.010in. center fixtures that can access both sides of an SMT assembly, these new and more expensive fixtures will be of limited use with compo-

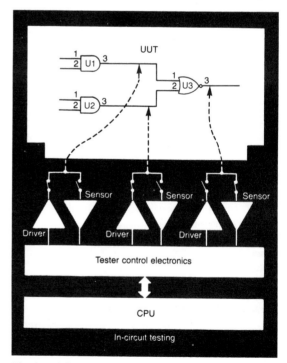

The schematic diagram of the in-circuit testing method shows how each electrical node within the circuit is contacted by a tester driver-sensor pair and how the individual circuits (gate U3 in this instance) are tested without testing the functions of other circuits (U1 and U2)

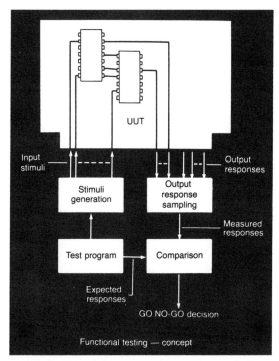

The schematic diagram of the functional testing method shows how the circuit is tested from the edge connector or normal input-input interface by the tester stimulus generation and response measurement electronics. This approach tests the functionality of the whole circuit as an entity

nents that feature 256 or more connections – all of them underneath the part.

In addition to the impact on fixturing, the actual in-circuit testing does not take into account such failure mechanisms as functional interaction between components, speed and timing related problems, and the 'soft' failures such as pattern sensitivities, timing sensitivities, and noise sensitivities that are more common with the LSI and VLSI devices the predominant components that are to be used on SMT assemblies.

Finally, with respect to in-circuit testing, the largest advantage of the in-circuit test techniques is its ability to detect multiple manufacturing defects in one testing operation. If, indeed, the number of defects on SMT assemblies is to decrease drastically due to increased automation, in-circuit testing will be highly inefficient because it will be used primarily for testing good SMT assemblies, while its strength lies in diagnosing defective assemblies. If, however, fixturing access can

be gained, the in-circuit tester will continue to be a very valuable tool for finding defects in the early stages of board production, and for providing the information needed for process improvement and the significant reduction in the total number of defects that the increased automation can provide.

Functional testing, on the other hand, does lend itself to SMT assembly testing. With functional testing, the actual performance of the assembly is verified, typically from a normal interface connector plus any other interface pins that may be available specifically for testing purposes.

Functional testing may be performed at high speed and will typically result in higher yield at next assembly test (e.g. more products assembled from multiple SMT assemblies or PCBAs will function the first time they are turned on). While functional testing is re-

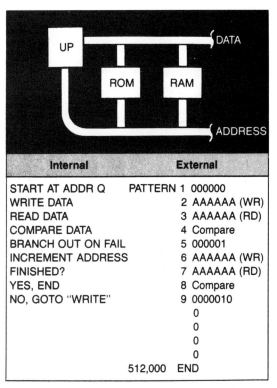

Internal	External	
START AT ADDR Q	PATTERN 1	000000
WRITE DATA	2	AAAAAA (WR)
READ DATA	3	AAAAAA (RD)
COMPARE DATA	4	Compare
BRANCH OUT ON FAIL	5	000001
INCREMENT ADDRESS	6	AAAAAA (WR)
FINISHED?	7	AAAAAA (RD)
YES, END	8	Compare
NO, GOTO "WRITE"	9	0000010
	0	
	0	
	0	
	0	
	512,000	END

Testing memory on a typical LSI/VLSI PCBA can be done with a small assembly language test program (left) which makes use of the intelligence already present on the PCBA under test or with a brute force approach (right) which could require hundreds of thousands to millions of test patterns

latively inefficient for diagnosing faults, fewer faults are expected with SMT assemblies. But the cost of functional test programming is very high, in many cases as much as five times more expensive than in-circuit test programming. The cost of functional test programming rises rapidly with increased UUT complexity, and high SMT assembly rework costs mean that diagnostic accuracy is very important.

Design strategy

What, then, is an effective testing strategy for SMT assemblies? The answer is both simple and complex. For the technical solutions are, if not simple in themselves, at least well known. But the implementation of the solutions is more complex as it crosses all organizational boundaries and requires both a great deal of human communication and cooperation as well as a thorough understanding by all parties involved of the total product costs and trade-offs needed to achieve the lowest overall cost.

In a nutshell, a successful testing strategy for SMT assemblies will involve the implementation of the following minimum set of critical elements:

- Careful attention to UUT design for testability to reduce programming cost.
- Increased use of built-in-test for both performance verification and fault diagnosis.
- Increased incoming component quality either via more incoming inspection or higher vendor quality levels.
- Increased use of test data, and fault isolation and rework data, to preclude

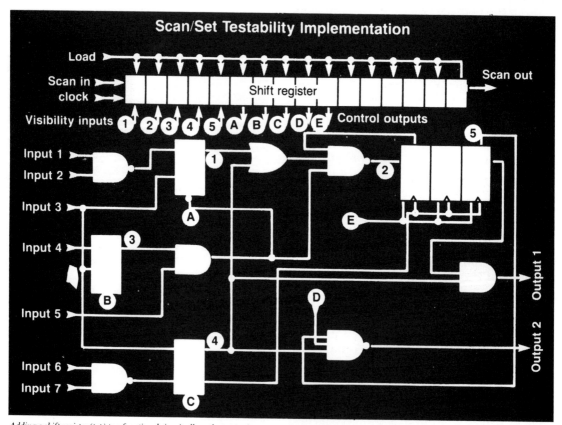

Adding a shift register (top) to a functional circuit allows the test equipment to gain electrical access to internal circuit nodes for both stimulus and response purposes. One might think of this approach as adding a portion of the bed-of-nails fixture to each unit under test

recurring problems at their source, thus reducing rework costs to an acceptable level.

A successful SMT assembly test strategy must begin with the initial electrical and mechanical design of the assembly so that it is optimally testable within the constraints of product size, price, performance, and reliability. The techniques for testable design are well known and widely published and have been proven to reduce test programming, go-no/go testing, and fault isolation costs by up to 50%, 30% and 150% respectively.

Over 50 companies that have committed to a design to test philosophy and educated their engineering personnel have estimated annual savings through testability averaging $1.2 million. Like any average, this $1.2 million figure must be carefully estimated within each organization. Some companies have only saved $100,000 per year while at least one company estimates an annual savings of $8.5 million. Thus the functional testing approach that is optimum for SMT assemblies can become affordable again – but only through commited cooperation between the design and test engineering disciplines.

One of the most efficient ways to implement this improved testability is to interconnect computer-aided design (CAD) and computer-aided test (CAT) equipment together. This interconnection makes enhanced testability faster and easier to achieve and also eliminates considerable duplication of effort between design and test in the areas of device and assembly modeling and test pattern generation.

As SMT assemblies continue to exhibit increased complexity and decreased access for visibility and control during testing, more built-in circuitry will have to be included on each UUT. Many forward-looking companies are allocating 3% and 5% (and more, in some cases) of hardware and software real estate just for built-in-test functions.

The built-in-test circuitry and software take advantage of the processing power usually present on a SMT assembly containing LSI/VLSI devices to reduce test program generation times and to make up somewhat for the lack of visibility and control that the new techniques tend to cause.

Consider as an example the testing of on-board random access memory. The testing of 256K bytes of RAM can be done one of two ways – with 8 or so instructions to the microprocessor, or with 512,000 discrete test patterns. The use of built-in-test for this type of performance verification provides huge leverage.

Another area where built-in-test is required is for fault isolation visibility once the bed-of-nails fixture can no longer be used. One way to implement increased visibility is by adding a shift register which can provide multiple internal point access with a minimum of extra edge connector pins.

The shift register approach allows internal data to be observed at any time for both on-line built-in performance monitoring and for fault isolation in the event of a UUT failure. There is a small price to pay in terms of extra parts cost but the small increase is far outweighed by the savings in test, troubleshooting, and (potentially incorrect) rework costs.

Catch faults early

The only absolute way, of course, to reduce the number of component defects on SMT assemblies is to reduce the number of defective components that are installed on them.

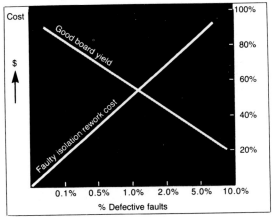

The graph shows the effect on fault isolation and rework costs, and on the percentage of circuit boards that will work the first time they are turned on, based on the percent of defective components that are installed in the PCBA. Incoming inspection reduces costs and raises yields at both PCBA and system test ends

This means more attention either to incoming inspection of components or to source quality control of the component vendors.

With the high cost of SMT assembly repair, it becomes critical to install only good components and to replace only bad components in the event of a failure. This ties back, of course, to testability and built-in test for efficient and accurate fault isolation.

It is also typically far less expensive to test components at the component level than it is to test them once they are installed on a complex assembly – regardless of the packaging techniques employed. Many past studies have pointed out the 'order of magnitude' relationship between the cost of finding faults at the device, PCBA, and system level. One such study points out that if it costs $0.50 to find a faulty device at the component level, it is likely to cost $5.00 at the PCBA level, and $50.00 at the system level. As the cost of fault diagnosis and rework rises at the PCBA and system levels due to the employment of surface-mount technology, this ratio is likely to rise well above the 10 times factor.

If we have designed the SMT assembly to be testable, included the appropriate amount of built-in-test, and taken steps to insure that only good components – including the substrate or bare pointed wiring board – have been used in the assembly, the assembly should work. How do we verify this?

The SMT assembly should be functionally tested to verify its performance. Whether the test patterns (stimulus and response) have been generated to (a) emulate the operation of the UUT in the next assembly or (b) to activate and propagate a preselected percentage of all possible faults with test patterns generated by design or test engineering personnel is of little importance.

Build right from the start

Since the fault diagnosis, isolation, and repair process is expensive regardless of the technology used for the assembly, but particularly so in the case of SMT assemblies, what can be done to reduce these costs? The answer is remarkably simple and, once fully under-

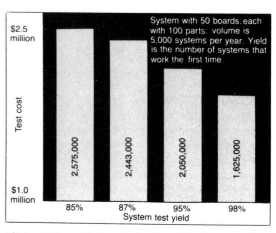

This is a specific example at an actual company of the effect on overall test cost that system first turn-on yield has. The 87% and 85% bars are for in-circuit testing only of PCBAs prior to system assembly and test, with and without incoming inspection, respectively

stood, quite obvious. The way to reduce or eliminate the expense of fault isolation, diagnosis and repair is to stop manufacturing defective assemblies – e.g. to build them right the first time!

Thus one of the key functions of the test equipment needed for testing SMT assemblies is the data gathering/analysis/reporting function now being actively promoted by most major ATE vendors. The data from the test equipment can and should be used to preclude problems at their source.

A simple analogy might be the difference between buying more and fancier fire trucks to fight an increasing number of larger forest fires (e.g. buying more expensive ATE and fixtures to deal with increased complexity and a large number of faults) versus finding the arsonist who is setting the forest fires (e.g. using test data to find and correct the source of the problem). It can be done. Many companies are realizing first turn-on yields of both conventional and SMT assemblies exceeding 95%. It takes work, discipline, and constant attention to detail, but it really pays off.

Testing SMT assemblies requires test involvement from product concept through design, manufacture and service. Coupled with discipline in testability and built-in-test, and an awareness of overall product costs, most manufacturers can realize considerable savings with SMT technology as long as they observe two simple guidelines:

- Test each element of the assembly as early as possible in the design and manufacturing processes.
- Make testing easier at each stage through adherence to testability disciplines and the connection of CAD/CAM/CAT hardware and software capabilities.

The philosophies of test earlier, test easier, plus defect prevention (as opposed to defect removal) and built-in-test, are the keys to success in testing SMT assemblies.

TEST AND INSPECTION OF HYBRID MICROCIRCUITS

The trend in hybrid microelectronics packaging technology is toward greater sophistication and greater density. Most hybrid circuit manufacturing facilities have adopted a somewhat 'standard' manufacturing technology, such as chip-and-wire, which is used to fabricate a variety of circuit types.

The one unique feature of each circuit type is electrical test. While each hybrid type may be manufactured with the same basic processing equipment, each circuit represents a unique challenge when testing the completed product. The challenge is presented to both the test equipment suppliers (to design equipment versatile enough to handle a variety of circuits) and to the user (to select the proper equipment for the application).

Inspection is a second area where technology must race to keep up with advances in hybrid circuitry. Approximately 20% of visible defects are not found by a simple visual inspection under a microscope. In some cases, over half the failures at first test are due to assembly errors which were not detected at inspection. The rapid growth of surface-mount technology places an added burden on inspection because many of the connections are under the components themselves and are not directly inspectable.

Electrical test equipment may be conveniently divided into that designed specifically for testing memory circuits and that designed for general-purpose analog, and digital circuits. The equipment for each is considerably different and, therefore, different criteria for selection apply.

The trend in microprocessors has been toward wider data paths and greater memory spaces. The most recent development is 32-bit microprocessors, many of which have address spaces as large as several megabytes. The 8k word by 8 bit static RAM and the 256k word by 1 bit dynamic RAM are among the larger memory integrated circuits currently available to accommodate the larger memory spaces required.

Memory testers

While the hybrid industry can turn out these high-density products, most test systems are directed primarily toward the IC and printed circuit board industries. Sometimes the test system can be adapted, but usually it is not that simple. Most hybrid microcircuits are designed to emulate the functional level of a printed circuit board in the space of an integrated circuit package. This can create some very unique test situations, such as a dual 64k word by 8 bit memory hybrid made up of 8k word by 8 bit devices. The circuit would require at least 18 clocks, 13 addresses, eight bidirectional data lines, and one power supply if it were designed in a minimum pin configuration.

The 18 clocks would consist of the 16 chip selects, one output enable, and one read/write line. If the configuration were changed to 64k words by 32 bits, this would require 32 8k word by 8 bit devices. The test requirement

First published in *Electronic Packaging & Production*, February 1986.
Authors: Jerry E. Sergent, Contributing Editor, and Howell H. Chiles and Robert Power, Litton Data Systems.
© 1986 Reed Publishing (USA) Inc.

for such a device could be 13 address lines, 10 clock lines, 32 data bidirectional data lines, and one power supply. With a mixture of 8-, 16- and 32-bit microprocessors on the market, the variety of memories is diverse, as are the test requirements for these memories.

In high-volume manufacturing operations, there are three main considerations. In addition to handling a large address space, a number of clock lines, and ample data path widths, the system must be capable of testing circuits in a minimum amount of time.

Secondly, if the system has the capability of multiplexing to two or more test heads, the overall test time can be minimized somewhat, but not in inverse proportion to the number of heads. Some time is required for the multiplexing operation itself.

Finally, if the production volume is sufficiently high, the test head or device under test (DUT) interface must be capable of interfacing with an automatic handler.

The test system must be capable of performing tests on the devices at their rated speeds. Accurate edge timing is required on the signals to the DUT, typically on the order of 10ns (10^{-9}s) edge placement, depending on the speed of the DUT. For a device that is rated with a 50ns or less access time, the edge placement will be more critical. In the 50ns device, the 10ns placement could lead to error as high as 40%, meaning the tested access could be as high as 70ns or as low as 30ns. Because speed has been one of the more important trends in memory devices, a test

SM2060 hybrid microcircuit test system manufactured by Scientific Machines Corp/

system should be capable of precise timing edges, possibly as low as the subnanosecond range, to handle the largest possible variety of memory circuits.

The physical size of the system is also an important feature. Many systems are physically quite large. Hybrid facilities must often produce analog, digital, and memory circuits in a wide variety of types. The range of test equipment for such hybrids is almost as varied as the hybrids themselves. A facility cannot afford to have a large portion of its test area taken up by one or two pieces of equipment.

Most circuits manufactured for the military require that hot and cold testing be performed. There must be a method for interfacing some type of instrument to vary the temperature of the DUT. This is usually simple if the DUT interface has a flat surface with no adjacent peripherals.

The user interface for test program generation must be kept as simple as possible for rapid test software development. In some cases, the user interface may be tailored to individual preferences, and this possibility should be explored.

The ideal memory test system for the hybrid industry would have a reconfigurable DUT interface, address generator, and clock generator, and as many features made flexible as possible. In this way, each time a new device is to be tested, the user could reconfigure the system. Certainly the degree of flexibility must be an important feature of any test system.

The DICE system (HP 81800S) offers high-volume testing of small- and medium-scale integrated circuits (Courtesy of Hewlett-Packard)

Midas-PR-830 hybrid circuit inspection system manufactured

General-purpose testers

Automatic testing of non-memory type hybrid circuits can also be quite difficult due to the complex electronic and physical characteristics of hybrid circuitry. The ATE market does not often address the hybrid testing problem directly. Therefore, the test engineer making a selection of automatic test equipment for hybrids must address several electrical and physical anomalies before considering particular aspects such as brand and price.

One of the problems associated with hybrid testing is that hybrids are often partioned for the convenience of the end user or system designer, which frequently results in odd combinations of analog, digital, power, and microwave circuitry. A test system should be easily reconfigurable for other circuits in the product line which must be tested on the same system.

Another consideration is the fixturing involved. Hybrids are generally larger in physical size than the standard JEDEC packages, such as the DIP or TO series. The locations of pin centers may fall on odd increments such as 0.050in. or even more exotic arrangements. These differences in size and pin locations tax the layout schemes for the test heads of typical component testers. These mechanical difficulties, however, may be circumvented through proper selection or custom design of

the test head, DUT boards, and sockets with respect to the user's product line.

Finally, hybrid testing is generally more stringent per input/output pin than that of printed circuit assemblies or large systems. This is because the hybrid circuit is still considered a component by most end users, and is therefore worthy of all the parametric and full-speed testing which an ATE system can muster. Requirements such as full parametric testing on all input and output lines, full input current characterizations, propagation delay measurements, and setup or hold times over the military temperature range are typical additions to the normal functional testing performed on modern high-reliability hybrid circuits.

These considerations, however, constitute the peculiarities of hybrid testing and are secondary to the other questions of throughput and total cost which are essential to any automatic test equipment justification. A serious organizational evaluation must be made concerning production volumes, overhead rates, and customer cost, and a balance struck between these parameters. Only after all of these technical and economic questions are answered can an automatic test system be correctly selected from the available vendors.

Despite the somewhat rigid constraints described above, a number of approaches are available for a variety of prices and through-

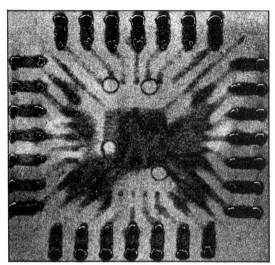

Defect in a thick-film print as observed by the MIDAS-PR830 system (Courtesy of Photo Research Vision Systems)

put rates. Several types of test systems are detailed below and represent a wide variety of costs and throughput rates.

One approach that should not be overlooked when considering more expensive test systems is the use of IEEE-488 bus compatible instrumentation, more commonly known as 'rack-and-stack' test systems. One of the more common systems in this category available from Hewlett-Packard is the HP DICE system.

The 8180A supplies a wide range of adjustment at the outputs from −5V to +17V, depending on the source impedance desired. Available with the data generator is an extender box which can provide up to 32 additional outputs.

The 8182A data analyzer has a powerful analysis capability such as a variable clock delay for delayed sampling, which is a requirement in all digital test systems, since all circuits have some delay. This feature lends itself to propagation-delay measurement quite readily. The analyzer may be configured in a single- or dual-threshold mode, with a wide range of comparison voltages. Furthermore, both the HP 8180A and the HP 8182A have binary pattern restoration modes which greatly accelerate the setup time required.

These digital instruments may be run safely

to clock frequencies of 50MHz at the HP 15425 test head with proper loading of the generator outputs. Local receivers are provided through the analyzer cables which contain active FET input drivers at the tips, thus augmenting the 50MHz driving capability with the same data reception rate.

The HP 15425 test head also runs on the IEEE-488 bus, with 84 input/output pins with a DC parametric ring which can be switched-in via individual relays. This brings the HP semiconductor parameter analyzer to all inputs and outputs for parametric measurements. The computer may be configured with a 1 Mbyte RAM card so a memory disk may be configured for the data files necessary in setting up the data generator and analyzer.

For those test engineering departments with higher throughput requirements, a turnkey approach to test equipment selection is generally required. One test system which falls into this category is the SM2060 hybrid microcircuit test system manufactured by scientific Machines Corp. of Dallas.

The modular test head of this system allows the user to modify a test fixture rather than use small DUT boards. Some of the more salient features of the SM2060 are a 10MHz data rate, up to 48 high-performance drivers and receivers, 32 dedicated analog pins, and up to 64 fixed-level digital pins. The pattern depth is presently 8 kbits per pin, with various types of formatting (NRZ, RZ, SBC, etc.) available, with four additional tristate formats as well.

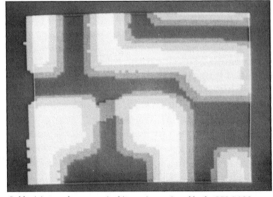

Solder joints under a ceramic chip carrier as viewed by the CSI-5100 system manufactured by IRT Corp.

All receiver pins have programmable windows for delayed sampling of devices, which may be used for propagation delay testing. Up to five programmable DUT power supplies are available, with DC parametric force-and-measurement testing on all inputs and outputs. An analog multiplexer is also available for switching signals of up to 200MHz to and from the DUT. Another useful feature is an available IEEE-488 bus along with a standard 19in. RETMA rack for the addition of other required instruments such as spectrum analyzers, function generators, counters, and voltmeters.

The controller for this system operates in a UNIX environment, so multitasking operations such as simultaneous test development and production test operations are possible with only a moderate cost increase for an additional terminal. With some additional cost, the system may be tied to a mainframe for detailed fault analysis and diagnosis for the hybrid, thus bringing the SM 2060 into competition with larger, more powerful ATE systems. This system is one of the first systems to address the problems of hybrid fixturing and test requirements directly, and represents a milestone.

For those organizations with high-volume production and/or high reconfigurability requirements, a more powerful test system may be considered. Two prime examples are the Teradyne A370 analog LSI system and the LTX77 with various extensions.

The Teradyne A370, being geared toward high reconfigurability, boasts the ability to test everything from codecs, filters, and modems to A/D converters (flash and otherwise), DACs, and digital circuits in various combinations. The operating system utilizes a custom language called Pascal/STEPS, which greatly reduces the lines of code required for test generation as compared with most types of structured BASIC. The test head is mechanically compatible with most handlers and wafer probers.

The other test system, the LTX77, has various extensions and slightly different capabilities. With the DX90 digital extension, the system integrates 96 high-speed I/O lines with the basic analog capability to test high-level mixed analog and digital circuitry. The CX80 consumer extension has RF and video source modules for testing radio and television products. The LX84 laser extension may be added for precision functional laser trimming of hybrid and monolithic resistors in production quantities. The operating system uses LTX extended BASIC for development of test programs.

Inspection equipment

The last few years have seen significant advances in automatic inspection equipment for hybrid circuits. It is now possible to detect faults in thick film circuits as small as 0.0005in. in size and to detect faults in solder joints under ceramic chip carriers.

Although primarily designed for inspecting PCBs and keyboards, the ScanSystem 1000 manufactured by Object Recognition Systems has been used to scan chip-and-wire hybrid circuits and SMT boards to ensure that the correct components are mounted in the correct places in the correct orientation. The system is capable of detecting part numbers on the top of ceramic chip carriers to verify that the proper part is in place.

The MIDAS-PR-830 series hybrid inspection system is manufactured by Photo Research Vision Systems. The most advanced member of the family, the PR-832, is capable of resolving 0.001in. over a 32in.^{2} single-circuit inspection area, with a resolution of 0.0005in. over an 8in.^{2} area. The system has a step-and-repeat capability to handle multiple prints on a single substrate. Able to resolve a 2 × 2in. substrate in one second with a 0.001in. resolution, the PR-830 series can be used off-line or on-line. The system may be programmed to detect missing or misshapen vias and defects in lines which are greater than 50% of the line width. The system is programmed by simply letting it view a 'good' part, after which it compares all other to the standard.

The CXI-5100 system manufactured by the IRT Corp. combines x-ray analysis with computer enhancement to perform non-destructive inspection of solder joints under surface-mounted components. This system can detect voids in solder joints under ceramic chip carriers. In addition, short circuits between solder joints, misalignment, solder height, and other defects can be detected. When silicon devices are mounted in a ceramic package, the system can detect voids in the bond material under the chip, whether it is bonded by solder, epoxy, or eutectically.

TESTING
COMPLETED
ASSEMBLIES

Once soldering operations are completed there is a need to test the full board. Divergent opinions on test procedures exist and must be examined.

IN-CIRCUIT AND FUNCTIONAL ATE

As printed circuit board technology improves, the automatic test equipment (ATE) needed to test PCBs is being presented with an increasingly tougher job. In particular, challenges are mounting from several fronts, including more and more use of VLSI circuitry on PCBs – with resulting demand for higher tester speed. Another challenge is the increasing use of custom chips having mixed signal technology – a combination of analog and digital circuitry on the same chip. In addition, surface mount technology (SMT) is pressing the mechanical ingenuity of test fixture designers. Test software costs are rising, too.

The two general types of loaded-board tester, which today must meet these and other challenges, are in-circuit and functional testers. Each represents a very different approach to the job of testing a PCB, and each approach can offer something the other cannot.

In-circuit testers

In-circuit testers locate faults by checking components and circuit paths locally, one-by-one. By contrast, functional testers stimulate the entire circuit under test (usually at the edge connector, but sometimes through a bed of nails) and compare the response to an expected response. With the in-circuit tester, circuit nodes are contacted by spring probes on a bed-of-nails fixture, and are isolated using electrical guarding techniques.

There are significant benefits associated with the use of in-circuit testers. They are particularly adept at finding manufacturing and workmanship-related problems, such as components of improper type, placement and value, etc., which have been estimated to represent as much as 85–90% of all board faults. Finding such faults on an in-circuit tester can be straightforward, while pinpointing the same faults on a functional tester can

First published in *Electronic Packaging & Production*, 1984 Automation Supplement, under the title 'In-circuit and functional ATE meets the challenge of the loaded board'.
Author: Robert Keeler, Associate Editor.
© 1984 Reed Publishing (USA) Inc.

take a great deal of time, necessitating complex diagnostics and the use of a guided probe.

With the in-circuit tester, the actual output function of the board at the edge connector is irrelevant. Instead, this type of machine performs a series of 'mini-functional' tests on the components, one-by-one. After the board description has been entered into memory, an automatic test-program generator can call up the needed test for a given component (IC, R, L, C, etc.) from a test library. Such libraries have to be upgraded each time new devices come along, and non-standard or custom devices will require original programming. Even so, compared to the functional tester, test program generation on the in-circuit machine is enormously simplified.

A major drawback to in-circuit testing, however, is that it offers no real verification of overall circuit functionality. In addition, the bed-of-nails fixture, required for accessing internal circuit nodes, can be expensive and cannot really handle the high test rates necessary for testing at speed. Nevertheless, lower programming development cost and highly efficient fault isolation are the hallmarks of in-circuit testers.

Pre-screening

Sometimes a tester called a 'pre-screener' is used prior to in-circuit test. DIT-MCO is a Kansas City-based ATE manufacturer which offers modifications to its bare-board testers to enable them to pre-screen loaded boards. DIT-MCO's Mike Mathews explains that these machines can be used to find routine opens and shorts in the board's copper due to plating and etching, or due to the reflow solder operation. These relatively routine, albeit undesirable faults, would waste the time of the in-circuit tester, which costs much more than the prescreener, says Mathews.

Computer Automation, Inc. offers what it calls a pre-functional tester as part of its Marathon line of ATE. Its Model 8100 is designed to do manufacturing fault screening prior to functional test. Warren Weinstein of

Computer Automation explains that although the 8100 does complete analog checking on all devices, as well as complete truth-table checking on all SSI and MSI devices, it has no dynamic test capability for LSI and VLSI boards. That is a role better served by a functional tester, says Weinstein.

John Fluke Co. offers its Model 3200A as a 'manufacturing defects analyzer'. Fluke claims it can do about 90% of the things a classic in-circuit tester can, but at about one-third of the cost and at much higher speed. Fluke's Jurgen Koch notes that programming time is drastically reduced with pre-screeners.

The Kryterion 550 from Everett-Charles is also intended for use in screening manufacturing defects. It features more than 8000 test points, and can test bare and loaded boards, coming with options such as a built-in bar-code reader and hard-disk storage. This product lends itself to fixture multiplexing, too. Everett-Charles' Model 55-175 is a smaller version of the 550, with a point-count capability of 2048 maximum. It can also test both bare and loaded boards.

Functional testers

The nature of functional testing is go/no-go, with the performance of the board as a whole being the parameter measured. Functional testers, especially those with the capability of testing mixed signal boards, are highly sophisticated machines requiring powerful computers and complex, expensive programming for test generation and operation.

Functional test, in the opinion of Craig Pynn, vice president of marketing for Zehn-

	Functional	In-circuit
Faults and problems which the tester is designed to find:	• ICs with inadequate drive current, or those that have excess leakage current or load down the driving IC. • ICs that are too slow or those with margin problems. • ICs with pattern, noise or timing sensitivities. • Problems due to interaction between chips.	Components (ICs, Rs, Ls, Cs, and discrete transistors, etc.) that are: • dead • wrong type • reversed in polarity • shorted between pins • open • missing • out of tolerance • improperly installed
General note:	When LSI/VLSI boards fail functional test, most IC problems not related to workmanship or assembly are found to be caused by timing or pattern sensitivities; in SSI/MSI technology boards, however, DC interactive problems are more prevalent.	In general, in-circuit testers find problems due to poor workmanship, component quality, assembly and incoming inspection technique; in-circuit testers can even pinpoint poor instructions or negligent operators. They can also find bad copper traces, although finding this sort of problem at the loaded-board stage is expensive.

Comparison of functional and in-circuit testers

tel, is intended to replace what most manufacturers refer to as 'hot-bed' testing. He explains that hot-bed testing amounts to placing a board, which has been successfully in-circuit tested, into an environment very close to that which the board will see in operation. A typical test setup might include everything in the system but the board in question, which gets plugged-in last. If the system works, the board is assumed to be good. If it fails, the board returns to the in-circuit machine or to a technician for diagnosis.

A major strength of functional testers is their speed. Although the programming effort that goes into them is much more involved, the go/no-go testing they do is significantly faster than the component-by-component test of in-circuit machines. Depending on the comprehensiveness of the test program, probably only 2–3% of the boards which test good on a functional tester will go on to fail in system operation.

When a functional tester does find a bad board, a guided probe is used to look for the fault(s) node-by-node. This can take a great deal of time. Functional testers have waned in popularity to some degree in the past years for this reason, and because the cost of programming them has risen sharply. This market shift has been to the advantage of

Wayne Kerr offers two benchtop test systems with built-in continuity, in-circuit and functional test capabilities

in-circuit testers, whose programming is simpler.

However, it may be that functional testers will be more suited to testing the new surface-mount technology boards which are now coming on strong. Probe access to closely spaced device leads and circuit nodes typical of SMT is not necessarily an issue with functional testers, which can connect to the PCB at the edge connector.

Combination testers

An alternative to buying two separate testers – one functional and one in-circuit – is to combine them into one machine. This was

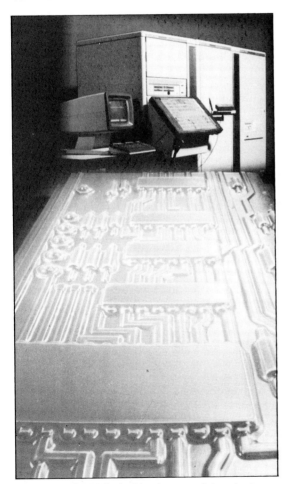

Hewlett-Packard's 3065 in-circuit board-test system features signature analysis capability with data compression to minimize the amount of memory behind each pin and speed test

done several years ago, when the first 'universal' tester, the L200, was introduced by Teradyne. The L200 has up to 1152 channels which can be multiplexed to 2304 pins. All channels and pins have both in-circuit overdrive capability and functional test performance.

Teradyne has recently upgraded its system with the introduction of the L280. It incorporates several new features, including several which Teradyne calls ABC emulation and D5 channel cards. These can be retrofitted into any L200 system in the field. With the D5 channel cards, Teradyne has tripled the density of its earlier high performance digital test D2 cards, allowing systems to be configured with up to 1152 10MHz channels or 2304 pins. Teradyne also offers what it calls 'MultiMode' testing, claiming that this approach gives the user functional test performance at the in-circuit test level. Teradyne explains that MultiMode testing allows the user to partition the board into functional subsets or clusters when doing functional test, by using in-circuit backdriving techniques. It also offers functional test performance when doing straight in-circuit test.

There is a strong interest now in obtaining a functional look at the PCB through a bed of nails to get better visibility of the board by accessing all of the internal nodes simultaneously. This is because there are limits to what can be determined at the edge connector. Access to internal nodes means that test engineers can simplify their programming task and shorten the time to run a functional test.

Dual-level fixturing is a relatively new concept which addresses this problem. In-circuit and functional test probes come down at two different stages of application, usually under vacuum or pneumatic activation. OB Test Group (formerly Ostby and Barton) and Evans Corp. and Pylon Corp. are among the manufacturers of dual-level fixtures.

Joe Burns, national sales manager for OB Test Group, claims that this development will give impetus to the use of the combined tester. OB's product, in particular, has simplified much of the control mechanism needed to activate the two sets of probes, requiring only that the user be able to provide two vacuum levels.

Some models offered

Zehntel Corp., long an advocate of in-circuit test, is now moving into the sphere of functional test. With its recent acquisition of a British firm, Columbia, Zehntel will be offering the Columbia 2000, a performance tester. Following this move, Zehntel now owns a company in each of the three principal areas of test: stress screening, in-circuit and performance (functional) test, says Zehntel's Craig Pynn.

The Columbia 2000 makes use of a technique called bus timing emulation for microprocessor boards, which takes control of the bus to test the microprocessor functionally. It is basically a guided-probe technique, although in a high-throughput environment, a bed-of-nails fixture would be called for, notes Zehntel's Mike Carroll. The company's Model 850 in-circuit tester, which was introduced in the fall of last year, is now its best-selling product. Carroll points out that this machine is especially suited to testing mixed signal technology boards.

Meanwhile, GenRad is positioning their new 2720 performance test system in the

A sequence processor and 4K 5 memory behind the pins on Fairchild's Series 30 Model 333 E in-circuit tester provide substantial vector depth for both algorithmic and linear pattern requirements

same market slot as the Columbia 2000. About the concept of 'performance' test, Bob Marchetti of GenRad claims there is no way that a classic functional tester can really simulate the end-operation environment of a PCB. He points out that it can duplicate operating speeds, but not, for example, the noise that comes in through cable harnesses. Marchetti claims that performance test, as on the 2720, really tests the board as it performs. For example, the system offers a technique called 'memory emulation' to exercise a board by putting its own microprocessor under test-program control.

GenRad also offers, the 227X in-circuit tester line. The new 2276 represents the newer generation of in-circuit testers, typical of an increasing trend to provide more functional test capability on an in-circuit tester. The 2276 possesses 5MHz in-circuit operation capability, and 8K of memory depth behind its hybrid pins.

Fairchild Instruments Co. is marketing both in-circuit and functional testers. The 30/333 series of hybrid in-circuit testers features a sequence processor and 4K × 5 of memory behind the pins to provide virtually limitless vector depth for both algorithmic and linear pattern requirements. The system can test LSI and VLSI ICs on boards with advanced, bus-oriented designs, and it can also handle mixed-logic boards such as ECL/TTL.

Fairchild is also offering a new performance system – the Series 700 System 720 – which has been on the market for about half a year now. The Series 700 incorporates all the experience Fairchild gained from their earlier Series 70 machines. It can test at 30MHz and its ability to make decisions at high speeds lends itself to running unlimited length, non-linear programs at full system speed. The system features 512 hybrid pins.

Hewlett-Packard introduced the Model 3065 about a year ago, a system which has in-circuit analog and digital, as well as functional test capability. HP has been steadily adding to the system's power since.

HP's Rod Parks says the 3065 has a unique system architecture that allows it to offer very high digital test throughput. The number of test vectors that have to be downloaded from the CPU has been reduced. The 3065 uses data compression, both to minimize the amount of memory required behind each pin and to speed test operations.

Wayne Kerr offers benchtop test systems. Its two models – the 8000 and the Combat 8315 – have in-circuit, functional and continuity test capability built-in. The British-owned firm is able to offer this combination of test capabilities by doing away with the scanner – a very large reed-relay matrix which switches signals from the unit under test to the test instrument. Eliminating the scanner, says Wayne Kerr's Phil Parsons, means that the tester can do without the typical tester's three or four thousand relays, the power supplies to run them, and the fans to cool them. In place of the scanner, Wayne Kerr uses a measurement op amp at every in-circuit test pin. Among other things, this reduces the serial resistance in each line. Eliminating the scanner also means that the functional modules present can use switching systems optimized for their own particular function.

Computer Automation offers its Model 8200 functional tester, along with its Model 8100 pre-functional machine, as part of its Marathon System, which is intended to set up a test environment that includes test engineering, quality control and manufacturing. CA sees it as a step toward the factory of the future.

The Model 8200 functional test station is aimed at complex boards and high-volume testing. It includes a memory pattern generation (MPG) feature which enables the user to test embedded memories – which are normally difficult or inconvenient to test. Since it is a real-time system, it can emulate a system environment so that the boards under test can be tested at speed. This facilities the pinpointing of timing errors.

Fluke's functional tester, the Model 3050B, is capable of producing yields of up to 98%. The system learns from a known-good board, with the result that programming is much simplified. Responses provided by the reference board in real time allow the 3050B to operate at rated microprocessor speeds. The system provides pass/fail totals and failure

GenRad's 2720 Performance Test System automatically tests and diagnoses complex, bus-structure digital VLSI printed circuit boards at speed in their native environment

data, and test programs are completely defined in hard-copy data file outputs. Test effectiveness is documented by an automatic fault-emulation output summary.

Marconi includes four models of in-circuit tester in its System 80 line, and all accept a wide variety of test fixtures. The System 80 line offers two methods of actuation – pneumatic and vacuum – as standard features. The company claims that virtually any board size and shape can be handled. Test programming is done in the INCITE language, a variation on BASIC developed by Marconi in 1973. Automatic program generation includes software for program generation, on-line editing, and analog and digital guarding.

Thinking about the future

Bob Broughton, a product marketing manager for logic analyzers at Tektronix, feels that the two most challenging issues for loaded board test now are growing circuit complexity and the pressure for tester speed to keep up. Broughton points out that, because it is not possible to probe the inside of integrated circuits, the only way to check ICs having a high level of integration is to do what amounts to a functional test on them. Thus, with the trend toward getting the entire function of a board onto one chip, in-circuit testers will have to have a greater functional test capability.

Speed of test will be an issue, too, says Broughton, because with increasing density, connections between circuit elements in the chip no longer behave like transmission lines but rather as small lumped constants, and circuit performance increases. The performance of a tester must increase too, so that it can test a device functionally at the device's own speed of operation. The appearance of an increasing number of custom VLSI chips on the market now will also be a problem that ATE manufacturers must address, says Broughton.

Peter Hansen of Teradyne agrees that the trend to higher levels of integration is a big problem, and he feels that no current test system or programming method adequately deals with it. Hansen claims that a ray of hope exists in Teradyne's aforementioned Multi-Mode testing feature, a technique offering the ability to divide a board into bite-sized chunks – circuit segments whose function the test programmer understands and for which he can write a program to exercise the device or cluster of devices functionally. Hansen says this sort of technique may be one answer to dealing with SMDs, where device nodes might otherwise be too dense to probe, or even completely inaccessible.

Surface-mount devices, says Allen Davis of Everett-Charles Co., are so far a very 'unruly' area – short on part and application standards. SMDs are sometimes leaded, sometimes leadless, and component heights will vary from almost flush with the board's surface to almost 80 thousandths above the board. In this vein, Everett-Charles currently offers a device it calls the Series 22 for probing SMDs which, says Davis, typically has to be modified to suit customer requirements. Davis foresees strong progress ahead in the area of probing SMDs, though the cost from an ATE standpoint will initially be high. He sees these rising costs ultimately forcing companies to adopt guidelines in design for testability.

FUNCTIONAL TEST

Modern commercial functional testers, especially those with the capability of testing mixed-signal boards (which contain both analog and digital electronics) are highly sophisticated machines requiring powerful computers and complex, expensive programming for test generation and operation.

Approaches to functional test, other than the use of a turnkey piece of commercial ATE, includes the use of an in-house built tester, use of a 'rack-and-stack' method (buying a group of individual test instruments and power supplies, then assembling them together with a computer in one rack), and hot-bed testing (adding a board to be tested to a known good system which lacks only the board in question). Though these other alternatives offer cost savings, they sacrifice either convenience or diagnostic power.

The evolution of functional test

Although the programming effort that goes into digital functional test is much more involved than in-circuit test, a major strength of functional testers is their speed. Functional test is by nature go/no-go, with the performance of the board as a whole being tested. Depending on the comprehensiveness of the test program, probably only 2–3% of the boards which test good on a functional tester will go on to fail in system operation.

Specifically, the functional tester detects faults arising from ICs with inadequate drive current or excessive leakage current, or those ICs that are either too slow or too sensitive to pattern, noise or timing discrepancies; it also detects problems due to interaction between chips. When LSI/VLSI boards fail functional test, most IC problems not related to workmanship or assembly are found to be caused by timing or pattern sensitivities. In SSI/MSI technology boards, however, DC interactive

First published in *Electronic Packaging & Production*, September 1985, under the title 'Economy and product complexity drive functional test'.
Author: Robert Keeler, Associate Editor.

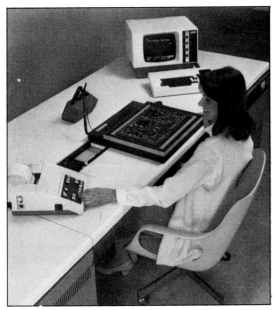
Factron's Model 720 functional tester

problems are found to be more prevalent. The term 'functional test', which is coming to be taken to mean test at less than full speed, is often associated with simulation as well as random logic and SSI/MSI; the expression 'performance test' is frequently taken to mean testing at full speed and is associated with emulation, as well as LSI/VLSI technology and bus structure.

Functional test sequentially applies test vectors to the PCB's inputs and compares the outputs from the board with predicted outputs. As Warren Weinstein of Computer Automation points out, a functional test program may contain over a thousand test patterns, each one corresponding to a single step within the test program.

Writing these programs consumes an enormous amount of time. In response to this problem, automatic test-program generation is often used during digital simulation. With the help of these tools, a test engineer can create a mathematical model of the circuit board and observe whether or not his test patterns toggle the individual logic units. How well the patterns do this is expressed as a percentage of coverage for each individual circuit board.

If a functional tester does find a bad board,

use of a guided probe is the prevalent method used to look for the fault(s) node by node. However, this process can take a great deal of time; functional testers lost popularity in the past years both for this reason and because the cost of programming them skyrocketed.

Notes Warren Weinstein, although digital simulation and automatic pattern generation have shrunk down the programming time, they have their limitations. For example, the difference in size between an automatically generated program and an optimum program shoots up rapidly as the percentage of logic units that are toggled rises above 60–70%. Even if the programs do fit into memory, they may still present problems for real-time testing because of the size limitation on the memory behind each driver pin. The development of emulation techniques has been one response to these problems.

Emulation is a tool that has greatly upgraded the test capability of functional testers. The technique is most often associated with digital functional testing of microprocessor boards. Technically, it involves creating a controllable stimulus source to stand in for a particular circuit under test.

Emulation offers a way of getting into the middle of the board for test at full speed, as opposed to working through the card-edge connector. As Roger Boatman of Computest points out, the latter approach requires a lot of high-speed electronics behind each pin, and does not offer an efficient way to verify that any particular test vector has uncovered a fault. Thus, it is difficult or impossible to tell if a microprocessor board is good or not.

Enrique Farne of Zehntel Performance Systems, a division of Zehntel Inc., notes that there are three approaches to emulation: in-circuit emulation, in which the microprocessor is emulated; memory emulation, in which the microprocessor's memory is emulated; and bus-timing emulation, in which timing cycles are emulated. Bus-timing emulation is an emulation of the bus cycles for every data-transfer instruction.

Farne states that with the first two approaches, control of the board still remains with the microprocessor itself. With bus-timing emulation, however, says Farne, the board is under the control of the functional tester; the microprocessor is not physically removed from the board in Zehntel Performance Systems' bus-timing emulation since it is tristated to eliminate it electrically.

Says Computest's Roger Boatman, emulation-based testing will probably be the front runner for microprocessor and bus structured boards for the next decade, simply because it uses a known good board as its own simulator and allows the user to greatly simplify the functional programming required to test a board.

Functional and in-circuit

The market preference until recent years has been weighted toward in-circuit testers, whose programming is simpler than for functional. The in-circuit testers, whose programming is simpler than for functional. The in-circuit tester looks for manufacturing process-induced defects such as components (ICs, Rs, Cs, discrete transistors, etc.) that are dead, wrong type, reversed in polarity, shorted between pins, open, missing, out of tolerance and improperly installed. It therefore has a well-deserved reputation as an excellent process diagnostic tool.

In-circuit test works on the assumption that if all the individual devices test out good, the

Teradyne's L200 combination tester

board as a whole is good. The functional tester, on the other hand, works on the assumption that if the board as a whole tests out good, its parts must be good.

It has been pointed out that because manufacturers are finally getting control of their processes, there are less consequent manufacturing defects to find. Ability to find these types of problems has always been the great contribution of in-circuit testers. With fewer problems to find, manufacturers are beginning to see mostly interactive and performance-type problems as the remaining barrier to the high yields they want. This is the sort of task at which functional excels. Thus, manufacturers are realizing that either more functional test capability must be added to their in-circuit testers or more functional ATE must be purchased.

In-circuit test is facing another great challenge in surface-mount technology boards. Probe access to closely spaced SMD pads and circuit nodes is difficult on a bed-of-nails tester, but not necessarily a problem with functional testers, which can test the PCB from the board's edge.

However, Tom Newsom of Hewlett-Packard makes an interesting observation in that surface-mount technology, being still fairly new, will likely offer many process-related defects and faults until its users have mastered it. Despite the fixturing problem, this might mean there will be a great demand for in-circuit test for surface-mount technology.

Newsom feels the line of distinction between the in-circuit tester and the functional tester is beginning to blur somewhat. On logic device test, the in-circuit tester essentially is already performing a minifunctional test device by device. In addition, many in-circuit testers have grown in diagnostic capability in recent years. Some in-circuit models, such as certain systems offered by GenRad, now include memory test capability based on functional techniques.

Furthermore, many functional testers are now offered that can test through the bed-of-nails, fixturing that once was strictly the hallmark of in-circuit testers. Use of bed-of-nails fixturing with a functional tester generally permits faster throughput and lends itself to automatic loading. It also gives greater visibility into the nodes.

The advent of the combination testers, able to do both in-circuit and functional test on the same machine, also helped to blur the functional/in-circuit distinction to some degree. For years Teradyne has been the pioneer in the field of combination testers with their L200 series. However, the ATE East shown early this summer saw the Factron Division of Schlumberger Inc. introduce a new combination tester called the System 750. Wayne-Kerr introduced a combination tester, their Model 8315, at the same show. Meanwhile, Zehntel Performance Systems does not rule out that they will investigate the idea of combination testers. Neither does GenRad, although a spokesman for GenRad hastens to add that they feel these testers will not be the mass-market solution.

The marketplace

Manufacturers of commercial ATE for functional test have concentrated heavily on serving the steeply priced, high-performance tester market. Don Crassas of Support Technology notes that the functional testers' price tags have gotten out of hand in recent years. He adds that while manufacturers need functional testers, they don't want to pay half a million to one million dollars for them. A price range of under $100,000, he feels, is more reasonable. Crassas sees a definite, growing demand for downsized, economical functional testers.

Crassas points out that technology has gotten to the point now where it is possible to put a great deal of computering and test power into a fairly small box. Thus, the advent of standard bus structure and the proliferation of the use of standard microprocessors has lessened the need for the general purpose pin-system necessary for random-logic boards. This sort of system, running at megahertz rates, demands a lot of memory and a lot of computing power, a factor which in the past has added to the cost of functional testers.

Warren Weinstein of Computer Automa-

HP's Model 3065 digital/analog board test system interfaces to automatic board test handling systems for high throughput

tion notes that another push toward the burgeoning lower cost functional tester market is the need to conserve floor space. He adds that lower cost models often approach the price range of an in-circuit tester. Thus, smaller companies that formerly only had an in-circuit tester will be able to buy a functional tester as a second machine and profit from the improvement in yield arising from having both machines. Regarding test capabilities of the lower end models now appearing, Jerry Hartman of PDI notes that, while it is true the larger systems traditionally have been able to range from 330 up to 500 I/O pins, the more economical systems are catching up here.

When shopping for a functional tester, the obvious consideration is equipment cost. However, this factor alone is not the whole story. Other factors to be weighed are the cost of program generation, the cost of operation and the cost of maintenance. The buyer should consider the volume throughput expected and the complexity of the PCBs as well as any new products planned for the future.

Also to be considered is the complexity of the tester itself and the corresponding skill level required to operate it. Will there be sufficient technical support from the vendor? How much floor space will the equipment take up?

Mike Renneberg of Eaton Corp. notes that other major considerations are the tester's capability for high speed and the presence of sophisticated software. Fault diagnostic capability is also an important criteria. Renneberg says the buyer should be aware that a high clock rate does not necessarily mean a high pattern rate across the test pins. He claims that this is probably the most misleading issue with regard to tester purchase today.

Pat Prather of Wayne Kerr states that the number of driver/sensor pins available should especially be considered when the intention is to do functional testing of random-logic circuits through a bed-of-nails fixture and perform diagnostics. Mike Renneberg adds that sufficient test points must be available to stimulate the board at the edge card. A capacity of 256 pins should be adequate for most boards.

Other questions to ask: Does the tester offer analog as well as digital functional test capability? Is an emulation technique or high-speed signature analysis available? What approach does it take and what are its strengths and weaknesses? Can the test points be programmed as input or output at full speed?

Both Prather and Renneberg agree that ease of programming is an extremely important consideration. Is the programming language involved easy? Does the system allow the testing program to be self-learned? Software should readily allow the user to learn the proper response of the board at full speed. Is a menu-driven program generator or a high-level language offered – or both? Some functional test products offered today follow in the next section.

Some testers offered

Computer Automation's 8200 is a high end, multicapability functional tester. C/A says it will allow a user to configure a system from the highest capability memory-pattern generation, with adjustable logic-level interface pins, down to static-level test. The machine has a 5MHz test-cycle rate and a 20MHz data rate.

C/A's 9200 Compact Series tester, on the other hand, has somewhat less flexibility than their 8200 though it still provides real-time functional test. It is priced in the same range as a midrange in-circuit tester, says the company.

Watkins-Johnson Model 1530-1540 digital/

Computer Automation's 9200 Series compact functional tester

analog hybrid test systems runs on their proprietary Colt programming language. The tester has 512 input-output pins and internal switching capabilities. There is a selection of programmable measurement and stimulus options and digital test capabilities.

The Eaton Model 800 uses guided probe software to lead an inexperienced operator to the exact fault location, point by point. Sophisticated learn-mode software here eases the programming task. The 800 can go full speed on all the commonly available microprocessors to exercise the circuit under its real-world conditions.

The Series 80 functional tester from Computest features advanced test-oriented emulators for such devices as the 80286 and 68020, as well as the 2900 family of bit/slice micros. Since these have been coupled with a high-level programming language, at-speed functional testing can be achieved in a matter of days.

Everett-Charles offers the Kryterion 800 functional test system. This is a modular system that the user can configure to his own test needs. The IEEE-488 bus is an available option for additional external equipment. The programming language here is a menu-driven version of BASIC. Use of a bed-of-nails is possible through the E-C 33-100 adapter which accommodates the E-C test heads for bed-of-nails.

Marconi has newly introduced its System 8080 dynamic functional tester. It is a 512-pin bipolar system, 10MHz parallel data rate. The System 8080 can be used in four functional modes: stimulus-response-compare (SRC), level-sensitive scan design (LSSD), functional stimulus mode (FSM) and ultrafast analog functional. There is 96k RAM behind each pin. Marconi's menu-driven software for the 8080 is called Climate.

The Factron Division of Schlumberger offers the 720 functional tester as part of their 700 family of ATE. Maximum data rate is 30MHz. Formatted modes can be done at a speed of 10MHz and signature analysis can be run at 20MHz data rates. Programmable cards are available to accommodate a large array of telecommunications products.

TDS (formerly Trade Data Systems) offers their Model TL8B, a console model with both digital and analog test capability. It has the ability to digital testing with up to 256 I/O pins and pattern rates of up to 10MHz. As an option it has the capability, with an IEEE 488 bus and a fully automatic matrix switching system, to handle up to 30 IEEE instruments.

Support Technology Inc. has recently introduced a portable functional tester for the field-service test market. It is capable of testing any board that has a microprocessor or a standard bus structure present. It runs on the Unix operating system and features 10 Mbyte of hard-disk memory. The machine is packaged in a 20lb, briefcase-sized package. The electroluminescent flat-panel display helps to cut down on the weight, and eliminates the need for a CRT and its associated big power supply.

John Fluke Co. offers its 3050B functional tester to the marketplace. The 3050B offers functional testing of both digital and analog circuitry when combined with the 3051B off-line programming station and the 3053B analog test station. The system includes guided fault diagnostics for fault isolation even with lower skilled operators.

Digalog's DMST 2020 Series functional tester is a 68000 microprocessor-based system with complete analog and digital test capability for microprocessor-based boards and systems. Able to work either through the card edge connector or a bed-of-nails, it features bus-timing emulation, an internal clock rate of 100MHz, and a 10MHz data I/O rate.

Microcontrol offers both high-speed memory board testers as well as logic board testers in its machines known as the M-7, M-10AT, M-10B and M-12. Both the M-7 and the M-10 systems have programmable clock systems, and the pattern rate is 10MHz, with stimulus and response being done at the edge of the board. All application programs are written in BASIC.

Instrumentation Engineering is offering their Model 4301, which is functional tester capable of both digital and analog test. It has digital capability of greater than 400 pins, 160 of them being tristatable, a data rate of 10MHz and an automatic diagnostic capability via a guided probe. Analog instrumentation includes a 1553 bus tester, pulse and function generators, power supplies, DMM, and timer-counter, among others.

Teradyne's L200 is the original combination tester. The functional capability of this machine offers timing resolution accurate to within 1ns. It has the capability for direct tie-in with LASAR software, a tool for helping the programmer generate test patterns. Within the L200 family there are two testers: the L210 and the L280. The L210 can be configured with up to 576 functional channels and the L280 can have 1152.

Greenbrier Electronics offers its 4400 portable universal tester, a benchtop/field service machine which can be used for functional screening of boards as well as component-level diagnostics. The base unit is under $10,000 and has the built-in ability to measure both analog and digital phenomena. There is a built-in signature analyzer, a 40MHz frequency counter/timer, and an autoranging

Everett-Charles' Kryterion 800 PAD functional test system

digital multimeter with voltage/current/resistance modes.

Hewlett-Packard's HP 3065 board test system is an in-circuit and functional board test system. It provides as standard equipment a number of analog functional test capabilities. These include frequency counters, sources, ramp and sine-wave generators. On the digital side, the 3065 can run at megaherz clock rates. The 3065 can perform cluster testing by defining a given section of the board and its stimulus and response with megahertz rate signals.

Wayne-Kerr's new functional tester is the model 8510, a combined functional and in-circuit tester which uses a single test fixture for both kinds of test. The 8510 is capable of both analog and digital functional test and includes emulation capabilities. For analog test, there is a frequency counter and the ability to measure both voltages and currents. It is possible to datalog, storing test results by serial number and performing statistical analysis.

Zehntel Performance Systems manufactures the Columbia 2000 PCB functional performance tester. It combines two approaches to functional testing that provide users with the field-proven conventional technique for testing SSI/MSI logic and a new bus-oriented approach for testing bus-structured LSI circuits. Bus-timing emulation allows the tester to address a circuit in the same manner as its own microprocessor and exercises the circuit in every possible mod. The system also performs a complete test to the microprocessor itself.

The Beaver Major MDS 9000 functional automatic test system from ATE Systems International offers a master program in combination with an automatic assembler/compiler to form a fully integrated, bug-free, resident operating system. Some options include a guided probe for easy DUT accessing and an external qwerty keyboard for prompting messages and instructions.

Computer General Corp., a division of EC&P, offers the Sleuth, a 10MHz real-time machine. There are 288 active driver/detector pins, all software programmable as either driver or detector pins. The system operates under the control of a test program which may be compiled either from the Sleuth's own high-level language or from an existing test program written in the high-level language of other ATE suppliers.

AUTOMATING THE TEST FUNCTION

In automated production, two problems face the production engineer. One is how to program the automated equipment. The second is how to bring the material to be tested to the test equipment and how to dispose of this tested part according to the test output data.

TEST PROGRAM DEVELOPMENT AND PROCESSING POWER

Flexible automatic test equipment (ATE) must work together with flexible manufacturing systems (FMS) and computer-aided engineering and design (CAE/CAD) in a factory expected to respond to myriad, rapidly changing marketing demands.

As a product takes shape, reaching the module, subassembly, or complete system phase, an automatic test system may be called upon to accommodate a high mix of low-volume products. Product life cycles, too, are generally becoming shorter. Testing new products in this business environment can require that existing test systems be quickly reconfigured and reprogrammed.

Manufacturing a larger variety of products with smaller volumes requires test flexibility. Test programming and distributed processing within individual instruments are crucial elements that must work together with reconfigurable hardware to assure that test flexibility is not constrained.

Flexible ATE

A flexible method must bring together into an ATE system individual instruments such as signal generators and digital multimeters in response to the specific production test requirements of the moment.

The IEEE-488 bus standard, now an established concept, is generally the core of flexible test configurations. Individual test instruments with increasing computing power of their own, and in some cases a higher level of integration within the instrument or measurement package, are linked by a bus conforming to the standard. The hardware configuration and test programs are typically established by individual ATE users to match their unique requirements.

Automatic test equipment established around the IEEE-488 bus fundamentally consists of a controller, multiple test instruments, and a signal switch. The bus itself is a cable that interconnects all devices in parallel so that any one may transfer data to one or more of the other devices.

A controller manages operation of the bus system. It designates which devices are to send and receive data. Keyboards, peripherals, and displays may be incorporated in the controller for entering data, storing and displaying test results, and monitoring system operation.

Each compatible test instrument has a standard 24-pin connector for interconnection to the bus. Up to 14 instruments and a controller may be connected in a linear or star configuration. Cable length can be up to 20m, with a device load for every 2m of cable. The maximum data rate under this condition is 250Kbytes/s.

Flexibility extends beyond the capability to link various test instruments. Flexibility encompasses the computing power and features of individual instruments, test program generation, and interfacing integrated test systems with their own internal bus to an IEEE-48 bus.

The IEEE-488 standard specifies mechanical, electrical, and application-independent functional interfaces and their logical descriptions. It does not include operational specifications that encompass allowable test instruments, nor their specific functions and logic descriptions. With a standard GPIB control message, there are slots to accommodate a device-dependent message.

Applications software for specific production test requirements must be developed to tell the instruments what measurements to make, where to make them, when to make them, and what to do with the results.

A controller works with an associated software language. It is frequently a high-level language such as BASIC, FORTRAN, or Pascal. Within the framework of the controller's language, the commands associated with each specific instrument's functions are sent over the bus. These commands may vary across different types of instruments, and

First published in *Electronic Packaging & Production*, September 1985, under the title 'Computer power shapes systems' test'.
Author: Ronald Pound, Executive Editor.
© 1985 Reed Publishing (USA) Inc.

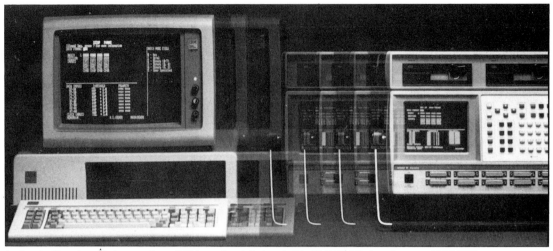

A subroutine library is used for programming an IBM personal computer to control Gould's K105-D logic analyzer over a IEEE-488 bus

across the same type of instrument from different manufacturers.

On the one hand it is the capability to custom-tailor test methods with software that assists in providing flexibility to an ATE system configured from IEEE-488 compatible test instruments. On the other hand, it can be software that limits flexibility. Inherent flexibility is constrained if, because of the basic elements and structure for developing test software, applications programs require too much time to develop or coding complexity promotes programming and testing errors.

With the framework established by the IEEE-488 standard, and its implementation in products such as the general-purpose interface bus (GPIB) or Hewlett-Packard's HP-IB, the focus today is on specific controllers, test instruments, and signal switches that can be placed on a bus to form an automatic test system.

Test programming

Test engineers conceiving a system for production testing face the task of designing procedures for unskilled operators to run. They are frequently interested in a specific measurement value, sometimes a go/no-go test. Minimizing test-development time, actual test time, and test-system costs are constraints faced by the test engineer. These issues have shaped the development of spe-

cific GPIB products in the marketplace today.

Intended principally for lower user-skill environments such as production-line testing, Tektronix' IEEE-488 compatible 4041 system controller is an 'execute only' controller. The operator on the factory floor cannot tamper with programs, nor even list them.

The unit was designed to be non-intimidating, says Tektronix. Operator interaction is limited to reading prompts on the alphanumeric display, inserting a tape casette, and pressing a small number of keys. In addition to system keys, there are 10 user-definable function keys that can be assigned subroutines by the applications program.

For program development by a test engineer, an RS-232-C CRT terminal attached to a 4041 with a program-development ROM option is used. This provides an interactive workstation for a test programmer.

Most test engineers, though, are not highly sophisticated, trained programmers. Many may write programs only occasionally. The activity of a test engineer should be focused on fundamentals of test strategy and test methods. He should not be consumed by programming complexities.

BASIC was chosen by Tektronix for the 4041. This is a good language for the occasional programmer with its English-like commands, simple syntax, and line-by-line interpreter implementation. Test program-

Test system is formed from Hewlett-Packard HP-IB instruments placed in a rack

generation software enables a test sequence that is manually performed once to be done automatically thereafter. Programs developed in the lab can thus be used for testing products in the manufacturing stage.

Test instrument control requires a system that is interactive with an easy to learn and use programming environment, notes Hewlett-Packard in describing the background for their new HP 9000 Series 300 for instrument control as well as technical workstation applications. These computers support three operating systems. BASIC and Pascal are available for single-user configurations. The manufacturer's version, HP-UX, of AT&T's UNIX operating system provides a multitasking, multiuser environment.

Flexibility is provided not only by IEEE-488 bus compatibility, but also by the compatibility of software for HP's earlier Series 200 with the new Series 300. Modularity of the concept, too, provides flexibility in upgrading the computer. Two different CPUs are offered. The 10MHz Motorola 68010 is available for entry-level and midrange systems. A faster 16.6MHz Motorola 68010 can be substituted if applications change to require high-speed processor performance.

Test software development packages alone, independent of controller and instrument sales, can be supplied too. Production Automation's UNITEST menu-driven software is designed to simplify programming of HP 9000 Series 200 computers for instrument

control. The UNITEST software will work for any manufacturer's GPIB test instrument under control of the system. Instrument libraries, supplied by the vendor or user-developed, include the GPIB capabilities of specific devices.

A user first picks from the instrument library those devices that will be used in the test. Then the control program is written in the normal manner except for the points where there is an instrument call. Here, the UNITEST software is entered. The specific instrument for that call is selected from a menu. Subsequently, the user specifies the task to be done for each call to an instrument and UNITEST generates the appropriate code.

As the control program is developed with UNITEST, checking of syntax and validity of the code assures that the test software will be error-free when completed.

Personal computers

Along with their penetration into engineering design systems, personal computers are appearing as test system controllers too. Personal computers used as GPIB controllers and turnkey packages interfacing instruments to a PC are approaches that differ markedly from the traditional GPIB control-

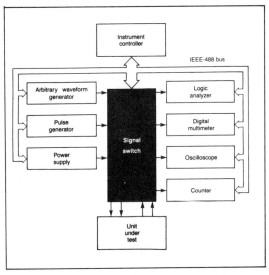

Automatic test system formed around the IEEE-488 bus includes an instrument controller, individual test instruments, and signal switch

ler approach, indicates Dolch Logic Instruments. Linking several instruments together in this fashion and adding some automatic test routines becomes relatively simple and inexpensive.

An IBM PC can be used to control GPIB instruments with Tektronix' GURU hardware and software package. Hardware includes a GPIB interface board that plugs into the PC's main board and a GPIB cable.

A test procedure generator written in BASIC allows users to generate a program that runs a specific test sequence without writing a single line of code. Details of the test to be performed and the equipment to be used are the only facts needed by the user.

Test routines are generated with Tektronix's GURU using a menu-driven program. At the heart of developing a test program is the function menu. This menu contains all individual steps that are elements of any test procedure.

Transforming a logic analyzer into a simpler tool for use on the factory floor is the central concept of Gould's Design & Test Systems Division's K105 Smartpak. This package's applications software runs on IBM or compatible personal computers.

Gould's Smartpak software for use in controlling the firm's K105-D logic analyzer is intended to relive the test engineer of GPIB programming burdens and permit him to concentrate on creating test procedures and instrument setups. In addition, less technical people in production can use advanced logic analyzer capability.

Each logic analyzer function supported by the control software is associated to a subroutine call with a certain number of parameters. A library of predefined routines written in 'C' language eliminates the need for test engineers to write device-specific code. A single subroutine call may represent up to 200 transactions over the GFPIB. The high-level subroutines contribute to reducing programming time and bugs due to typing errors.

Programming of Keithley Instruments' IEEE-488 interface for IBM PC or XT computers is done with statements taking the form of BASIC calls. Statements are included to program instrument functions, obtain data, and for IEEE-488 control functions. A number of operation programs for use with various Keithley GPIB instruments are included.

Price positioning between IEEE-488 instrumentation systems and manually operated test equipment is the strategic thrust with Vistar's PC1000 system based on an IBM PC. Small to medium manufacturing operations are one of the firm's target markets.

Along with the personal computer, a system consists of test instruments in an expansion chassis operating over the PC bus. A GPIB

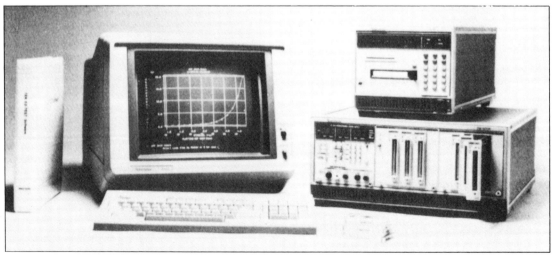

A GPIB instrument controller, Tektronix' 4041, sits on top of a rack holding a programmable power supply and modules

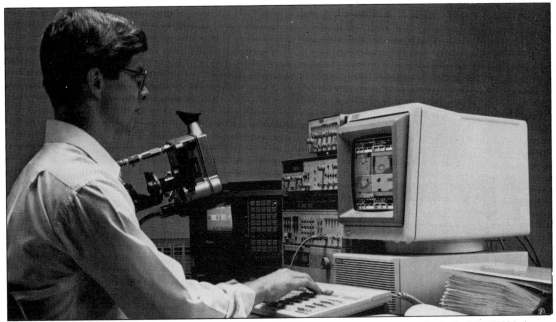

Three operating systems — BASIC, Pascal, and the manufacturer's version of AT&T's UNIX — are supported with Hewlett-Packard's HP 9000 Series 300 computer for instrument control

interface, however, permits blending of the features of personal-computer control over its bus with the features of an IEEE-488 bus. Automatic operation of test routines written in BASIC can be established on the PC1000 system.

Instrument features

With increased computing power being incorporated in the test instruments themselves, their programmability as well as their measurment specifications is crucial.

Programmability and processing within an instrument provide the user with increasing testing fleixbility and can take some of the burden off the system controller and the bus. With programmable functions, the scope of what any single instrument can do is increased. Processing within an instrument, such as finding the mean of an ensemble of measurement values, can reduce traffic on the bus. This can, for instance, enable the controller to work with other instruments in the time period when it would otherwise have to collect and process data itself.

Signal sources

Programmable sources place in the hands of a test engineer the capability to excite a unit under test with a signal that is unique to his particular application, but still doing this process with a standard test instrument.

Production applications for Wavetek's 275 programmable arbitrary function generator, the firm says, include testing of pacemakers and other medical equipment and simulation of heartbeats, nerve responses, and EEG brain-wave patterns. Simulation of solenoids and relays also require special waveforms which simulate contact bounce.

GPIB compatibility enables users to incorporate the arbitrary waveform generator into ATE systems. GPIB instruments such as the 275 arbitrary waveform generator are used by a variety of firms in ATE, says Wavetek. These include Medtronics, GTE and Honeywell, as well as ATE manufacturers such as Fluke who integrate the units into their products.

For ease of programming, front-panel entry on the firm's GPIB-compatible instruments is identical to the GPIB entry sequence.

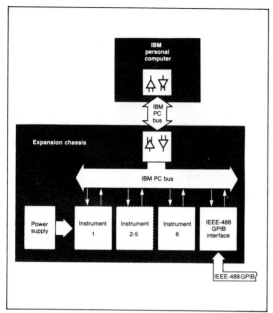

An IBM PC is used to control instruments in Vistar's PC1000 system

The ASCII character for GPIB programming appears on most keys on the front panel. This aids in transferring a manual setup to a controller program, or vice versa. To help the operator also, command recall can display previous characters entered either at the front panel or over the GPIB.

User-friendly design and logically positioned touch-sensitive front-panel controls make it possible for a casual user to more easily use the 9087 signal generator, says its manufacturer, Racal-Dana. The microprocessor-controlled, synthesized signal generator is used primarily in RF and microwave system testing. ATE compatibility is achieved with GPIB control of all functions.

Fast frequency switching, or frequency agility, speeds certain test procedures using the signal generator. Checking a filter for flatness, for example, or testing a receiver for spurious responses can be performed rapidly.

Measuring signals

Final test in manufacturing is one of the application areas for which Hewlett-Packard's HP 3457A digital multimeter is intended. The unit, with measurement capability for frequency and period, was designed with such

users in mind as manufacturers of mobile, two-way radios, says HP.

Programming can be accomplished through the HP-IB, the manufacturer's implementation of IEEE-488. A digital multimeter language conceived by HP, called HPML, is used to program the instrument. The programmer constructs a command by simply specifying the type of measurement, range, and accuracy.

Optional plug-in scanner cards permit multiplexed access of up to 10 signal channels. This capability eliminates the need for a large scanner in an automatic test system. The multimeter is in an enclosure compatible with 19in. racks used for building up test systems. It has a service-request (SRQ) output that can be used to actuate mechanisms to remove from the manufacturing line assemblies that fail testing.

Managing the flow of measurement information in a test system is aided by the instrument's capability to store up to 1050 readings that could be entire, extended measurement sequences for fast throughput on the bus at a convenient time.

Built-in math routines include statistical functions such as mean and standard deviation. Other routines include a pass-fail limit test, which can be useful in production testing.

Software for Keithley Instruments' Model 193, a 6½-digit digital multimeter, permits the user to define language with English structure rather than cryptic alphanumeric commands. Keithley's Translator software also allows device command strings to be replaced by user-defined mnemonics. These more compact commands can reduce traffic on the bus as well as making programming simpler.

The Model 193 DMM can emulate most other manufacturers' instruments by translating their device-dependent commands. This minimizes software changes if they are placed into an existing ATE system.

Internal processing is prominent with Solartron's 7071 digital voltmeter. Production testing and quality assurance are target applications for this instrument. Routines include maximum and minimum values, standard deviation, and high and low limits with alarm outputs and display of data in relation to the limits (also shown on the display).

This digital voltmeter can be controlled via an IEEE-488 bus or RS-232-C line. A serial RS-232-C link operating at speeds up to 9.6k baud can be used over longer distances than a IEEE-488 bus. Commands for passing instructions over both types of links are in an English-language style. An SRQ button can be actuated by the user to flag the controller for attention in an interactive system test operation.

More products are being designed with fast ECL and high-speed TTL and CMOS parts. These fast systems demand correspondingly fast, wide-bandwidth instrumentation for production testing.

Handling high-speed logic families with a rise time less than 1ns is the marketing basis for Hewlett-Packard's HP54100A/D digitizing oscilloscope with 1GHz bandwidth. Firms that are faced with production test of high-speed products include manufacturers of computers and peripherals, terrestrial and satellite microwave systems, and guidance systems. The HP digitizing oscilloscope can be combined with a controller such as the manufacturer's HP 9000 Series 200 or Series 300 controllers to form a test system.

This oscilloscope is programmable with English-like mnemonics. Uncomplicated syntax and command hierarchy contributes to

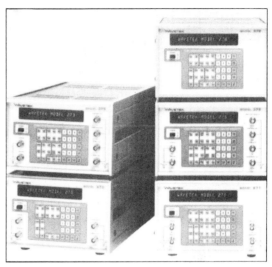

Programmable GPIB compatible test instruments from Wavetek include sweep/function generator, function generator, synthesized function generator, arbitrary waveform generator, and pulse/function generator

simplifying software development for the test engineer. Parameters determined by internal processing for a waveform captured by the oscilloscope can be compared to go/no-go limits entered into the test sequence.

Universal counters are common measuring instruments in test systems. For example, GPIB talker/listener control is optional on Racal-Dana's 1900 Series of universal counters to incorporate them into a test system.

These instruments may be used to measure frequency, period and time interval or phase and to totalize. Functionally, they are used in

Measurement package includes Tektronix' 4041 GPIB system controller, 2465 DVS oscilloscope, and 4150 color-display terminal

test systems to examine frequency accuracy and stability (frequency or period); rise and fall times and propagation delay in digital systems (time interval); phase offset in avionic control systems (phase); and bit error rates in data transmission systems (totalize).

Integrated packages

Integration of GPIB systems is at the level of linking separate test instruments. At the next level of integration, but still lower than a fully integrated ATE system, is the linking of essentially different test modules within a single test instrument.

Dolch's COLT and ATLAS products are examples of these systems. Plug-in modules for digital word generation, logic analysis, and in-circuit emulation offer an integrated, though flexible, digital stimulus and measurement system. From one to four plug-in modules can be used, depending on the instrument model and whether an expansion chassis is used. Internally, modules are linked over the instrument's own buses. The instruments have, however, an interface so they can be controlled on GPIB.

These integrated test systems are used by engineering for development and then passed on to production with turnkey test programs. This approach requires little learning by the production user and offers commonality between the test equipment used in engineering and production, observes Dolch.

Another style of test package is based around the marketing concept of selling together (though not necessarily in a rack or single enclosure) a test instrument, GPIB system controller, and display terminal. Tektronix's package, with their 2465 DVS oscilloscope, 4041 GPIB system controller, and 4105 color display terminal, plus test program generator software is an example of this approach. One advantage of this marketing package for the user is that it can provide fast implementation of an automated, portable test system.

ROBOTIC AUTOMATED-TEST WORKCELLS

Operating automatic test equipment for printed circuit boards requires material handling systems with intelligent control capabilities.

Automating this task also requires the flexibility to change rapidly and easily as board runs change, and the ability to realize a cycle time that makes automation economically feasible, usually thought of as approximately 20 seconds per board.

Robots meet most of the requirements mentioned, provide flexible automation, and the dexterity of a six- or seven-axis arm enables one to adapt to the existing dimensions and design of testers and the existing layout of a test station, thus minimizing fixturing cost.

Robot's work envelope

Robotic board handling with ATE involves a minimum of two workstations, or testers, per forbot to be economically feasible. This places certain requirements on the robot to facilitate implementation by the user.

The robot should have a large enough work envelope to service two testers without modification to either robot or tester. It should also be able to easily access the other equipment required in the workcell, such as delivery and takeaway conveyors. The robot should be the adaptive piece of equipment in the workcell.

Forcing applications engineers to adapt to a robot with a more limited work envelope lengthens the development time. Everything is more tightly packed in the workcell, and tolerances are tighter. The robot may have to be mounted directly on the tester. This may be acceptable if the robot and tester are

First published in *Electronic Packaging & Production*, 1984 Automation Supplement, under the title 'Robots handle PC boards in ATE workcells' Author: Rollie Woodcock, Intelledex.
© 1984 Reed Publishing (USA) Inc.

delivered as an integral unit from the manufacturer, but certainly more expensive to implement if it is done on a custom basis.

The Intelledex Model 705, used in workcells with ATE, is a six-axis arm that is rotated with a seventh axis, or waist joint. This has two advantages. First, it gives the robot a work envelope that is roughly a circle with a 10ft diameter, enough to service two testers and associated equipment. Second, because the rotation is accomplished with a seventh joint, a six-axis arm is available at the point of work. Six axes are the minimum required to realize the dexterity of a human arm. This dexterity minimizes the requirements for specialized fixturing that occur when the workcell must compensate for the more limited movement of a four- or five-axis arm.

The robot must be fast enough to move over these larger distances and achieve the required cycle time of approximately 20 seconds while meeting the repeatability requirements for picking up a board and placing it on a test fixture. Speed is the more critical concern here. Unlike other applications, for example loading odd-shaped components into printed circuit boards, repeatability is not one of the premium requirements. Although the boards must be placed very precisely on the test fixture for the actual test, the robot can position the board much less accurately over self-centering alignment pins on the fixture, and use a gripper with compliance to seat the board at the bottom of the pins.

Gripper design

End-effector, or gripper, design is a key part of any robotic application, and automated board testing is no exception. In order to minimize cycle time, a gripper that can handle two boards at once is desirable. This enables a robot to remove a board from a test fixture and then, without moving from the area, load the fixture with the next board. It does not have to move first to a drop-off point for the just-tested board, and then to a pick-up point to retrieve the next board.

Two such grippers have been developed by Intelledex for use with its robot. The first is a

A robot handles PCBs in an automated functional test workcell at the Electronic Manufacturing Division of Xerox Corp. Clockwise from behind the robot are: tote boxes for failed boards, one of the functional testers, the board pick-up point with the robot's gripper above, the second tester, and a conveyor with totes for passed boards

two-sided paddle with vacuum pick-ups on both sides. Using this gripper, the robot picks up the next board to be tested, moves to the test fixture and picks up the just-tested board, rotates its gripper 180° to load the next board, then delivers the tested board to the appropriate take-away conveyor. A second gripper, better suited for larger and heavier boards, features two gripping mechanisms opposite each other that are rotated in the X-Y plane. These mechanisms have metal clamps with notches that are closed on the board with stepper motors. The sequence of movements with this gripper is the same.

In addition to being able to handle two boards at once, the gripper needs to be able to be fitted with sensors. These can take the form of micro or optical switches, which sense whether or not a board is present in the gripper; or force sensors, which can indicate possible problems in locating or seating boards on test fixtures. A load sensor might also be required to measure the weight of the board currently being handled.

Intelledex robots have an integral end-effector system to support these varying requirements. Twenty internal electronic lines and two pneumatic lines which terminate at the tool interface are controlled by an auxiliary computer. This computer accepts plug-in electronics in open board slots which allow the electronics and pneumatics to be configured for the particular sensors required.

Machine vision

Other sensors, such as vision, may be required for some applications. If the boards are not registered in a repeatable location by fixturing, vision could be used to locate randomly oriented boards on a conveyor belt. The method of presentation used depends on the volume of boards to be tested. Fixturing is faster than vision. On the other hand, for on-line vision applications, the speed of the vision system may not be as important as its other capabilities because the pacing item in the workcell tends to be other automation equipment.

Vision could also be used for hole detection or identifying faulty boards. If vision is required, a robot with an integral vision system eases workcell design and implementation. The integrated vision systems of Intelledex robots is controlled with the same software program that controls the robot. It

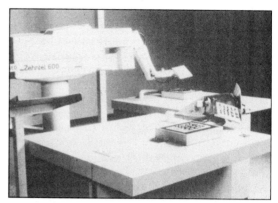

Two automatic testers are within the work evelope of the Intelledex 705S shown here in Zehntel's robotic board handling system. The robot's gripper can grasp two PCBs at the same time with vacuum pick-ups on both sides of a paddle

eliminates the need for converting from camera, or pixel, coordinates to the workcell coordinates used by the robot. It also eliminates the need for data conversion and management of communications protocol that arises when the robot and vision system are from two different manufacturers.

Board presentation and take-away systems are another key part of an automated board test station. The most important feature of the board presenter is that it be able to deliver boards to the robot in the same position every time. Although sensors can accommodate a degree of inconsistent presentation, the processing of sensory data necessarily lengthens the duty cycle. A stable and accurately machined fixture which can present boards consistently to the robot is not one of the more expensive parts of the workcell, but it pays off in consistently keeping the more expensive parts of the workcell running.

After the board has been tested, the easiest way to remove it from the cell is to drop it on a conveyor which would automatically carry the board to the next station. In most cases, however, the automated board test station is not part of such a continuous flow line. This means boards have to be batched in tote boxes. Once again, this is where a robot's large work envelope is helpful. The robot could have three, or possibly four, such totes from which it works. The size of the totes would depend on the space available in the workcell and the robot's reach. One tote would be for

The dexterity of a seven-axis robot provides flexible automation that can be adapted to the existing dimensions and layout of ATE and the test station

boards to be tested, one for boards that tested good, one for boards with minor failures, and one for boards with major failures; the difference between these latter two being the level of technician required to do the repair.

Task management

The premier requirement of a robot for an automated board test system is that it incorporate a communications capability. The information that it gathers from its own sensors is data from which decisions are made, and those decisions must then be communicated to other equipment in the workcell. This means, for example, querying testers to find out which is available next, receiving signals from the testers when the test is completed, signaling a conveyor belt to move after a board has been placed on it, or signaling for operator intervention when required.

The robot can also log tested boards into a data base that contains the test history of each board. This can be accomplished by placing an optical sensor somewhere near the board pick-up point. After grasping the board, the robot passes a bar code label affixed to the board over the optical sensor. Data read by the sensor is converted to ASCII and communicated via an RS-232C link to a central computer. The board identity can also be sent to the robot so that it can make a decision about which test to run without demanding the central computer's time.

Communications in the workcell is normal-

A PCB is seated on the test fixture of a functional tester at Xerox. Edge fingers on the board must be placed between test pins

ly handled on standard RS-232C or parallel buses. Safety equipment such as pressure mats and light curtains often communicate directly with the robot via TTL lines.

Indigeneous to the communications task is a requirement for the robot to prioritize responses to the communications received. An interrupt from a safety device is a higher priority than responding to a signal that a board is ready to be picked up. The robot must also make decisions about the sequence of loading and unloading the two testers so that each is used to the maximum degree.

The management of all this is in a software program. To accomplish it, the robot needs the ability to be programmed in a high-level computer language. Intelledex uses Robot BASIC, an enhancement of Microsoft BASIC. The power of the BASIC language enables the robot to receive communications, and make decisions based on the priorities assigned in the control program. Robot BASIC contains specific commands for control of vision, end-effector, and ancillary equipment, so that the control of all devices is integrated into one program.

In addition, a Robot BASIC program can be developed off-line, enabling program development or modification without tying up the robot. This is a necessity for complex programs such as those controlling a fully automated board test station, which require considerable time to develop.

Workcell control

The automated board test station comprises a system in itself. A possible configuration of the work process is to have the robot controller in charge of the entire workcell, including the ATE. The robot would be responsible for initiating tester operation, controlling throughput of the system, reporting test results, monitoring the station for error or fault conditions, and providing summaries of daily processing.

In total, this implies a robot that is really a computer with an arm functioning as a peripheral. The newer generation of intelligent robots are designed from this perspective. Even if the robot controller does not

At the pick-up point, boards are grasped by the robot for seating on a functional tester at Xerox

function as the main controller for the entire workcell, its capabilities still ease the implementation of the total job by providing localized intelligence for all of the aspects of the task which are specific to the robot arm and associated sensors.

Robot applied

Manufacturing operations at the Electronic Manufacturing Division of Xerox Corp. in El Segundo, CA, include an automated functional board test station designed by Intelledex. The workcell comprises an Intelledex Model 705 robot, an input station with a buffered board feeder, two stand-alone functional testers controlled by personal computers, a tote box or shipping box conveyor for 'pass' boards, a table with two tote boxes for 'failed' boards, and an auxillary output buffer.

A tote box for passed boards is loaded by the robot in Xerox' ATE workcell

According to Ken Zablotny, manufacturing engineer at Xerox, "The robotic test station provides early feedback on assembly quality. We know very early if there is an assembly problem upstream. It's also a very cost-effective solution. Our payback on the investment is 1.2 years. The workcell has also eliminated two days from the work-in-process cycle. A final advantage is that the efficiency of the robot has enabled us to move one employee per shift to another job. Additional reductions in manpower were achieved through the re-layout of the board handling and inspection operations."

Xerox chose robots, Zablotny said because of both increased throughput and yield. The workcell is expected to handle nearly 300,000 boards annually. The goal of reducing operator involvement was achieved through judicious use of sensors. An operator assures proper setup at the start of each shift, then workers adjacent to the cell respond to the call lights.

Intelledex robots were chosen, according to Zablotny, because of their large work envelope, the ability to network within the plant operating system, the ability to use the Intelledex controller to control the entire cell without the use of an add-on programmable controller, and their previous experience with Zehntel in board test applications.

As with many robotics applications, Xerox' part in implementing the workcell was critical to its success. The Xerox engineering team defined the workcell concept and goals, defined the proper location for sensors to achieve minimal operator involvement and to provide self-diagnostics, provided facilities changes, helped with the design of the material feed and takeaway systems, and had their personnel attend Intelledex' training and maintenance courses.

In a typical cell cycle, the robot controller polls the entire cell status and determines the first action. If both testers are occupied, it acts on the first 'tester ready' signal, removes the board, places it in the appropriate pass or fail tote, then reloads. If one tester is occupied it loads the other.

At the input station, boards are presented on a conveyor. The board buffer, capable of

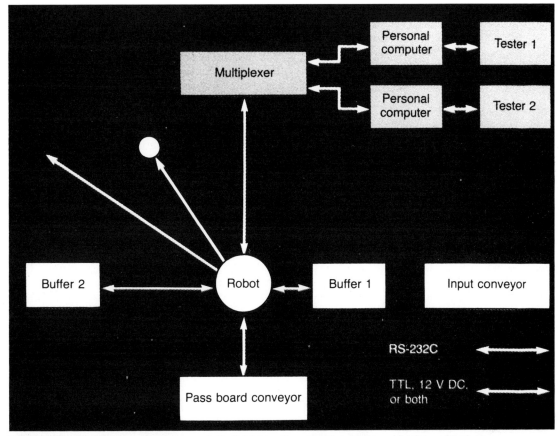

Task management in the ATE workcell requires communication links between the equipment to gather data and distribute resulting decisions

holding 20 boards, is just prior to the board pickup point. This buffer serves two functions: it prevents the entire line from shutting down should the robot or a piece of equipment upstream from the workcell be temporarily put out of service. It also assures continued during operator breaks and between shifts.

If there is a board occupying the pickup point during normal operation, the next board which flows to the testing workcell from the assembly line goes into the buffer. The buffer sends a board to the pickup point only upon receiving a command from the robot.

The end-effector is a parallel jaw V-gripper using a rack and pinion. It has a 90° air-actuated 'wrist' to reduce the cycle time involved in grasping the boards parallel with the ground, and placing them into totes perpendicular to the ground. Sensors verify that the 'wrist' is open or closed.

Picking up a board is more complicated than it looks. The robot is equipped with a sensor that tells it how wide the gripper jaws are open. It uses this to tell if a board is actually present. If the jaws close to a spacing less than 'hat of the board's width, the robot knows it has not successfully grasped the board.

If the grasp is unsuccessful due to board warpage, the robot moves in a search pattern until a successful grip is achieved. It has been found that the additional five to eight seconds to complete a search is a better investment of time than waiting for an operator to correct the condition.

Once the robot has a board, it sends a 'query' byte to the testers. There are two testers, each controlled by a Xerox 820 personal computer that is interfaced to the robot via an RS-232C link. These personal computers act as slaves to the robot controller, iniating no action or data transfer without first receiving a command from the robot controller.

Two testers are utilized to maximize throughput. While one is testing a board, the robot is removing a board from the other and inserting a new board. The cycle time averages 22 seconds.

Each tester has a photocell to detect the presence of a board on the tester, and a switch to verify that the board is fully seated. This switch eliminates the false failures that occur when the board does not make proper pin contact. When this condition is encountered, the controller will restart the vacuum until the board seats properly. This simple software routine eliminates five to seven operator interventions per eight-hour shift.

The testers respond to the query byte with a status byte indicating the state of operation each is in. If the tester has a board present and the test is complete, it will send one more byte,

with the code defining the type of failure, if any, the board experienced.

In loading the boards into the test fixture, the most critical concern is an edge connector requiring two-sided pin contact. This requires the robot to insert a 0.065in. thick board into a 0.180in. jaw opening that contacts both sides of the board. Commands in Robot BASIC were used to implement smoothing of a straight line, enabling insertion on all but the most severely warped boards.

There are three output stations: a 20-board buffer similar to the input buffer, the tote box conveyor for 'pass' boards, and the table for two totes for failed boards. The fail totes are removed by an operator since the volume of boards handled at this station is considerably less than that handled at the pass station. The robot counts the boards put into a pass tote, and when it is full, it issues commands controlling devices which move the box from the loading conveyor to the take-away conveyor.

Excluding the gripper, the robot controls 17 output devices to manipulate the conveyor

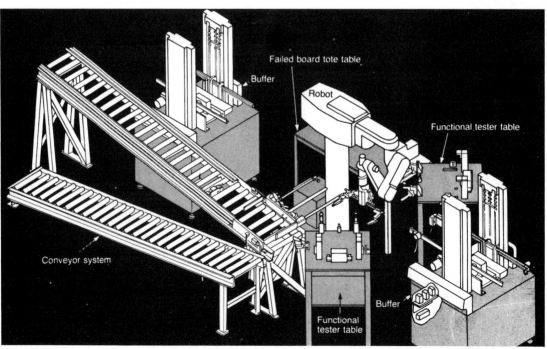

Two testers are used in the functional test workcell at Xerox to maximize throughput

Xerox personal computers control functional testers in the workcell. The robot is seating a board on the test fixture above the computer

system and board input and output. There are 11 input sensors to indicate items such as buffer status and conveyor status.

The controller also makes decisions about tester performance. If a single tester fails five boards for the same reason (a tester problem), the tester is automatically taken off-line and operation continues with one. If six consecutive failures occur from both testers, the cell will stop, suspecting a board assembly error.

CHAPTER 8

ENVIRONMENTAL CONCERNS IN ELECTRONICS ASSEMBLY

Production engineers in their task of specifying, procuring and implementing the tools of electronics assembly must be constantly aware of the effects of the environment on the quality of their products and the impact that their chosen processes have on the atmosphere, on their manufacturing staff and on society as a whole.

The concerns which impinge on the health of the workers will usually come from solvents used to clean circuit boards, remove fluxes and from adhesives used in SMD processes. These are discussed in the appropriate chapters and are usually the subject of stringent government regulations concerning the emission from and the disposal of such material.

This chapter addresses itself to those environmental concerns which affect the quality of the assembled product. These concerns fall into two areas, cleanliness and the prevention of electrical charges and discharges detrimental to electronic components such as integrated circuits.

CLEAN ROOMS

Many of the processes of electronic manufacturing must be done in clean rooms free from atmospheric contamination. This freedom from air box particles is not easy to achieve and the cost of operating such rooms and maintaining this environment through personal discipline can be a most significant part of factory overhead costs.

CLEAN ROOM TECHNOLOGY

How clean must clean room facilities and equipment be, and how can their cleanliness be effectively measured? These questions have not been answered to the satisfaction of many clean room users and manufacturers, although clean rooms have been in existence for over 40 years.

Over the years, many attempts have been made to define and standardize different areas of clean room technology. The most widely recognized standard in the field in the USA is Federal Standard 209B, "Clean Room and Work Station Requirements, Controlled Environments." It is also widely misunderstood and misused. For example, there is confusion over exactly what the standard can be applied to. Thus, despite claims made to the contrary by some manufacturers, there is no such thing as a Class 100 or a Class 1000 garment, wiper, table, etc.

What 'class' means

Four classes of air cleanliness are defined in the standard: Class 100, Class 1000, Class 10,000 and Class 100,000. The class numbers indicate how many particles are permitted per unit volume. For example, the particle count in a Class 100 room should not exceed a total of 100 particles per cubic foot of a size 0.5m and larger.

Stephen Tucker of Marshall Kinatechnics, a manufacturer of clean room workstations, believes that there is a general lack of knowledge on the part of the specifier as to applications needs versus standards applied. "Most of our requests for clean room equipment are prefaced by the comment, 'I need Class 100 or 1000 or better.' Many of these individuals are not even sure what Class 100 or 1000 really means. Researching these requests usually reveals the need is for Class 10,000."

At the root of the problem in understanding clean room standards is that, except in a

handful of companies, the responsibility for contamination control is often a secondary function of a design engineer, an operating foreman, a maintenance supervisor, or the like. The rapid turnover of people involved in contamination control further contributes to its second-class status. But the need for clean rooms is growing, so the technology will keep developing and many of the current problems will be overcome.

Current application areas

In the electronics industry, clean rooms are used mainly in the manufacture of semiconductor devices and in the manufacture of thick and thin films for hybrids.

Hybrid fabrication processes which are taking place in clean rooms include vacuum deposition, screen printing and chemical deposition. Vacuum technology processes include vapor deposition, sputtering, and other vacuum techniques for depositing metals and dielectric materials on substrates.

Thin-film circuitry can be deposited and etched to tolerances less than 0.001in., while thick films are limited to about 0.005in. as screened, and 0.002in. if etching techniques are used after screening. Contamination as small as 1–10 m, or 0.001in., would probably result in an open circuit in a thin-film substrate.

Terry Francis of Burroughs Semiconductor feels that "clean rooms are getting a lot of attention right now because higher densities have brought us to a transition point." Francis, who has been involved in contamination control for about two years and is currently vice chairman of ASTM's subcommittee F-1.10 on process chemicals, feels that control of particles will have to be at the parts per billion level in order to meet future needs, as opposed to today's parts per million levels. He is looking more closely at what happens in the immediate product environment, rather than at specific clean room classifications. The approach that Francis is taking to contamination control may well be the wave of the future.

First published in *Electronic Packaging & Production*, May 1984, under the title 'Change sweeps through clean room technology'.
Author: Susan Crum, Assistant Editor.

The sputter metal deposition process for thin-film hybrid circuits requires Class 100 conditions. A particle as small as a few micrometers settling on the substrate could result in a defect in the pattern (Courtesy of Tektronix)

The search for new technology

Stuart Hoenig of the University of Arizona emphasizes "I don't believe in this word class." It is a concept that he believes will eventually disappear. Hoenig headed up a three-year study at the University of Arizona, begun in 1982 and supported by the National Science Foundation, of new technology for clean rooms. Traditionally there has been a low level of technology in contamination control. Besides the advent of laminar-flow and the increase in equipment sterility, the technology of contamination control has been much the same as it was in 1960. Since 1960, Hoenig contends, there has been a tendency for people "to do more of what they have been doing;" what he calls the 'brute force approach' to contamination problems. However, he feels that now there is "a

movement to look at the problem and see what it is and try to solve it."

Hoenig spent part of his time in 1983 travelling around the USA trying to stimulate interest in a 'clean room center', supported jointly by industry and government and based at the University of Arizona. This center provides ongoing development of clean room technology after the NSF grant ran out. A similar center has also been launched by the Japanese at the Tokyo Institute of Technology.

Revision of FS 209B

In response to suggestions from industry and government that the existing standard needed to be brought up to date once more, the Institute of Environmental Sciences, with the cooperation of the Association of Contamination Control Manufacturers, began the task of revising FS 209B for the federal government in early 1982.

Initially, general meetings were held for interested persons to discuss how the standard should be changed. Because these gatherings were too large for detailed consideration and evaluation of recommendations to take place, a small working committee was appointed in early 1984. This committee represents various areas of interest in contamination control, and includes 10 users, six government employees, three consultants, and 14 suppliers.

During a February 1984 meeting of this committee, it was tentatively decided that the standard would continue to deal only with particulate air cleanliness. The need for air cleanliness classes of less than 100 and for particle size reference of less than 0.5m was discussed and assignments were made for further study. A system of nomenclature and definition of classes, which would be compatible with the existing standard but would allow for future expansion of both class and particle size designations, was tentatively adopted.

Criticisms of the existing standard are widely varied. Nino Cerniglia, vice president of Systonics, feels that Federal Standard 209B is a good standard which has served the industry well and should be built upon, but that "the set guidelines, which aren't really appropriate to today's technology, should be detached from it." Marshall Kinatechnics' Stephen Tucker believes that "the industry could best be served by publishing a set of standards in 'layman's English'. Descriptions of typical applications and the personal protective clothing required at each class would also be helpful."

The core of the controversy, however, is the things that are not covered in the standard. Requirements for the laminar flow air velocity are in the appendix, not in the standard itself. Classifications for clean rooms are valid only for particulate contamination of 0.5m and above. Laser-type particle counters and UV/visible particle counters, which could measure particles of 0.1m, are not standardized. The air cleanliness classes are merely statistical averages, unrepresentative of any specific type of facility. Only particulate contamination is recognized in the standard. It ignores such diverse yet significant types of contamination such as radiation, vibration, magnetism, heat, and static.

People in the clean room

Even if the revised standard should take all these considerations into account, there still remains another set of problems with the clean room's single greatest source of contamination: people. People compromise the cleanliness of the facility just by being in it. Every time someone scratches a mosquito bite or a sunburn they release particles of their skin. Sneezing releases particles. Just brushing a hand across the sleeve of a clean room uniform can raise the level of particulate contamination in the work area 1.5–3 times over ambient levels.

Furthermore, people that smoke, even though they do not smoke in the clean room itself, emit more particulates than non-smokers. According toa University of Arizona study, as long as 10 minutes after finishing a cigarette a smoker will emit 35 times as many particles in the 0.2–0.4m range as a non-smoker.

Not only are people themselves a source of particle contamination, their presence in the clean room disturbs the laminar air flow, creating areas of turbulence which pull contaminated air from the room into the work area.

Aside from complete automation of the clean room facility, which does not appear to be imminent, there is not much that can be done about particle contamination by personnel besides offering training programs in clean room procedures and choosing appropriate clean room garments.

Those responsible for specifying clean room clothing should look for garments that are comfortable to wear, available in a variety of sizes and designs, durable, attractive, of good quality, and cost effective. These are characteristics that are not difficult to find. The nagging problems for garment specifiers are the absence of guidelines as to the appropriate garment cleanliness for the different clean room classifications, the lack of an

	Clean rooms	Clean room equipment	Clean room uniforms	Clean room assembly/ subassembly
American Air Filters		●		
Angelica Uniform Group			●	
Blue M, Unit of General Signal		●		
Clean Room Products	●	●	●	●
Contamination Control, Inc.		●		
Controlled Environment Equipment Corp.	●	●	●	
Conwed Corp.		●		
Interlab, Inc.		●		
Marshall		●		
Mobot Corp.		●		
Nacom Industries		●		
Oak Medical Supply			●	
Panel Controls Corp.		●		
Solder Absorbing Technology		●		
Static, Inc.		●		
Superior Linear Weld	●	●		●
Systonics	●	●		
Techni-tool		●		
Veco Intl., Inc.	●	●		●
Weber	●	●		

Suppliers of clean rooms, clean room equipment and assembly services

adequate definition of the initial particulate count on the garment, and the possibility of particle migration through the garment negating the cleanliness of the outside. While there is a long list of sophisticated tests to measure these parameters, there is no standardization.

Air showers are a part of most clean room designs. While they were designed to blow dust off the surface of uniforms before wearers enter the clean room, military studies suggest that 1m particles may actually be driven into the fabric by the shower. A number of alternatives to conventional air shower systems have been proposed in the course of the clean room study at the University of Arizona. Among these is a ring jet system that would be pulled downward by the wearer, over the garment, progressively removing particles. Residual air and dust would be drawn through a floor grill. Use of a vacuum wand was another suggestion. A garment to which an air line could be connected for blowing low pressure air from the inside out through the suit was also proposed.

A more unconventional approach was suggested in the deliberate use of electrically charged fabrics that would attract dust, thereby continuously removing dust from the clean room environment. The effect this electric charge would have on products manufactured under such conditions was not explored, however.

Building a clean room

A number of companies exist for the purpose of providing engineering analysis and design services to upgrade existing facilities or to construct new clean rooms (typical room configurations are shown in the sidebar). Clean Room Products Inc., of Bay Shore, NY, is one of these. To meet the varying technical and economical needs of its customers, it offers a program in four phases. Phase I is an on-site engineering analysis of overall manufacturing and production from as far back in the manufactuing chain as raw material procurement and subcontractor processing. This phase results in a conceptual design and basic specifications for a new or upgraded facility. Phase II takes the concept and specifications developed in Phase I and produces complete working drawings and detailed specifications that can be used for contractor bidding or in-house construction. Phase III supplies major materials and equipment with drawings and specifications and on-site supervision. During Phase IV the turnkey operation goes on-stream after certification.

Ray Schneider, engineering manager, explained that people often come to Clean Room Products when they are experiencing a yield problem, or quality control problem, and think that having a clean room might help. "If the client has a well-defined idea, it simplifies the problem for the supplier," he said. It is not unusual for the company to design a clean room without knowing exactly what processes will take place in the completed facility. Customers often consider such information proprietary.

Systonics, another company that designs

and constructs clean rooms, routinely establishes the criteria of the room with the client according to the process, not just according to what the customer tells Systonics they need. Vice president Nino Cerniglia says that design problems are best treated by up-front planning. Since most customers lack a person whose sole responsibility is contamination control, Systonics works mostly with the director of operations, often with a financial officer, and sometimes with facility managers in designing the clean room. They build Class 100,000 to Class 100 rooms; using particle-scattering techniques, they have developed what they feel would qualify as a Class 10 clean room.

A considerable amount of debate exists as to whether a Class 10 should be included in the revised federal standard. Theoretically, Class 10 is possible. Practically, it is close to impossible. The main problem in achieving Class 10 is in particle counting.

Measuring 0.1 particles per cubic foot (or 1 particle in 10 cubic feet), would require a long sampling time, making it difficult to detect short-term trends. Even supposing that Class 10 cleanliness is achieved in the unoccupied clean room, there is a high probability of contamination at the product.

Just because a room was once tested and certified to meet the air cleanliness requirements of a particular class is no guarantee that the levels of contamination are constant. There are several sources of contamination, mentioned previously, that cause these levels to change. In addition, the condition of the ambient air outside the clean room can affect the cleanliness inside. On hot, muggy days, for instance, a higher amount of contamination seeps through the cracks and crevices of the room.

Regular monitoring, not only of particulate contamination, but of pressure differentials, temperature, humidity, and the frequency of people entering and leaving the room, will point to contamination problems as they arise.

Typical clean room configurations

The three basic types of clean room configurations are non-laminar flow, vertical

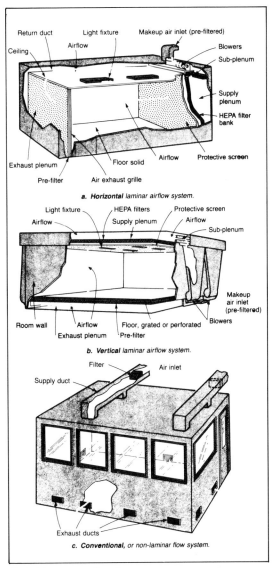

Typical clean room configurations

laminar flow, and horizontal laminar flow.

In the horizontal laminar-airflow system, high efficiency particulate air (HEPA) filters constitute one entire wall of the room, moving filtered air across the room to be either recirculated or exhausted to the outside.

This design can provide Class 100 conditions, but the cleanliness level drops as the air moves away from the filters. As the air moves across the room it picks up contaminants from personnel and equipment in its path. In situations where many people are present, the air becomes turbulent and particulate control is lost.

The use of ceiling filters in vertical laminar airflow systems allows more flexibility in clean room design, allowing the whole room to be used for critical operations. The air either exhausts or is recirculated through floors or sidewalls.

Conventional, or non-laminar flow clean rooms are still in use but are no longer being built. These rooms, which can usually achieve no better than Class 100,000 conditions, have

air which enters from the ceiling and takes a random flow pattern exiting near the floor. Frequent cleaning is required to maintain clean conditions in conventional clean rooms.

Workstations are available in both laminar and non-laminar flow. Laminar-flow benches can provide a clean environment for small items providing, for example, Class 100 conditions within a Class 10,000 clean room. However, contamination can be introduced into these environments by just walking past them, therefore constant maintenance is required to keep them clean.

MODULAR CLEAN ROOMS

Lowering capital costs and stabilizing the process has always been a problem area when planning a Class 100 manufacturing facility (no more than 100 particles of 0.5.160m or more in diameter per cubic foot). Typically, a one- to two-year construction cycle is involved during which the process normally changes. Commonly, the end result is large, unexpected cost overrun necessitated by process changes.

Ideally, what is needed to adapt to changing requirements and prevent cost overruns is a flexible factory layout which can provide Class 100 environments to only those parts of the assembly process requiring such cleanliness. Unfortunately, many of today's clean room designs take an opposite approach. In anticipation of an increase in the production schedule requiring additional space within a short time, most clean rooms are constructed to encompass the entire process including those portions not requiring stringent contamination control.

The MCR approach

One way to keep costs under control while still providing an extremely clean manufacturing environment to critical portions of the process which require such control is to integrate

modular clean rooms (MCRs) and the normal factory into a single working unit. Such an approach does challenge the engineer making the layout. However, in creating such a design, it is possible to provide Class 100 clean-room conditions only where required; not wasting such conditions on portions of the manufacturing process which do not require them.

In designing a modular clean room, several objectives should be met. The ideal MCR is adaptable to process changes, flexible, self-contained, highly reliable, easily installed and maintained, energy efficient and cost competitive.

Module clean room

An example of prototype modular clean room (MCR) is illustrated. This clean assembly room unit can be prefabricated and installed in a relatively short time. As shown, the MCR provides conventional Class 100 environment as specified in US Federal Standard 209B, by using 0.3m HEPA filtered airflow at 100ft/min. The temperature of the MCR will be controlled to approx. 22°C, and the relative humidity of 35% can be adjusted to 55%.

The size of this MCR is 10ft 8in. wide, 16ft 6in. long, and 14ft high. This size has been determined as being ideal for robotics and advanced material handling systems. Each MCR unit can stand alone or be combined

First published in *Electronic Packaging & Production*, May 1984, under the title 'Modular clean rooms promise flexibility at lower costs'.
Authors: G.J. Horky, B.O. Williams and J.R. Block, IBM Corp.
© 1984 Reed Publishing (USA) Inc.

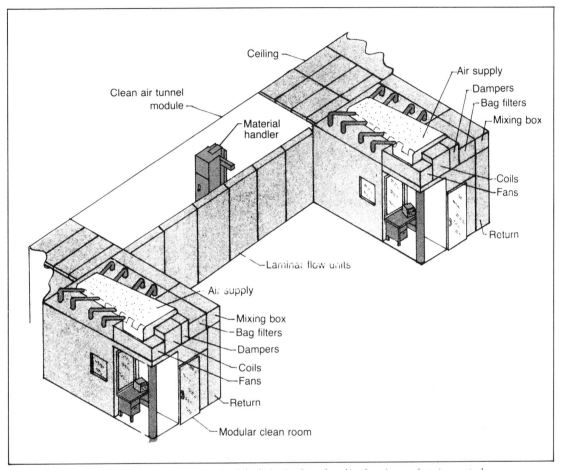

Two modular clean rooms are connected by a clean air tunnel module, eliminating the need to subject the entire manufacturing area to clean room conditions

with others in multiunit assembly areas. The layout of such units can be planned so that future units can be added in the locations not needed when the factory was first planned.

In practice the factory will consist of MCRs placed wherever the process dictates a Class 100 environment, either singly or in multiunits. Whenever the units are placed alone, individual MCRs can be joined together by a clean-air tunnel. Automatic material handlers can transport work from one MCR to another or to external workstations.

In cases of multiunit assembly areas constructed from several MCRs, individual change areas can be deleted and a common entrance constructed. When the process dictates a series of operations that can be

performed outside the MCR, a blow down-air lock or an in-liner cleaner can be installed before parts reenter the MCR.

Auxiliary functions

The MCR has been designed so the window/ pass throughs provided on each side of the room will have the lower half panel removable to permit the installation of doorways when the modules are to be joined together. One wall has been designed as a perforated blank wall down to the workstation height. This will allow 50ft/min clean air to pass through it, maintaining a clean air flow over the product being stored in the vertical bin storage rack. The room allows a 2ft space to be reserved in

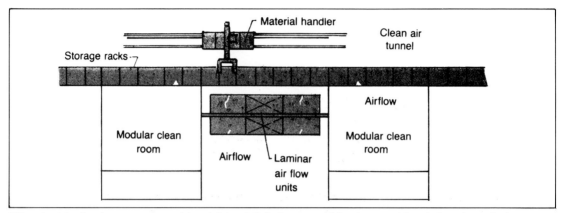

This overhead view shows how parts can be passed through the rear wall of a clean-room module and transported down the clean-air tunnel to the next clean-room module

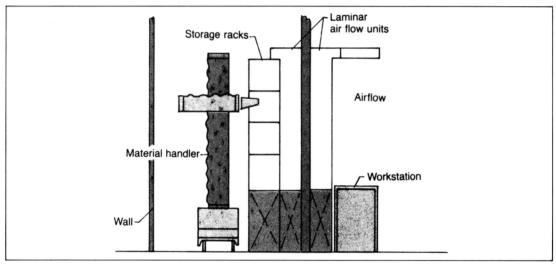

This side view of the clean-air tunnel outside the modular clean room illustrates how the tunnel can be used as a storage area. Parts can be kept here under Class 100 conditions without the necessity of controlling temperature and humidity. Parts can also be passed out of the tunnel to Class 100 workstations without temperature and humidity control

this area, which provides vertical storage for the more automatic material handling systems. The space can be included as part of the clean room when no material handling or storage space is required.

The raised floor in the MCR will vary in height from 8in. to 2ft. The minimum of 8in. is required for air recirculation back into the MCR air-conditioning unit. The entrance/change room area has pushbutton sliding-glass doors for visibility and ease of entry. The area outside the entrance, 4 × 6ft, can be used for operator desk, terminal, tooling controls or electronic cabinets required by the process.

Design features

The MCR has been designed to utilize lightweight honeycombed walls of module construction with a self-supporting external frame that will support not only the walls, but the air-conditioning unit as well. The raised floor will be standard 2ft squares, with adjustable perforated louvers to adjust the air flows according to the tooling sizes and shapes that are installed. A full HEPA-Filtered ceiling with hypobolic lighting and the required sprinkler and loud speakers to meet all safety codes are part of the MCR.

PERSONAL DISCIPLINE

It is obvious that more than an intitial investment is necessary for effective clean room use. Absolute discipline in entering the environment must be enforced to maintain the projected levels. It will also be necessary to monitor the levels of contamination.

CLEAN ROOM AIR MONITORING

There is no typical clean room. Because each clean room is set up differently, with its own individual, often proprietary, manufacturing process, there are no easy, standardized solutions to testing air cleanliness.

Nonetheless, solutions must be found. Adequate monitoring of clean room air is important because there are so many variables that can affect air cleanliness and, ultimately, product reliability. A change in ambient weather conditions, the introduction of people into the room and failed filters are a few of the things that can bring about changes in the level of air cleanliness. Regular monitoring can point to some of these problems before major product damage occurs.

Initial testing

Initially, clean rooms are certified either by their builders or by independent testing firms to conform to the requirements of a specific class of air cleanliness.

Systonics, a clean room manufacturer, is one of the companies which brings in an independent firm to test their products. "It keeps us honest," says vice president Nino Cerniglia, while adding that he does not feel that this practice is the only way for a clean room manufacturer to operate.

"If the room is constructed well, there should be no problems with particle counts," he continues. "Also, customers are increasingly concerned about the air immediately around the product rather than the air in the room conforming to a federal standard. They want the air to be clean enough to avoid defects in their products."

After the initial certification, however, the customer is on his own in finding the test configuration and test frequency that works best in keeping the room clean. Federal Standard 209B offers little guidance to the novice in this area. It says only that "particle counts are to be taken at specified intervals during work activity periods at a location

which will yield the particle count of the air as it approaches the work location. The preferred location for the particle count is at work level height with the sampling probe pointed into the airstream."

The bottom line is that monitoring airborne particles in a clean room is a learn-as-you-go process, best taken on by someone who is familiar with the manufacturing processes which will take place in the room. If the test configuration is not working, "the product will tell you," says Dave Lupo, president of B&V Testing Inc., an independent agency that verifies air quality by measuring and counting airborne particles. B&V Testing Inc. will trouble-shoot rooms for customers, and will suggest workable solutions to test problems.

There are two accepted methods of particle counting. One is by using a light-scattering system, the other is by manually counting particles trapped on a membrane filter by examining them under a microscope. Because the latter method is a tedious process, adequate only for monitoring air in the Class 10,000 to 100,000 range, it is rarely used today.

Types of particle counters

Light-scattering systems are available in two basic types: those using high-intensity lamps as their light source, and those that use lasers. Both types measure the light scattered from a particle when it interacts with the light source. This scattered light is collected and focused on a light-detecting element (a photomultiplier tube, photodiode or phototransistor). The pulses of light are then converted to pulses of electrical current. The light intensity and the electrical current it produces are therefore proportional to the size of the particle.

High-intensity lamp type, 'white light' particle counters are generally reliable; however, they have their limitations. Because the intensity of light scattered by a particle depends on shape, color and index of refraction, various instruments will produce data varying as to particle size, even though they

First published in *Electronic Packaging & Production*, September 1984, under the title 'Clean room air monitoring challenges users'.
Author: Susan Crum, Assistant Editor.

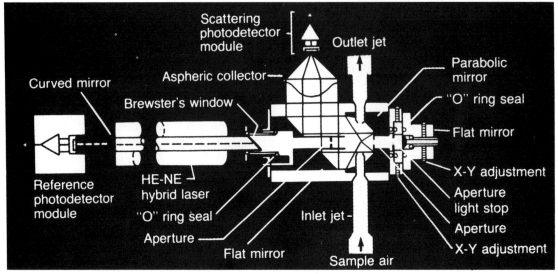

The optical system of a laser particle counter, Particle Measuring Systems Inc.'s Model LPC-101

can be calibrated perfectly with polystyrene latex spheres. Deceivingly low counts may result when some electronic pulses are not counted because simultaneously occuring particles may exceed the instrument's counting rate capability. This same peculiarity results in the counter's interpretation of the pulse as being that of one larger-sized particle rather than two or more smaller particles. Falsely high particle counts can be the result of either noise from the electrical supply or noise radiated directly from the light source.

Currently available counters

Climet, a manufacturer of particle counters, claims to have beaten the problem of the

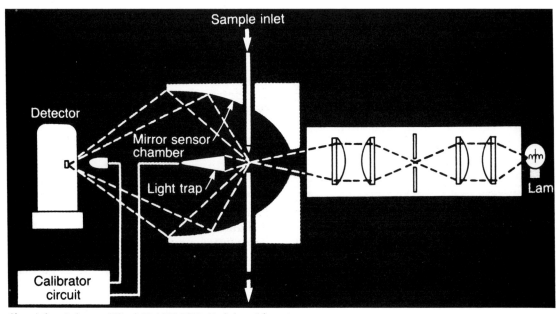

Shown is the optical system of Climet's Model C1-208C white-light particle counter

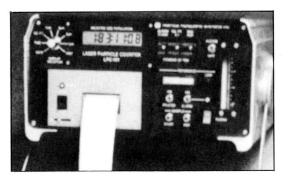

Particle Measuring Systems Inc.'s Model LPC-101 laser particle counter features a high sample flow rate and the ability to size particles down to 0.1μm

sensor's sensitivity to properties other than particle size. Its sensors use an elliptical mirror to collect scattered light over very wide angles around the particle, thereby providing a relatively large particle signal which is essentially independent of particle properties other than size.

Climet manufactures white-light particle counters exclusively, citing problems in laser counters' alignment, resolution and repeatability as the reasons why it chose not to enter that market.

Climet's counters have a sheath flow system which focuses the air stream at a precise location in the sensing zone, and additionally prevents particles from settling on the lens of the optical system, thereby assuring that only airborne particles are counted, not dust on the surface of the lens as well.

The company's microprocessor-controlled continuous monitor, Model CI-8040, can handle up to 40 simultaneous inputs from any combination of sensors, with a capability to expand to 128 sensors with the addition of up to eight Model CI-298 multiport air samplers.

The basic sensor used with Climet's monitoring system is the Model CI-226 remote airborne sensor. It will sample air at either 0.25 or 1.0ft^3/min (CFM) and detect particles as small as 0.3μm. A more compact model, the CI-222, is available for bench or wall mounting. It requires only 0.45ft^2 of space, and has a 0.5μm sensitivity. The system includes an optional RS-232C computer interface and keyboard terminal, allowing operation and data collection from any connected in-house terminal.

Met-One (Grants Pass, OR), claims that the high sample flow rate (1 CFM) of their Point 5R particle counter prevents large particles from collecting in the sample tube. The Point 5R uses a quartz-halogen light source and measures particles 0.5μm and larger. It is part of the company's System Four particle-counting system, which consists of a central monitor, recorder, and remote sensors. The basic System Four can monitor 40 inputs simultaneously, and can be expanded up to 1040 channels.

Pacific Scientific's HIAC/Royco is a manufacturer of both laser and white-light particle counters. Their Series 4100 system is illuminated by a tungsten-halogen lamp. An RS-232C interface is standard.

The 4100 system is used with either the Model 1100 or 1200 sensor. The Model 1100 sensor includes 0.5μm particle-size sensitivity with a 1.0ft^3/min sample flow rate. The Model 1200 sensor includes 0.5 or, optionally, a 0.3μm particle size sensitivity, with a selectable sample flow rate of 0.01 or 0.1 CFM.

Counters used with the 4100 system are the 4100 and 4150 models. The Model 4100 includes front-panel-mounted digital thumbwheels for user setting of channel-size thresholds. The Model 4150 contains factory-set size thresholds that cannot be changed by the operator.

Laser-type counters

Particle Measuring Systems Inc. (PMS) of Boulder, CO, manufactures laser particle counters exclusively. Its LPC-101, introduced in January of this year, features a high sample flow rate and a high sensitivity for sizing small particles down to 0.1μm. The laser light source is a hybrid helium-neon tube.

Aluminum extrusions are used in the LPC-101 to provide mechanical stability. Laser and detector alignment are achieved through springloaded X-Y screw adjustments. The reference voltage for size determination is derived from the laser's mirror, providing automatic compensation for variations in laser intensity.

PMS claims to have the largest collecting solid angle in the industry, which is responsible for its counter's high sample flow rate of 0.1 CFM. Scattered light from particles in the sample stream are collimated onto a 45° flat mirror. The light reflected from the mirror is refocused onto the solid-state photodector by an aspheric lens located at the top of the sample block.

Each particle encountered generates a signal from the photodetector which is compared with a reference signal in the pulse-height analyzer to determine the size of the particle. The reference signal is derived from a similar photodetector which monitors laser intensity by measuring the leakage of gas from the laser through one of the mirrors. This provides automatic compensation for variations in laser intensity, supplying accurate sizing of particles throughout the lifetime of the laser.

The laser's advantage

The advantage of using laser particle counters lies in their high sampling rates and ability to count particles down to 0.1m in size, for Class 100 and Class 10 applications. The ability to measure Class 10 brings the industry a step closer to actually achieving it.

B&V Testing Inc.'s Dave Lupo believes that, even with the ability to count small particles rapidly, clean rooms must be completely automated before Class 10 rooms become feasible. "People's presence makes rooms a minimum of Class 100," he maintains.

"It is definitely the wave of the future," says Jim Daniels of Pacific Scientific's HIAC/Royco Instruments Division, in speaking of laser counters. Daniels is aerosol product manager for the company. Laser counters are for people who want to see a better statistical sample, he said. "We realize that to expand our (counter) sensitivities at a higher flow rate, we need to have a very bright light source over a large view volume." White light is just not adequate for this application, he said.

While admitting that some older laser counters were difficult to use because of their sensitive alignment, Daniels says that newer systems are more reliable. Additionally, he points out that lasers have a life of from 17,000 to 20,000 hours as opposed to approx-

imately 1100 to 1200 hours for white light. The longer life of the laser cuts down on the number of times it must be replaced and the system realigned.

"With the new optical designs and automatic electronic gain adjustments developed in the last few years, the laser systems are able to attain a very high degree of repeatability," said Daniels. "Once the laser is allowed to stabilize for a minimum of 15 minutes, the laser light becomes stable over an extremely long period of time."

HIAC/Royco's Models 4000 and 4130 are microprocessor-controlled, eight-channel laser counters. Their sensors count and size particles in the 0.3–15.0μm range using a helium-neon laser for a light source. Optical components of the sensors are continually bathed in clean air to prevent particle deposition. The 1 CFM sample flow is continuously monitored and can be viewed by a front panel flow meter and adjusted by an internal valve. Model 4000 contains a continuously charged battery to maintain the clock memory and operator-selected parameters, for use in remote locations. This model can be used for up to one week without AC power.

Reference particle size

Some users prefer white-light particle counters over the newer laser counters because of their firmer reputation for reliability and because the current reference particle size given in Federal Standard 209B is 0.5μm, which is well within the reach of white-light particle counters.

Although FS 209B is being revised, it appears that the revised standard will also use 0.5μm particles as the reference. There are two major reasons for maintaining this size. One is the fact that even the highest efficiency filters cannot trap particles smaller than 0.3μm. The second is an uneasiness of the group revising the standard concerning the reliability of the newer laser particle counters.

Because of the widespread debate over the adequacy of the standard's requirements concerning test times and locations, this group is considering setting up new requirements. Revision of Federal Standard 209B was planned to be complete by mid-1985.

PROTECTION AGAINST ELECTROSTATIC DAMAGE

The use of modern semiconductors in electronic manufacturing is particularly sensitive to damage caused by exposing these components to electrostatic discharge or magnetic fields or pulses.

ESD CONTROL

The time is past when ESD control was not considered a serious problem. It is a malady that affects the electronic manufacturing industry, resulting in losses of billions of dollars each year. The purpose of this article is to bring this problem to light and discuss what can be and is being done. First, to take a look at the culprit.

Static electricity is an imbalance of electrons on a surface. Nature prefers equilibrium, so imbalanced charge makes the attempt to redistribute itself over the charged area. If two charged objects are touched together then separated, a discharge occurs. Since this is the case, an excess or a deficiency of negative and positive ions must be remedied here, rather than waiting for a discharge to occur – an expensive event where sensitive electronic parts are involved.

The reasons for static are easy enough to see, but it must be recognized that the potential for damage to sensitive parts is enormous. Less than 100V of generated charge can literally melt the circuits of a semiconductor chip. As geometries and layers get smaller, less voltage is necessary to destroy a chip. However, the damage is on such a microscopic level that only an electron microscope will reveal the extent of such a catastrophe.

The setting then for the battle against static is the manufacturing environment where these precious devices are made and used.

All materials fall into one of two categories: conductors or non-conductors. Static can be eliminated from a conductor simply by connecting it to ground, the main idea in many static-protection products, as will become evident. A non-conductor, however, will not move charge with a change in potential so a different solution is in order: the electrical imbalance is corrected simply by introducing enough positive or negative charges to neutralize the static.

It is important to remember that the word

A complete workstation must include comprehensive protection for static-sensitive devices (Courtesy of 3M)

antistatic implies a particular class of material. An antistatic material relies on an increase of humidity inside itself to eliminate the problem, moisture being the means of dissipation. A conductive material does what its name indicates: it conducts the electric charge away from the device being protected.

Different types of electrostatic charging are differentiated by the way they are produced. One is triboelectricity, which means the electricity is produced by friction. Another is induction from external fields, wherein an outside source of electrostatic energy induces a current in a previously uncharged device. The source and the device need not touch one another for this process to take place. Both are a common danger in the factory or manufacturing environment.

It will be useful to follow a surface-mount device or some other type of component on its journey through the factory, and see where the hazards lie.

First published in *Electronic Packaging & Production*, August 1985, under the title 'ESD control: When 'no charge' means profit'.
Author: John Harris, Technical Editor.
© 1985 Reed Publishing (USA) Inc.

Suppose a device in a reel, tape or tube is being transferred to its destination, the factory, where it will be incorporated into a tiny circuit on a printed circuit board. Let's follow an electronic component on this trek through to final packaging.

Everyone who will be working with the component is aware that it is coming. Receiving got the word yesterday and contacted Incoming Inspection, just in case, the bags, tubes, sticks or boxes do not allow their personnel to do a quantity count without opening the ESD protective packaging. The people in Incoming Inspection will take the shipment to a protective work area if need be. Once this is done, the shipment is sent to the stockroom where attendants will unload or kit the materials, if necessary, and then only if it is done at a static-free station. The unloading of the tape, reel or tube is the first encounter with dangerous static electricity.

For the moment, no charge is generated, because the component manufacturer has placed the surface-mount device on a reel with a conductive vinyl backing and covered it in a transparent, antistatic sheet. If the component is a DIP, it has been brought in a tube with a metallized plastic frame which protects the component during the trip to the automatic handling machine or assembly person. The tube is then unloaded or the tape

Even in an automated handling environment there is danger of static discharge (Courtesy of Xerox)

pulled apart, but static is still not generated since the machine is grounded or the person doing the unloading job is wearing a wrist strap.

If the assembler is an automatic robot, it has been grounded and the 'fingers' of its hand have been coated with an antistatic material to prevent discharge. An air ionizer is also in operation here, since the anvil transport portion of an automatic handling mechanism can generate very high levels of static electricity as a result of the continuous flow of devices over and then separating from the anvil.

Perhaps this component and the other parts must be marked in some way. This operation probably is being done with laser scoring or a similar method so that no personnel, possibly carrying static charge, need touch them.

If the assembler is a person instead of a robot, the workstation is the next stop. In many operations, the board will pause at a workstation in any event, and most certainly will be touched by human hands at some point. Of course, people turn out to be the greatest cause of static dissipation. One person can easily generate 30,000 volts. Because of this, the necessary steps must be taken to avoid costly loss of inventory. A number of control devices are available, starting first with the factory employee and his or her workstation.

In an assembly situation, it will be necessary to use conductive or antistat parts bins. Mats and table tops that are conductive or static dissipative must be used to drain off charge. Wrist straps or heel grounding devices are two common and effective solutions to prevent charge buildup on people.

Conductive floor waxes are another option in controlling plant-wide static generation. Charleswater Products, for example, developed a colloidal-film wax that drains charge produced by walking – a charge that is present even if the floor or tile is conductive.

The board on which the device travels throughout the manufacturing process is now found in the hands of a factory employee. A wrist strap is being used; for the moment, there is no danger. But be careful – a destructive charge could be generated if any of these common on-conductors are present:

Wrist straps must always be used to eliminate the possibility of static discharge when handling a PCB (Courtesy of 3M)

coffee cups, candy wrappers and cigarette cellophane. Air ionizers are employed to balance ions on surfaces that cannot be effectively grounded. Even though the factory has prohibited food and similar items at the workstations, there is always the possibility that someone either does not know the rules or does not realize why they have been implemented.

Personnel control, then, is one step toward the static-free facility. Handling of the part beyond the workstation must be equally free of static. It will be beneficial to take a quick look at what is available for the manufacturer and user here, following advice on its travels through this factory.

Handle with care

Static-protective bags are of value to the manufacturer of sensitive components or the user of such items because they provide protection for individual boards or other parts being stored, shipped or handled away from the protected workstation. Tubes for DIPs or chip carriers provide a similar static-free environment. The component and board moving through the factory are placed in a protective container at a protected workstation.

When choosing a bag or tube for control of ESD, one must consider the application for which the item is being used. Does it matter if the bag or tube is transparent? What electrical properties are desired, and to what tolerance are they to be chosen? These needs must be evaluated in order to make a logical and economical choice of a protective device.

The available bag materials are numerous, and differ in cost to the user in as many amounts. Pink or blue poly are antistatic bag materials that are inexpensive and have good tensile strength. They do not however, shield devices from external fields as do conductive materials.

Conductive carbon does shield from these fields but it is not transparent. There is also concern that a carbon-loaded bag will lose this layer during handling, degrading the protective properties.

Metallized film bags combine excellent conductive properties with transparency, but are not extremely pliable and cost more than antistat products. Care must be exercised to see that the bottom edges are not creased, as loss of continuity might result.

Properly marked conductive bags form a protective 'cage' for PCBs and are transparent to allow easy handling (Courtesy of Xerox)

Some bags have a grid of conductive material rather than being metallized, thus providing strength, flexibility and transparency. Scaled Air Corp., for example, markets a conductive carbon-grid bag that can be stretched to twice its length or width without destroying electrical continuity. Cost is higher than antistatics but these items are also conductive, providing shielding from external fields as well.

Some speculate that since charge is only drained through the grid itself, there is a decrease in effectiveness in comparison to a metallized product which has a continuous conductive layer about the bag's surface. It should be remembered, however, than an electric current will seek the path of least resistance; something that a conductive grid should provide.

When any kind of bag is employed to protect against static discharge, it must be tested for continuity at each reuse to ensure that it is still an effective tool.

All the eggs in one basket

Totes are the multipurpose carriers designed for the transportation of PCBs loaded with ESD-sensitive devices. If a device needs to be moved from the worktable to a shipping facility, chances are that it will travel in one of these conductive totes.

It is important that boards be placed in a sturdy, protective container if they are to be handled. Some totes provide adjustable racks so that boards of different sizes might be stored and transported without touching one another. The tote has a lid which should be used to assure protection. An unsealed tote could still permit a free charge to get inside. The lid also ensures that the boxes can be stacked on top of one another rather than being nested, a process that generates charge.

Durability is also of importance here since the industrial setting may indeed be hard on both the boards and the boxes they are stored in. By being hard on these totes, the production setting also presents a 'carbon problem' to the user. Wear in handling, such as sliding boxes, tends to remove carbon from the box, affecting its protective integrity.

Everything goes smoothly for the component and it is placed in a tote and moved to the shipping area. The last stop for a component is in this department. It is part of a board that will become (or perhaps already is) the central brain for a computer or other useful electronic device. The printed circuit board is put into a protective case for the trip to wherever it will be used.

A cardboard case manufactured by Conductive Containers Inc., for example, has a conductive layer bonded onto the case surface itself, with an antistat foam cushion inside on which the part is placed. As with any such container, its lid must be in place to provide a 'Faraday cage effect' that will protect the device inside. Were the lid not used, a complete protective circuit could not be established.

Such a shipping container must ensure that parts or boards cannot move around during transport. If there is movement inside the carton a charge can accumulate, making a discharge possible when the products are unloaded.

Another thing to be on the lookout for: a tote box placed on an unprotected surface. If the box should become charged and a person touches it, the devices inside could become part of an electrical circuit, causing an electrostatic discharge to course through them.

Conductive foams are also used to provide a protective layer for parts being transported. This protection method, known as shunting, provides some protection, but static charge on

Shipping materials must provide a protective Faraday cage since people will be handling them

the lip of a DIP or waffle pack can cause an arc between the lid's inside surface and the chip's top surface. The main value of foam, then, is for cushioning and lead protection.

Proper use of these containers and bags is essential, regardless of the material used in construction. According to one manufacturer of static-control products, the most common improper use of these items is the failure to properly drain off any static charges on the container or bag prior to opening. This is important because it applies to both the user and the manufacturer of the devices that are being protected. Not only are such procedures essential to use in dealing with static, but periodic checks of equipment and personnel must also be made to ensure that all the static-protective devices are doing their jobs.

Static control

One of the easiest ways to control static is simply to prevent it entirely; give it no chance to be generated. Education of employees, then, is one very essential part for a successful operation.

At one time, people might have denied that static could be such a problem, and could not in their own minds justify the 'expense' of taking the necessary steps to control this profit thief. But as boards that were perfectly fine at quality inspection began to fail for no 'good' reason, eyes began to open.

This story is set in a factory that places components on boards, but a component manufacturer needs to take the same precautions. Even a few defective components can amount to a product failure rate that is surprising. For example, if a business makes DIPs and only 0.05% of the inventory is degraded by static, that's only one of every 200. If the users of these DIPs install them at an average rate of 20 per board – that's a potential failure rate of 10%. With five boards per system, that rate skyrockets to 40%.

It is now probably clear that the elimination of an electrostatic discharge problem from a manufacturing facility is indeed an advantage. But how does one implement these profitable measures in one's business?

Several companies offer their services in a consultation capacity to assist customers in choosing methods and devices to control ESD. 3M Static Control Division, for example, has a program wherein static-control analysts check an operation from station to station, on the lookout for potential causes of static. Solutions are then recommended to correct any static problem that might exist. 3M also conducts classes on the static problem for their customers' employees and managers, to make sure they are aware that damage due to static is a reality and a program must exist to deal with it.

Just what should you have in mind when it's time to go shopping for products to control static? Cost, of course, is a major factor, but choosing a protection system that is not sufficient for the operation can prove to be more expensive. Materials are the key factor in this decision and it pays to have or seek expertise in choosing the right equipment for a static-control job.

Customizers also seem to be coming into the picture more often. The customer supplies the resin or design for a static product and the firm does the job of molding or assembly from the design or specifications. Note that the user may also ask for cetain tolerances from a custom product, which might be a useful aid in developing a static-control program, especially in an established factory.

A last consideration is what the static-control industry believes is needed as far as the products that are available. The key word in this puzzle seems to be materials. One manufacturer of static-control products would like to see the development of a static dissipative or conductive tote box that is carbon-free. Another expresses the desire for a new molding resin for tote boxes other than polypropylene, since it has a high shrink rate.

Further, standardization of stress and test methods is of great concern to everyone. Of course, the on-going search for improved ESD protective products and standards benefits the user directly, in that a more effective product is the result. For the manufacturer of components and boards alike, a successful static-control program means a lower failure rate and increased customer satisfaction.

STATIC SHIELDING TECHNIQUES

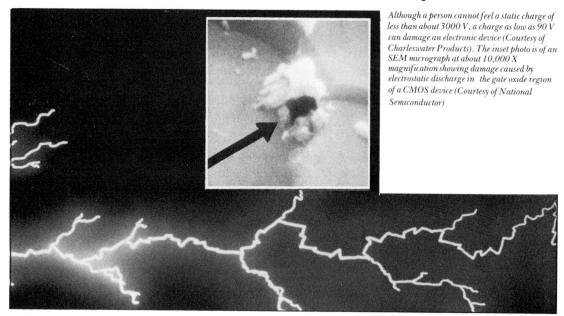

Although a person cannot feel a static charge of less than about 3000 V, a charge as low as 90 V can damage an electronic device (Courtesy of Charleswater Products). The inset photo is of an SEM micrograph at about 10,000 X magnification showing damage caused by electrostatic discharge in the gate oxide region of a CMOS device (Courtesy of National Semiconductor)

With the trends in electronics toward more closely packed devices, higher density circuit boards, faster response times and more capacity per chip, static protection of CMOS devices is becoming more important and more difficult to achieve. At the same time, more people are becoming aware of the damage electrostatic discharge can cause. These two forces combined are steering the static protective packaging market toward more effective products.

According to Dave Swenson, of 3M in Austin, TX, a container can protect sensitive electronic devices from three types of static electricity: triboelectric charge, caused by friction; electric discharge, which is simply a spark; and an electric field. Anti-static material shields devices from triboelectric charge because the material itself is unable to generate a static charge. Devices can be protected from discharge with non-conductive insulation. To protect against an electric field, however, requires a conductive container that will drain the charge to ground

before it can reach the device. The trend in static protection is away from simple anti-static and insulation protection and toward total shielding of sensitive components from all three kinds of damage.

Bagging the components

One type of product designed to provide this kind of shielding is the layered bag. Several companies manufacture a variety of this bag, which generally consists of a conductive carbon or metal layer sandwiched between static dissipative or anti-static material. The advantage is that, when the bag is sealed, the conductive layer provides a Faraday cage effect, preventing any charge, including one from an electric field, from reaching the inside. The anti-static or static dissipative layers prevent sparking between the components inside and the conductive layer. For conductive layers that are carbon-filled, the outside layers prevent sloughed-off carbon from contaminating the working environment.

One such bag is made by Sealed Air Corp. A sheet of nylon is reverse-printed with a carbon grid, and then laminated with anti-static

First published in *Electronic Packaging & Production*, January 1985, under the title 'Static protection calls for new techniques'.
Author: Ruth Berson, News Editor.

polyethylene. The contents of the bag can be seen through the transparent interstices of the carbon grid. Moreover, says Sealed Air, the bag can undergo a certain amount of stretching without loss of conductivity, and one can tell by looking when the grid is broken and a bag is no longer functional.

A similar bag, the XT-7000 from Quality Packaging, has the grid laser-embossed onto an anti-static film, and covered with a static-dissipative, abrasion resistant coat. According to sales manager Franz Heldwein, the trouble with printing and then laminating is that, between the two steps, the material is rolled up and carbon particulates can contaminate the anti-static protective layer. The Quality Packaging bag is claimed to have no particulates in the outer bag and inner layers and like the Sealed Air bag, is translucent and can withstand some stretching.

American Converter seals its product's conductive layer between anti-static polyester on the outside and anti-static polyethylene on the inside. According to Creighton White, the polyethylene is stretchy and can accommodate pricks from leads and pins, while the polyester is durable and will withstand the

The open conductive grid in the Sealed Air shielding bag allows a person to see into the bag

wear and tear of being handled. "More and more people have gotten away from other types of static-control bags and into a static-shielding bag," says White. "Our bag was introduced this year, and it's been going crazy for us."

Shielding and cushioning

Two thicker bags that protect from physical as well as electrostatic damage are available from Seco. The 2100 Stat-Zap offers a layer of anti-static cushioning in addition to the conductive and anti-static protective layers. The Pack 99 has the cushioning layer as well, but the conductor is a carbon-filled poly, rendering the bag opaque.

Another product combining shielding with cushioning has been developed by Colvin Inc. A wrap-around box, known as the Cady-Pack, is made out of corrugated cardboard laminated with a few angstroms' layer of aluminum, and, on top of that, static-free cushioning material. When unfolded, the box looks like a cross, with the circuit board in the center. When it is wrapped up, says Colvin's Mark Dwyer, the box provides static shielding. The circuit board can be sent from workstation to workstation wrapped up in the Caddy-Pack. When it arrives at a given workstation, the operator can unwrap the box, snap his or her wrist strap to it, and work using the unfolded box as a static-protective surface. When the board is ready to be shipped, it is folded into its package once again and inserted into a shipping tube made of the same material.

Alternatives to carbon

While many companies are moving toward conductive shielding, others are continuing to make static-dissipative products but trying to get away from the use of carbon. Carbon has several disadvantages, according to most manufacturers. First, as mentioned previously, it tends to slough off when exposed, endangering the devices it is meant to protect and making it unsuitable for use in a clean room. Second, carbon can be too conductive and must be shielded from sensitive parts to prevent damage from sparking. Third, unless it is laid down in a grid, as in some of the

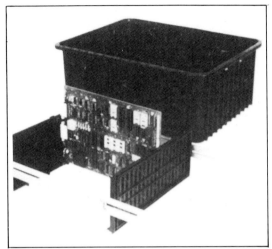

Fancort's adjustable Karry-All is designed for boards with components running up to the edge

shielding bags, it creates an opaque layer in the protective package that precludes visual inspection of the contents.

Among the companies developing alternative static-dissipative material is Fancort Inc., West Caldwell, NJ. The company is working with an anti-static plastic, ABS, with a resistivity of about $10^9\Omega$/sq. According to Ron Corey, Fancort's sales manager, until now only General Electric (among resin manufacturers) has been able to produce static-dissipative material that is acceptable to the electronics community. The material is expensive, however.

The ABS, from Borg-Warner, costs about half as much, says Corey, and has the advantage of being highly impact resistant. In addition, unlike carbon, the ABS can be colored. "You get so much black around the plant," Corey says. "No one likes the look of it and everyone worries about carbon coming off."

Fancort is currently testing the ABS to see how well it molds their circuit board rack parts, and what the rate of static decay is. If the results prove promising, they intend to shift out of anti-static polypropylene and into ABS.

Another alternative, developed by Charleswater Products Inc., is colloidal film technology. This technique uses materials in which are suspended conductive ionic polymers. The material has a resistivity of between 10^7 and $10^9\Omega$/sq. Use of colloidal film allows the company to make anti-static work surfaces in keeping with their philosophy of "incorporating static-safe properties into the materials that are going to be used anyway, as opposed to making products that are additive," according to president George Berbeco. Charleswater has introduced an anti-static floor wax which can be used in place of an anti-static floor mat or anti-static tiles. According to Berbeco, they are also about to bring out a flexible colloidal film material which can be used to make static-shielding bags.

Quality Packaging is about to introduce a material that, according to sales manager Heldwein, is made permanently static-dissipative through radiation. The material can be used in clean rooms since it does not outgas, and the static-dissipative qualities do not decay with time, adds Heldwein.

Ease of handling

A trend in static protection that is not new, but continuing, is to make products more convenient to use. Fancort is marketing a conductive rack for holding circuit boards that have components running to the very edge. The rack is a conductive, mica-filled polypropylene held together by aluminum brackets covered with molded card guides which hold the overhanging components. The guides are

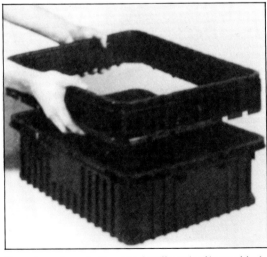

The snap-on collar on Buckhorn's tote box offers optional increased depth

wide enough not to put pressure on the components, says Fancort; the width also makes loading boards on the rack easier for the operator. The rack is designed to fit into a 15in. tote box with a lid so that the boards can be protected during transportation.

Henry Mann offers an anti-static rack intended for use in dealing with circuit boards that have been assembled but not gone through soldering. "The idea is to keep the boards horizontal so the inserted components won't pop out of the board," says advertising manager David Belshaw. The rack is designed so that it can be picked up by hand or loaded on a transport dolly.

Rob Rothfuss of Buckhorn feels that flexibility is an important aspect in protective products. "One of the problems with tote boxes is that a company may buy a box of a certain dimension, and then the size of the board they make changes," explains Rothfuss. "We try to design a lot of flexibility into our boxes." Buckhorn's tote boxes come with a 3in. collar that can be snapped on to increase the box's depth. A divider system on the inside allows a box to accommodate various sizes and numbers of boards. The lid is made of the same conductive material as the box, so that once covered and sealed, the tote box with circuit boards inside can be shipped as is.

Colvin Inc. has designed static-protective component storage trays that can be kept in an automated warehouse and fetched by a robot for placement on an automated assembly system.

While there has been increased awareness about static-protection in the workplace, people have also begun to realize that boards can be damaged in the field as well, if they are removed for repair. The Colvin's Caddy-Pack, mentioned above, provides one solution to the field repair problem. Another is offered by Sealed Air. Its field service kit consists of a fold-up mat made of a conductive layer between two anti-static layers. The edges are sealed so that the conductor is not exposed and there is a fastener for grounding a wrist strap. The mat has pockets in which the repair technician can keep tools. A similar mat is offered by Simco.

CHAPTER 9

THE ROLE OF CAE/ CAD/CAM IN ELECTRONICS MANUFACTURING

Although this book is focused on assembly rather than component manufacturing, one must not forget that the success or failure of any manufacturing process is often totally dependent on product design that is compatible with the chosen assembly process.

Computers are increasingly used to design the product and control the processes of fabrication and assembly. Some familiarity with the terminology of computer-aided manufacturing is essential to an understanding of today's electronics manufacturing processes. This chapter exmaines the use of CAD/CAM in all types of circuit board use.

COMPUTER CONTROL OF MODERN FACTORY OPERATIONS

Although the number of computers used in finance, business management, medicine and other office environments far outnumber those applied to manufacturing operations, computers will eventually play an essential role in increasing productivity and product quality on the factory floor. Many observers believe that American industry may even have an early lead in the application of computers for production processes. This advantage is said to be the best opportunity for overcoming the technological inroads made by other industrial nations.

CAM (computer-aided manufacturing) and CIM (computer-integrated manufacturing) are today's popular acronyms for describing the role of computers in the manufacturing process. In CAM, the computer is often not in direct communication with all aspects of the production operation. An operator interface is required. CIM, however, implies that the interface between the computer and the manufacturing process is direct. The computer is in direct control and is receiving data directly from all segments of the production operation such as process sequences, parameter monitoring and control, bar-code reading, robot and machine control, materials handling, lot sizes, etc.

The final goal of CIM is the implementation of the fully automated factory or 'factory of the future'.

Digital Equipment Corp. (DEC) says that factory-floor control via computers offers significant flexibility in communications. By direct link, or by disk transport, complex, accurate and complete data sets can be easily transferred to internal functions such as quality control, manufacturing engineering or finance, and to vendors to communicate new specifications, or to rapidly resolve quality problems involving incoming parts. This becomes especially important when the 'Just-in-time' (zero inventory) method of materials handling is used.

Advances allow flexibility

A number of recent advances have made possible the adaptation of computers to the manufacturing environment and for achieving a CIM system. These include:

- The introduction of quality personal computers that are highly flexible both as stand-alone devices or networked as smart terminals of larger computers.
- Hardened terminals and suitable data collection devices designed to endure shop-floor environments.
- Powerful but affordable minicomputers.
- New interfaces between computers and automated shop-floor equipment.
- New networking concepts and products that can link all these components together, allowing them to share a common database and providing opportunities for fully computer-integrated manufacturing.
- Cluster systems linking multiple computers to multiple shared on-line storage devices for reliability and speed, allowing con-

Production monitoring
 by plant
 by department
 by workstation
 by machine

Machine comparison reports
 uptime/downtime
 idle time

Maintenance management
 scheduled maintenance
 recalibration
 tool changes

Equipment cycle reports
 equipment wear
 energy use

Programmable devices support
 programmable logic controllers
 process controllers
 NC machines
 robots
 bar code readers

Integration/Engineering
 manufacturing engineering
 industrial engineering

Quality control functions
 receiving inspection reports
 vendors communications
 process "capability" calculations
 process parameter plotting
 process control adjustments
 fault identification
 faults analysis

Factory alarm annunciation
 dwindling WIP inventory
 machine failure
 process limits exceeded
 other bottlenecks

Materials handling
 materials used
 parts required
 critical parts levels
 scrap/rework reports

Integration/Management
 manufacturing management
 marketing information
 financial data
 human resources management

Energy management

Factory operations applicable to computer integration (CIM)

First published in *Electronic Packaging & Production*, March 1985, under the title 'Computers control modern factory operations'.
Author: Howard W. Markstein, Western Editor.

figurations with more than 100 Gbytes of disk capacity.
- A rich menu of innovative software.

According to DEC, these advances have allowed computers to be configured for satisfying the requirements of a given manufacturing application. For example, a company may only have a present requirement for a single stand-alone system. However, this can be a single start-up unit of an expandable system, allowing for future add-on and growth. Eventually, the computer system can be networked to a larger computer for communication or to acquire greater storage and computing capacity, then later networked to cluster systems where numerous computers are linked to many large disks and can independently use a common database.

The goal of CIM can then be approached by having the system directly interface with bar-code readers and other sensing devices, programmable controllers, NC devices, robots and process controllers that will provide integrated manufacturing with computer-controlled shop-floor automation. The computers can also be linked in distributed processing systems with mainframes, other computers and outside networks without hierarchical limitations in fully integrated manufacturing enterprises.

This PRO-350 computer from Digital Equipment Corp. is shown in use on the factory floor at Lanx Corp., a disk manufacturer. The unit represents a start-up for an expandable CAM system that will eventually evolve into CIM

Computers in action

An example of computer control in manufacturing can be seen at Lanx Corp. (San Jose, CA), a manufacturer of high-quality hard disks. The computer-controlled operation began by using a single Digital PRO-350 desk-top computer located on the factory floor. The system uses RS/1 software (BBN Software Products, Cambridge, MA) and acts as an indirect shop-floor quality-control instrument to form an expandable CAM system that can eventually evolve into a CIM system. The PRO-350 has since been networked to a VAX-11/750 mainframe and two other PRO-350 computers, one for the QC manager's office and the other for the firm's financial officer in corporate management. Two VT-102 terminals are also part of the system. One terminal provides parametric testing inputs to the VAX-11/750, and the other is for system programming.

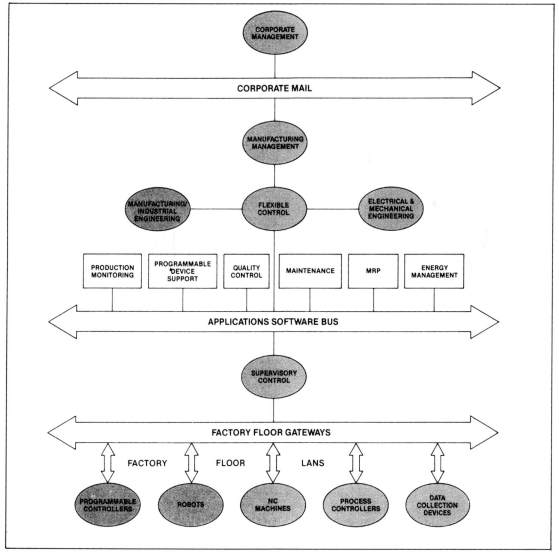

Full computer-integrated manufacturing (CIM) is achieved when direct two-way data flow exists on the shop floor via local area networks (LANs) under supervisory computer control and linked to a hierarchy of management systems through appropriate software (indicated by rectangles)

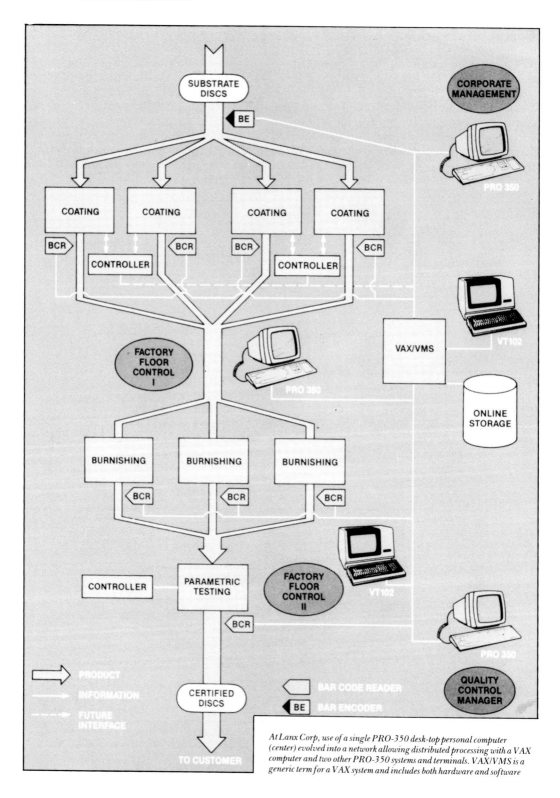

At Lanx Corp, use of a single PRO-350 desk-top personal computer (center) evolved into a network allowing distributed processing with a VAX computer and two other PRO-350 systems and terminals. VAX/VMS is a generic term for a VAX system and includes both hardware and software

The manufacturing process at Lanx involves three main steps: coating of the substrate disks in several automated coating units run by stand-alone process controllers; burnishing; and parametric testing.

The PRO-350 computer on the shop-floor is convenient to all of the controls and readouts of the coating units. The operator periodically inputs in-process parameters on the PRO-350. These are then processed with RS/1 statistical analysis software and graphically displayed and evaluated. At any moment during a day's production run, personnel can command the production status, resulting in a display. A direct interface from bar-code readers on each production line to the VAX system allows tracking of each individual disk and lot. This permits statistical correlation of defects with the operating conditions relevant at the time of production of a particular lot or disk.

Using a personal computer on the shop-floor to emulate a VAX terminal provides the ability to continue operations when the larger computer is not accessible. Enough current production data is stored in the PRO-350 to sustain production operations if and when required.

The VAX computer provides each PRO-350 with the greater storage and computing power required to handle larger databases and to speed computations. Other terminals can be added as needed to accommodate system growth. Lanx says that eventually the VAX/PRO network will interface directly with the existing process controllers that operate the automated disk coating system in each line, marking the transition from CAM to CIM.

Flexible automation

James F. McDonald, general manager of IBM's Industrial Automation unit, says that CIM provides the flexibility in automation required by today's manufacturers. He warns that in the present environment where a product's life cycle may be as short as six months, and where customers are making demands for customization, investing in a 'hard' automated manufacturing system may be foolhardy and expensive. Flexible automa-

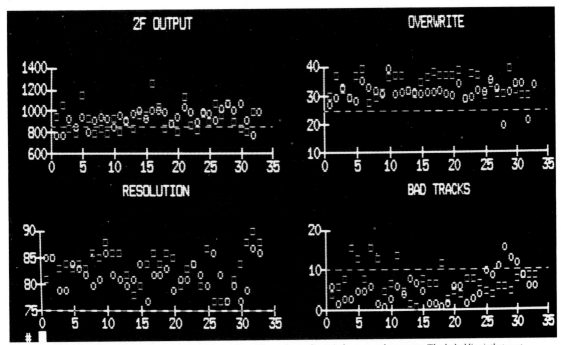

The PRO-350 displays four major quality parameters. Abscissa is sample number; ordinate is the measured parameter. The dashed line is the target maximum or minimum value. Squares relate values on the front side of the disk; ovals, the reverse side

tion, however, will allow rapid product changeover.

"Flexible automation offers an attractive alternative," explains McDonald. "Because it is reprogrammable, you can use the power of the software and simple, inexpensive tooling to accommodate design changes. And that translates directly into greater opportunities in the volatile, competitive market place, whether you're building cars or computers, pens or printed circuit boards."

As defined by IBM, computer-integrated manufacturing, or CIM, is a complex and creative blending of a variety of automation devices – robotic systems, material handling systems, fixed automation, programmable controllers, parts feeders, test stations – all operating under the control of a computer, or, more likely, multiple computers.

McDonald says that as a large spectrum of automation devices are installed, computer control becomes a matter of necessity. "If you visualize a factory where you have 250 robots, the first thing you think of is, well how do you turn the system on? What makes everything come on in sequence? How do you get all the materials to the right system? If you shut down the system for some reason, how do you know the status of the individual components in order to restart the system? It's difficult to imagine people trying to manually start a system having that many devices. So, a key element in building an automation strategy is the ability to drive all the pieces from a host computer. And the 'glue' that ties this type of automation system together is the software, the programming language," concludes McDonald.

CIM at IBM

IBM is in the process of assembling a 'showcase' of computer-integrated manufacturing at its Lexington, KY, plant, a major location for producing IBM electric typewriters. When completed, the plant will have more than 250 robotic systems for manufacturing subassemblies and keyboards.

The main computer area has no walls and is located in the center of the manufacturing floor. These manufacturing lines communicate with each other and with a higher level

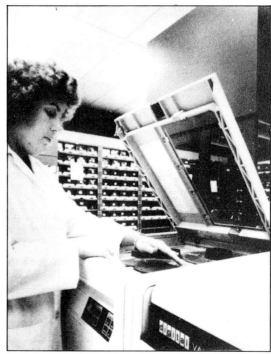

The VAX-11/750 is networked to three PRO-350 computers and provides access to increased computing and storage capability. The desktop computers are therefore converted into powerful workstations

plant system through a hierarchical network of computer systems.

The production system architecture consists of control systems at the plant level, manufacturing area level, production line level and direct machine interface level. The plant level system is primarily concerned with logistics such as inventory management, parts ordering, production reporting, and overall manufacturing module scheduling.

The manufacturing area system handles the individual manufacturing module scheduling and performs production analysis. At the line level, part identification, error analysis, and program loading (foremen) is performed, while the interface level system provides specific control and instruction for robot and conveyor movement, error recovery, and tool start-up and stop commands.

Each of the system levels, plant, area, line and interface, consist of multiple manufacturing modules, with each module controlled by an IBM Series 1, a mini-processor specialized for sensor-based processing.

The plant level IBM 4341 is interfaced with the distribution and shipping function, and with engineering. The latter provides the bill of materials, quality specifications, test algorithms, and defines robot/tool control. IBM says the computer-integrated manufacturing system has enabled the company to become a low cost, high-volume producer in a competitive worldwide market.

Programmable controllers

Programmable controllers are one of the most popular devices for direct machine control on the factory floor. Gould Inc., said to be the largest supplier of programmable controllers, defines the device as a 'blue-collar' computer packaged to survive on the production floor. It represents an electronic approach to machine control. In essence, a programmable controller is a solid-state electronic replacement for the banks of electromechanical relays formerly used for machine control. Its function is to ensure that each individual machine operation and each cycle occurs in the correct sequence. The device can be easily reprogrammed to alter the operation of the machine. Modern programmable controllers, armed with memory and linked to a factory communications network, form the basis of factory automation.

Richard E. Morley, director of advanced technologies at Gould, and credited as the inventor of the modern programmable controller, says, "A programmable controller can be likened to a PDP-11 in 'drag,' and its presence on the factory floor is to replace or augment sequential control, monitoring control, quality control and manufacturing overview." He adds that it also performs logic and arithmetic functions and can be programmed with special algorithms for optimization of control.

IBM's Lexington, KY plant is a prime example of computer control in manufacturing. The main computer area is located in the center of the manufacturing floor, thereby emphasizing the integration of computers within the factory environment

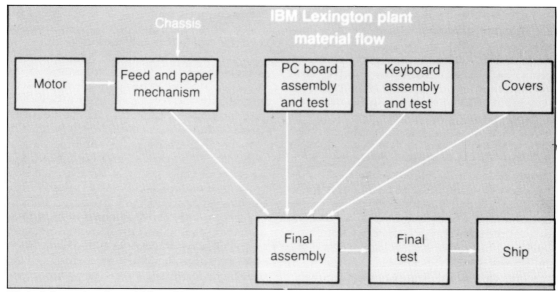

This simple diagram shows the material flow and the major manufacturing operations at the IBM plant for the production of electric typewriters

In one figure a programmable controller from Westinghouse Electric is shown controlling a robotic assembly station at Tektronix. The robot, under the direction of the programmable controller in the adjacent rack, is assembling the rear section of a CRT for a Tektronix oscilloscope. The technician is shown receiving and monitoring prompting messages from the programmable controller relating to the robotic assembly sequence.

A total of three programmable controllers are used in the Tektronix CRT assembly area, each controlling a specific operation. Westinghouse says the three controllers are connected via a serial communications link in a daisy chain configuration. This arrangement allows each programmable controller access to a section of common memory with at least one other controller, thereby creating the effect of multiline communications while using only four wires per pair of controllers.

Since programmable controllers are used on the factory floor in proximity to machinery, heat sources, contaminants, etc., they must be designed and tested to withstand hostile environments. Another figure shows a number of Gould programmable controllers in a burn-in chamber where they are operated under extremes of temperature and humidity.

Programmable controllers also form an integral part of the environmental test operation. A total of four burn-in chambers are interconnected by a control system consisting of a programmable controller for each chamber, with all four reporting to a central programmable controller. The system provides automatic control of test start, stop, environmental conditions, unit failure reporting, time of failure and documented error readings.

The importance of programmable controllers, and also programmable process controllers, in the overall scheme of computer controlled factory automation cannot be overemphasized. These units form the first link in the control system chain and have a primary influence on manufacturing productivity and quality.

Tying it all together

Networking and software are the key elements in tying together a computer-integrated factory. The communications network provides synchronization of the process by allowing discrete sequential events to become part of a continuous process, all communicating in a common language provided by the software. Networking allows a master computer at the plant level to drive data down to the production equipment which, in turn, feeds back status information to monitor, control and adjust the entire plant as a single system.

Networking has also been defined as the means of achieving a vertical hierarchy of programmable controllers, minicomputers and mainframes to work together in a concerted manner acceptable to management. When properly implemented, networking simplifies growth and eases the transition from a stand-alone computer to CAM to CIM. It can also maintain the relative autonomy of workstations while providing access to the greater computing power and storage capacity of the system's mainframe computers.

Requirements of CIM

According to Gould, an effective computer-integrated factory automation system must facilitate the entry of CAD and CAM into the production flow, support the needs of manufacturing planning, provide the database for machine and facility optimization programs, keep maintenance records, chart machine trends, control inventory, and monitor and control machine operation on the factory floor.

Implementing such a system requires expertise in computers, communication networks and control systems. Gould says computers are the most visible elements, and hence receive the majority of attention, but, in practice, the ability to balance all these elements "makes the difference between just talking about factory automation, and actually making it work."

IBM defines CIM as the integration of business functions, engineering, and plant floor operations. Not only are the production machines in direct communication with the computer network, but so are engineering and distribution, thereby forming a fully integrated manufacturing facility. According to IBM, there are four basic requirements for developing a CIM system: database, com-

An IBM Series 1 processor (rear right center) controls an automatic PCB component inserter at IBM's, Lexington, KY, plant. The terminal at right is used for entering commands

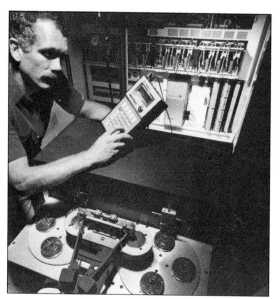

A technician holds a programming and monitoring unit for verifying proper communication between a Westinghouse programmable controller and the robotic assembly station. The programmable controller is the two-tone enclosed unit at the lower center of the equipment rack. Adjacent PCBs at lower right contain input/output circuits

munications, information access, and network management.

Database includes the data needed and generated to drive plant operations and involves the programmable controllers, production machinery and amsterials handling systems. It also includes the data captured from the plant floor such as parts counts, machine utilization and quality monitoring, which is all then fed to the plant level system.

Communications means that the machine controllers and various levels of computers can electronically converse with one another through a systems network architecture. It allows and provides an exchange of data between the production machinery and support equipment, even though the various systems may be supplied by different vendors. Basically, communication occurs across multiple levels of the plant architecture to different kinds of equipment using different communications protocols.

For example, communication exists between the interface level IBM Series 1 and a programmable controller, even though the latter has a vendor's proprietary network protocol. If the programmable controller, for instance, is a product of Allen-Bradley, then it has a proprietary network called the 'Data Highway', and IBM has a programmable controller software product to communicate across this proprietary network.

Access to information implies the capability, through appropriate software, of allowing people to have access to data across the entire plant, via CRT terminal or hard copy. The goal is to offer a means of providing human evaluation of plant operations such as yield or productivity analysis and the evaluation of cost factors. In effect, access to information provides the means to assess results.

Network management is the capability of having a plant-wide view of the instantaneous operations, or status, of the production machines and support equipment. Maintenance engineering can then determine potential problems by noting any irregularities in equipment performance. Network management, therefore, means monitoring the functionality of the plant.

The use of computers to control productivity on the factory floor, and to integrate with other manufacturing and management functions, has been an anticipated improvement in plant operations. From the products presently available, both hardware and software, the technology of computer-integrated manufacturing is slowly being applied, initially as a start-up version by the smaller companies such as Lanx Corp., and as a full-scale pursuit by the larger firms such as IBM, capable of making the required investment.

Gould programmable controllers undergo a series of environmental tests to simulate conditions on the factory floor. The testing is performed under the automatic control of other external programmable controllers

ELECTRONICS CAE/CAD/CAM

CAE is a term given to a collection of tools that assist electronics designers in specifying and simulating electronic circuits. The term has been generally associated with low-cost engineering workstations. CAD tools assist designers in implementing circuit designs, but they do not design automatically. Printed circuit board and integrated circuit layout digitizing systems are examples of CAD tools. Design automation tools, on the other hand, perform design work of their own. Examples of design automation tools include automatic placement and routing. For the purposes of this discussion, all three are combined within the overall CAE/CAD/CAM function.

Electronic CAE/CAD/CAM systems typically support the following electronic design functions or activities:

- Printed circuit board design.
- Schematic design.
- Logic design and simulation.
- Circuit design and simulation.
- Integrated circuit layout design and verification.

These activities speed up printed circuit board design and verification. They also help prevent physical and electrical design violations from entering into the design process. They also expedite the transfer of design information between the printed circuit board (PCB) designer and engineer. CAE/CAD tools are used in an estimated 50% of all new PCB designs. By 1990, they are expected to be used in 90% of all new designs.

As electronic products become more complex, their development time in some cases threatens to exceed their useful product lifetime. This makes the use of CAE/CAD/CAM tools extremely important because they can enhance productivity, reduce design errors, and improve design document management and control.

CAE/CAD/CAM overview

The design process starts with the development of a schematic of the circuit to be produced. A board plan is developed which shows the physical outline of the board and general placement locations for the functional blocks of the circuit. The next step is to actually place the individual parts on the board. This depends on board layout and routing considerations. When parts placement is complete, interconnects are added. While routing interconnects, it may be desirable to alter some of the parts placements made previously. After the layout is completed, it must be verified. Circuit connectivity is verified by checking the layout against the original schematic (or netlist). Design rule checking is also performed to ensure that the layout conforms to the various rules of spacing, minimum trace width and so forth.

The layout is corrected if any errors are discovered in the verification process. The artwork used to fabricate the printed circuit

First published in *Assembly Engineering*, May 1985.
© 1985 Hitchcock Publishing Co.

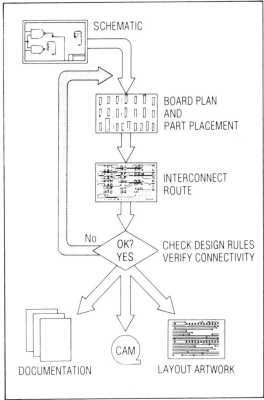

The basic activities involved in PCB design

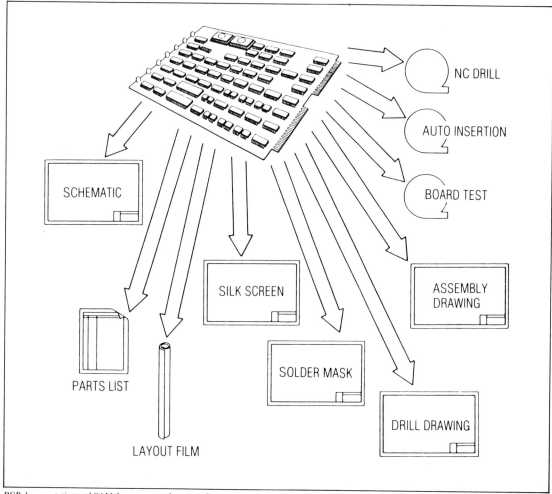

PCB documentation and CAM data are normal outputs of many design systems

board is then generated. A documentation package consisting of a bill of materials, assembly drawings, netlist and so forth are also generated. The original schematic is back-notated to update parts designators and pin assignments reflecting final parts placement.

PCB design/engineering analysis

The schematic is developed by the engineer responsible for developing the board and shows circuit topology and components. Special requirements could include critical parts placements, special layout considerations for analog circuits, etc. The PCB designer uses the schematic to help develop a bill of materials which lists each component type and quantity for each design. If the CAE/CAD system is used for layout, the designer also develops a netlist which describes the components and interconnects that comprise the circuit. The netlist is the basis for all placement, interconnect routing and verification of PCB layout.

The CAE/CAD system makes it easy to enforce design standards. But the most important reason for using the CAE/CAD schematic capture is the netlist generation. The netlist is a list of all interconnects or 'nets' connecting the different components in the circuit. The netlist in turn is used to drive automated PCB placement and routing algorithms, and for automatically verifying connectivity during design. Manual entry of the circuit netlist would be both error-prone and time-consuming.

A number of CAE/CAD systems and workstations offer logic simulation which enables the designer to simulate and observe the behavior of the circuit. This software modeling of circuit behavior can drastically reduce the time and cost of constructing an electronic breadboard to check circuit behavior.

Before the board can be laid out, the physical board size must be determined. This must include dimensions, location of special connectors or other hardware, location of tooling holes, identification and logo information. This is done quite simply using the graphics capabilities of a CAE/CAD system.

The PCB designer must also gather information about the size of each component and pinouts before beginning board layout. In a CAE/CAD system, these types of information are usually stored in the parts library resident in the system. CAE/CAD also provides for manual, interactive and automated placement of components, depending upon the particular system. The designer, however, must usually specify the board positions into which the components can and cannot be placed. Once specified, the placement routines can determine the best fit for placing components into the appropriate spaces.

One way in which a designer can receive feedback about the choice of placement is through use of a graphic 'rats nest' that shows all of the interconnects between the components. Automated placement systems use

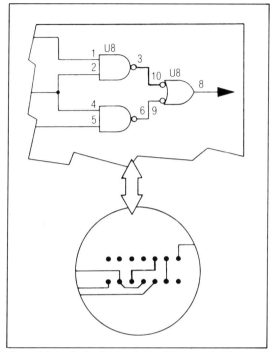

Connectivity verification involves checking the layout against the schematic

computer power to determine optional component placement schemes. Automatic placement algorithms are driven by interconnects, restrictions on component placement, and other weighted factors determined by the designer.

Interconnect routing begins after component placement is determined. CAE/CAD systems offer solutions ranging from manual to fully automatic routing. Systems that incorporate both interactive and automatic routing allow the practical experience of the designer to be coupled with the analytical power of the computer. But before routing can begin, the designer must specify the trace widths, minimum spacing, and other information influencing routing.

After the PCB is laid out and before production artwork is generated, the design must be verified. Design verification usually includes at least a connectivity check and a design rule check. As mentioned previously, connectivity is checked by comparing the physical board layout against the original schematic. CAE/CAD systems can save a lot of

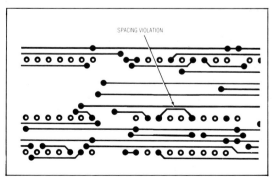

An example of design rule violation

time in this area since parts placement and connectivity algorithms are driven by netlist data, and therefore accurately reflect that data. Similarly, it is unnecessary to check for design rule violations since this is built into the system and can be programmed to define such things as minimum spacings between pads and traces, minimum trace widths, and so forth.

Supporting documentation is an important output of CAE/CAD systems. It is essential in the development of full-scale production plans for printed circuit boards, including final board assembly. After completing PCB layout, the original schematic must be updated to include reference designators and pin assignments. Back annotation is performed automatically by most CAE/CAD systems.

Essential documentation that must be generated as part of the PCB design process includes:

- Layout artwork or film for manufacturing the PCB.
- Parts list or bill of materials required for board assembly.
- Solder mask drawings showing the areas which should be solder-free.
- Drill drawings defining the size and location of holes to be drilled.
- Silk screen drawing.
- Assembly drawing and parts lists used during component stuffing and assembly of the printed circuit board – including component location, orientation, polarity, etc.

All of these are natural and necessary outputs of a comprehensive CAE/CAD system.

Electronics CAM

Electronics computer-aided manufacturing (sometimes called ECAM) has its roots in the CAE/CAD process. CAE/CAD provides many interfaces to the computer aided manufacturing (CAM) process. These tools often provide for the automation of both the printed circuit board itself as well as the printed circuit board assembly.

If automated drilling of the PCB is required, an NC (numerical control) drill tape may be output automatically as part of many CAE/CAD manufacturing documentation packages. The drill tape contains data about the locations and sizes of the various holes to be drilled in the board. This is a required feature in high-volume production of PCBs where high-speed automated drilling machines are used in the manufacturing process. CAE/CAD systems create the files for automated drilling, extract the necessary information from the database, and format it for the tool to be used.

Automatic component insertion is now used in many high volume PCB production requirements. If automatic insertion is used, data must be supplied to instruct the machine which components are to be placed and their respective locations. Component location, orientation and related data can be extracted from the database of many CAE/CAD systems and formatted to directly drive various insertion machines.

Board test and inspection information is a natural complement to other types of CAM interface output data. Furthermore, it is becoming increasingly essential if automated testing of either the PCB or final assembled board is to be accomplished. Many testers use a 'bed of nails' for applying electronic stimulus and collecting response data. In such cases, the test documentation should show the physical location of all nodes to be tested. Some CAE/CAD systems are capable of outputting mechanical and electrical data which can be formatted for conducting board tests.

In addition to the above, both plotter and photoplotter interfaces are normally provided to obtain hard copy of layouts or artwork for board production.

CAE/CAD/CAM equipment configurations

CAE/CAD/CAM tools come in a variety of configurations. Once the tool requirements are defined, a decision must be reached on what type of configuration is best for the particular user's situation. There are three basic options: a standalone turnkey system, generally in the form of a workstation; a central computing facility, either mainframe or minicomputer, capable of running specific CAE/CAD/CAM graphics software; or a hybrid system that combines features of both.

The workstation is a self-contained system that does not ordinarily use an external computer for support. Workstations usually support a single user, although many of them are also capable of being networked into larger systems sharing a common database and peripherals. This capability will be discussed further in the next section.

One of the main advantages of the workstation is cost – generally less than half that of a mainframe or minicomputer. This may enable smaller organizations to purchase several workstations rather than one larger, central-

Digital Equipment Corp. (DEC) and Silvar-Lisco have announced availability of a computer-aided engineering (CAE) electronic software package that runs from workstations to host systems. The new software enables VAX station 1 and 500 systems to be fully integrated as CAE workstations into a network of VAX computer systems. Using Digital's DECnet/Ethernet communications facilities, with a uniform VAX/VMS operating system environment, the Silvar-Lisco software establishes a uniform applictions and files continuum across the entire network of integrated engineering systems. It provides users with full performance employing one operating system, one protocol and one integrated network. Within the Integrated Design Environment created by the Silvar-Lisco software, the user can perform design capture, verification, semicustom place and route, and layout of PCBs

ized computing system. Another advantage is that when a workstation goes down, it only affects the user of that station and not others on the network.

Centralized computing systems, based on mainframes or minicomputers, have the obvious advantage of raw computing power. New workstations based on 32-bit architectures in some cases come close to offering equivalent power, but they are not quite there yet. Further developments in the application and use of very large scale integration (VLSI) micro chips are required to close that gap. This will make it possible to incorporate the power of a mainframe in stand-alone workstations. Until then, computation-intensive analytical problems are best run on mainframes or super mini-computers.

Another advantage of centralized computing is that the incremental cost of adding users is generally lower. The cost of adding an additional user to a networked system of stand-alone workstations means the purchase of another complete workstation. Adding a user to a centralized system generally means adding an alphanumeric CRT terminal and perhaps a graphics terminal. The cost of adding a user to a centralized system can be one-fifth to one-tenth that of a complete workstation.

In practice, the combination of a central mainframe/minicomputer networked to various workstations may be the optimal approach for many situations. This arrangement can support different users and meet a

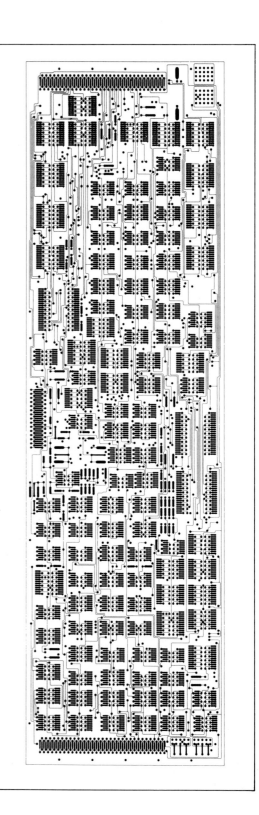

A new enhanced version of the Calay V03 CAD system for PCB includes improved routing performance, high resolution graphics, output for automatic insertion machines, new interactive features, and true automatic component placement software. The Calay RPR 300 Router, known for its ability to achieve 100% automatic routing of most circuit boards, now routes 30% faster, thus providing shorter turnaround time. The automatic component placement package automatically places components in the optimum position for routing, resulting in faster process completion. The software employs constructive placement algorithms which will allocate space proportionately for each component. This method of automatic placement consistently checks each component placement to all other components on the board. This differs from the commonly used pairwide interchange method in which only the connectors between two components are checked at any given time. Output for automatic insertion machines and sequencers, in the form of punched tape, mag tape, or floppy disk can now be generated by the Calay V03 CAD system. It is also reported to handle chip carriers, flat packs, leadless components, surface mount components and off-grid components fully automatically

variety of computing needs. Workstation users can perform most of the work locally. larger processing jobs such as autorouting, on the other hand, can be sent via network to the mainframe or minicomputer. This approach takes advantage of the workstation's ability to offload most design tasks, and the central computer's ability to handle highly analytical 'number crunching' functions.

CAE/CAD/CAM networking

The networking concept is a communications scheme allowing users to communicate with each other, share information, and share peripherals and other resources. Networking is most effectively used in a distributed computing environment where each node or user station on the network performs some computing functions of its own. Large com-

puters can also be networked so that they can share data and other resources as well as communicate with one another.

Networking is a critical topic for all functions within the automated factory. There are various approaches to the architectures and protocols involved in structuring local area networks. It is not possible to deal with them in great detail here.

It should suffice to say, however, that the main advantage of using networked workstations is that system throughput is relatively independent of the number of users on the network. Each user is supported by a separate workstation. System degradation occurs only when users share peripherals such as printers, mass storage devices, and so forth on the network. This may cause delays while users wait for access to a peripheral.

Telesis has introduced a design capture system for use on IBM Personal Computers. The EDA-1000 Design Capture System provides electronics engineers with the ability to create hierarchical or flat schematics using IBM-PC, -XT, or -AT computers. Fully integrated with Telesis' PCB CAD workstations, the Design Capture System serves as a low-cost extension to the physical design process. The hierarchical schematic capability provides up to 99 levels of drawing segmentation. High-level diagrams to gate-level schematics can be organized in drawing trees, enabling the engineer to efficiently manage complex designs. Changes made at one level are automatically incorporated into all other levels. The system's logic checking utility allows the engineer to identify and correct circuits before board layout begins. Five types of errors are detected: unconnected nets and pins, multiple use of circuit designators, improper use of pins, and incorrect signal name cross-reference. An online symbol library contains more than 1200 of the most frequently-used schematic symbols.

If a common database is used in a networking environment, management controls can be exercised so that user groups always work from the same design revision levels. This helps prevent errors and delays in PCB layout that may occur when engineering makes changes that are not passed all immediately to the design group. All users working on the same project network receive updated information relevant to that project simultaneously.

CAE/CAD/CAM trends

Most of today's newer workstations feature 32-bit microprocessors with at least 0.5 Mbytes of main (RAM) memory and 40 Mbytes or more of disk type mass storage. Workstations utilize bit-mapped high resolution displays that are available either in monochrome or increasingly in color. The advantage of using color in PCB designs is that it can be used to represent various routing layers, thereby aiding design conceptualization.

The UNIX operating system developed by AT&T's Bell Laboratories is being used in an increasing number of new CAE workstation designs. It offers: program development utilities, transportability, multitasking capability and virtual memory support. Program development utilities can often be combined to perform increasingly complex functions. Transportability simply refers to the fact that software programs written under UNIX can be run on all other machines that use the same UNIX operating system. Multitasking is the ability to run or execute several different programs or functions simultaneously. And virtual memory loads pages of information into physical memory as they are needed, thus allowing microprocessor-based workstations to run programs normally executed only on a mainframe.

Recently, a number of software programs have been written enabling the IBM PC and similar microcomputers to be used in CAE/CAD/CAM applications. As computing power increases and costs decrease, it is reasonable to expect this trend to accelerate the move to networked microcomputers – especially for the less computation-intensive functions.

Relationship between CAE/CAD/CAM and electronics assembly

Printed circuit board technology is moving from discrete wiring and wire harness; single-sided, two-layer designs; analog circuitry; discretes to cans; and low speed, low power. Its evolutionary direction is towards multilayer designs; digital circuitry; DIP and SO packaging; surface mount technology (SMT); and very high speed and power.

The implications for PCB design are clear. These factors will contribute to increased board fabrication complexity, finer lines, denser conductor layouts, more holes per square inch, and modified layouts for alternative assembly technologies such as SMT.

Until recently, the most profitable aspect of using CAE/CAD/CAM systems in the design and development of printed circuit boards has been in the generation of tapes for automatic drilling and component insertion. To a lesser extent, this same type of information has been used to assist in computer-aided test and inspection as well. Systems are also being developed that, in the future, may employ a postprocessor in the CAE/CAD/CAM system that will evaluate board thickness, trace spacings, and hole sizes plus other PC board parameters, determine initial settings for a wave solder machine, and transmit the settings to the machine via a data link.

But the influence even newer design approaches will have on the electronics assembly industry are critical. Robert Rowland of Sperry corp., speaking at the recent NEPCON West conference, said, "The electronic assembly industry is basically still in its infancy. Until the last four or five years electronic assembly, for the most part, was done by hand or with dedicated equipment such as DIP inserters and wave soldering machines. Not much thought was given to automation. The problem with many of the package styles used for the past decade arises from the fact that many of them were not designed for an automated production environment. When these packages were developed the prevailing method of production was by hand assembly. For the marketplace of a decade ago this method was adequate. To manufacture in the USA today a company must be able to compete in the world market, an environment in which most of the foreign competition enjoys an advantage in labor costs. The key to competing against this advantage is to automate the manufacturing environment."

"Surface-mount technology (SMT) will not only improve product functionality, size, and quality," he said, "but it will dramatically affect the methods used to manufacture PCBs. The reason for this? Surface-mount devices (SMDs) are designed to be used in an automated production environment. Thus the main obstacle to automation has been overcome. Many conventional components that are difficult or impossible to auto insert, such as transistors, are easy to auto place when using the surface-mount component."

Since most CAE/CAD/CAM systems originated before the development of SMT, they lack some of the software tools needed to help design and produce those assemblies. The majority of CAE/CAD/CAM systems for auto-routing and autoplacement, for example, are directed towards printed circuit board technology. The use of these autorouting and autoplacement methods for surface mount hybrids, for instance, generally yields good results since the technologies are similar in terms of board structure. The main difference is in the use of printed resistors versus discrete resistors. No conductor run passes underneath a printed resistor unless it is on another layer underneath the resistor. For chip and wire hybrids, however, most auto-routing and autoplacement systems are not very satisfactory, and serve only as a starting point for interactive layout. These software

systems generally have difficulty in handling wirebond jumpers on the same layer, a commonly used technique in chip and wire hybrid circuits.

A flexible grid system is required on CAE/CAD/CAM systems to accommodate the fine lines needed to take advantage of the higher packaging densities offered by SMT. These boards are designed using fine-line rules consisting of 0.008in. or smaller trace etch and 0.008in. or smaller minimum clearance between features. CAE/CAD/CAM systems also must deal with components mounted on both the top and bottom of the board, and the typically higher pin counts – not to mention the need to process larger amounts of data due to greater board density.

CAE/CAD/CAM systems require the flexibility to create new components and store them in a component library, and modify them as necessary. They must also be able to accommodate different pin-to-pin spacings, routings, grid choices, and special design considerations such as heat sink parts. Component placement of SMDs also becomes a three-dimensional problem rather than a simple two-dimensional auto insertion process. And automatic routing procedures must be developed to accommodate SMDs as opposed to current DIP-oriented board designs.

Implications for electronics assembly

The CAE/CAD/CAM tools designed to accomplish automated routing, drilling, component placement and testing of PCBs utiliz-

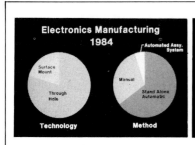
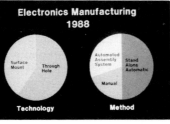
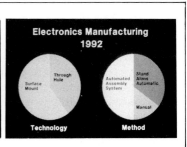

Universal Instruments projects that surface mount technology will become dominant by the early 1990s, and that the accompanying assembly methods will tend toward integrated, automated systems as opposed to standalone or manual assembly techniques

ing through hole technology are fairly well developed in theory, if not in practice. According to Jim Baudanza, electronics product specialist with Computervision, CAD data can generally be post processed to feed CAM hardware for automated component insertion, wire wrapping, routing, and so forth. There are also CAD interfaces to bed-of-nails testing equipment and similar devices. However, in most cases standalone automatic equipment is still being programmed individually through keyboard entry or similar means. Manual assembly still remains a large part of the assembly process; while fully automated assembly through CAE/CAD/CAM interface represents probably no more than 5% of current applications.

Baudanza sees this picture changing gradually as the technology and communications interfaces to various devices are further refined. He also noted that CAE/CAD/CAM software designed to drive SMT placement equipment is now coming on the market. This could help accelerate the move to fully automated assembly as well.

Universal Instruments forecasts that automated assembly will gradually increase until it represents about 50% of all assembly applications in 1992. Manual assembly, however, will account for only a slightly smaller percentage of applications than it does currently. Therefore, the biggest gains in automatic assembly are expected to come through linking the great number of standalone automated devices in use currently.

There are, for instance, three principal techniques for programming automated component insertion devices: walk-through teaching, keyboard terminal entry, and computer entry. In the teach mode, a programmer walks the machine through the required sequence of steps using a hand-held portable terminal or teach pendant. In direct entry, a keyboard terminal is used to manually program or enter the sequence of actions to take place in the assembly process, and to locate the coordinates where the components are to be placed. This can be a tedious, error-prone process. Remote computer entry allows an assembly program to be developed offline from the robot or pick-and-place machine and downloaded over a communications link.

The use of a CAE/CAD/CAM system to generate the component style and location data, and then transfer it directly to the pick-and-place machine is a special application of this latter mode. Similar to producing automatic insertion tapes from a CAD database, this procedure generally begins with the CAE/CAD/CAM vendor serving as the point where programs are developed to match the CAD data format with that required by the specific assembly machine.

In summary, as surface-mount technology, CAE/CAD/CAM techniques and related communications technologies develop and receive wider acceptance, they will be integrated to reinforce one another in the creation of the automated electronics assembly plant. Where it occurs, this synergistic effect could produce a marked improvement in the competitive capability of US electronics manufacturers.

CAD/CAM FOR PCB MANUFACTURING

To date, computer-aided manufacturing for printed circuit boards (CAM for PCBs) has primarily meant numerically controlled drilling and component insertion. Numerical control is a technique where a computer-based controller reads a program, which is a

First published in *Electronic Packaging & Production*, March 1985, under the title 'CAD interfaces with CAM for PCB manufacturing'.
Author: Robert L. Myers, Omnimation Consultants.
© 1985 Reed Publishing (USA) Inc.

sequential list of detailed codes, and translates the individual steps in the program into commands for a machine tool. When a PCB is designed on a computer-aided design system (CAD) a database exists which is a description of the board. CAD/CAM is having a computer extract information from the CAD database and generate the NC program.

The database in a CAD system is layered. This layering can be compared to the ability to overlay many sheets of very fine vellum with no stretch and exact registration, while being able to see the information on all the sheets. Graphic and alphanumeric information may be added, removed or changed on any layer individually or in any combination of layers.

One of the most significant benefits of CAD is accuracy. A computer can only understand and manipulate graphic information if the graphic information can be transformed into digital format (numbers). PCB CAD systems are based on an orthogonal X-Y grid. Locations can only be described by X and Y coordinates; therefore, they must be on grid.

A CAD system will contain in its memory a library of predefined symbols. CAD systems create drawings in much the same way as using stick-on decals. Instead of taking a decal from a sheet and positioning it on a layout, an operator tells the system which library component is wanted from the library, and uses an interactive graphics terminal to indicate its position and orientation.

There are three layers of a typical library component for a resistor with 0.5in. spacing. Layer three is the pad master and contains the pads, to scale and at the proper spacing. Layer four contains the component outline. As can be seen from the registration marks, the component outline is exactly registered over the pads. Layer five contains the drill holes. Different geometric shapes are used to denote different hole sizes, and the hole size for this component is represented by a square. The pads are defined to be on the same grid points as the drill holes.

When a design has been checked and released to manufacturing, all the coordinates in the CAD database will be exact. When extracting a photoplot program and a drill tape from the database, the exact same coordinates will be given to the photoplotter to flash a pad and a drilling machine to drill the hole. This is the most accurate input possible for these machines, and annular ring will be a function of the accuracy of the photoplotter and drill.

In a database created using library components with layering similar to that described, all the drill holes will be on layer six. A computer program called a postprocessor in

the CAD system can use layer six to produce an NC drill program.

The postprocess will sort the drill holes so that same size holes are drilled together. This eliminates unnecessary changing of the drill bits. The holes are then sorted in a serpentine pattern that will produce the fastest drilling sequence for the board.

Automatic component insertion

Once set up, fully automatic insertion equipment can insert components into a PCB with no operator intervention. Insertion machines consist of a fixture that can hold a PCB and be moved in two dimensions underneath a fixed insertion head. The insertion head holds a component and can move downward, inserting the component into the board. Some inserters can rotate the component 90° before insertion. DIPs require different insertion machines than do discrete components.

DIP inserters have long vertical tubes that

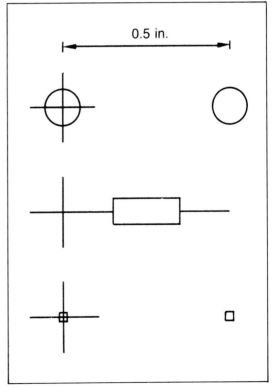

Three layers of a typical library component for a resistor with 0.5in spacing

hold many of the same type DIP. The DIPs are gravity-fed onto a conveyor or similar mechanism that moves them to an insertion head where the leads are bent to fit the standard pattern of holes in the PCB. The insertion head then lowers and inserts the DIP into the board. Underneath the board, a 'cut and clinch' mechanism trims and bends the DIP leads. The inserter program must tell the controller which DIP to have in the head when the board is positioned to insert that component. Some inserters want to know the location of the physical center of a component, and others use pin 1. If the inserter has the ability to rotate a DIP 90°, the inserter must know whether or not to rotate the insertion head prior to moving downward.

Axial component inserters use and the same type of fixture moving in X and Y under a fixed insertion head, but have different requirements. When inserting an axial component, the leads are trimmed and bent to the proper spacing before insertion, and a cut-and-clinch mechanism is under the PCB. Components are arranged in a continuous array by long strips of tape attached to their leads and wound on reels. Some discrete inserters require that a reel of components be mounted on the inserter which has all of the components to be inserted in the correct sequence. Components are fed directly from this reel to the insertion head. This requires an intermediate step of taking components from reels with all the same component and creating a new reel with components in the insertion sequence.

Insertion machines that insert radical lead components are also available. Regardless of what they insert, the programming requirements can be generally described as having

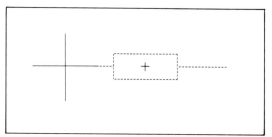

An insertion offset is used for locating a component which is actually on a different layer

the right component in the insertion head when the board is correctly positioned underneath the head.

Manually programming insertion equipment can be accomplished by typing in all of the component location coordinates in the correct sequence, and associating a component with each location. This is a tedious, error-prone process. Another approach is to mount a bare board in the insertion fixture and move it under the head in the desired insertion sequence using manual controls. A fixed position light shines on the table, and when the operator optically determines the PCB is in the correct position, a button is pressed. The coordinates are recorded for later use, and the operator positions the table to insert the next component. A more convenient and accurate method is an off-line programming console which allows an operator to locate components with a probe or digitizer.

A postprocessor can extract information from a CAD database and create an insertion program. When creating a CAD library, a layer is reserved for insertion offsets. In the example, layer eight is to be used for insertion offsets. A library component can contain whatever offset with respect to the origin of the library component is required. The second figure shows what would be on layer eight of the resistor library component. The middle of the component would be located by a point. The outline would not be on layer eight and is used to show the point is in the middle of the component.

When creating an insertion program, the CAD computer can find the location of the component and add its offset, producing the coordinates of the component's center. If the part number had been added, it could be extracted and associated with the coordinates. If the CAD computer were told which part numbers were in which component holder location, it could create the entire program.

Manually programming insertion equipment to insert components into boards made with hand-taped artwork and bomb-sighted drill tapes is an inexact process. The artwork has a random pattern of errors. The drill tape has a random pattern of errors. The program

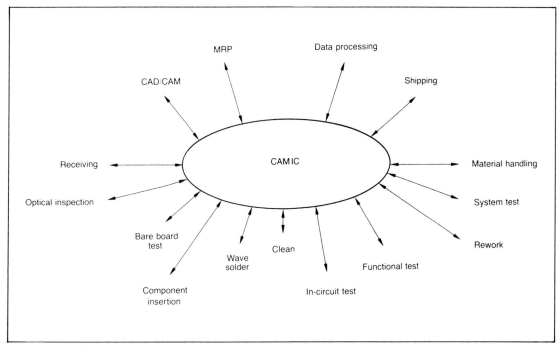

The author's view of a PCB automated factory depicts CAM at the core of production

generated by moving a bare board under the insertion head and optically selecting a location also contains random errors. These three random pattern of errors generally combine to make insertion programs very inefficient. When in production, the insertion machine often must be stopped to realign a board in its fixture. Furthermore, when an inserter is being used to program itself, it is not in production.

CAD-generated insertion programs, like CAD-generated drill tapes, are accurate. When the controller reads a command to insert a component, the location coordinates are exact. The bare PCBs will be more accurate because they were made with photo-plotted artwork and a CAD-generated drill tape. When the inserter is told to insert a component, the holes will be in the right place. This allows the inserter to run more efficiently and more often, since it is not being used to program itself.

Design for insertion

Designing for manufacturability should be a concern anytime, but the advent of CAM makes it a major consideration. Some of the issues are:

- Reducing the number of different component types can have a significant impact. If the design calls for 51 different DIPs and the DIP inserter can handle 50 at one time, the last DIP will require changing DIP sticks and another run, or more likely manual insertion.
- Components should be oriented at only 0° and 90°.
- Since most insertion equipment requires a different fixture be made for each board outline handled, standardizing the board outline or reducing the number of board outlines can lower fixturing time and costs.
- When inserting a component into a pattern of holes, the board must be positioned underneath the insertion head. There is a positional tolerance stack-up due to fixture accuracy, machine X-Y positioning accuracy, positional and dimensional accuracy of the drilled hole, accuracy of the tooling hole and variation in lead wire diameter. Holes must be large enough for automatic

insertion, but not large enough to cause soldering problems.

- DIP, axial and radial inserters and must have room to grip components and therefore require a minimum spacing between components. The cut-and-clinch mechanism underneath the board also has minimum spacing requirements.

Many other aspects of PCB layout that are of minor interest without CAM can become crucial with CAM. Having manufacturing engineering sign off on PCB layouts before they are released for manufacturing could allow problems to be found and fixed before a board is in production.

Computer usage is increasing in the PCB manufacturing process, and computerized wave soldering is an example. Prior to computerized wave soldering, the control settings for preheat temperature, conveyor speed and other parameters were estimated and entered into the machine. The conditions inside the soldering machine may have fluctuated, but the control settings remained constant. With computerized wave soldering, a microcomputer monitors temperature, speed, flux and other variables and continually adjusts the control settings to ensure the best soldering possible.

The first machines require control settings to be input by an operator, and the microcomputer controls soldering. In the future, a postprocessor in a CAD system will evaluate board thickness, trace spacings and hole sizes plus other parameters of a PCB, determine the initial settings for the wave-soldering machine and transmit the settings to the machine via data link. As a batch of PCB were being soldered, values of the key parameters of the soldering process could be sent to a quality control computer via data link. That computer will store the batch number and the operating parameters. If PCBs failed in the field and soldering were suspected as a cause, the data could be retrieved and analyzed.

DESIGNING PCBs FOR SURFACE-MOUNT ASSEMBLIES

The printed circuit board has been undergoing a gradual evolution over the past several years. With the introduction and acceptance of surface-mount technology (SMT), surface-mount devices (SMDs) now coexist with, and in some cases replace, dual in-line packages (DIPs) on circuit assemblies.

The origins of systems commercially available today for computer-aided design of PCBs, however, came before the advent of surface-mount technology. It is imperative, therefore, that those responsible for the automation of PCB design examine the challenge SMDs present to CAD systems, and how CAD vendors are changing their software to accommodate them.

First published in *Electronic Packaging & Production*, March 1985, under the title 'SMT challenges CAD-system speed and structure'.
Author: Mike Marsh, Racal Redac.
© 1985 Reed Publishing (USA) Inc.

SMDs characterized

Surface-mount technology is a packaging technique for integrated and discrete circuit that eliminates the need for larger plastic or ceramic DIP packages. Unlike DIPs, which are rectangular with leads along the long sides, SMDs may be square and have leads on all sides. Lead-to-lead spacing, too, is usually much smaller for SMDs than for DIPs.

These factors allow SMDs to be much smaller for a given number of pins than the DIP equivalent. An SMD may occupy less than 25% of the physical area that a DIP component with the same number of leads occupies.

In addition to saving space on a PCB, surface mounting does not require drilling of holes into the board for mounting components. SMDs are attached to one or both surfaces of the board, rather than inserted into holes from one side.

DIPs, however, have been the predominant packaging approach since integrated circuits were introduced. One of their advantages over SMDs, therefore, is that many current design and drafting skills, including CAD-software expertise, specifically center on creating DIP-type PCBs.

SMD challenges

Every CAD system is a compromise between the goals of maximum performance and maximum generality, with generality usually sacrificed for performance. Put another way, a PCB CAD system cannot be all things to all people; it usually is created to do a specific function very well and other functions less well.

Because most CAD systems available today originated before SMT and were sold into a market dominated by DIP-component PCBs, they have problems in designing SMD printed circuit boards. They lack some of the software tools required to help design these assemblies.

Sometimes the problems are only related to performance. For example, it may take 10–20% longer to design an SMD-type board

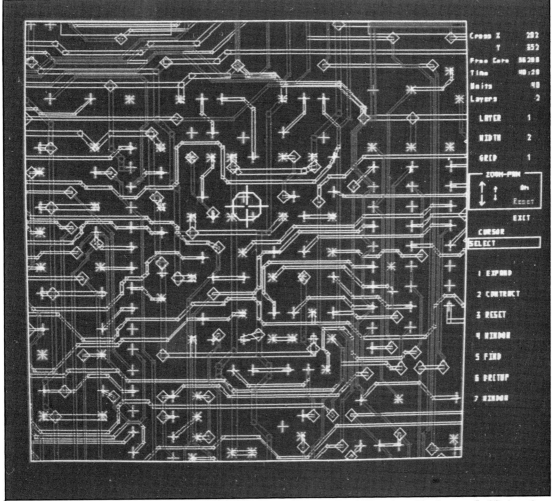

An SMT PCB's routing is shown on a CAD-system display with the two colors representing traces on different layers. Note the greater flexibility in routing the fine-line traces and shorter connections compared to the DIP routing in the last figure in this paper

than its equivalent DIP type because the automatic placement and routing functions work less well and require more designer intervention.

In other cases, however, the problem is fundamental in the CAD-system structure; it is simply not suited for SMT design.

The CAD system functions that are unique to SMDs can be divided into five areas: system database structure, system flexibility, component placement rules, autorouting, and manufacturing tools support (or CAM).

Database structure

The majority of PCB CAD systems have built into them assumptions about the data they will process and the way it will be processed. These assumptions are made to streamline the system for performance and a great many make excellent sense. For example, no commercially available PCB CAD system is able to design boards larger than, say, 48 × 48in., as very few PCBs are this large. Also, few CAD systems can design boards with more than 1000 equivalent ICs since this does not represent a major market segment of potential users.

However, important database requirements that are not found in DIP designs are significant in designs with SMDs.

An important example is the need for a flexible grid system or unit of distance. The use of fine lines on boards is necessary to take advantage of the potentially greater packing densities of SMDs. SMT boards are almost always designed using fine-line rules – 0.008in. or smaller trace etch and 0.008in. or smaller minimum clearance between features.

To be effective for SMT design, a CAD system must support a 0.001in. grid resolution. This will allow any possible grid to be used. The most common grid size used for DIP-type designs is 0.025in. (although even this is changing as well in an effort to maximize packing densities for DIPs).

SMDs can be placed on both sides of the PCB, unlike DIPs which are always on a single side. It is thus required that the CAD system support components on both the top and bottom of the board. Usually this means that the system should be able to mirror the

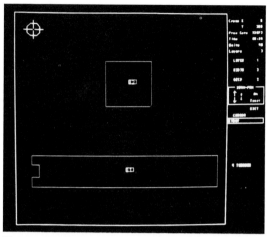

The component library of a CAD system shows an SMT chip carrier's smaller size and distribution of leads around its four sides compared to an identical circuit in a DIP

component shape on the graphic screen as well as display components on the different sides separately.

SMT designs normally contain many more equivalent ICs than DIP circuit designs. To cope with this greater density, the CAD system needs to be able to process larger amounts of data.

DIP-component lead pads are usually circular, square or oval. SMD pads are more complex and can vary from one component to another. Often a plated through-hole is included as part of the board's component lead pattern to allow electrical access to other routing layers. A CAD system must be able to easily support such complex pad shapes and patterns.

Also, SMDs typically have higher pin counts than DIP devices. The CAD system must be able to accommodate up to 250 pins on a single component.

Most of these requirements are unique to SMT and were not considered necessary in the design of most existing PCB CAD systems. If a CAD system is not designed to handle these database requirements, it probably is not adequate for SMT design.

System flexibility

One of the major traits of SMT is that it is rapidly changing. There are few standards on issues such as component size, pin count, and lead-to-lead spacing. Manufacturing skills and equipment are also changing rapidly, causing the rules for designing boards with SMDs to change as well.

To accommodate this rapidly changing environment, CAD systems must be flexible. There are different aspects of flexibility. First, creation of new component shapes must be easy, as the need for them arises constantly. A system must have the capabilities for creating a library of components, and quickly modifying shapes.

Second, the CAD system must be flexible in the routing grid choices available to the user. It should be possible to change the grid, interactively if necessary, in local areas of the PCB to accommodate SMDs with different pin-to-pin spacings.

Third, design of SMT assemblies often requires special design considerations such as heat-sink parts or specific routing rules. To allow the designer to implement these rules, the CAD-system's user interface needs to be responsive, powerful, and provide quick graphic feedback on the designer's action.

Flexibility will remain a key element in a PCB CAD system for SMT design. As the technology evolves, the system must be able to continually accommodate new parts and new design rules.

Component placement

Automatic component placement is a neglected area in CAD software because most user interest has been focused on the efficiency of the system autorouters.

Virtually all automatic component placement software in commercial CAD systems is designed only to help design PCBs with a large population of DIP components. With only a couple of exceptions, such as the Racal-Redac Maxi, available systems lack provisions even for placement of analog components such as resistors, capacitors, and diodes.

Very little work has been done on the specific placement requirements for SMDs. Since SMDs can be attached on both sides of the board, the process is more than a simple two-dimensional problem. None of the currently available placement software is able to

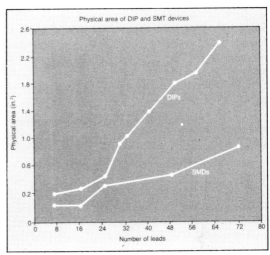

Larger amounts of data must be processed by a CAD system to handle the greater component density of small SMDs. An SMD may ocupy 25% of the area occupied by a DIP with the same number of leads

handle this new situation. Most of these placement programs are variations on algorithms that have been proven to work in only two dimensions.

As a result, for all PCB designs that are done using two sides for component positions, placement is done interactively by the designer, not automatically by the CAD system. Until algorithms are designed that work in a three-dimensional manner and are tested for efficiency, this problem will remain.

Thermal considerations play a greater part in the design of SMT assemblies. Typically, there is a much greater concentration of active parts placed in the given area and the devices are mounted directly on the surface of the circuit board.

There are means to help dissipate the heat, but the designer has to assure that the thermal problems are considered. This is an area where automatic analysis would significantly help designers and they can expect to see tools soon. Commercial CAD systems, at present, make no provision for integrated thermal analysis as part of the automatic component placement process.

Autorouting

SMT presents its greatest challenge to CAD systems in automatic routing. All commercial routers have been developed for DIP-oriented designs, and all perform rather poorly with SMDs. To see why, one needs to examine first how a DIP-type PCB looks to an automatic router.

Normally DIPs are oriented on a matrix, with rows and columns of components across the circuit board. The DIPs are almost always oriented in the same direction, except in a localized part of the board. Between the DIP rows and columns are channels, to allow routes.

In the major direction, parallel to the longest dimension of the components, DIPs offer little problem as obstacles to the router. In the minor direction, the components are semi-obstacles as an etched trace can go between the leads of a DIP.

For this type of PCB, a channel router gives optimum results in a reasonable amount of time. A channel router works in two layers of the circuit board, with a trace in one direction on one layer of the board, and a trace in the transverse direction on the second layer. There is a plated through-hole at the point where the direction of an interconnection is changed. Channel routers take advantage of the regular structure, give predictable results and are efficient in terms of CPU time.

Most CAD systems also provide a maze (or Lee's) router to complete the interconnections that are not handled by the channel router. A maze router has the property of guaranteeing a solution. It finds an etch path for a connection if a path is possible, but at a cost in CPU time and etch-path length. For this reason, it is almost always reserved as a cleanup autorouter.

SMT autorouting

A channel router makes several assumptions. It assumes routes are on convenient grids for its internal routing rules, usually 0.025in. or 0.050in. Also assumed is that an etch can exit from a component lead on any layer of the PCB. For SMDs, these assumptions do not hold and thus lead to poor results from a channel router.

SMDs routed using fine-line design rules make 0.025 and 0.050in. grids insufficiently precise. Channel routers that blindly follow an 0.025in. rule almost always add plated through-holes that cause short circuits with etched routes.

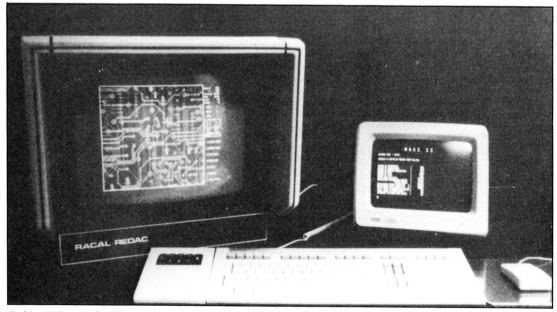

Evolving SMT requires flexibility in a CAD system to accommodate new parts and new design rules. Shown is a Racal-Redac MAXI workstation configuration

In addition, SMDs, particularly the high pin-count chip carriers, have leads on all four sides (unlike DIPs with leads on two sides only). From an autorouting perspective, a DIP represents a partial obstacle in one direction only, while a SMD is often a complete obstacle in all directions.

One of the major areas of difficulty for channel routers is the edge connector on a board. Because the edge finger pads are on both sides on the circuit board, they physically must be routed on two layers. But they should be routed on a single layer to satisfy the channel router's rule of one layer for the horizontal etch direction, and a second layer for vertical etch. If the edge connector is at the top or bottom of the circuit board, the etch should be on the vertical routing layer. If it is on the left or right sides, the etch should be on the horizontal routing layer.

This problem invariably leads to congestion, and so less-than-optimum autorouting results around edge connectors. While not desirable, this is acceptable because edge connectors are limited in number, and their location is confined to the periphery of the routing area.

A similar problem exists for SMDs, though worse in its impact because they are placed throughout the routing area and are numerous. All etches from a SMD must exit on a single layer, even though the preferred

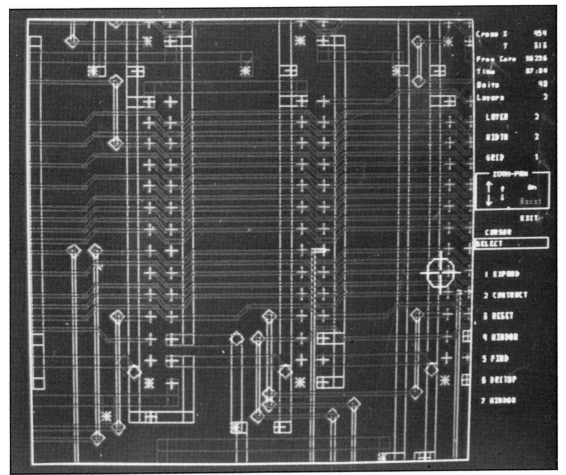

A DIP-type board layout is shown in a CAD-system's display with the two colors representing traces on different layers. Note the longer connections compared to those with SMT shown in the first figure

direction could well be on another layer. As a result, channel routers are usually poor when a large number of SMDs populate the board. Use of a maze router helps avoid this problem as well as the problem of assumed routing grids.

Many companies, among them Racal-Redac, are working on modifying the maze router algorithm to tailor it for SMDs. This usually involves providing the router some degree of heuristic capability for working in the area immediately around a SMD and modifying the 'costing functions' that govern the trade-offs the router makes (such as between a plated through-hole versus etch on a layer in the non-preferred direction).

These costing functions are usually quite complex, and sometimes need to be tailored to the SMD type and component placement in a specific PCB layout. This is impractical for the average user. Thus, CAD vendors are working to develop 'intelligence' in their autorouters. Intelligence means the router's ability to modify its run results if they are unsatisfactory, thereby improving the autorouting completion rate.

These intelligent router algorithms are known as 'rip-up-and-retry' routers and as a class represent the best opportunity for coping with the problems posed by SMT designs. These routers, which also permit designer assistance at critical junctions in the process, normally require several passes, or iterations, through the layout database before completion. At Racal-Redac, research into advanced router techniques has allowed close to 100% completion using rip-up routers.

Buying SMT CAD

It is difficult to know which CAD system is best suited for automating SMT design. A particular circuit board might be designed well with System A, and poorly with System B. A second board, with slightly different design rules, could give the opposite results.

The prospective buyer of a CAD system should look for flexibility in the system above all, both in the ability to interact with the design and the generality of the database. Furthermore, the buyer should investigate what SMT-specific software is being developed. The buyer should not expect to find the state-of-the-art in CAD for SMD work to be as advanced as for DIP-type circuitry. A compromise in performance is necessary today. In one to three years, however, one should look for systems with many features that are specifically SMT oriented.

CAD IN HYBRID PRODUCTION

CAD systems may be used for a variety of tasks involved in the design and manufacturing of hybrid microcircuits, including creation of the layout, analysis of the circuit, creation of the artwork, and material usage. Quality may be improved, while the overall cost of the design is reduced. Virtually every company which manufactures hybrid circuits uses CAD or CAE or CAM to one degree or another, either on internal systems or by subcontracting to an outside service.

General capabilities

The autorouting and autoplacement systems used in most CAD systems are directed toward printed circuit board technology, and the degree to which they can be utilized in the design of hybrid microcircuits depends on the packaging scheme incorporated. For surface-mount hybrids, autoplacement and autorouting methods used for printed circuit boards yield good results since the technologies are quite similar in terms of the board structure. The major difference is in the use of printed resistors versus discrete resistors. No conductor run may pass underneath a printed resistor unless it is on another layer underneath the resistor.

First published in *Electronic Packaging & Production*, March 1985, under the title 'CAD systems pervade every area of hybrids production'.
Authors: Jerry E. Sergent, Sergent & Sergent, Howell Chiles, Litton Data Systems, and Peter P. Molkenthin, Honeywell Avionics.
© 1985 Reed Publishing (USA) Inc.

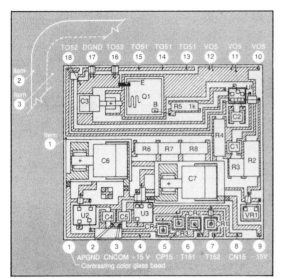

Master layout of a hybrid circuit created on a CAD system (Courtesy of Litton Data Systems)

At the other extreme, for chip-and-wire hybrids, autoplacement and autorouting systems are not as successful and generally serve only as a starting point for an interactive layout. Most software systems have difficulty in handling wirebond jumpers on the same layer, a common technique used in chip-and-wire hybrid circuits. In addition, many constraints must be placed on the layout parameters, increasing the time required and resulting in a less optimum layout. For example, a CAD system will generally use the maximum number of layers specified, while, in many cases, this number could be reduced if the layout were done by hand or interactively.

One of the most useful features available on some systems is the ability to perform a point-to-point check of the layout and compare it to the circuit schematic. Those who have performed a point-to-point check on a multilayer thick-film circuit can appreciate the amount of effort required and the potential for error. Further, certain specified lines, such as power and ground, may be highlighted for easier visual checking or for trouble-shooting. This facet must be considered when entering the raw data since each line on the hybrid layout must be identified in the same manner as on the circuit schematic.

All hybrid circuits are designed according to a set of design guidelines which specify dimensional constraints for such parameters as conductor line widths, wire bond length, minimum resistor area, via size, component interference, or via overlap from one layer to another. Some systems have the capability of comparing the layout to the design guidelines, permitting a more producible circuit with a higher probability of success in initial production.

The design of both thick- and thin-film resistors is a complex procedure. Termination effects and screen-printing variables complicate the design of thick-film resistors, while complex configurations create difficulties in calculating the resistance of thin-film resistors. For certain systems, software which will permit interactive resistor design to a high degree of accuracy is available. Once thick-film resistor materials are characterized for termination and geometric effects, the raw data is entered into the system, which calculates an empirical equation for each of the design curves.

The designer inputs the power and tolerance requirements to the system which then calculates the dimensions of the resistors. Geometric effect, such as the resistance of corners, can be implemented to determine the square count of complex thin-film resistor designs.

The newer systems are becoming more and more user-friendly, providing more on-screen help and minimizing the amount of operator training and experience necessary to obtain full utilization of the machine capability. Input interfaces range from direct digitizing from a keyboard, to a light pen, to a digitizing pad, to a mouse. Output in interfaces also vary widely, and must be an important consideration in selecting a machine. Some systems may directly interface with a mask maker and provide complete phototools for making screens or etching, while others provide only a drawing, leaving considerable work to be performed prior to obtaining phototools. A common technique is to cut and peel rubylith. While rubylith may be cut by the machine, with a slight modification of the plotter, it still must be peeled by hand, a very time-consuming task when many layers are involved. The driving factor is the initial cost of the equipment which must be weighed against the expected use.

A further consideration is the degree to which the system can be interfaced with existing mainframe equipment. With proper equipment and/or software selection, existing equipment may be able to perform many of the desired tasks, such as circuit analysis and logic simulation, which would add considerable cost if bought as a package in the CAD system.

Complete interfacing with hybrid assembly equipment, such as automatic wire bonders and pick-and-place systems, is still somewhat in the future, although some success has been realized in programming pick-and-place systems used in the surface-mount technology. While the technology is available, the demand has not been such as to justify the development and implementation cost.

Much useful manufacturing information, such as wire usage, paste usage, and other material requirements, can be obtained from CAD systems. In addition, parts lists and visual aids for assembly, test, and troubleshooting can be created. Visual aids can be created for less cost and time than a photograph, and process or layout changes may be implemented instantaneously.

Circuit simulation and modeling is important in an age where new products are introduced almost daily. In an effort to minimize the prototype development stage in new products, hybrid designers need circuit simulators such that the circuit design is more interactive, particularly for chip-and-wire designs where the active devices may have a wider range or parameter variation. Software packages work effectively for designs that are not extremely complex. For those whose complexity prohibits such an approach, other solutions must be sought.

One solution that works well for large digital circuits is the so-called simulation engine which can achieve over 10×10^8 gate evaluations per second. Circuits requiring hours to analyze with other methods can be analyzed by the simulation engine in a matter of minutes, allowing more time for redesign

and, in some cases, eliminating the prototype phase altogether. The logic engine is a specially designed piece of programmable hardware that will interface with a host computer and which can run a logic simulation in real time. The user need only download the required information from the host computer into the engine, let it do the simulation, and wait for the results to be returned to the host on completion. Most logic engines are limited in the types of circuits they can simulate because of the complexity in designing a universal engine.

For analog circuits, there are software packages, such as SPICE, which run on many popular minicomputers and mainframes. For personal computers, there are packages, such as Micro-CAP, which run on the Apple or IBM PC. One package particularly helpful in new products is the Analog Workbench, created by Analog Design Tools Inc., a new system specifically created to aid in the analog circuit design and analysis process.

This system does not require an expensive mainframe, but has its own computing power and mass storage that allows hundreds of designs to be stored for future use. The system uses the window approach, letting the designer create the schematic in one window

A computer is used to refine the design of a network layout (shown on the CRT display) in Allen-Bradley's computer-aided design department in the Greensboro, NC, microelectronics plant

and design the test in another. The test may then be executed and the analysis results will appear on the screen. If a change is required, it can be implemented and the analysis run again as often as needed. The interactive nature of this system decreases the time required for the design phase of a product, with greater assurance that it will work the first time.

Available systems

There are a variety of systems available at prices ranging from one thousand dollars to over half a million dollars. Four typical systems are described here: a low-cost system, an advanced general-purpose system, a more specialized but still general system, and a system designed exclusively for hybrid layouts.

Some companies market software/hardware packages which interface with personal computers of the type found in many laboratories. These systems are inexpensive, but are time-consuming to use and generally provide only simple drawings. Others, such as the system marketed by FutureNet for the IBM PC/XT family, can be used to draw schematics and can interface with other equipment to perform such functions as circuit analysis.

One of the most widely used general-purpose systems is the Computervision system, which has a number of software/hardware options. CADDS-II is a VLSIC design package which can also be used to design hybrid circuits. Some features as CADDS-II include high resolution, up to 1in., and design rules checking. CADDS-II may directly interface with a maskmaker, but some data manipulation, with the utilization of CADDS-III, is basically for designing printed circuit boards and has the autoroute and autoplacement capability.

The package may also be used for designing hybrid circuits, but may not directly interface with a maskmaker for such applications as a thin-film resistor array or master slice. CADDS-IV is for designing mechanical systems and may also be used for designing hybrid circuits. CADDS-IV suffers from the same limitation as CADDS-III in being unable

to directly interface with a maskmaker. The computervision system may interface with a mainframe computer for such applications as autorouting and circuit simulation.

A system directed more toward hybrid design is the Via SystemNode 200, which also has the capability of thin-film and VLSIC design. This system has a schematic capture tool, which enables point-to-point checking, circuit analysis, and logic analysis to be performed. Analysis programs such as HSPICE and CADAT are available for this system. It also has the capability of using Ethernet to interface with other computer systems or peripheral devices. The SystemNode 200 may drive either a photoplotter or maskmaker through the use of a tape or directly. Autoroute and autoplace are not yet available for hybrids but may be available at a later time.

One of the most advanced software packages is the HCAD package, marketed by Tektronix. This system, designed specifically for use in hybrid microelectronics design, interfaces with a DEC VAX 11-780 mainframe computer, which in turn interfaces with a Cyber mainframe for the more complex analysis programs. HCAD is an interactive system with no autorouting or autoplacement capability, but is capable of designing both thick- and thin-film resistors, capacitors and inductors. In addition, stray capacitance calculations and trimming simulations may be performed.

One of the most innovative aspects of HCAD is the capability of performing a three-dimensional thermal analysis on the finished layout. The temperature rise above ambient of both active and passive devices may be determined and compared to established guidelines. The output may be stored on magnetic tape for creation of artwork.

Future developments

Two recently announced items relating to CAD/CAM/CAE of importance to the hybrid microelectronics industry were the formation of a committee to discuss standards for user interface in CAD systems and the development of a chip designed to speed up graphic processing by a considerable amount.

In 1984, the American National Standards Institute (ANSI) and the Association for Computing Machinery's Special Interest Group on Computer Graphics (ACM/SIGGRAPH) carried on discussions on the subject of creating a standard for the Graphical Kernel System (GKS). The GKS is the computer code that forms the user interface in CAD systems and comprises the set of functions the system will perform in the creation of graphical entities. In February of 1984, a copy of the first draft of the proposed standards was distributed to members of ACM/SIGGRAPH for reading and approval.

Representatives of the CAD manufacturing industry were also given the opportunity for input in the process so that currently implemented commands would have a chance to appear in the proposed standard. The standard is still in discussion at this time.

The GKS standard will help to establish uniformity in CAD systems from one manufacturer to another to produce greater communications between equipment from various manufacturers. This will accelerate the development of totally automated manufacturing facilities controlled from a central system. The GKS system, when adopted, should help prorate the cost of a CAD system somewhat by permitting a wider usage.

Recently, NCR Microelectronics announced the development of a Systolic Array processor chip known as the Geometric Arithmetic Parallel Processor (GAPP). The GAPP was developed by NCR in conjunction with Martin Marietta Aerospace for use in real-time target recognition. The GAPP contains 72 single-bit processors with 128 bits of memory behind each processor, all contained on a single integrated circuit chip. The processor is designed to be used in an array structure where each chip need only communicate with its nearest neighbors to the north, south, east and west.

There is no limit to the number of chips that can be used in a system, and as the number of chips increases, the speed of operation increases linearly. When operating with a 10MHz clock, a 48 × 48 cell systolic processor, containing 32 chips, can grab a 48-bit word every 100ns. The array thus has a bandwidth of 480 Mbit/sec.

The implication to the CAD industry is one of reduced turnaround on new designs in the form of faster and better-handled graphics capabilities for design systems. The system could, if desired, have a processor dedicated to every pixel in the graphical display. This system would allow each pixel to be operated on simultaneously. Most of the work could be done in the design station with little interaction with the host computer until the completed design is to be stored.

The use of CAD/CAE/CAM techniques to design and manufacture hybrid circuits is a very useful tool. With proper selection and utilization of the available equipment and software, the design cycle for a hybrid microcircuit can be reduced by as much as 50%, with an additional improvement in accuracy and producibility.

The selection of equipment must be done very carefully with the end application and expected use factor in mind. While the equipment described here is typical, it is by no means comprehensive and a thorough survey of the available equipment, including a hands-on demonstration, is well worth the time. Otherwise, it is possible to spend a great deal of money and still not have the desired capability.

Companies which specialize in artwork generation are located all over the USA. Layouts are performed, or an existing layout or tape can be converted to artwork. This can be a less expensive alternative for the company whose need for a CAD system is minimal but still essential.

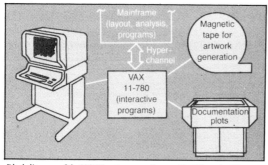

Block diagram of the HCAB system manufactured by Tektronix. The major output of HCAD is pattern data used to generate fabrication artwork and documentation plots for fabrication/process control

DISCRETE WIRED CIRCUIT BOARD PRODUCTION

Today's complex circuit board designs demand a flexible system packaging technology that allows fast changes, is highly reliable and readily accommodates high-density interconnects and high-speed logic signals. A computer-aided design system is also needed to fully integrate the design, documentation and manufacturing processes and to shorten the time from concept to finished product.

The finished circuit board design represents the designer's efforts to reconcile different mechanical and electrical specifications with required interconnection and component densities. Signal timing, critical noise levels, component placement and electrical constraints peculiar to specific logic types are all problems that tax the designer's ability to quickly and easily create the right circuit board for each application. These problems can be minimized if the right design tools are used.

Two technologies developed to solve these problems are now working together to offer designers an integrated system that can bring high-density boards, averaging 500 ICs or more, from schematic capture to production in as little as one week.

The Multiwire Division of Kollmorgen Corp. (Hicksville, NY), which designs and manufactures discrete wired circuit boards that use insulated wires instead of etched traces for signal interconnections, has found that Intergraph Corp.'s hardware, used in combination with Electronics Design System (EDS) software, improves turnaround time for circuit board design and production cycles as much as 10 to one.

Multiwire technology not only lends itself to both large and small designs, but also provides excellent signal quality, fast turnaround, density approaching 100%, and fast and easy modification and revision. EDS has reduced design turnaround by integrating the schematic and layout data bases and by minimizing, if not completely eliminating, the laborious task of routing circuit boards by hand.

Permanent libraries of component specifications help to reduce clerical errors in both design input and documentation, while the system files for each design contain all data necessary to automatically generate bills of materials, photo artwork, NC drill data, NC wiring data and silkscreen leg-ends.

Multiwire technology

Multiwire is a discrete wired circuit board. Like conventional multilayer circuit boards, it employs etched foil layers to handle power and ground interconnections. However, Multiwire replaces etched foil for signal conductors with computer-generated patterns of polyimide-insulated wire, laid down by CNC wiring machines to form permanent interconnections.

Because the insulated wires can be crossed and placed very close together, Multiwire boards routinely accommodate component densities of 2.0 ICs per in.^{2} and greater. These boards offer a more reliable alternative to multilayer and wire wrapping in applications that require tight tolerance, controlled impedance transmission lines and high-speed ECL circuits.

Multiwire's crossover capability, however, rules out most CAD vendors who design systems for conventional printed circuit boards based on the tenet that two signals cannot cross each other. While these CAD systems could be used to prepare data to be submitted to Multiwire for processing, they could not interact with Multiwire's discrete wired data base. Intergraph Corp. (Huntsville, AL), which has committed its equipment to run on Multiwire boards, specifically addresses the Multiwire CAD problem.

The CAD system

To accomplish the entire design process using a common database, Intergraph developed the Electronics Design System (EDS), used

First published in *Electronic Packaging & Production*, July 1985, under the title 'CAD accelerates circuit board production'.
Author: Tony Filippone, Kollmorgen Corp.

extensively for all in-house circuit board design. This schematic-driven, computer-aided design and manufacturing package includes features to route and check very large-scale system designs and minimizes the possibility for clerical errors through the use of on-line specification barriers. In addition, EDS documents schematics, provides bills of materials and other reports, and generates postprocessing output for Multiwire's computer-driven manufacturing process.

EDS is fully compatible with the standard VAX/VMS 32-bit operating system from Digital Equipment Corp. It allows design engineers to select from both interactive and automatic procedures for packaging and placement of components, and engage a variety of route methods for printed circuitry and Multiwire technology.

Menus allow access to a full repertoire of preprogrammed electronics design functions and a complete set of basic drawing components. To suit individual requirements, the designer can build additional element menus and use any three simultaneously with the main EDS menu.

Multiwire design technology, which complements the Intergraph EDS software, per-

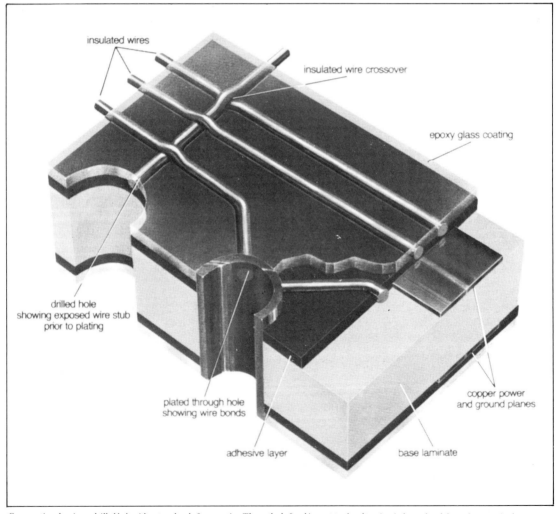

insulated wires

insulated wire crossover

epoxy glass coating

drilled hole
showing exposed wire stub
prior to plating

plated through hole
showing wire bonds

copper power
and ground planes

adhesive layer

base laminate

Cross-section showing a drilled hole with exposed end of severe wire. The method of making copper bond to wires in later plated through process is also shown

mits the designer to assemble and format signal lists and component holes data, and route high-density interconnect circuit designs.

The output data generates graphic plots of the routed conductors for documentation and checking, and magnetic tapes or floppy disks for NC wiring, testing and fabrication of a Multiwire circuit board.

CAD enhancements

In a typical design cycle, the schematic design process generally consumes the largest portion of time, depending upon the complexity of the board. In this initial phase, the designer works interactively with the Intergraph system to develop a detailed logic schematic, and performs traditional tasks, such as establishing general system architecture and selecting the components required to accomplish a specific function. All logical interconnect information can be derived from the completed schematic.

During schematic creation, the designer uses the system cursor to access and insert schematic symbols from the on-line symbol library. Logic symbols are placed in position, and the connections are created to link them into functional networks. As each logical connection is made, the design system keeps track of the pins that are being connected. Signal names are assigned to each connection so the designer will be able to link the

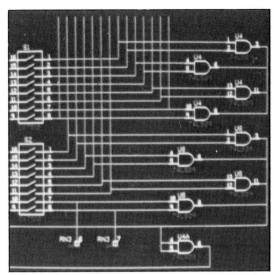

The designer works interactively with the CAD system to place logic symbols on a schematic

different components of a functional network.

While the schematic is being entered, the design system is establishing a database that will be used to create the from-to and parts lists. Multiwire design centers work from the same input material provided for wire wrapping or multilayer designs: schematics, component lists and net lists. Boards can be designed from a logic diagram or from-to list, with board geometry and special parameters described. The Multiwire design system will accommodate most types of interconnect formats.

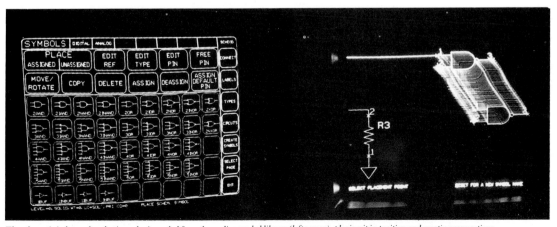

The schematic is drawn by selecting a logic symbol from the on-line symbol library (left screen), placing it in position and creating connections

Component placement

Once the schematic net list has been generated, components that are represented by the logic gates must be placed on the surface of the circuit board. Prior to component placement, the Multiwire design process is very similar to multilayer. While poor component placement will not degrade its design as much as it would conventional etched circuitry, Multiwire's tolerance to handle placement is enhanced by the design system's manual, semiautomatic and fully automatic methods for component placement.

The manual method allows the designer to place the components in any location or order. Semiautomatic placement tells the designer the mathematically correct placement, showing interconnects to components that have already been placed on the board. The fully automatic method is driven by user-defined parameters; the designer tells the system how to perform the placement, which it then does on its own.

Component placement on the circuit board takes less than one week for a 500-IC board. Using the CAD system, not only is the location of every component quickly established, but the system also locates the X-Y coordinates of every pin on each component in the design. As each component is placed, the majority of information needed to produce the silk screen, such as component outlines and reference designators, and descriptions of the power and ground-plane artwork, is automatically extracted from permanent system libraries and added to the design file.

Additional board layout functions include: options for interactive gate prepackaging and component preplacement; constructive initial placement algorithms to automatically compute the optimum placement location for each package; a check-off feature for rapid identification of unpackaged gates, placement improvement algorithms with rubber-banding for easy package manipulation; automatic generation of straight line interconnects (SLIs); user-selectable trace widths; and a rip-up feature for rapid deletion of traces back to SLIs.

Multiwire boards will accommodate many

Components can be placed manually in any location or order. Placement is tracked by the system's check-off feature (left screen)

different component and logic types. Chip carriers and pin grid arrays, as well as newer surface-mount components, can be mounted and interconnected on a multiwire board. When component layout is completed, the system generates an interconnection list that will be used during conductor routing.

Speeding routability

The task of routing interconnections between various points, a formidable task in multilayer design, is greatly simplified by Multiwire's use of wire rather than etched circuitry. The fact that insulated wires can cross each other raises the density of interconnections achievable on a single plane, whereas circuits that cross in multilayer must reside in separate layers.

ECL and high pin count devices, which make routing more complicated, increase the need for an interactive design system that can cope with a greater number of incomplete routes and also increase the necessity for designer-placed wires. Using stored schematic and board layout data and the Multiwire Design Technology (MDT), EDS can automatically generate wire routes for signal interconnections in a single day.

EDS software allows Multiwire's automated router to take over, and the designer to change or complete what the router cannot. A graphical presentation of the finished product helps the designer visualize obscured routing.

To route conductor paths, the designer must send the Multiwire router several pieces of information: the border of the wired area, which serves as the limit for wire routing; any obstruction such as cut-outs in the circuit area or locations where wiring is not permitted, so that these areas will be respected by the router; and all of the component hole locations that were identified during component placement.

Crossover capacity allows the router to place more conductor paths on each level, reducing the total number of levels that will be needed for a specific interconnection density.

Discrete wire conductors can be rerouted using the interactive CAD system. Trace movement in multilayer design is more difficult because of the interdependence of the levels and the associated interlayer connections by blind vias. With Multiwire each is independent. If the designer moves one, he has to move only that one; there are no blind vias to contend with.

If the router can not find legitimate paths for all of the from-to's it will show them in the design file. The unrouted from-to's will appear as straight lines between the points to be interconnected. Because insulated wire conductors can cross without shorting, the designer can use the reroute option to

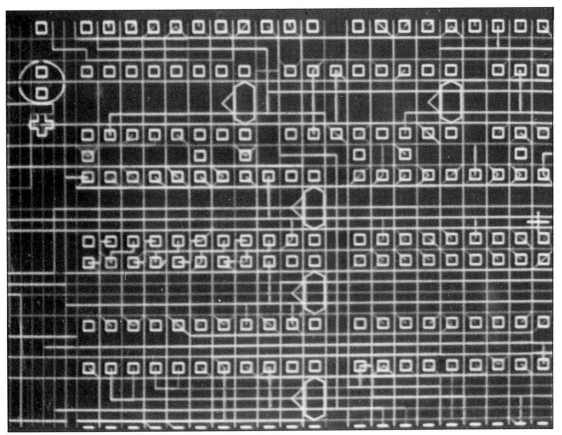

The design file contains all components and routed wiring at different levels

manually place the unrouted from-to's on any level, using any route or path desirable.

The method for placing the interconnects is very similar to that used during the schematic generation phase of the design process. As the designer manually routes each wire, the coordinates of each inflection point will be saved in the design data base.

Special routing requirements

Some board designs require that conductors be routed in a certain way, such as clock signals, digital signals that must be routed to avoid an analog area, or signal lines that have high current requirements. Multiwire conductors are 34 or 38 gauge polyimide-insulated copper magnet wire which has been manufactured to a high degree of accuracy.

The designer can make a predetermined, repeatable pattern a parameter for automatic routing or, after automatic routing, use the interactive routing option to place these wires in acceptable locations. Changes are incorporated in an updated design file.

Parallelism analysis

In some circuits, the amount of cross-talk interference that can be tolerated is minimal. Boards that require specific noise margins are designed using wire pattern analysis and interactive routing techniques.

After wire routing, Multiwire is able to generate a report that will tell the designer how close the wires are to each other, for how long a distance. This report can be used to determine which of the wires are more likely to couple and produce noise levels in excess of specification.

The maximum allowable coupled length is calculated based on the electrical characteristics of the devices being used and the electrical characteristics of the board itself. The report that is generated shows the signal name and the locations of segments that are in violation of the selected parameters. With this listing, using the reroute option, the designer can move the segment of the wire that is in violation to a safe distance or location. The

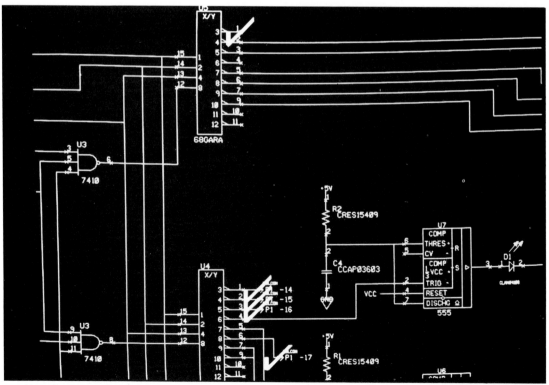

Board layout functions include a check-off feature for rapid identification of unpackaged gates

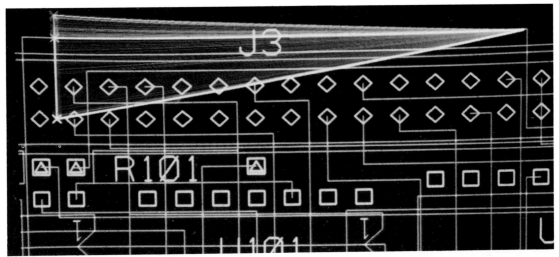

Wires are easily rerouted using the system's rubber-banding technique, which leaves one end point connected and moves the other to the new connection point

parallelism test can be run again to make certain no new violations have been created.

Wire length analysis

In circuits where signal timing is a critical factor, the delay time of some signal lines must be kept to a minimum. Multiwire can tell the designer how long each of the wire segments between connections is. The wire length report lists the shortest orthogonal length of the wire, the actual length and the difference between the two.

The typical propagation delay for multiwire circuits is 1.8ns/ft (less than stripline and about the same as microstrip). Using the wire length report and the known propagation velocity, the designer can determine the total delay time for different parts of the circuit.

Nets that have critical timing parameters can be rerouted to meet specifications using the reroute option. The system's interactive routing function lets the designer shorten wire segments to decrease time delay or lengthen segments to increase delay. After rerouting those wires that violate signal timing requirements, the designer can rerun the analysis to make sure no new violations exist.

After the design has been completed and all parameters are satisfied, the complete documentation necessary to generate wiring and drilling machine tapes and photoplots, and to generate power, ground and silk-screen legend artwork, is available within EDS. This data will be used to run the CNC manufacturing equipment.

The system also generates bills of materials formatted to user specification and data that will be used for electrical testing and other inspection tools.

With Multiwire manufacturing facilities available in-house, prototype fabrication generally takes less than a week. If manufacturing facilities do not exist on-site, the same step can still be accomplished inside of four weeks using outside vendors.

Simplify design changes

Since multiwire's point-to-point wiring interconnections are laid down on each board by a computer-controlled wiring machine, signal interconnections can be rapidly changed by making simple computer command entries.

Using the system's trace editing option, the designer can rip up and replace particular interconnections, or completely reroute the entire board. Assuming no additions to the bill of materials, major rerouting of a multiwire board with 1000 equivalent ICs can be routed to 99.99% completion in under four hours. When design changes occur, they are incorporated in an updated digital data base,

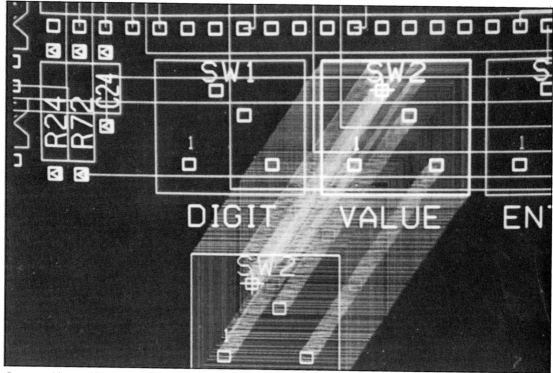

Customer revisions requiring movement of components are accomplished dynamically using the system's manual component placement option

which is used to write a new floppy disk.

Debugging a high-density prototype takes from two to 12 weeks, or about half the time normally required to trouble-shoot a board. This substantial reduction in time is feasible using routing software which extracts logical hookup information for routing directly from the schematic. Thus, a designer can be completely assured that all pins are properly interconnected, since illegal pin-to-pin connections become an impossibility.

Signal corrections to the schematic, usually ranging in number from 50 to 100, and corrections to the component layout, are accomplished using EDS. Based on the corrected schematic, EDS deletes wires that no longer exist and produces SLIs for the new connections.

The amount of time required to route new wires is dictated principally by the density of the board. For the average 500-IC board, typically only two hours are necessary to interactively place new wires resulting from corrections to the schematic logic.

Using EDS, new wires can be routed without disturbing wires that have already been placed. EDS executes design rule checks that ensure proper tolerances have been maintained and verifies that these new wires accurately reflect the changed schematic logic.

Using the manual component placement option, the designer can also move a component to a new location and easily reconnect it, rerouting only those wires that have to be modified.

EDS provides board checks and refinement options to ensure that designs ae production-ready and error-free. The system's back annotation feature, for example, updates the schematic to correspond precisely with the completed board design, while numerous quality assurance checks enhance final design reliability. Board check and refinement features include: design rule checking to ensure geometric tolerances are maintained; graphic highlighting of design rule violations; and quality assurance checks to ensure accuracy of the final design.

CHAPTER 10
AUTOMATING ELECTRONIC MANUFACTURING

A substantial number of the preceeding pages are devoted to automatic machinery, even programmable machinery. In a limited sense this can be labeled as automation.

The word 'automation', however, is more encompassing. It involves the *integration* of discipline and the internal ability to determine failure, identify cause and take corrective actions.

Ideally, automation begins at the design stage. Early stages of implementation will come through use of CAD and CAM. On the factory floor, islands of automation will have to be tied together, first into workcells, then total systems.

This tying together will have to be achieved in two ways: control integration (through LAN?) and physically by integrated material handling devices; automatic storage and retrieval, and automatic transfer by conveyors and guided vehicles.

After examining these tools of integration, it is worthwhile to examine a selection of case histories of attempts at full system integration. Unfortunately, it seems that the majority of these attempts have been made as research and development programs, often by companies with vested interests in broad-scale implementation of such integration. Others, with defense orientation, often seem to be developed without cost considerations.

The thrusts are not mere labour savings. They concern product quality, miniaturization accompanied by improved capacity, batch lot manufacturing and product consistency. Whether this can be achieved with identifiable returns on investment remains to be seen.

AUTOMATING THE ASSEMBLY OF ELECTRONIC PRODUCTS

Electronics is changing the face of most industries today. This is no more so than in those industries concerned with information transfer such as telecommunications, office systems, computers and the like. A company at the forefront of this revolution is Plessey. In particular on its Beeston, Nottingham, site Plessey Office Systems Ltd (POSL) produce digital switching systems, terminals and office automation systems. The Transmission Division of Plessey Telecommunications (PTL) is also located on this site and is primarily responsible for digital transmission systems for voice, data and video displays. The company's manufacturing managers and engineers continually need to adapt current production techniques and devise new ones to meet the electronic industrial revolution.

About 4500 people work on the Beeston site and at a major satellite unit for engineering about four miles from Beeston. This compares with 9000 or so of some 14 years ago, reflecting the development in both the product and equipment used in its manufacture. Plessey has been able to improve productivity by careful planning of capital investment and through natural wastage.

Now, there is little to be seen of the old fashioned equipment, such as mechanical dial telephones, relay exchanges, and the like. Although these are still produced in small quantities, the major production is of advanced digital computer controlled systems and terminals. Plessey's standard telephone is designed for simple assembly, with a single PCB and elastometer keyboard. Components are minimized with a single integrated circuit dealing with both transmission and signal functions. Simple transducers of the moving iron type are identical for both the transmitter and the receiver in the telephone head.

Functions

Exchanges are now totally solid state with microprocessor control of the functions. This

equipment has changed dramatically as it is no longer designed just for voice transmission but can transmit data and therefore can be used as the basis of the fully automated computerized office. Plessey's IDX system is the first third generation exchange for voice, text, and data communications. It is designed to be the heart of the IBIS (integrated business information service) electronic office. With this equipment desk-to-desk transfer of information in all forms can be achieved.

Similarly, transmission products have been undergoing a steady revolution. These products are also digital. The main change taking place on this side is a move to fiber optic transmission. This is because of the many advantages such as noise immunity, security, safety, the inherent isolation of terminal electronics, ease of multiplexing and so on. With Plessey being the largest UK manufacturer of optical fiber connectors coupled with its expertise in signal processing and transmission, it is an excellent position to manufacture complete fiber optic transmission systems, from the transmitter to the receiver.

The type of printed circuit assemblies produced for private exchanges and telephones varies enormously. Although each has some subassemblies that are very similar, they are all configured to meet customers' requirements. So there are boards within the base system such as switching blocks that are needed in large quantities. CPU boards are also standardized, but only one board may be required for each exchange. Then there are the customer requirements and these can be selected from a number of preferred varieties such as 'conference calling' on telephone terminals. Such items are only required in small quantities and typically would be built in batches of 20.

Transmission equipment also needs hundreds of different products. Normal production is 10,000 professional boards per week, chosen from 500 different codes (or part numbers). The manufacture of these pro-

First published in *Assembly Automation*, February 1984.
© 1984 IFS Publications

Plessey use the Dynapert dual head machine for inserting axial lead components

ducts encompasses the complete spectrum of methods from fully manual to fully automatic production.

With the exception of a board prototyping unit all PCBs are manufactured off the Beeston site principally from Plessey's subsidiary in South Shields. The main concern with board manufacture therefore is assembly, primarily component insertion. Inevitably because of the variety of products manual assembly is used extensively. But there are the larger volume boards for which automatic insertion is justifiable. Plessey have opted for Dynapert insertion machines because each factory has found these machines to be well suited for its type of production. The machines are very fast and capable of inserting up to 9000 components per head per hour, have a very quick change over from code to code, a matter of minutes, and can insert in two directions.

The telephone shop produces about one million standard digital telephones a year. A Dynapert twin head VCD-F machine is installed here together with a 40 station sequencer which are assembling about 10 different codes on boards. It will insert standard axial lead components and represents the first stage in the automation of board assembly at POSL. The next stage includes the extension of the sequencer up to 80 stations (this machine is on order) and the installation of machines for both radial lead components and dual-in-line packages (DIP).

The sequencer produces bandoliers of any number and combination of axial components which are stored on drums at each of the machine stations as well as BTC (bright tinned copper) links which are stored on a drum at the end of the machine. The completed bandoliers are stored on drums ready for the insertion machine. Normally three days' stock is held between the sequencer and the insertion machine.

Verification

The sequencer also carries out verification of the resistor, diode and so on. As each component reaches the end of the track prior to bandoliering it is automatically lifted and held between two contacts where polarity and value are checked. Any error or out of tolerance component fault stops the machine and summons the operator who removes the faulty part and inserts an identical one from a loose store. This component is also checked before being dropped back into its track groove so that the machine can continue. This check is carried out within the normal operating cycle of the machine which can be as much as 20,000 components per hour. A further check is carried out to ensure that there is no missing component on any track groove. Centering of the component is carried out mechanically before the final station where the two leads are trapped between two drum fed color coded tapes which identify polarity. Spacing, fixed at 0.2–0.4mm, can be used but this needs a major tool change.

The twin head VCD-F is the latest microprocessor controlled inserter from Dynapert. The two heads work in duplex and produce two identical boards at a time. The table on which the boards are mounted accommodate jig plates for four boards per head, two of which are available for loading and unloading. The head remains fixed and the table manipulates the board in the X and Y directions. The third (rotation) table axis, allows components to be inserted in two directions (X and Y) and indexes the boards to the load/unload stations.

As each component is fed from the bandoliering drum it is picked up by the head, where the leads are cropped to the programmed length, bent, and inserted into the two

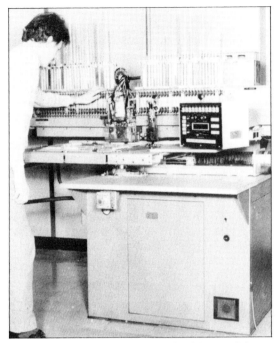

The Dynapert inserter being set up to assemble boards for transmission products

holes on the board. Having pushed the leads through they are bent over to secure the component in place prior to soldering.

Optical alignment

The VCD-F machine is fitted with programmable automatic optical alignment of the board holes. As a hole comes under the head a light is shone through and sensed. The board is micro-manipulated to obtain the optimum output from the sensor and to align the head with the exact center of the hole. Technically this could be done for every hole but this would be very impractical as it would slow the insertion rate down to an unacceptable level. It is more normal to program for one particular hole position to be sensed, either on each board or more likely on each batch of boards. This hole acts in effect as a datum and the program for all hole positions is adjusted to compensate for any error measured at this stage. Programmable pitch is achievable between 0.3 and 1.3mm.

The changeover from one board code to another is very quick, a matter of minutes at the most. All that has to be done mechanically is for the bandolier drum and the board mounting jig plate to be changed. Then the board code is typed into the machine to select its program automatically and the machine is then ready to go on the new board.

Currently the machine is working flat out on two shifts and very often on three. This is good news for Plessey and means that shortly a second double-head machine will be needed to meet the demand.

Generally transmission products require heavily populated boards and utilize many DIPs. Consequently, the automatic insertion equipment for these includes a 120 station sequencer and a Dynapert single-head DIP inserter as well as a twin axial lead inserter.

The DIP inserter is very similar to the axial lead machine except that it takes standard integrated circuits which are fed from tube magazines. There are 60 magazines with each station dedicated to one component type. A second 60 station magazine is being purchased which can be loaded off-line and rapidly exchanged with the one on the machine. The machine can accommodate six to 20 lead 0.3mm center DIPs.

Each DIP component is selected in the program order and fed to the inserter head

A 120 station sequencer, an axial lead inserter and DIP inserter are centrally controlled by Dynapert's System 8000, which also provides inventory and quality report print-outs

Several Electrovert inclined track-soldering machines are employed on the Beeston site

where its leads are pressed inwards to the correct pitch ready for insertion. Also, this gives electrical contact with the leads and an electronic logic test is performed. The programs for these tests are stored, for each component, on floppy disk. Plessey writes its own programs based on the IC manufacturer-supplied specification, but as not all specifications are available 100% checking is not possible at the present time. After insertion the leads are bent over in the direction of the PCB tracks. Plessey achieves 3500 insertions an hour from this machine which compares very favorably with the manufacturer's claimed absolute maximum of 4000 per hour.

Although both the DIP and axial lead machines can insert in two directions, generally boards are of single-axis design. This has been achieved by persuading engineering to design for automatic insertion. Eliminating two axis designs avoids the need for a 90° rotation of the table. Also engineering are encouraged to design boards with single direction polarity – if the polarity direction has to be switched the table will need to be rotated through 180° to insert the components in the correct direction. A third area

where production is influencing design is the avoidance of components close to the edge of the boards. Such areas are inaccessible to the insertion head and then manual insertion would have to be used. These design guidelines are helping the company to increase the degree and rate of automatic insertion.

Flow soldering of the boards is a well standardized technique and again there is an unwritten policy to standardize on one supply of equipment. There are six Electrovert flow soldering machines in use on the Beeston site and many more in other Plessey factories. Typical of these is an Econopack 229 which is an inclined chain track machine, with platen load at one end and unload at the other. The sequence is foam flux, preheat, and lamda wave solder. Standard carriers are used capable of taking different jigs to accommodate different numbers and sizes of boards. Currently the private telephone systems operation solders 5000 boards per day with an average number of 150 joints per board.

Transmission products use the solder-cut-solder technique. This dual-soldering technique is dictated because of the high density of components which means that there is little room for bending leads. It is also a means of minimizing the amount of component preparation prior to insertion.

Cleaning station

The installation consists of two flow soldering machines with a cutting-cleaning station inserted between the two. Boards are loaded onto a twin belt conveyor which transfers the work carriers to the first soldering machine. It then passes to the cutting station and finally to the second soldering machine before being cleaned, dried, and returned underneath the three stations back to the front end of the machine for unloading. Thus the loading and unloading station are housed together at one end of the machine.

The first soldering machine is a horizontal chain driven machine. Rather than using a foam flux, a spray flux system has to be used because it was found that the long dangling leads from the uncut components interferred with this operation. Similarly, it was not possible to use a wave-soldering technique

An eight port extension line board about to be flow soldered in the CDSS (Computerised Digital Switching Systems) area at Beeston

because of the lead length and in this station the boards are lowered into a drag bath. The level of the solder bath is critical in this instance having to be maintained to a half board thickness tolerance of $\frac{1}{16}$in. A tool setter keeps a monitor of this situation and tops up the tank manually whenever necessary.

The second stage of cutting is performed with two vertical axis tungsten carbide blade disks followed by a pair of vertical axis brushes and a horizontal brush. The latter is a standard item on the Electrovert machine and Plessey installed the two other brushes to meet its own requirements.

The third stage is a standard inclined chaind riven machine which uses foam fluxing, preheating and lamda wave soldering. At this final stage the ends of the wires, that had been cut previously, are tinned. Also, Plessey has found that this has led to a better quality of joint. However, it finds that a poor joint after the first stage is not improved by the second soldering stage.

Automatic insertion is just one of the areas that Plessey is using to improve its productivity and product quality. Such areas operate in isolation at the present time but the long-term objective at Beeston is that they should be integrated. David Williams, Business Execu-

tive in the Standard Systems Division sees the need to create a database accessible to engineering and manufacturing. "I see the creation of a database as the way to go to achieve computer-integrated manufacture. There are many problems that we are tackling including the avoidance of database corruption for which solutions are available. We have standardized on VAX computers for engineering to be linked by high-speed highway communication. We can link several computers using the VAX Cluster, DECnet and Ethernet links. This will provide us with a distributed engineering database." The links between design and production databases are still being examined.

Meanwhile plans for further automation are going ahead. The policy is ultimately to link 'islands of asutomation' such as automatic insertion, asutomatic assembly and robotic testing. A project nearing completion is the automatic assembly and test of telephone transducers. Another project likely to come to fruition during 1984 is a sheet metal FMS. In the view of John Leeming, Industrial Engineering Executive in the Transmissions Divisions, "We hope in future to be able to build bridges between these islands but the solutions are not yet clear. If we waited for such solutions to become more concrete we would never start out." It is not possible, either practically or economically, to install completely automated factories at the present time, and the island approach seems to be the only practical one according to both Williams and Leeming.

Plans for automation are being accelerated by the emergence of a new technology in electronics, that of surface mounting of components or 'onsertion'. This is evolutionary rather than revolutionary. Already there is some surface mounting being done. This will certainly increase until boards are virtually 100% surface mounted. Plessey at Beeston has recently ordered its first onsertion equipment from Dageprecima who is now part of the Dynapert organization. This will be used for surface mounting of components onto PTL's professional boards.

Surface-mounting technology allows the use of substrates other than the conventionally epoxy-glass. These can help in improving product quality, reliability, and ease the problems of testing. The testing problem is getting more and more acute because of the increasing use of VLSI. These are powerful and high component integrity is needed. Given that assembly must be right first time, the high quality of surface-mounting technol-ogy will be of great value.

With ceramic substrates all components will be surface mounted, but this will still leave the odd large and difficult components such as transformers and transistors which cannot be handled by conventional surface-mounting equipment. It is here that Leeming sees the need for orbots. Accuracy and speed have prevented robots from being used for conventional component placement because specialised machines inevitably do it better. But now there are several robots available which are ideal for handling larger components. There are plans to link the placement machine with the robots in the cable television contact. This integrated system should be up and running at Beeston by next October.

The other area where Leeming sees robots being of value is as an interface between the board production line and asutomatic test. Robots can be used to handle boards from the end of the line and into automatic testing. This integration of production and tests will enable Plessey to keep a closer control of its production. Stored test data can be related back to design and production and forward to performance in the field, leading to improved design performance and better customer satisfaction.

TEST, REWORK AND INSPECTION MANAGEMENT COSTS

The quest for quality has always existed in electronics assembly operations. The problem has been at what cost. An acceleration of both quality improvements and reduced inspection and rework costs has become apparent over the last two years. The trend toward better products at lower prices does not stem exclusively from the competition in the marketplace.

Recognized Japanese quality plays its part but higher quality also becomes a major

First published in *Printed Circuit Assembly*, February 1987, under the title 'High yields at low cost'.
Author: Charles Henri Mangin, CEERIS International Inc.
© 1987 PMS Industries

ingredient of the corporate culture at the best managed board assembly operations. Reducing the cost of assembly is also becoming a matter of survival for many US operations facing the competition from low-cost overseas circuit board assembly shops. The cost of quality can be reduced to as little as 10% of the cost of assembly through proper test, rework and inspection management (TRIM).

New technologies

Major technological and management changes are occurring whose full benefits can only be achieved when high yields are

	Baseline context	High yield operation
Incoming component yields	●	●●●
Component and PWB storage	●	●●●
Autoassembler yields	●●	●●●
Preventive maintenance	●	●●
Autoassembler setup inspection	●	●●●
First-article inspection	●●	●●
PCB inspection before solder	●●	●●
Soldering process monitoring	●●	●●●
Solder joint inspection	●●●	●●
Design for producibility	●●	●●●
Work-in-process	●	●●●
Human resources dedication and training	●	●●●
Low	●	
Medium	●●	
High	●●●	

Major management concerns

maintained all along the assembly process.

Flexibility is becoming a major issue for electronics assembly operations for several reasons: market shifts and shortened product life imply versatile assembly resources; the introduction of SMCs modifies the assembly operation sequence steps; and smaller batches reduce the work-in-process (WIP) and its associated financial cost.

New technologies are available which allow for the implementation of flexible electronics assembly systems. These include computerized monitoring and control at the supervisory level; automated guided vehicles, PCB conveyors, WIP automated storage and retrieval systems (AS/RS); automated board loading/unloading modules; and bar-code based board tracking systems.

The need for flexibility and the availability of new technologies leads to drastic modifications in the management of electronics assembly operations. Fully automated assembly systems are spreading and pull or Just-in-time organizations are being implemented. Integration of assembly into the whole operation, e.g. board design, final test, component

purchasing, final product marketing is becoming the state-of-the-art.

Fully automated assembly combined with integrated management exists today at a few large electronics companies. These operations tend to be high volume/ low mix and consistently achieve above 90% board yields at first-pass automated test equipment (ATE) with scrupulous process monitoring and minimal in-line inspection costs.

These achievements will become the norm for lower volume and higher mix environments in the near future. Quality is also becoming of major concern to all vendors. Their near term objective is to reduce their defect rate to 200–500 ppm for incoming components, automated assembly equipment performance and the soldering process.

Baseline vs high-yield

This analysis compares two 'typical' environments. It is recognized that there is no typical assembly shop, no typical boards, etc. For the purpose of clarity the *baseline context* represents a composite assembly operation designed on the basis of our experience with

Process step	Result
Bare boards	99.9% good boards released to assembly (defect rate 1000 ppm)
Components	99.9% good components released to assembly (average defect rate 1000 ppm)
Populating	99.9% good insertion/placement (defect rate 1000 ppm)
In-line QC	Release 91% good analog boards and 90% good digitals soldering
Soldering and cleaning	99.99% good solder joints (defect rate 100 ppm)
Pre-ATE inspection	First-pass yield –90% analog boards –85% digital boards

Results achieved through high-yield process implementation

		Baseline context		High yield operation	
Defect origin	Early 1980s	Analog board	Digital board	Analog Board	Digital board
Incoming components	25	40	34	61	48
Assembly	30	40	34	27	28
Soldering	45	20	32	12	24
Total	100	100	100	100	100

Origin of defects identified at ATE (%)

island-of-automation operations. It is deemed representative of the most common and quality-conscious existing facilities.

First-pass yields at ASTE are 71% for 'typical' analog boards (150 components and 636 solder joints) and 60% for 'typical' digitals (200 components and 1762 solder joints). This implies total TRIM costs of $1.3 million or more per 100,000 boards.

The *high-yield operation* is deemed capable of achieving first-pass yields at ATE of 90% good analog boards and over 85% good digitals. Consistently high-yield components at incoming, good assembly practice and proper use of automated inspection equipment are essential. This permits the reduction of TRIM costs to below $750,000 per 100,000 boards.

High quality process

High-yield assemblies stem from good management practice. High quality incoming components, proper maintenance and adequate operation of the assembly and soldering equipment combined with automated board inspection after assembly and soldering will raise the yield at first-pass test to 90% for analog boards and over 85% for digitals. Selected recommendations for implementing a high-yield process are:

- *Bare boards* – Select vendors who deliver consistently high-yield boards with zero non-repairable defects, and few repairable defects: incoming yield should be >95%. Repair repairable defects. Manage stock

carefully; use FIFO to avoid shelf contamination; develop JIT delivery with vendor progressively. If captive shop, test and screen bare boards before transfer to assembly. Process and plating specs: very high standard to ensure solderability. Pre-bake before assembly and soldering to reduce blow-out of water at wave solder.

- *Components* – Buy prescreened/tested components wherever possible. Test and burn-in those types which have infant mortality

	Analog board	Digital board
First-pass yield at ATE in % good boards		
baseline context	71%	60%
high-yield operation	90%	87.5%
TRIM costs in $/board		
baseline context	$12	$13.50
High-yield operation		
● Functional then in-circuit testing	$7.50	$7.80
● Combined testing when yield and volume permits	$2.50	$2.55

Baseline context vs high-yield operation

problems, e.g. VLSI, memory chips. Buy properly tinned components. Employ sound, JIT/FIFO stocking procedures. Employ careful component handling everywhere to avoid bending leads or damaging component bodies.

- *Populating process* – Acquire most rugged, most easily maintained/repaired autoinserters. Limit on-machine testers to off-line sequencers and verifiers on autoinserters. Electrical head tester not needed because of high component yields. Use good setup aids; write perfect insertion programs. Employ strict and thorough preventive maintenance and parts replacement programs. Implement setup inspection for first article verification of accurate setups. Design boards for autoinserted components wherever possible.

- *In-line QC* – Use computerized-vision inspection wherever possible for 100% inspection of top and bottom before solder.

- *Soldering and cleaning* – Buy best soldering and cleaning system with as much preheat capability as possible. Avoid additive dross suppression. Factor in solderability at board design level. Set careful wave height techniques. Carefully monitor flux specific gravity and cleanliness. Test flux activation time and temperature for each board type. Coordinate cleaning and flux chemistry to ensure perfect cleaning. Train solder technicians to oversee operation. Delegate responsibility.

- *Pre-ATE inspection* – Use low-cost shorts and opens continuity tester for solder joint inspection.

- *General recommendations* – Compute maximum board yield achievable based on process and number of operations/components affected per board type. Implement statistical QC and make information available to assembly shop in close to real-time. Train and motivate all personnel involved with the operation.

Optimum cost strategies

An improved process combined with new test strategies allows drastic TRIM savings.

High-yield operations are deemed to be implemented because they are economically justified. The functional-in-circuit sequence in a high-yield environment generates TRIM cost savings of 40% over the baseline context, which is not a poor quality environment.

When applicable, combined testing reduces TRIM costs by five times over the baseline context and by three times over the functional-in-circuit sequence TRIM expenses.

FLEXIBLE MANUFACTURING

The flexible manufacturing system (FMS) is a manufacturing philosophy. In the broadest sense, FMS describes the capability to respond or conform with change in the factory environment, particularly with respect to high product mix and low volumes.

Flexible manufacturing applies broadly to both fabrication and assembly, and in environments ranging from totally automated to completely manual.

Theoretically, flexible manufacturing provides complete flexibility in the quantity and type of product manufactured by the factory. A totally flexible printed circuit board assembly and test facility, for example, would produce any size board in any quantity ranging from one to thousands.

Perhaps the most versatile and flexible of all operating systems, the robot is ideally suited for printed circuit board assembly workstations. Although slower than dedicated systems, robots can insert or place non-standard as well as standard components; and in multiple configurations in a work cell, they

First published in *Electronic Packaging & Production*, 1983, under the title 'Flexible manufacturing increases productivity'.
Author: S. Leonard Spitz, Eastern Editor.
© 1983 Reed Publishing (USA) Inc.

With up to 320 inputs available, this flexible automated workstation picks up a chip package for placement on a substrate as large as 10in. (Courtesy of Universal Instruments)

can pick components from feeders, straighten leads, and cut and clinch.

Most robots use some form of transmission, gears, harmonic drives, chains, stainless steel cables, to convert lower power from several stepper motors to high torque where needed at the articulated joints. This permits a low center of gravity by housing the motors, in the machine base. However, transmission systems can degrade accuracy, repeatability, and reliability, and care must be exercised to limit backlash. The alternative is to mount motors so that they are connected directly to the joints while compensating for the raised center of gravity with a stiff, relatively heavy structure.

Robots come with a broad range of capability, from Microbot's benchtop Alpha 2 to Adept Technology's heavy duty, speedy Manipulator. With 4, 5, or even 6 axes of articulated motion, robots have a reach of as much as two to three feet, and can slue to programmed points in space with a velocity of up to 30 feet per second. Typical accuracy and repeatability range from a low value of ±0.015in. to a more precise value of

±0.001in. Cycle times for a given operation depend on path length and payload which may be as much as 15lb or as little as a few ounces. Over short paths and bearing light loads, one second cycle time is not unusual, conversely, with long paths and heavy loads, cycle times may go to 20 seconds.

Sensors

Robots 'feel' and 'see' using force, proximity, and optical sensors, as well as vision systems. While adding capability, the optional inclusion of vision systems also adds complexity and cost with some commensurate degradation in speed. Especially suited to high speed, force sensor technology incorporates strain gauges into the robot's gripper, enabling it to feel components coming into the workstation from an assembly line, parts feeder, or conveyor. Proximity sensors in the gripper use air backpressure to detect whether the gripper is closed or open. Optical sensors include an infrared emitter detector pair also installed in the tip of the gripper. This device differentiates between reflective objects,

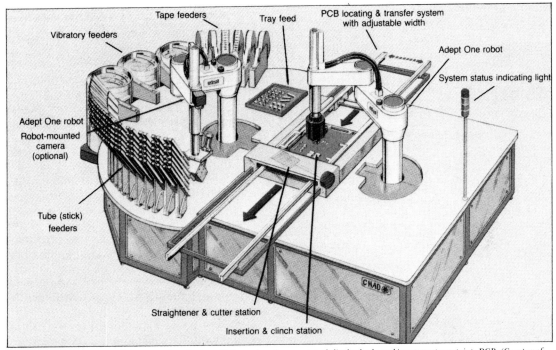

Labels on figure:
- Tape feeders
- Tray feed
- PCB locating & transfer system with adjustable width
- Vibratory feeders
- Adept One robot
- System status indicating light
- Adept One robot
- Robot-mounted camera (optional)
- Tube (stick) feeders
- Straightener & cutter station
- Insertion & clinch station
- CHAD

Automated workcell is comprised of two coordinated robots which, as a team forms, cuts and clinches leads, and inserts components into PCBs (Courtesy of Adept Technology)

usually white, and non-reflective objects, usually black or absent, giving the robot a primitive scanning capability when suitably colored objects are used.

Three levels of flexibility characterize the movement of parts between workcells in a flexible manufacturing system. At the lowest level, Very Limited Route Flexibility refers to early generations of powered belt conveyors with divert arms for removing product upon reaching the appropriate workstation. The returning belt at the underside of the drive pulley is often used to deliver empty tote cartons to the dispatch station. In this transport system, the route between kit inventory and field workstation is fixed; only delivery times and product volume are variable.

The next higher level of material transport flexibility, Routing Flexibility, defines the delivery of a particular part from a staging point to any one of a number of alternative workcells as a function of the start of the part and the state of the cells. For example, a printed circuit board may be delivered to a robotic station for component insertion. If there are several identical assembly stations, the specific destination can be selected to balance the workload. This approach requires the ability to make destination decisions in real time rather than as a predetermined part attribute.

Total Route Flexibility, the highest level of material transport flexibility, refers to the movement of a part or subassembly from any staging point to any workstation, and from any workstation to any other workstation. Such route flexibility can be achieved with automated storage/retrieval systems (AS/RS), automated guided vehicles (AGVs), smart monorails, and carts-on-tracks. Route flexibility results from the existence of more than one possible path from a specified origin to a specified destination, combined with the controls that permit an intelligent selection from among the alternatives. The route selection must be made in real time, as opposed to predetermined on the basis of average congestion.

In addition to being characterized by degree of flexibility, material handling sys-

Vision-guided robot arm finds component leads and places high pin count packages on surface-mount boards (Courtesy of Adept Technology)

tems can be identified as either synchronous or non-synchronous. Utilizing continuous moving conveyors and indexing conveyors, synchronous handling systems transport workparts together, all at the same time. Operators have no control over the movement of parts, and in performing assembly tasks are paced by the speed of the parts on the conveyors.

By contrast, non-synchronous handling systems move workparts independently of other parts, permit operators to control part movements between workstations, pace the assembly line to the speed of the slowest operator, and easily provide for in-process inventory of workparts.

Supporting non-synchronous material handling methods, automated storage/retrieval systems automatically replenish parts to workstations, remove materials from workstations, move materials from one workstation to another, and store work-in-process.

System control and communication

Acting much like a symphony conductor, a central host computer orchestrates the various and diverse functions of the flexible manufacturing system. Through a network which includes local process control and material handling computers, the host accesses information, otherwise unavailable, for the purpose of making real-time decisions such that operational harmony and true integration are achieved.

At a printed circuit board assembly facility put in place by Universal Instruments Corp. and operating under the FMS philosophy, a real-time process controller (RTPC) automates assembly activities at three distinct operational levels, operating individual assembly machines; managing part movement and associated processes on the assembly floor; and integrating neighboring manufacturing functions.

With respect to operating individual machines and depending upon the machines' intelligence and communications ability, the RTPC identifies incoming materials; downloads the machine program; assists in machine setup and extracts management information.

With reference to assembly part movement, the RTPC makes WIP routing and scheduling decisions, communicates with the transport system to move WIP tracks and reports work flow and traffic status, identifies processing errors and assists in recovery and generates system performance statistics.

System architecture

At the bottom level of the RTPC's modular system architecture, machine concentrators link to a group of machines with similar interface and control characteristics. These machine concentrators are further integrated with the main process computer which ties together other design, planning, and production computer.

The machine concentrator acts as an intermediate data buffer between the main process controller and the machine workstations, regulating communication traffic by selecting and packaging information going from the machines to the main process controller. The machine concentrator receives instruction from the main process controller to direct machine activities.

The main process controller accumulates workstation information received from the machine concentrators and makes real time assembly process decisions to direct machine activities. This capability requires the main process controller to keep a database containing items such as 'what to make', 'how to make', and current and historical assembly flow information.

COMPUTER SIMULATION

The careers of engineers and their managers can be at stake as they attempt to bring a sophisticated, automated manufacturing facility into profitable existence. Considerable financial and market risk for the firm, too, are associated with automating a business – as well as with not automating.

Reducing the risk of implementing an unsatisfactory system that does not meet expectations of the firm's top management, nor its shareholders and customers, is just one problem that can be addressed with computer simulation. It is a tool increasingly used to evaluate the performance of a manufacturing system before significant capital is invested in implementation or change. Scheduling and planning of day-to-day operations, too, pose issues where simulation can point the way to making profitable decisions.

Modern manufacturing

Simulation is used extensively by The Confacs Group to support facility layout and design of automated material handling systems, as well as providing a decision support tool, says Kerim Tumay, industrial enginer specializing in simulation. Computer simulation models implemented by this Tempe, Arizona-based, facility-systems consulting firm have included multilayer printed circuit board fabrication as well as PCB assembly.

The firm claims their simulation models eliminate many of the risks associated with the more traditional approach of statically designing a facility and hoping the integration of

dynamic system components will perform to the required specifications.

Using simulation, questions can be answered on issues such as equipment utilization, production queue times, throughput, productivity, and adequacy of manufacturing capacity into the future. As a decision tool, simulation permits users to pose questions in terms of 'what if' a specific situation were to be encountered, then find what will be the response of the manufacturing system to that situation.

Manufacturers' move to surface-mount technology (SMT) provides an example of applying simulation to look ahead at the risks of manufacturing decisions made in the face of product mix uncertainty. Confacs has been working with companies on the introduction of SMT, states Tumay.

Most people do not know what percentage of their boards are going to be based on SMT, he observes. They have questions about how changing the mix of through-hole component assemblies and SMT assemblies will affect the size of a facility needed, and how one arranges the work positions.

Though applicable to manufacturing with an in-line configuration of machines committed to running one product in relatively large volume, simulation is being viewed with increased interest as manufacturers consider flexible manufacturing systems.

Frequently, a high degree of complexity emanates from myriad combinations of assembly and process machines, software commands that control them, and jobs to be run from the mix of products produced in a modern electronics manufacturing facility. Routing and scheduling complexity is certain

First published in *Electronic Packaging & Production*, March 1985, under the title 'Simulating the factory of your future reveals problems early'.
Author: Ronald Pound, Senior Editor.
© 1985 Reed Publishing (USA) Inc.

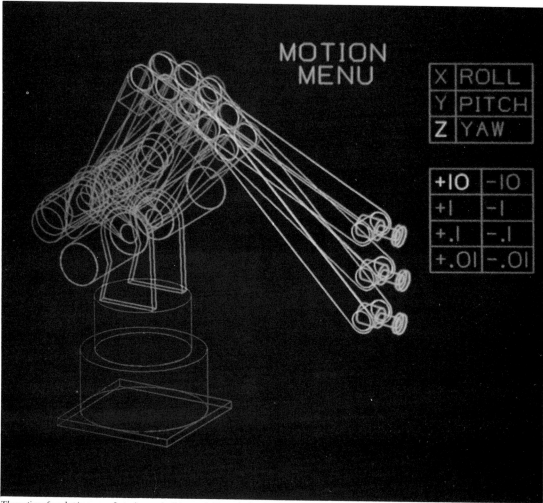

The motion of a robot in a manufacturing workcell is revealed by computer simulation (Courtesy of Calma)

to increase with flexible manufacturing systems using robots or automatic guided vehicles (AGVs) to move low volume products, from a high product mix, between uncommitted machines in workcells.

Simulation process

Computer simulation is a technique for examining performance of the model of a real system as it operates over a period of time. The state of the modeled manufacturing system can be observed while the computer program runs and statistics measuring its performance can be automatically collected.

The capability to address system interac-

tions dynamically is one characteristic distinguishing simulation from other problem-solving models, such as queuing theory and linear programming. Simulation, too, is particularly appropriate in situations where the size and complexity of the manufacturing problem make other analysis techniques difficult to apply.

On the other hand, simulation does not directly provide an optimum solution to a problem. Results of a simulation only show what will happen for the specific situation considered.

While general-purpose simulation languages are available on the market, setting up

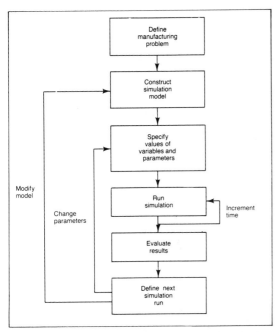

The simulation process requires development of a model and specification of data for the specific system to be investigated

to run the simulation still requires development of an individualized model of the specific manufacturing system to be examined. It is also necessary to collect or determine the values of variables and parameters associated with the system and incorporated into the model.

A system model consists of logic and arithmetic processes that can be executed on a computer, or manually if sufficiently simple, to mimic the dynamic response of a real system. Variables are the properties of a system that change during a simulation run, such as those that are explicitly a function of time. Parameters are properties that are constant during a run, but may be changed by the user to simulate various systems with the same model configuration but different specifications.

Simulation does not tell one exactly what the real system does. The model is a representation of a real system, and deviations between the two occur. Confidence in the simulation results depends on the validity of the assumptions and the accuracy of the data. It is generally necessary to use approximations of the real system to make the simulation

problem tractable, and it is frequently difficult to nail down the values for parameters of systems that do not yet exist.

Modeling defects are not the only source of differences when real system performance is compared with simulation results. The real system may be constructed with its components not conforming to the specifications derived on the basis of simulation. It may, also, be subsequently operated in a manner different from that specified for the simulation.

Applied simulation

General-purpose simulation language packages available include SIMAN, marketed by Systems Modeling Corp. SIMAN is the acronym for Simulation Analysis. SLAM II and MAP/1, available from Pritsker & Associates, are also well-known simulation language packages.

Pritsker's SLAM II is fully implemented on the IBM Personal Computer, as well as being available for mainframe and minicomputers. If the computing power of a large machine is required for running the simulation, SLAM II allows the user to develop models on a personal computer, then run them on the larger computer. The package provides three different modeling approaches – network, discrete event and continuous. Combinations of the three can be used in a single simulation model.

MAP/1 discrete manufacturing simulator enables users with a relatively low level of computer skills to model and analyze a system, Pritsker claims. Recent enhancements to the package, the firm says, have increased its applicability to automated systems. These new modeling capabilities for representing manufacturing systems include conveyors, fixtures, and conditional parts routing.

SIMAN, a FORTRAN-based package, has a lot of capabilities for simulating manufacturing system elements such as tote stackers, AGVs, and robotic systems without writing a lot of code, says Kerim Tumay of Confacs. SIMAN, too, executes on personal, mini, and mainframe computers. Models can be developed on a personal computer, loaded into a larger computer to get faster running time

Simulation of automatic guided vehicles, such as these shown here moving between automatic component insertion machines, requires specification of time to charge the battery as well as speed (Courtesy of Universal Instruments)

when required, and the simulation results loaded back into the personal computer for further analysis.

Confacs' SIMAN simulation models are constructed on three levels. Level 1, explains Tumay, is basically what happens between processes. Level 2 is what happens within processes, which also relates to the material handling system as well as how a particular process operates. Level 3 takes the detail of Level 2 and makes it user friendly for shop-floor people.

Tumay says they have found that generic models can be designed that require minimal changes for simulating a specific client's system on Level 1. Starting here, they can get an assessment of which areas in a manufacturing system are going to cause problems. Then they expand the detail of investigation in these areas by going to Level 2 and Level 3 models. At Level 3, the model can be used as a supplement to the user's current planning methods, such as MRP.

An extensive database characterizing various models of manufacturing machines has been developed by Confacs for use in their simulations.

Database include, for example, component insertion rates for a variety of specific automatic insertion machine models. Also included are parameters for preventive maintenance and setup times, as well as efficiency factors

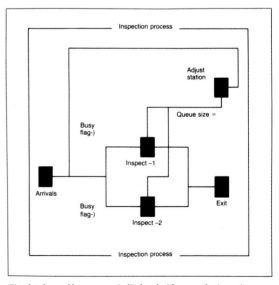

Simulated assembly movement is displayed with an overlay image in Simulation Software System's PCMODEL. This representation of an overlay is for TV sets moving through inspection

applied to the maximum insertion rate. Using this data, simulation has been applied to evaluate the impact of using different automatic insertion machine models under various conditions of production volume and component density on the printed circuit boards, remarks Tumay.

Another example, continues Tumay, is the simulation of a robotic system handling PCBs in an automatic test equipment (ATE) workcell. Here, they draw on a database for robots which includes work envelope, speed of the arm, minimum and maximum reach, and preventive maintenance.

Software packages

Following the pervasiveness of personal computers onto every engineer's desk, Simulation Software Systems has developed a screen graphics modeling system running on IBM personal computers. The firm's PCMODEL, says David White, has been developed to aid manufacturing and industrial engineers in their task of planning and analyzing production facilities.

This simulation package is designed to model the movement of manufactured assemblies through the assembly process. Objects, time, tools, and route are the fundamental elements through which users individualize models of their specific systems. Assemblies, such as PCBs, are referred to in PCMODEL as objects that are processed for specific lengths of time at one or more processing tools (such as automatic insertion machines) located along a defined route.

Complicated networks of routes involving bypasses, loop-backs, and crossovers can be modeled. The program's logic can sense pending collisions, then force the movement of a simulated assembly to be delayed until a free path is available.

Movement of assemblies is shown on the video display terminal in simulated time. An overlay screen can be defined and calibrated for the display to show an object's location with reference to features of the real system. The speed of the simulation can be slowed from its maximum to close to real time.

In addition to viewing an animated display as the simulation proceeds in time, tool locations may be specified for the collection of statistical data on machine utilization.

Production floor simulation is aided with a job release and tracking feature that allows the user to define up to 100 different jobs, with up to 65,000 assemblies per job. Each job can be assigned a priority of release to give a predefined sequence of assemblies flowing into the system. This is an example of a decision rule used in simulation.

Another decision rule in the model is specification by the user of maximum work in progress. Failures or path blockages can also be simulated.

AutoSimulations' InterFaSE (acronym for Interactive Factory Schedule EValuator) permits simulation of existing manufacturing operations in a manner that can be used by production or operations planners. Market introduction of the package is imminent, says the firm's national marketing manager Sam Kershaw.

Its emphasis, explains Kershaw, is on the factory processes and machine contention, not so much on materials handling. The concept addresses decisions that have to be made, for example, when loads arrive at a machine, and a choice has to be made on which load to accept based on predefined rules.

In use, InterFaSE consists of a model of the manufacturing facility, and an interface to the plant's manufacturing database. This allows the simulation to access the current state of all production orders and machines.

An automated system's programmable controller provides inputs to HEI's real-time simulator

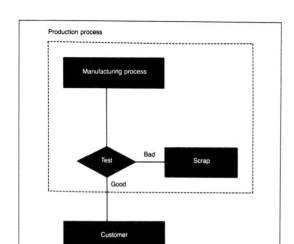

The production process is divided into manufacturing and test in Navone Engineering's Process Quality Simulator

With InterFaSE, planners can establish in the model actual conditions of materials, inventory, machine status, and personnel availability for a forecast period. Then, running the simulation with various operational scenarios, the best plan for conducting plant operations can be developed.

Robotics design, simulation, and analysis is the focus of ROBOT-SIM. AVailable from general Electric's Calma Co., this package is intended for a user to evaluate the performance of robots and their workcells. The simulation provides animation for motion path verification, cycle time analysis, and interference checking. It generates graphs of the robot's dynamics during a simulation run, including position, velocity, and acceleration.

A library of robot models includes kinematic and dynamic data. Though ROBOT-SIM comes with a library of several robot models, the user can also create library entries of his own. Libraries with different end effectors and workcell equipment are also used.

The simulation packages PCMODEL and ROBOT-SIM do not directly interact with the computers or machines of the real manufacturing system that is simulated. InterFaSE, however, interacts with the manufacturing database of the real system to obtain factory status and data for the simulation run.

HEI Corp.'s Real Time Computer Simulator, on the other hand, interacts with an automation system's programmable controller. The programmable controller executes the same program which it uses to control the automated system. Outputs generated by the controller are used as inputs to the simulator. They are processed in the simulation to imitate the system response.

Simulation in this case is used to assess software, in particular the program residing in the programmable controller, as well as hardware associated with the real manufacturing conveyor system and process machines.

Each manufacturing system element is described by the user through fill-in-the-blanks tables. These simulation tables are stored in a host computer, which may be an IBM PC or compatible machine. The tables are loaded into the non-volatile memory of HEI's simulator whenever a new simulation run is to be performed.

The simulator, through an interface, accepts commands from the programmable controller output, responds with input state changes, and graphically displays the system condition. The programmable controller inputs may be changed from the operator's console and the controller's response observed.

Modeling quality control processes as well as manufacturing is the object of Navone Engineering's Process Quality Simulator, indicates Alan Reddin. This software package uses Monte CArlo simulation, which is the generation of random numbers for the simulation's variables. The firm claims this is

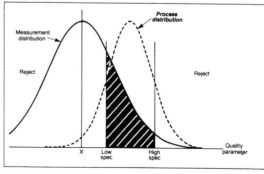

Monte Carlo simulation with Navone Engineering's software handles random variations in manufacturing and test. The shaded area is a measure of the fraction of assemblies manufactured with quality parameter X that the tester indicates are good, though they are bad

the first package using the technique that is available for running on small computers. IBM PC and Apple computers will support program execution.

The basis of the Process Quality Simulator is to assumme a production process consisting of two operations: the manufacturing step (where the assembly is actually made) and the inspection or test step (where the assembly is classified as being either good or bad). Furthermore, it is assumed that one quality parameter at a time is measured to determine whether a part is good or bad.

Production processes exhibit random variations in their quality parameters and test measurements have error associated with them. This simulator assumes both are normally distributed.

The user must specify the process mean and standard deviation for the quality parameter considered. Experience with the manufacturing process, statistics of field failures, and laboratory failure analysis can provide the user with clues for the quality parameters to consider and their quantitative characterization.

In addition, the standard deviation of the tester's random measurement error must be specified. The mean of the random measurement error distribution is assumed to be zero. It is also required that measurement bias error – or the systematic, constant measurement offset from the true value – be estimated.

After giving the computer information that defines the difference between a good and bad assembly, and specifying the type of inspection, it runs the simulation. Results give the percentage of the manufactured assemblies that the test measurements said were good, but were in fact bad products that may be shipped to customers. They also reveal the percentage of assemblies categorized as bad, but that are really good and are thus incorrectly scrapped or sent to rework.

LOCAL AREA NETWORKS

Factory automation is receiving a great deal of attention in the engineering community. Sometimes referred to as the factory of the future, the automated factory has become the focal point of American industries' efforts to improve quality and productivity, and to regain cost-competitive footing with its competitors. A key element of factory automation and improvement in quality and productivity is expeditious transmission of crucial information from one location to another.

Today, many production operations in the electronics manufacturing facility are performed by computer-controlled automated systems. Unfortunately, while much useful information is generated at these islands of factory automation, it is frequently not shared with other portions of the manufacturing operation which could use it. This is, in part,

due to the fact that few direct communications links currently exist on the manufacturing floor through which information can be readily shared.

Recognition is growing that a computer network can provide a direct link between these various intelligent positions, simplifying the exchange of information. This link can eliminate waste of physical movement and time to obtain information such as picking up a magnetic tape or disk, a paper tape or computer printout with the desired information. Such networking schemes, which are typically restricted to a single manufacturing site, are often referred to as local area networks (LANs).

Currently, however, very few manufacturing operations possess such a totally integrated system of information sharing. Nevertheless, some inroads have been made. Some manufacturers have taken it upon themselves to develop a network appropriate to their needs. GM has been actively pursuing factory

First published in *Electronic Packaging & Production*, 1984 Automation Supplement, under the title 'Networks bridge the gap between islands of automation'.
Author: Tom Dixon, Senior Editor.
© 1984 Reed Publishing (USA) Inc.

networking in automotive production. Automatic test equipment manufacturers have developed specialized networks with which a user can tie together similar test systems.

The establishment of a network permits access to information, previously unavailable or too costly to process, with which the user can make better decisions. Information from a CAD system can be transmitted to a CAM system, instructing it as to what operations need to be performed to build the product. Part movements can be tracked so that materials requirements and shipping dates can be established. Manufacturing data can be collected so that the designer can determine how good a design is made and what improvements could be made. Yield information can be collected which could help accounting and industrial engineering departments to determine manufacturing cost and plan future material purchases.

Network topology

Topology, the way individual stations are physically linked together, is a crucial issue in selecting a network appropriate to the application. An ideal network would permit each of the attached intelligent devices to obtain information from any other attached intelligent device on the network on demand and without much of a delay.

One way to accomplish this would be to have a direct link from each unit in the operation to every other unit. Unfortunately, such a distributed network scheme becomes cost prohibitive as the number of stations grows. Imagine the number of additional transmission lines required as the number of operating units changed from five to say 25.

One way to minimize the number of transmission media links is to have all the outlying systems connected to a central unit which acts as a switching center, directing communications between units. Such a network topology is sometimes referred to as a star network. One of the shortcomings of this system is that, if the central control unit fails, the entire network is shut down.

Another approach to networking is the ring scheme. Instead of going through a central control unit, a ring is established by linking one unit to another which in turn is connected to another unit and so on. Messages are passed from unit to unit until they reach the appropriate unit. A commonly cited shortcoming of such a topology is that the addition of another node (system) on the network requires that the entire network be shut down. Furthermore, if one node of the ring fails the entire ring may shut down. However, special ring schemes that remain operational even after their structure is severed were successfully implemented several years ago.

To get around these problems, a hybrid of the star and ring configurations can be used. The star-shaped ring runs each of the links through a central hub. If a node failure is sensed it can be bypassed on the central hub, permitting the network to continue to operate.

One topology that addresses the problem of having to shut down the network is the bus construction. In the bus system each device is attached to a common cable by an individual link. The premise is that each device can be attached and detached without reconfiguring the network.

The network medium

A variety of materials can be used as the transmission medium on which the network carries signals from one node to another. Among these are fiber-optic cables, twisted wire pairs, ribbon cable, coaxial cable, CATV cable and several others. Signals can also be transmitted in the form of IR and microwave radiation.

Several factors come into play when selecting the medium from which the physical network will be constructed. Among these factors is the amount, and cost, of the cable that must be strung to connect network systems.

Unlike the case of tying together test instruments, in which relatively short distances are involved, local area networks in a factory normally involve much greater distances between attached units and therefore significantly more cable.

As a consequence of cable costs, instead of employing interconnection techniques similar

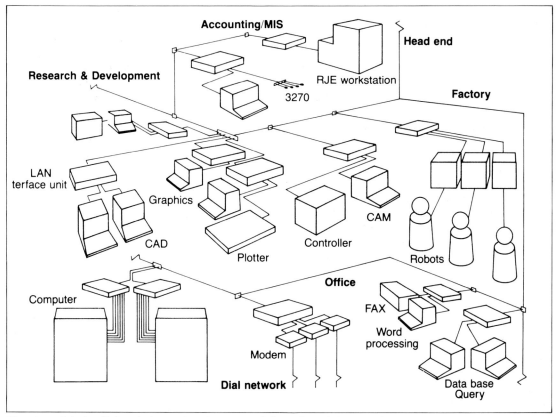

Networks permit sharing of information between the various systems on the factory floor and other support functions. This ability to share information holds out the promise of improved efficiency for business operations by permitting network users to make better decisions in a more timely manner

to IEEE-488 with its multiple conductor cables that permit parallel data interfacing, local networks used in manufacturing usually involve sequential data transmission and the associated cable interconnection techniques.

Environmental considerations such as electromagnetic interference and the operating temperature range over which the network is expected to work, can also influence the medium selection. For instance, a fiber-optic cable which is immune to EMI may be appropriate in some manufacturing facilities. On the other hand, a shielded electrical cable which is less expensive than the fiber-optic cable may be totally adequate for shielding out unwanted EMI in another case.

There are other technical considerations to medium selection, such as the data rate of the signals to be transmitted and the path length over which the signal must be transmittable. If

the required data rate is 10 megabit/sec (10 Mbs), a coaxial cable medium might be appropriate, however if the required data rate is 50 Mbs, fiber-optic cable might be a more appropriate medium. Along a similar vein, if the distance over which the signal must travel is significantly long, fiber optics or interconnection techniques such as those used by the cable TV industry are appropriate.

Two other terms are frequently mentioned when describing a network medium. These are baseband and broadband. Baseband simply refers to a medium in which only one channel of information exchange exists. Broadband refers to the medium over which several channels of communication exist simultaneously. Baseband mediums usually involve some sort of signal pulsing such as alternating voltage levels or transmitting bursts of different frequencies to represent data. Broadband transmissions typically involve some form of signal modulation to represent data and usually use established CATV techniques.

Accessing the network

Aside from selecting the networking medium mechanism, the mode in which the system will gain access to the network must be determined. Without some control scheme the network will not function.

In a star-type configuration where a central unit controls the network, each station can be polled by the control unit. When one wishes to send a message, the central unit simply informs the control unit. Unfortunately, such a scheme ties up the control unit in administrative functions.

One approach, which eliminates the need for a central administrative unit, is to schedule a fixed time slot for each unit to access the network. However, such an approach can waste valuable time. If a station has no message the network must wait until the allocated time slot has expired before turning control over to another unit.

A method to eliminate wasting time by tying up the network with a station that has no message to transmit is to permit each system connected to the network to monitor the network traffic. If there is no activity on the network it is permitted to transmit. This scheme is sometimes known as carrier-sense multi-access (CSMA).

A shortcoming of such an access method is that two units on the network might try to place a message on the network simultaneously. In order to work, the transmitting stations must be able to detect when a collision has occurred. If a collision occurs, the colliding stations must back off and try again after an appropriate wait.

One criticism of the CSMA method of accessing the network is that there is no assurance that a unit will gain access to relay a message when needed. This could be a significant problem where process control is being performed and a delay in relaying a message is caused by a crowded network. Proponents are quick to point out that in most instances network loading is not high enough for this to be a problem.

The token passing access control scheme offers a means of assuring that each unit on the network is guaranteed access to the network on a predictable basis. A code known as the token is given to one of the stations on the network. Only the station with the token is permitted to transmit. Once the holder of the token has finished transmitting his message or the allotted transmission time frame has expired, the token is passed on to the next position.

If a station does not have a message it simply passes the token on to the next station. Through the use of multiple token configurations, it is possible to give more frequent access to the network to more crucial positions. Such stations may be allowed to access the network on an every other time slot basis (i.e. first, third, fifth . . .). Less important stations may be allowed to access the network less frequently on a second, sixth and tenth basis, for example.

Some 300-plus types of networks are presently available on the open market. A major portion of these are targeted for the office environment. However, many of these claim

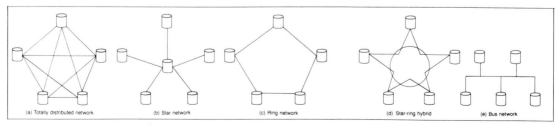

Network topology, the way devices are physically linked or connected together, can take on a number of forms. The illustrations are simple representations of commonly used network topologies. Often it is hard to determine what topology is actually being used; factory systems connected to the network might not be as neatly organized as our representation. (a) totally distributed network, (b) star network, (c) ring network, (d) star-ring hybrid, (e) bus network

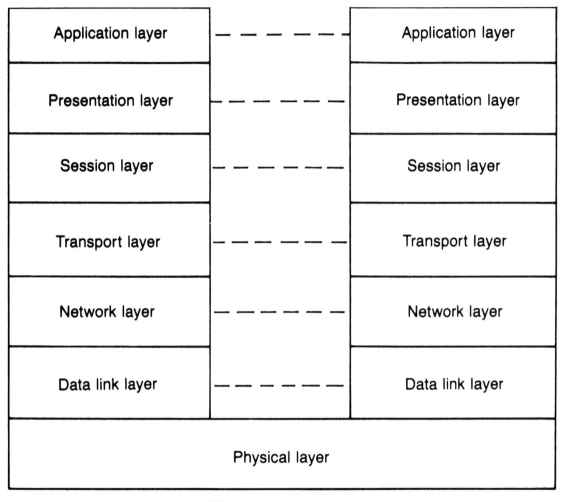

ISO model of network

Unfortunately, the networking is not just as simple as establishing the physical link between the computer based systems. A number of other hardware and software solutions are required to make the network fully functional. The ISO has developed an ideal layered model of the tasks which must be performed to achieve a functional network

to be adaptable to the needs of the manufacturing operation. Such claims must be examined with some degree of scepticism. What is more significant, however, is that many of these products are incompatible because of the numerous possible design approaches that can be taken.

IEEE-802 standards

Fortunately, there seems to be help on the horizon that might alleviate some of the confusion. A few years back, the IEEE formed

the 802 committee specifically to develop standards for networking. Interestingly, atypical of the normal mode of operation of computer systems manufacturers in which everyone goes off and does his own thing, they have for the most part participated and supported the efforts of the committee. Even the International Standards Organization (ISO) appears ready to adopt the IEEE standards as its own with few modifications.

The committee has been divided up into various subcommittees, some of which are developing standards for various networking schemes. What the committee has done allows some flexibility in the choice of the networking method while affording a degree of uniformity and compatibility to manufactured products.

Three of the IEEE 802 subcommittees are developing networking standards. These standards cover such issues as access methods, topologies, physical mediums and data transmission schemes.

The 802.3 group is developing standards for a carrier sensor multi-access network with collision detection (CSMA/CD), including one based around Ethernet. It has been working on both a broadband and baseband bus system.

Meanwhile, the 802.4 has been developing a standard for a token bus network. Here again the option exists to use either broadband or baseband medium. It should be noted that GM has chosen the broadband token bus network method for use in its operations.

A standard for baseband token bus ring networks is being developed by the 802.5 group. This standard specifies a point-to-point ring topology which is accessed by token passing and requires operating stations which are not transmitting to repeat transmissions from other stations.

One criticism of IEEE 802 committees has been a lack of network standards on which fiber optics are used as the transmission medium. However, IEEE 802 appears to be addressing this problem with the establishment of a fiber-optics advisory group.

ISO network model

Unfortunately, the physical link provided by the networking methods being addressed by IEEE 802.3, 802.4 and 802.5 are just a part of the networking problem. It only provides the means of getting information bits from one node to another.

These bits of information still must be made useful to the receiving system. The ISO network model represents the types of hardware and software functions that are necessary to make information exchange usable. This idealized model is divided into seven layers: the application, the presentation, the session, the transport, the network, the data link and the physical. Each layer performs a number of tasks. IEEE is mainly concentrating its efforts on the physical and data-link layers of the ISO node.

The bottom layer is the physical layer that connects one node to another and gets the bits of information from one node to another. This is primarily what is addressed by IEEE 802.3, 802.4 and 802.5.

The next layer is the data link layer. This layer puts the data into block format, adding information such as who is talking to whom as well as source and destination address. IEEE 802.2 is addressing some of the aspects of the ISO model.

Above the data link is the network layer. One of the functions of this layer is to concern itself with routing the messages.

Next comes the transport layer which is concerned with end-to-end reliability of the network communications. It provides the feedback mechanism which assures that what was transmitted from one node is received and understood at the other node.

The session layer is concerned with establishing the dialogue between the nodes. One of its responsibilities is to initiate and terminate tasks within the node.

The presentation is responsible for making sure that communicating nodes are speaking the same language. If one node is using a different data format it would translate the data into a form which the application layer could understand.

The top layer of the ISO model is the application layer. It is concerned with what the user is doing. It is the interface between the user and the network.

Standards for the various layers of this ISO model are currently being developed by the X committees, a group of the Computer and Business Manufacturer's Association as well as the ISO.

PHYSICAL INTEGRATION

The concept of linking islands of automation is more than control linkage. It includes the physical means of carrying production from material storage, fabrication, assembly, joining, testing and packaging.

Just as insertion machinery depends on parts with known orientation of location, so do all other elements of manufacturing.

These systems must prevent physical, electrostatic and electromagnetic damage to the products. It must provide buffering between islands of manufacturing. It should provide for a minimum of inventory in process and still provide for immediate response to customer order. The following papers discuss material handling appropriate to electronics assembly.

MATERIALS HANDLING ACCESSORIES

The effectiveness of materials handling systems involved in PCB fabrication, such as automated storage and retrieval, and computerized transporters and conveyors, depends in large measure upon how well the various accessory items are integrated into the overall operation. These important but often-neglected accessories are usually overshadowed by the major emphasis placed upon the automation and control aspects of the materials handling system. However unsophisticated they may appear, accessories such as tote boxes, work organizers, handling fixtures, carts, static protective containers, etc., are necessary to the efficient operation of the total system.

In addition, many of these items occupy an important niche in materials handling and storage of parts and subassemblies for manual PC assembly. They also play an important role in the manual routing of materials. Tote boxes, carts, roller conveyors and manual workstations all combine to form a well-organized assembly line.

According to Marshall Industries, the wide variety of materials handling accessories and production aids available provides a fully integrated approach to the assembly environment, ensuring that each part or component within the work area has a dedicated location. Knowing the exact location of each assembly item offers motion economy for the operator, efficient plant housekeeping, ease in identifying materials on the line, and ease in materials handling to and from the line.

In essence, the economies realized by using well-design materials handling accessories are achieved through: reduced number of handlings of individual subassemblies or components; reduced product damage in transportation or storage; reduced direct labor through organized job kitting and workflow; and reduced assembly reject rate through accurate identification and organization.

First published in *Electronic Packaging & Production*, May 1984, under the title 'Materials handling accessories govern production efficiency'.
Author: Howard W. Markstein, Western Editor.

Static protection in demand

Ensuring the efficient routing, containment and protection of assemblies and components along a production line, whether automated or manual, together with associated storage and inventory requirements, is the task of the materials handling accessory. What are these accessories and how do they perform this vital function? Suppliers say that the high-demand items are anti-static or conductive tote boxes, parts bins and PCB storage/carrying devices or fixtures. Other important items include kitting trays, static-free bags, PCB carriers, dollies, pallets, transfer carts and mobile racks.

In a typical storage and dispatching area where PCB assembly kits are routed to various workstations via conveyor or carts, the parts are often housed in static-protective containers which provide efficient work organization and materials handling.

3M Co. says that all tote boxes, containers, bins, etc., should be statically shielded to protect today's sensitive devices from electrostatic damage. Tote boxes, especially, must be made of materials that protect both against electrostatic fields and electrostatic discharge, plus have the capability of being grounded. In addition, tote boxes and containers used in routing work through automated or manual assembly lines should meet mechanical requirements. These include vibration and shock resistance, impact strength, abrasion resistance and friction.

Polypropylene is the most common material used for fabricating tote boxes. Other materials include polyethylene, ABS and fiberglass. Totes can also be of open-frame metal construction. This allows possible use as a holding fixture during some cleaning and temperature testing operations.

It is now well accepted that some form of static control is also needed in PCB materials handling and assembly operations. Damaged devices on assembled boards are a threat to profits, and wakened or partially damaged devices installed in a customer's system pose a potential threat to the supplier's reputation.

Assembled PCBs are shown inserted into a conductive plastic tote box equipped with a cover for maximum protection against electrostatic fields. The inside walls of this box are adjustable to fit different board sizes. Another technique to provide size flexibility is by use of inserts (Courtesy of 3M)

Two simple rules are advised for developing an effective static control process:

- Handle all static-sensitive assemblies or components at a static-controlled work area.
- Transport all static-sensitive items in static-shielded containers.

Degree of conductivity controversial

Static control products can be categorized

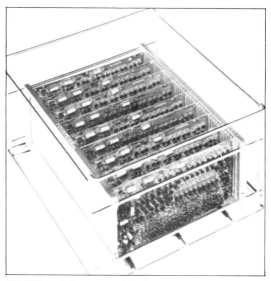

This PCB carrier features a metal wire frame with adjustable aluminum sidewalls for size flexibility, light weight, durability and conductivity. It can be used through chemical cleaning processes, in temperature cycling ovens, and can act as an insert for a plastic tote box (Courtesy of Protecta-Pack)

depending upon their resistivity. Materials handling accessories are therefore either electrically conductive, static dissipative or antistatic. There is much debate among users and manufacturers as to which materials and resistivity levels provide optimum protection.

Alacra Systems reports that the controversial range generally runs from $10^{12}\square$ surface resistivity for anti-static materials to $10^5/\square$ for conductive materials. Doubt exists as to whether anti-static offers adequate protection, and many claim that conductive materials dissipate charge too quickly.

3M explains that an anti-static tote box or parts container is formulated with a hygroscopic agent that attracts humidity, giving it a surface resisistivity of about $10^{12}\square$. This resistivity prevents triboelectric charging but does not provide shielding from electrostatic fields. A conductive tote box, however, is constructed of a plastic impregnated with a conductive element. This type of structure is volume conductive rather than surface conductive. 3M emphasizes that materials handling accessories made from conductive materials will not allow external electric fields to penetrate, thereby preventing a buildup of electrostatic charge on parts within the tote box or container.

The Simco Co. claims that containers for PCB handling and storage should be electrically conductive – carbon-loaded polymer – with a resistivity of $10^5/\square$ or less. For best protection, the containers, or totes, should be sealable with lids and be designed for use with automated storage and retrieval systems – a concept predicted for future wide use. The containers should also be of modular design, with appropriate inserts, to prevent becoming obsolete due to changes in PCB sizes.

Shielding bag popular

The static-shielding bag is also an important accessory item. According to 3M, the transparent static-shielding bag is the accessory item of highest demand in PC assembly. Once an assembled PCB is placed inside the bag, it may still be transported in a tote box, tray or other container. An advantage of the shielding bag is that the assembly may be removed from the tote box and handled without the fear of causing damage through static charge. 3M emphasizes that the transparency of the bag permits visual inspection of the contents, while the low resistivity of the surface creates an envelope of static shielding.

There are a host of other products available to provide protection against harmful static discharge, although not directly related to materials transport. These accessory items include floor and table mats, seat covers, wrist straps, garments, shoe grounders and ionized air blowers.

Varied PC handling accessories

The well-designed, conventional plastic tote box is ideal for use with manual or conveyorized work flow and storage systems. There are also many other items used for the transport of PC assemblies and bare boards.

An aluminum carrier with adjustable bulkheads to handle varied boad widths can be fitted with conductive or non-conductive board guides, and the entire assembly is stackable for reduced floor space. Aluminum carriers offer light weight, sturdiness, and provide inherent static shielding.

Instead of metal, a PCB carrier or handling tray may be made of a conductive plastic held together by metal end pieces. The supporting sides are adjustable in span to accept varied board sizes. When filled, the boards are secured by a metal bar fastened over each open side. This design typifies many of the simple, low-cost alternatives devised to provide support and to transport PCBs. Another version of this principle is one of all-metal construction.

Perhaps the simplest method of board handling and transport is in using the correct type of tray. The crosscut recessed slots allow inserting boards in two directions. The base can be of conductive plastic for static control. Although this type of tray can be used as shown, it is also meant for insertion into a tote box, thereby becoming the base and PCB holder for the tote.

Carts and dollies also form an important aspect of a materials handling system. They provide batch movement of product throughout the work area and also act as a natural storage medium. Carts are used not only as

Static-shielding bags are an important materials handling accessory, finding wide use in the protection of assembled PCBs during transport and storage (Courtesy of Simco)

carriers for tote boxes but are also used for the handling and storage of finished products. The cart forms an integral part of the assembly workstation, whereby stuffed boards are inserted into open carriers for transport to the soldering station. This cart is often referred to as a mobile rack.

Work holding fixtures, although usually not used as a transport item do provide a

All-aluminum PCB carrier can be transported by hand, cart, conveyor or stacked on dollies. The unit has adjustable ends and internal dividers and is available in varied-length top and bottom panels (Courtesy of Multi-Tool & Manufacturing)

PCB handling rack is formed using two handling trays joined by metal end supports. The trays are of conductive plastic for ESD control. Spacing is adjustable to accommodate different board sizes (Courtesy of Calmark)

This all-metal carrier features ease of adjustment and inherent static control (Courtesy of Inter Metro Industries)

materials handling function at the workstation. PCB fixtures used at workstations are accompanied by a means of having components readily accessible to the assembler, such as open-top plastic bins.

Another common method of presenting components to the assembler is by use of an array of circular open-top plastic bins employing the 'lazy Susan' tray concept. For semiautomatic component insertion, programmable assembly directors provide built-in board fixturing and integral bin sequencing for component selection.

Flexibility and durability

Suppliers of the various accessories discussed here advise that accessory items be evaluated as stringently as that of major materials handling equipment purchases. Specific factors to evaluate include function, durability, space utilization and static control effective-

ness. A major consideration is in evaluating the accessories with the goal of maximizing the efficiency of large automated materials

PCB tray has a conductive plastic base and removable carrying handles. The tray can be used as shown or as an insert for a tote box. Cross-cuts allow use with different board sizes (Courtesy of Conductive Containers)

Pallets loaded with finished products rest on rollers awaiting transport by pallet carts or dollies. These particular pallets can interlock and be stacked for greater capacity. The cantilevered shelves are adjustable for different heights (Courtesy of Herman Miller)

handling systems while increasing the efficiency of manual materials handling.

Multi-Tool and Manufacturing advises the user to determine the requirements for each production or assembly step. The accessory that fits most of the requirements is a first choice; secondary requirements can then be met by further evaluation. In general, those accessories that are adjustable to accommodate size changes and possess durability are usually the best choice.

The user is advised to consider the following questions when choosing materials handling accessories:

- Will the tote, tray, cart, etc. accept the work geometry?
- Is it adjustable and adaptable to varying sizes?
- Is it durable enough to withstand the expected environment?
- Has static control been incorporated?
- If necessary, is it compatible with automated materials handling and robotics?
- Is space utilization (such as stacking and nesting) maximized?
- Is the cost compatible with the quality?

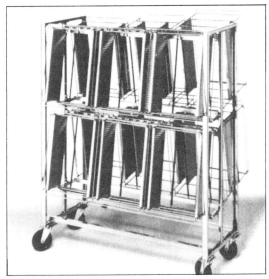

This car serves as a mobile rack for accepting PCBs before and after assembly. Carts of this type feature high-capacity batch movement and storage (Courtesy of Protecta-Pack)

PCB assembly fixtures provide a secure mounting for manual component insertion. This fixture has an integral parts bin that rotates for access to the opposing bins (Courtesy of Henry Mann)

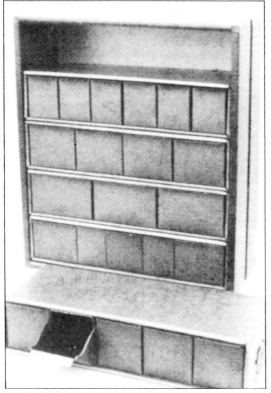

Parts bins hold components for PC assembly and can be positioned at the workstation or used for compact storage. These bins are of static-protective material (Courtesy of Alacra Systems)

System integration

Many materials handling systems integrate the various accessories with each other and with the workstation. The intent is to provide a non-dedicated system adaptable to the work being produced at any particular time. For example, Herman Miller inc. is marketing a materials handling system of standardized modular design that can be easily reconfigured by the use of various inserts and subcontainers, totes doubling as workstation drawers, carts with adjustable and removable shelving, pallet/cart combinations where totes are cantilevered from a center support to allow easy access, and easily configured workstations constructed of metal framework and hanger rails for accepting totes and individual subcontainers. Tote box console assemblies can also be moved about via dollies, brought to the workstation and hung on a support rail within the operator's reach.

According to Herman Miller, an integrated, flexible system of this type can be configured to fit the customer's process rather than modifying the production process to fit the materials handling system. The meshing of the materials handling function with the bench operation leads to an efficient approach to closing the production/materials-handling process loop. Herman Miller advises

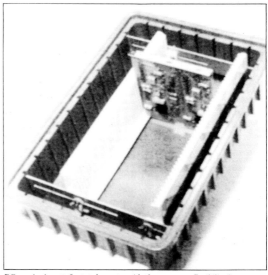

PC carrier inserts for tote boxes provide the necessary flexibility for accommodating changes in board sizes (Courtesy of Fancort Industries)

the user to consider only those materials handling systems and accessories that are flexible, durable, non-dedicated, human factored and readily available.

In general, for any materials handling system the proper choice of accessories is vital to the function of the total system. Accessories can determine whether or not a system becomes functionally efficient.

"To ignore the details of container size, style and function is to negate the value of the materials handling system being installed," says Alacra Systems. "Material handling accessories should be specified as an integral part of the total handling system design, rather than as an afterthought."

AUTOMATED PCB HANDLING

As the electronics assembly industry continues to rise from the slump of 1985 and 1986, we find ourselves confronting a new wave of terms, phrases, acronyms, and philosophies concerning manufacturing. Among these are CIM (computer integrated manufacturing), Just-in-time, pull-system, factory integration, MRP (manufacturing resource planning), et cetera, all promising to solve the problems of manufacturing.

Although there are merits and truths in the explanations of these concepts, the fact remains that market position depends on the ability to reduce costs, adapt to changes, and improve quality. With regard to this, manufacturers are scrutinizing every aspect of their electronic assembly operation. And an area deserving special attention is board handling.

Addressing the issue

The phrase board handling refers to the fact that in any electronic assembly system, boards must be moved from point 'A' to point 'B' to point 'C'. In the process of assembling a printed circuit boad, each board must be moved from a stack of empty boards, through an assembly process, to a location for finished boards.

Traditionally, this chore has been performed manually. Boards were carried, by hand, from assembly machine to assembly machine, where they were automatically populated with electronic components. This method of manufacturing, typically referred to as islands of automation or off-line processing, was the first step towards a larger vision of automating manufacturing.

For years, the off-line production method was sufficient. Manufacturers were busy reaping the benefits of automatic assembly machinery, and any problems concerning board handling and process integration were a back burner issue, at best.

But the times have changed. Automatic component placement is no longer a luxury, it is a given, and manufacturers are looking for any back burner issue that may give them an edge. What they are finding is that many of the same problems solved by automating board assembly can be solved by automating board handling.

But automating board handling, board transfer time is decreased, board quality increases, and board routing problems are eliminated. Automated board handling installs a discipline in the assembly process that immediately provides the following benefits:

- Board transfer time is pre-determined and guaranteed. Consider this, especially with regard to surface mount technology; after an adhesive or solder paste has been applied, components must be placed within a specified time. After component placement, curing must occur within a specified time.
- The potential for handling errors is eliminated. A properly installed board handling system will not drop or mishandle a circuit board.
- Overall process times can be recorded and governed. Problems in the factory, such as

First published in *Assembly Engineering*, July 1987.
Author: Bob Osterhout, Universal Instruments Corp.
© 1987 Hitchcock Publishing Co.

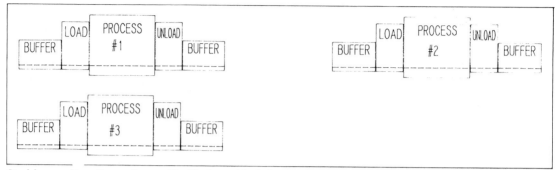

Standalone assembly machines require manual handling between islands of automation. This normally means quantities of PCBs accumulate at the input and output of each machine

disabled or slow machines, can be detected easily by examining the entire process flow.

Having witnessed the results of automating board assembly, it is easy to perceive board handling as the next target for improvement. But as with assembly automation, board handling presents a new field of problems and questions.

Incorporating automatic board handling

One of the main problems with traditional off-line processing is the inherent ability of the system to accumulate work-in-process (WIP). WIP, as we know, is simply inventory – money tied up in the factory.

When analyzing the efficiency of an assem-

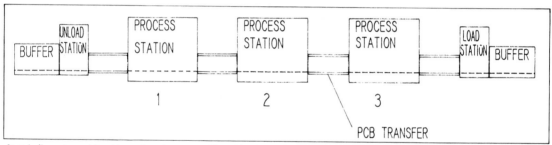

In an in-line system each board passes through every assembly machine. Manual handling is eliminated, but the slowest machine or a breakdown can limit system speed

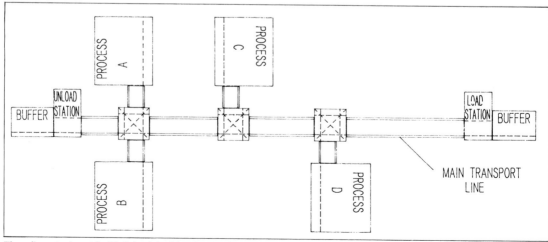

The on-line system is most flexible. Printed circuit boards visit only those machines necessary. Programming the transport system is more complex, however

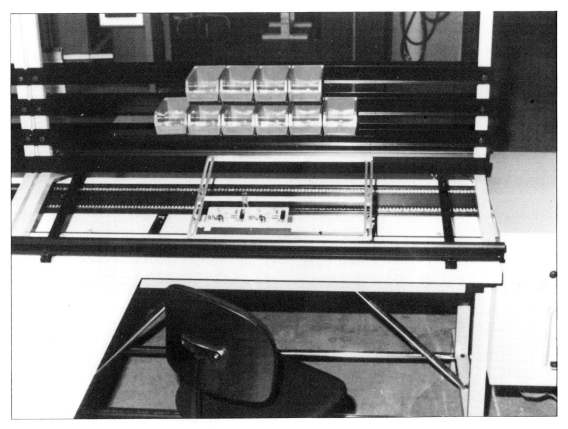

A slide rail and fixtures carry PCBs from assembly station to assembly station in a manual assembly system (Courtesy of Isles Industries)

bly operation, the WIP serves only as a cover for problem areas. Because of this, automated board handling is best addressed as a single board process. Certainly, off-line, batch-oriented operations could be automated via magazines and automated guided vehicles, but the potential for accumulating WIP remains.

Single board processing, however, lends itself easily to the Just-in-time philosophy. By allowing the transfer of individual boards, a series of assembly machines can be set up as a true, Just-in-time pull system; that is, a system in which boards are transferred from machine to machine on demand, rather than in batches.

This pull style of manufacturing decreases the potential for WIP. When a machine is disabled, the entire process stops, and rather than allowing the other processes to continue to build WIP; all attention is focused on the problem area. Not only does this discipline cause reduced WIP, but when coupled with an effective program of preventive maintenance, it ensures that machine down times will be less frequent and shorter in duration.

Single board processing, then, becomes a method of streamlining the assembly operation and removing the cushion of WIP. The details of incorporating such a system, of course, still depend on the goals of the manufacturing facility.

Designing a board handling system

One of the benefits of the off-line processing method is that the process worked the same regardless of product mix or volume. Tote bins were carried from machine to machine, and a new product simply meant storing the bin for as long as it took to set the machine up for the new product.

Where lot sizes are sufficient and a certain amount of WIP is allowable, magazines can accumulate a number of PCBs before transfer to the next machine

A single board processing system, however, presents new issues. For example, if a group of machines is to be connected with a single board handling system, how can we ensure that it will be flexible enough for high mix, low volume jobs, yet fast enough for low mix high volume jobs?

In response to these issues, two fun-

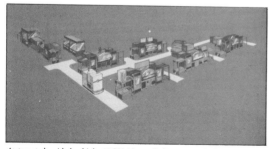

Automated guided vehicles (AGVs) instead of conveyors tie this assembly system together, carrying magazines of PCBs from machine to machine

damental system designs have evolved: in-line processing and on-line processing.

In-line processing systems

In an in-line system, assembly machines are connected in a continuous line, via board transfer hardware. In such a system, there is a single, fixed routing path for every board entering the system, and every board passes through every machine.

An in-line system is typically set up for low product mix, high volume operations. The system is fast because of short travel distances and the absence of board routing decisions.

Initially, in-line systems were viewed as inflexible because of the fixed routing path, but further investigation revealed that these

This conveyor board inverter will automatically, in-line, turn over a PCB 180° for additional operations on the reverse side (Courtesy of Dynapace)

Most automatic insertion equipment such as this Universal Instruments SMT placement machine can couple by conveyor directly to the next or previous assembly station. Here the connection is to a reflow solder system

systems could, in fact, be configured to handle more than one product type. By installing machines with larger component inventories and sequencing capabilities, varying product types could be handled by changing the software setup only. And with the advent of bar code capabilities, the process of machine software setup and changeover has become an automated reality.

Although an in-line system can be efficient and flexible, complete product flexibility requires an on-line processing system.

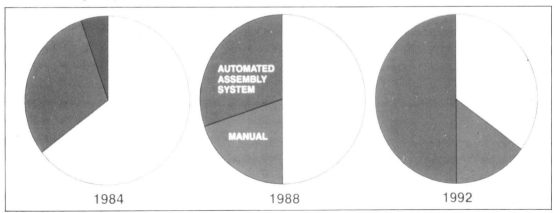

The nature of electronics assembly has changed and continues to change from manual, to automatic machine, to automated assembly system

On-line processing systems

In an on-line system, assembly machines are connected to a central board handling system via single board transfer mechanisms.

In such a system, a board is transported through the central board handling system until it reaches a board routing intersection. Then, depending on the needs of that product type, it can be transferred into a machine connected on-line at that point.

This method of board handling allows a system to be designed with a wide combination of machines, for a wider range of products. Since there is no fixed routing path, boards only visit the machines required for assembling them. Such a system also allows selective routing for rework and line balancing.

In addition, a correctly configured on-line system can increase floor space and maximize machine usage by allowing the system to process multiple products simultaneously.

Overall, the on-line system provides the greatest manufacturing flexibility. When this flexibility is not required, the problem of automatic board handling can be addressed through an in-line system.

Conclusion

The benefits of automatic board handling are obvious. By automating yet another process of electronic assembly, we continue to increase quality, while decreasing production time and cost. Although traditional off-line processing served its purpose, it is time to rearrange the machines and add board handling hardware. And for new systems, any design that excludes board handling is starting out with an established flaw.

AUTOMATED GUIDED VEHICLE SYSTEMS

Today, automated guided vehicle systems are one of the most exciting and dynamic areas in material handling. Through the years, developments in electronics have led to the rapid advancement of these systems. These ongoing changes may have given AGVS more flexibility and capability, but market acceptance has really given the technology application variety to allow it to expand into the standard accepted material-handling method it is today.

As with any new and relatively unknown commodity that experiences such a growth in popularity, there is a dire need for more information and knowledge about how these systems works, what makes them tick and how to get the most out of them. No doubt, if you plan a conveyor system or a tow-chain system, you probably have many knowledgeable people within your organization to call upon for advice and assistance. However, when it comes to planning an AGVS, you'll quickly find there are not nearly as many people who know how they work or how to use them properly.

Functions

To begin with, automated guided vehicle systems have five specific functions. The proper method selection for each function and its ability to work with the others determines, in great measure, the degree of success of a system. These functions include:

- Guidance – this allows the vehicle to follow a predetermined route which is optimized for the material flow pattern of a given application.
- Routing – the vehicle's ability to make decisions along the guidance path in order to select optimum routes to specific destinations.
- Traffic management – vehicle ability to avoid collisions with other units at the same time increasing vehicle flow and load movement throughout the system.

First published in *Machine & Tool Blue Book*, June 1985, under the title 'A primer on AGVS'.
Author: Gary A. Koff, Mannesman Demag.
© 1985 Hitchcock Publishing Co.

- Load transfer – the pickup and delivery method for an AGVS which may be simple or integrated with other subsystems.
- System management – method of system control that is used to dictate operation.

Vehicle types

To carry out these fundamental operations, a number of vehicles are available. Most of the units in operation across the country fall into one of six basic categories:

- *Towing vehicles.* These were the first vehicles introduced and are still popular today. Towing vehicles can pull a multitude of trailer types and have capacities ranging from 8000 to 50,000 pounds.

- *Unit load vehicles.* Unit load types are equipped with decks which permit unit load transportation and automatic load transfer. The decks can either be lift-and-lower type, powered or nonpowered roller, chain or belt decks, or custom decks with multiple compartments.

- *Pallet trucks.* These trucks are designed to transport palletized loads to and from floor level positions, eliminating the need for fixed load stands.

- *Fork trucks.* The fork truck is a relatively new guided vehicle which has the ability to service palletized loads, both at floor level and on stands. In some cases, these vehicles can also stack loads in racks.

- *Light-load vehicles.* These carriers have capacities in the neighborhood of 500lb or less, and are used to transport small parts, baskets, or other light loads throughout a light manufacturing environment. They are designed to operate in areas with limited space.

- *Assembly-line vehicles.* Assembly-line types are an adaptation of the light-load AGV for applications involving serial assembly processes.

Guidance

A common misconception is that guided vehicles can be made to do anything. With so many new vendors on the market, it is very difficult to differentiate what is practical from what is not. In order to appreciate what you can and cannot do with AGVS, an understanding of vehicle control would be helpful.

AGVS steering control allows vehicles to manoeuvre in different ways. There are two basics types: differential speed and steered wheel.

Differential-speed control uses two fixed wheel drives and varies the speeds between

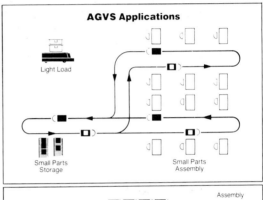
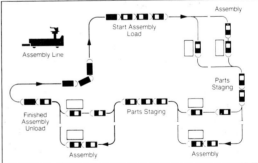
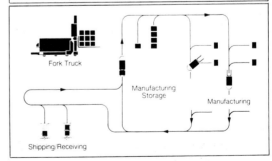

Typical light-load applications (top) of AGVS distribute product from small parts storage areas to workstations where light assembly is accomplished. Employing assembly-line systems (center) provides individual tracking of assemblies and measured work rates. Fork truck applications (bottom), new to the material-handling industry, include added flexibility in integrating other subsystems

AGVS pallet trucks, like this one, are generally used in distribution functions

the two drives, on either side of the guidepath, to permit the vehicle to negotiate a turn,
similar to the way a tank or a tracked vehicle
turns. An amplitude-detection guidance sensor is used to signal the steer system and is
based on balancing the signal received from
the left and right sensors in front of the
vehicle. When one or the other signal is
greater, the steering system compensates by
correcting the steering until the amplitude of
the left or right signals are equal.

Steered-wheel control uses automotive-type
steering in which a front-steered wheel turns
to follow the guidepath. Phase-detection
guidance is usually used for this type of
steering, and incorporates a sensor which
differentiates whether the vehicle is to the left
or right of the path by detecting the positive
or negative phase of the signal received from
the guidepath wire. This information causes
the vehicle to correct its steering so that there
is no phase difference.

Steered-wheel control is used in all types of
vehicles; however, differential-steer techniques are not used in towing applications or on
vehicles with man-onboard controls.

In space-limited applications not requiring
high-throughput volume, differential-
controlled vehicles are used in pivot steer
modes. In situations like this, the vehicle stops
on the main line, rotates 90° and proceeds
into, or out of, a station. If sufficient room is
present, a normal radius curve is often better
from the standpoint of flow, and controls
simplicity.

Steered-wheel-type vehicles have excellent
guidance tolerance along the guidepath.

Differential-steer vehicles normally have less
tracking tolerance and therefore require
external guides or positioning devices at load
stations when used in automatic load-transfer
applications.

Routing

Routing, another fundamental control function of AGVS, includes two approaches:
frequency select or path-switch select.

Frequency-select routing. In frequency-select

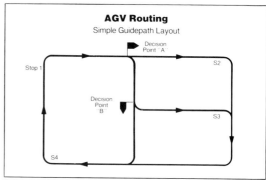

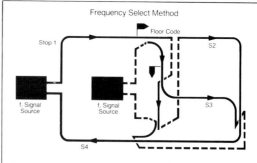

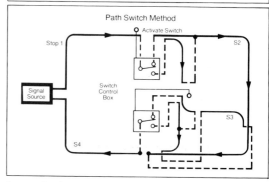

*In comparing the two routing techniques, frequency select and path-switch
select, the frequency methods relies on the vehicle itself to choose the proper
guidepath. Floor controls cause the vehicle to take a specific route when
path-switch select is used*

systems, the vehicle approaches the decision point and reads a marker in the floor which identifies the vehicle's location. This marker is a passive code device which is usually in the form of buried magnets, metal plates, or other code devices. When the vehicle is approaching the decision point, it is following a single frequency. At the decision point, two frequencies are present. The vehicle, depending on which direction it wishes to go, selects a frequency to follow and the routing is automatically accomplished.

This method requires that at least two different frequencies be present at a decision point for a two-path decision. If there were three paths to select from at a decision point, then three different frequencies would be required. The multiple frequencies are only required at the decision point or at a convergence point where multiple paths come together.

In frequency select, the frequencies loop through the system in a continuous wire and are used over and over wherever they are required. They are always active, resulting in reliable, long-life operation.

Path-switch select routing. In the path-switch select method, when the vehicle approaches a decision point and passes an activation device, it causes one path to be turned on while the other paths at the decision point are turned off. The vehicle has one live path to follow and the routing is then automatically accomplished. In this case, the vehicle has to communicate to the decision point switch control which path it wishes to follow, based

Towing AGVS, the earliest designs available, involve movement of product in warehouses

on the stop locations to which it was routed.

The guidepath is divided into segments which are switched on and off by separate floor controls. Normally, these systems use one frequency, and the paths for divergences and convergences are switched in and out as required by the vehicle in the area. The routing requires external floor controls to switch paths on and off, which usually has an impact on reliability and maintenance. The controls necessary for path layouts using this method can become quite complicated.

Traffic management

Traffic management, another element of control, concerns the movement of vehicles from one part of the system to the next. There are three types of this management in general use today: zone control, forward sensing and a combination of the two.

Zone control. Zone control is the most widely used and permits one vehicle in a given zone at a time. When a vehicle occupies a zone, the closest a trailing vehicle can get is into the next completely unoccupied zone behind the lead vehicle. The lead vehicle must then proceed into the next zone before the trailing vehicle can move ahead. If a vehicle is allowed to occupy a zone, it can proceed to any stop in that zone. Zone control is accomplished by three general methods.

The 'distributed zone control' method uses individual controls for each zone, linked together in sequence. A zone control box for each zone is connected to a hold beacon which, when energized, will stop a guided vehicle from entering a given zone. When the traffic ahead has cleared into the next zone, a signal is sent from that control box to the box in the rear, forcing it to deenergize the hold beacon and allow the vehicle to move forward. It can only proceed until it encounters the next zone with an energized hold beacon.

Distributed control requires many zone control boxes and is normally used in smaller systems where only a few zones are necessary. The more control boxes in the system, the more maintenance required. In small systems, however, this approach has the least cost.

'Central zone control' accomplishes the same zone sequence action as distributed control, but uses a central controller to control each block zone through a communication point. Instead of each block zone having an individual control box linked in a distributed fashion, the central controller regulates the entire network of zone communication points. When a vehicle approaches a zone entrance, it communicates through the zone point to the controller. If the controller determines that a vehicle can enter a zone, it allows the vehicle at a point to proceed; otherwise, the vehicle waits.

Central control is used in systems where traffic flow is complex and several vehicles are in use. If the controller should go down, however, there is no ability for the AGVS to operate. Advantages of the system are greater flexibility, blocking sophistication and increased vehicle movement for higher throughput rates.

In the third method of zone control, 'onboard control', vehicles communicate without the need of a central zone controller or distributed zone control boxes. By passing over floor codes at the start of each block zone, the vehicles know where they are and communicate that information by radio frequency through the guidewire or air to other vehicles in the area. When these vehicles receive the information, they then decide for themselves, based on their location, whether or not they can enter a given zone. When the lead vehicle passes into the next free zone, it transmits its new zone position. This new information is transferred to all vehicles in the system and the appropriate blocking actions are taken.

Onboard control is the newest form of zone control and includes a microprocessor on each vehicle with sophisticated software logic. Since there is no central or distributed control, failure can be limited only to specific vehicles and not to a zone controller. In addition, vehicles can be removed from the path without requiring manual resetting of zone control boxes to allow other traffic to move.

In any form of zone blocking, the more zones in the system, the greater the degree of vehicle movement. Fifty-foot-long zones permit much more vehicle movement than 200 foot-long zones because in a given length of path, you can fit four times as many zones, permitting vehicles to move with more freedom and yield higher throughput.

In distributed- and central-controlled zone blocking, each zone usually costs a fixed amount of money. The more zones, the greater the cost of the traffic management system. Onboard vehicle blocking systems have fixed prices due to traffic control logic onboard the vehicle. Typically, these systems have the optimum number of zones installed for the greatest possible vehicle movement.

Forward-sensing control. Traffic management by forward-sensing control uses a sensing system onboard each vehicle that detects the presence of a vehicle in front of it. Three types of sensors are used: sonic, which works on a radar principal; optical, which uses infrared light sources; and bumper, which uses a physical contact of vehicles to cause traffic control. These methods are used on straight sections of path.

In general, forward sensing is useful where there is a lot of straight path in a system and where paths are not interrupted by intersecting curves. In these applications, whenever the sensing system senses a vehicle in front of it, the vehicle goes into a hold pattern. When the lead vehicle moves out of the range of the sensing device, the trailing vehicle automatically restarts.

Advantages of forward sensing for traffic management include the fact that vehicles can come quite close to one another and there are no fixed holding points as in zone control methods. These conditions permit a greater density of vehicles in a given area.

System management

System management ties the AGVS together. Like traffic management control, a number of different types are available to suit user requirements.

Onboard dispatch selector management involves a control panel on each vehicle which is used by an operator to dispatch the vehicle to a single or series or stop stations. The operator may also select the function he wishes the vehicle to perform at the stop station. This is the most common form of system management and is the most flexible.

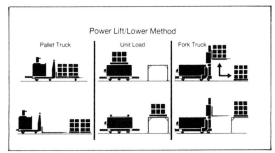

Power Lift/Lower Method

Pallet Truck   Unit Load   Fork Truck

AGVs include a variety of load transfer techniques that give vehicles added versatility in their material-handling tasks. Power lift-lower is a commonly employed method

An offboard call system can also be used for management. Simple call systems involve only a push button at a call station in order to stop the passing guided vehicle. More complex call box controls not only can call a specific vehicle, but can also remotely dispatch that vehicle to other destinations without operator interface. Operators can input new destinations into the panel at their station which will communicate those destinations to the vehicle after it leaves its stop location. This is useful in systems where load transfer is automatic.

A remote dispatch terminal can also be used in systems to provide centralized dispatching control. The remote terminal approach allows an operator to control the individual guided vehicles from a central location. In order to do this, the operator must have some visual status of the vehicle's location and condition so that he can effectively dispatch vehicles to areas in the system where they are needed. This is usually accomplished with a CRT graphic display or locator panel.

Many times this approach is used in systems which cannot justify totally computer-controlled solutions and where a high degree of selective movement with automatic transfer is required. For example, if loads are received at a receiving area and taken to selective locations in a storage area with automatic load transfer, then a centralized control may be effective. A central operator would use the keyboard to send the vehicles to the receiving area to automatically pick up loads and then dispatch those vehicles to selected storage locations where the loads would be automatically dropped off.

Central computer control for system man-

agement is the highest level of control possible. Many integrated material-handling systems incorporate central control for AGVS system management. Most of these systems interface with an automated storage and retrieval system (AS/RS), taking loads into and out of storage. The computer directs the vehicles in coordination with the loads moving into and out of the AS/RS system and permits automatic tracking of every load in the system. Where the volume is high and automatic load transfer occurs, the computer form of system management is very efficient.

It is also possible to mix these various forms of system management together in the same system. For example, if your system is computer controlled and the central computer fails, then resorting to remote-terminal or onboard-vehicle control would allow the AGVS system to continue to operate in the integrated mode.

Monitoring system management is an important consideration in automated guided vehicle systems. One approach for a simple system is a locator panel which merely indicates vehicle locations. CRT color graphic displays provide real-time monitoring capabilities which can instantly detect problems, identify specific vehicles and show location of failures. Central logging and report capabilities are also helpful and develop historical data on the system's performance. Periodically, performance reports can be printed, indicating such things as how often the vehicles have moved on how many loads a given vehicle transported in one particular period. Performance data is helpful for keeping efficiency at the highest level.

Hints on planning AGVS

System planning begins with developing the scope of the system and involves an analysis of system requirements. Planning requirements include identifying load characteristics, floor conditions, interfaces and flow requirements.

The AGVS must handle loads with specific sizes, weights and stability. The footprint of the load is also important for automatic load transfer systems.

Although vehicles travel slower than a fork truck, stability of the load is important. Loads which are stacked too high or loosely can tip on turns or during emergency stop situations.

Engineers must also be careful about floor conditions during the planning stages. Most systems operate indoors; however, outside operation is possible. In these cases, attention must be made to weather and path conditions and the necessary adjustments in the vehicle and control system design. In addition, floors may require special treatment such as antislip floor coatings.

The floor surface is another important criteria. AGVS systems normally run on concrete floors, but are capable of operating on asphalt, wood-block, plank, and tile materials as well. Wood-block floors tend to move and increase the probability of a wire breaking and asphalt tends to sag during summer months.

When planning vehicles to negotiate inclines (not to exceed 6%), care must be taken in rating the load capacity. Vehicles must be derated from their level of load capacity. Special traffic management restrictions generally apply, such as not having vehicles pass in opposite directions on an incline.

Another condition that is very important is discontinuities in the floor. Guided vehicles can pass through expansion joints, trenches, rails or other metal in the floor, but these conditions may require added cost.

System planning also must include an evaluation of interfaces. The AGV may have to pass through automatic doors or fire doors, and must be capable of actuating the door for both open and close functions. Special controls are provided so that if the door fails to open, the vehicle will not proceed. Likewise, the vehicle must have priority when passing through a doorway. These special controls block out manual closing methods while the carrier proceeds through the door.

If the AGVS utilizes pick-and-drop stations, then the mechanical and electrical interface with these stations is very important. Mechanical clearances for vehicles must be planned and executed properly.

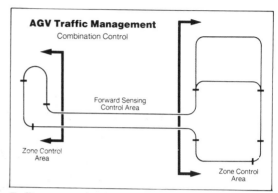

Traffic management is maintained through zone control and forward sensing configurations. This system benefits from both, using the zone method where guidepaths converge and forward sensing for straight runs

Conclusion

An AGVS is not a system unless it meets your flow reuqirements. The optimal combination of routing, traffic management, system control and load transfer methods dictate the flow capacity of the system.

In the planning process, the from-to movements must be examined carefully, along with average and peak rates between points in the system. A guidepath that optimally services those points, giving priority to the higher-volume areas, can then be planned more effectively. It is also important to look at area activity periods. In certain systems the work flow varies considerably on different shifts in different areas of the system. Consideration has to be given to where activity rates are at given periods of time.

An efficient system is one in which the waiting time for vehicle service at a given station is minimized. This is achieved by blending experienced appraisal of flow requirements with knowledgeable understanding of AGVS controls capabilities.

A successful system is a result of a partnership between a customer and vendor. To achieve this, vendors must involve themselves in a customer's application needs, not just his own product technology. At the same time, customers must understand a vendor's AGVS product technology, not just their own application needs.

CASE HISTORIES IN ELECTRONIC ASSEMBLY AUTOMATION

The following papers are representative of attempts to integrate the tools and processes described in previous pages.

As such they are pioneering ventures, perhaps never fully achieving all the goals originally envisioned. They clearly show the complexity of the task and the variety of skills that must be joined together.

BATCH-OF-ONE ASSEMBLY

In 1982 Martin Marietta began a technology modernization (TechMod) program to reduce hermetic chip carrier and assembly costs through the implementation of state-of-the-art manufacturing technology. The original concept was that of an automated production system to assemble the HCCs and other components on PCBs for certain military systems. Shortly thereafter, a proposal was made to implement a second program which would take advantage of the latest computer-aided manufacturing concepts. It could be shown by cost benefit analysis that each step would be well justified by reduction in direct labor, the major cost to be addressed.

Martin Marietta defined the requirements of a fully integrated hermetic chip carrier PCB manufacturing system as part of its proposal for production of LANTIRN (Low-Altitude Navigation and Targeting System for Night). Comprising four work cells: hermetic chip carrier (HCC) assembly; HCC test; PCB populating; and assembled board test, it was found that a system could first be automated and later be fully integrated with computer-aided data management, robotic machine load and unload, inter-workstation transport and bar-code data collection.

Computer-aided data management enhances equipment utilization, tracks quality, and offers continuous management data availability. The data management software supplied, Shop Floor Data Collection System (SFDCS), by Digital Equipment Corp., offered the following functions: on-line tracking of individual parts; work cell and workstation reporting of work-in-process; change control and tracking; lot tracking; and data collection and transaction recording. Modifications for full integration require changes to the SFDCS including: communications to robots; inventory and cycle counting at SMC placement equipment; and bar-code and NOR level tracking. Several other modifications such as work cell start-up and shutdown,

First published in *Assembly Engineering*, June 1985, under the title 'Martin Marietta integrates towards batch-of-one assembly'.
Author: Walter H. Schwartz, Senior Editor.
© 1985 Hitchcock Publishing Co.

maintenance scheduling and work order scheduling can be added later.

System control is down loaded from a host computer for work cells 1 and 2, and another for cells 3 and 4. Machine level programming is accomplished at the host.

The importance of system integration can be seen by considering the activity of the robots in the first work cell which assembles the HCCs. Robot 1 is positioned to perform these operations at the beginning of the line and also has access to the end of the line where it unloads fully completed HCCs from boats and places them in tubes. Robot 2 at the wire bond station loads and unloads the wire bonder and automatic video inspection equipment. Robot 3 can load and unload up to 3 wire pull test stations. Robot 4 loads the sealing furnace conveyor and unloads the laser marker conveyor. Robot 5 unloads the sealing furnace and loads the laser marker. Movement between cells is via a programmable conveyor; there is buffer storage at each station. Bar-coded boats of chip carriers are tracked at all times; the system knows the relative and absolute position of each boat at all times.

Cell 1 – HCC assembly

Cell 1 assembles hermetic chip carriers. Assembled in boats of ten of a type, the system must be able to handle more than 200 different types and assemble approximately 6000 HCCs on a five shift per week basis. The operation of five robots and automatic transport are integrated by the host computer.

Robot 1 loads chip carriers into boats for die bonding. Bar-codes ask the host computer to down-load the proper die bond program. The die bonder places the chip in the carrier and makes the bond. The robot then removes the boat and loads it onto the conveyor for delivery to the wire bonder.

At the wire bonding station, an automatic video inspection station checks to see that the chips are oriented and centered correctly. Bar-code scanners note the number of the boast to request the correct wire bond program. The inspection system notes the

The LANTIRN package, the two pods hanging beneath the air intake, allow this F-16 to fly safely at low altitudes in any weather and at night, and to locate and target armor under such adverse conditions

position of any badly placed chips; the wire bond system will ignore these. Robot 2 loads and unloads the boats from the wire bonder while a bar-code scanner tracks them. The proper wire bond program is down-loaded upon demand, from the host computer.

Robot 3 can load and unload up to three wire pull testers. The correct program is again down-loaded from the host computer in response to the bar-code identification on the boat. Failure information is collected in the host computer and may be called up at the rework terminal via a bar-code scanner.

After wire pull test, the boats enter a vapor degreaser via a programmable conveyor before going to a semi-automatic inspection station where a split-image comparison system is used. Again, bar-code tracking informs

the host computer of arrivals and departures. Defect data is entered into the system by the inspector's terminal.

At this point, rework can be done based on the data available at the rework station terminal and the operator's judgment. Bar-code scanners track movement in and out of rework.

Robot 4 puts lids on the HCCs, loads and unloads a vacuum oven and places the boats on the sealing oven conveyor. Bar-code scanners continue to track all boats. After sealing, which is a gold-brazing process, the boats go on to a marking station.

Robot 5 removes the boats from the output of the sealing oven and loads them on the laser marking system conveyor. A bar-code scanner identifies the boat and the host

The Mini-Cartrac conveyor system from SI Handling Systems ties the robotic workstations together. Its control system will accept down-loaded instruction from the host computer

computer down-loads the proper marking information. The assembly system layout is such that Robot 4, the sealing station robot, can transfer the boats at the exit of the marking system to the conveyor where Robot 1, the die bond robot, unloads the boats and puts the completed HCCs in tubes.

Cell 2 – HCC test

HCC test is time rather than labor intensive; little automation is used except in data collection.

Sticks of hermetic chip carriers are manually loaded and unloaded from a 150°C bake at a rate of approximately 40 sticks per day, always tracked by means of bar-codes. Sticks are similarly handled at temperature cycling, acceleration screening, and hermeticity screening stations.

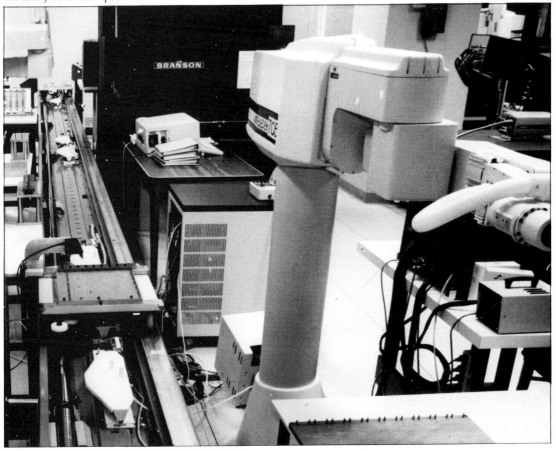

Robot 2 loads and unloads the wire bonder. Boats are stored temporarily, precisely positioned on the holding jig in the foreground

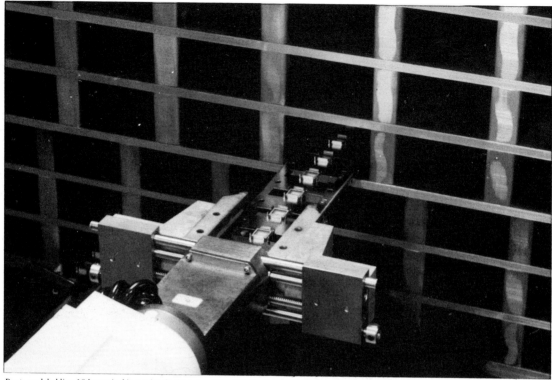

Boats, each holding 10 hermetic chip carriers facilitate handling. The boats are stored in a precise and sturdy rack

Automatic equipment will load and unload sockets at burn-in. All HCCs are subject to manual visual inspection after electrical test and are transferred to plastic storage sticks. Bar-coding is used for data tracking at all times.

Cell 3 – Circuit board assembly

Forty-one circuit board types using 314 different HCC part numbers and 48 chip capacitor types must be assembled. Individual boards use an average of 50 HCCs and 70 passive components. Circuit boards are standardized at 8.0 × 11.7in. or sub-multiples thereof, to reduce fixturing and handling problems. Bar-code scanning tracks all cell activities. Analog signal data from soldering and cleaning equipment monitors those processes and generates an alarm if preset limits are exceeded. Historical data is available if problems arise. The circuit boards first have solder paste applied, currently stenciled rather than screened because of the precision of the application. Loading and unloading the screening equipment is manual at present but is convertible to automatic in-line operation if required. Following solder paste application, a buffer magazine can hold up to 25 boards temporarily in a dry nitrogen storage box. Boards are manually transferred to the component placement machine.

The placement equipment can stock up to 64 types of passive components on tape reels and approximately the same number of HCCs in metal feeder sticks. The system can dispense adhesive but adhesive is not required at this time. In operation, the host computer down-loads the required programs to the machine controller's floppy disks. It also tells the operator how to stock the machine. A loaded magazine of boards is placed on the input platform and the machine is directed to start. Boards are automatically fed and a bar-code label on each board identifies the program required. A sensor recognizes any board marked as bad and prevents its being populated.

The machine places passive components at

Robot 4 loads and unloads the vacuum oven, places covers on the HCCs, and places boats on the sealing furnace conveyor. Robot 5, its arm just visible at the right, unloads the sealing furnace conveyor and loads boats into the marking system conveyor

a 0.6 second rate and HCCs at 1.2 second rate. When populated, the boards are stored in a buffer magazine and then delivered to an oven where the solder paste is dried.

After drying, the buffer magazines are placed at the input to the vapor phase soldering machine. If the vapor temperature and all other parameters are correct and the in-line cleaner is also ready, the boards are reflow soldered. Soldered boards move directly and immediately to the cleaner; after cleaning they accumulate in buffer magazines to wait for visual inspection.

A color video comparator alternately superimposes the image of a known good board and the image of the board under inspection. Differences appear as motion on the screen. Rejects are routed to rework; good assemblies go to test. Rework information from the comparator is available via a RS232 port.

Cell 4 – PCB test

The initial test on completed PCBs in Cell 4 is a video inspection. Boards are compared with a known good unit. The operator identifies the board to the TRACS (GenRad's Test,

Repair, Analysis Control System) by bar-code and inspects for gross manufacturing defects, communicating results via a touch-screen menu on the video terminal. Following video inspection, solder joint inspection is performed manually under a stereo microscope; results are entered into the TRACS via another touch screen video terminal.

In-circuit and functional test of complete PCBs are performed by a Teradyne L-200 Test System under the control of programs generated on and down-loaded from the VAX 11/730 computer. Data is collected by the GenRad TRACS installed in a 2294 computer.

Bar-code scanning of the board under test informs the VAX computer to down load the appropriate test program. When ready, the operator issues a test command on his terminal. Test data, diagnosis down to the component level, is transmitted to the GenRad TRACS computer.

Rework data is available to video terminals at off-line rework stations and all test results and rework data are stored in the TRACS for management reporting.

Current status

The system operates as predicted; software remains to be developed for full integration. Cells 2 and 4, HCC and PCB test are essentially complete. Software for start-up and shut-down of the work cells, for performance monitoring, report generation and work order scheduling has not yet been implemented.

Trial run production has been completed in Cell 1 to verify that each process will produce acceptable results whether operated automatically or manually. The soldering and cleaning processes in Cell 3 have been developed to the point that assemblies meet LANTIRN environmental test requirements.

FMS FOR PCB ASSEMBLY

FMS is a philosophy that has traditionally been thought of as applicable for metal cutting and metal forming. However, there is increasing awareness, within the industry that has made advanced manufacturing technology possible, namely electronics, of the advantages to be gained from FMS.

This paper gives some background information relating to the Factory Automation Systems Technology (FAST) Division within GEC, the motivation within the electronics industry for adopting FMS and gives an outline description of a typical system highlighting some of the more innovative areas.

Background

FAST Division is a trading division of GEC Electrical Projects Limited, which in turn is a part of the General Electric Company plc with a turnover of £5500 million.

The Division consists mainly of experienced application engineers in the various aspects of factory control. Supporting this and other divisions of the Projects company are specialists in systems design, software, microelectronics, robotics and materials handling. There is additionally a large worldwide field service organization which can supply 24 hour cover where required.

The FAST Division is specifically concerned with factory automation projects for the manufacturing industries, covering a variety of fields including production control systems, automatic guided vehicles, transport systems, stores control and full CIM projects.

FAST is a commercial organization within GEC and has not only provided a focal point for flexible automation schemes in-house but is also involved with companies outside of GEC who require such expertise for their own applications.

FAST has very close links with other companies in the GEC group with related interests, such as GEC Industrial Controls, GEC Mechanical Handling, GEC Computer Services, Marconi Research Laboratories and the Hirst Research Centre.

FMS – The requirements

The medium-scale PCB manufacturer's business may be characterized by a number of what are often conflicting pressures: a fast moving and competitive market place; increasing complexity of board designs and manufacturing techniques; a wide range of products, batch sizes and production volumes; stringent quality demands and a large number of production stages.

Each one of these characteristics may be readily addressed in isolation. However, to meet all their conflicting demands requires a very flexible manufacturing philosophy such as FMS.

It is important to note that the very nature of PCB assembly, where a wide range of products may be manufactured from a relatively small component base, is in fact the ideal environment for truly flexible manufacturing.

Author: Andrew Freer, GEC Electrical Projects Ltd.

Average weekly production rate	No. of board types in this range	Total weekly production rate of all board types in this range Actual	%
0.4 – 1.5	1175	470	26
1.5 – 3.0	34	50	3
3.0 – 7.0	67	200	11
7.0 – 14	11	75	4
14 – 50	18	250	14
50+	15	760	42
	1320	1805	100

Average production rates of boards

Further motivation for adopting a flexible manufacturing policy is found in the way boards are assembled. This has traditionally been a manual process which although slow and prone to error is very flexible. The benefits in speed and quality realized from the use of automatic and semiautomatic machines has so far only been available to the volume producer. Programming and setup times together with severe inflexibility has restricted their use in small and medium batch size applications. In order to open up the benefits of automation to the small and medium batch size applications, a control policy is required which is inherently flexible and sufficiently forward-looking to enable sensible planning decisions to be made.

Having established that the nature of PCB assembly is suitable, and in fact in many ways ideal for FMS, it is worth looking at some of the more specific requirements which its implementation must meet.

Companies with a wide range of products and production volumes face a major management task in maintaining an efficient and effective factory. The scale of this problem may best be highlighted by the analysis of a medium-scale PCB manufacturing plant. Tabulated below are the average production rates of boards output over the year with an indication of their variety and of their batch sizes.

The extremes of these figures show the extent of the problem i.e. 26% of the total output of the plant comes from nearly 1200 different board types, each with an average weekly production rate of between 0.4 and 1.5.

A large variety of boards presents the additional problem of requiring a proportionately large number of different manufacturing routes (or process plans). Although the events required to assemble and test a typical PCB may be reduced to the following sequence:

Prepare kit – insert components – flow solder – inspect joints – fit components not flow solderable – visually inspect components – electrically test – thermal cycle/ burn-in – system test.

The actual production processes required by the range of different PCBs means that the process plans have to be specified in greater detail, typically containing between 20 and 30 operations. This situation poses a difficult scheduling problem which is further compounded by the variations in batch sizes and

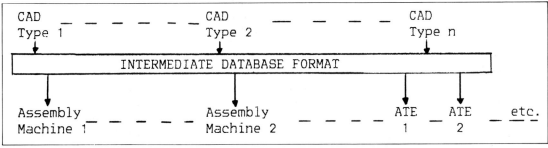

Intermediate database format

production volumes already mentioned. There is then a range of unpredictable problems such as inspection failures, component shortages, design changes and priority changes which further complicate the scheduling requirement.

In order to remain competitive in a highly mobile market, the PCB manufacturer must not only match new products as quickly as possible but must also introduce innovative ones with a minimal period between design and production. It is therefore necessary for the design process to be as quick as possible and for data capture done at that stage not to be duplicated later. It is also a concern that there typically exists a 40 or 50 to 1 ratio between elapsed time and operation times when manufacturing a PCB. The control system has much scope for improving this ratio.

In any move towards FMS, it is vital to consider the impact that will be made on the quality of the product, firstly from the point of view of systems that may be installed to check the PCBs. Secondly, the actual manufacturing processes must be considered in an attempt to eliminate those which are inherently error prone, replacing them, where appropriate, with new or improved automation.

An important change that is occurring in the PCB manufacturing process is the advent of new technologies such as surface mounting. This reinforces the requirement for a flexible manufacturing system since it brings with it the prospect of new and increasing automation. Furthermore, the production control system must be able to cope easily with a continually changing set of processes and operations as these new technologies are implemented.

The implementation of automation can do much to improve the manufacturing process, however many of the problems will still exist. It is therefore vital to look much wider than just the individual operations in order to obtain real benefit. The PCB assembly process as described is an ideal environment in which to implement FMS. However FMS itself is just part of a business strategy vital for a company to obtain a predominant position in the electronics market.

Implementing FMS

In implementing FMS it is important to consider the philosophy in its widest context. As mentioned earlier it is not simply a matter of installing automation; what is more important is the analysis of each operation in relation to all the others. This means considering the type of operation and whether or not it should be performed by a machine, robot or human operator. It means looking at the resources required by each operation and how they can best be integrated with other relevant systems on site. The production control system itself must be considered in conjunction with other site systems such as CAD, stock control, material requirements planning, stores control and manufacturing requirements planning. If there is inadequate integration of these systems and of the manufacturing operations then the level of benefit realized from implementing FMS will be severely limited and this must be an area where change must be considered.

In all aspects of FMS, but especially so in the area of production control, the requirement of flexibility remains paramount. Towards this end, it has been found that the control system should be biased towards letting production happen rather than stopping it. This may seem obvious, however it is easy to design a multitude of, what in isolation are very sensible, checks into a control system, such that the amount of data required to be entered before a job can be released for manufacture is so daunting that it discourages the system's use.

The well-known policy of top-down design and bottom-up implementation has proved successful in the past and leads to the most essential requirement, that of plant scheduling and control. The principal features of the control system are now discussed:

- Long-term plant loading and scheduling. The plant loading enables policy decisions to be made concerning plant and operating practices before a difficult situation arises. The scheduling provides the detailed plan of work to be carried out on a day-to-day basis.
- Tracking of work-in-progress closes the

Assembly technique	Typical insertion rate (cpts/hr)	Maximum number of component types	Maximum number of component styles
Manual	20 – 600	Unlimited	Unlimited
Semi-auto	500 – 1000	80	Unlimited
Robotic	600 – 2000	25	10
Automatic	2000 – 30000	40–240	1

Assembly techniques which should be considered

control loop and allows comprehensive management reports to be generated detailing the state of different orders, products or production plant or areas.

- Adaptation of the schedule to take account of local unforeseen events.
- Monitoring and analysis of test and inspection results to spot trends in failures allowing action to be taken for their prevention.
- Direction of transportation to ensure correct and quick progression of all work.
- Identification and monitoring of resources required for manufacture. This includes materials, programs, fixtures, manufacturing and test instructions.

The above features either use or generate large amounts of data. It is important, therefore, to pay particular attention to the way in which it is collected, stored and disseminated. At the shop-floor level this means DNC of machines (where justified), operator information, user-friendly data screens and simple data-collection methods such as bar-code readers.

At all levels within the FMS it is an absolute requirement that the data be sound and consistent. It is therefore essential that it be generated and maintained wherever possible by the CAD function. The manufacturing data comprises the following information:

- Parts lists – for stock control and purchasing.
- Circuit diagrams – in NC form for test equipment.
- Circuit diagrams – for all manual operational requirements.
- Layout diagrams – for on- or off-site bare-board manufacture.
- Component positions – for sequencing and insertion machines.

- Component positions – for building of bed of nails test fixtures.
- Track and component layouts – for repair functions.

By using CAD as the primary source of much of the manufacturing data, the overall timescales and costs of bringing a new PCB to production is reduced and a 'more standard' product results. It is also our experience that it helps produce a better product with fewer errors and requiring less setup and debug time. The first prototype boards can also be readily produced in the factory since much of the manufacturing data is generated at the design stage.

The use of CAD systems to generate data for a variety of manufacturing and test machines does present a problem of data format. The potential number of post-processors required to generate the programming data could be huge. To overcome this potential problem a neutral intermediate database format has been used to provide a standard interface for machine or CAD data type.

The distribution of data to and from the various systems and machines used for PCB manufacture is achieved through a local area network. This provides good data integrity, a high data capacity and flexibility with regard to positioning of nodes. Currently using this network are the ATE machines, the in-circuit test machines, the QA database, the CAD systems, the production control systems, the stores control system and all the shop-floor data collection devices and printers. The installation of this network has in fact enabled a number of integration goals to be reached which would otherwise have been more difficult and costly to achieve.

In preserving the long-term flexibility of the plant it was decided that the transport

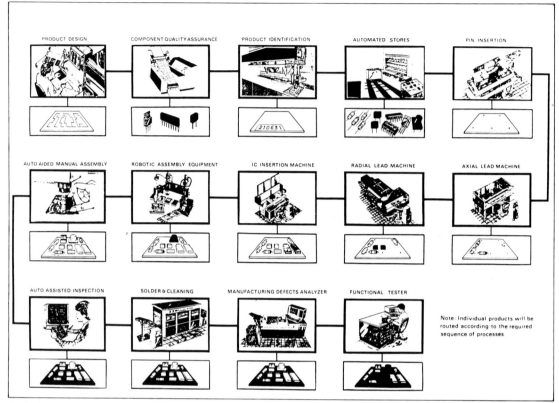

Production processes for electronic assembly

system could not be based either on wire guided AGVs or on conveyors. It is therefore the intention of using either self-guided vehicles with on-board navigational equipment or a manual system or a combination of the two. In either case it is the control of the system that is important, ensuring that work is progressed without delay.

The major features of the FMS presented so far are essentially common to all other flexible manufacturing systems:

- Production control system.
- Flexible transport system.
- CAD data management.
- DNC.
- Integration of manufacturing support functions.

The major part of the system that is peculiar to electronics is the plant used. This paper does not attempt to look in any detail at the assembly operations since they are essentially peripheral to the operation of the FMS philosophy. There are, however, four basic assembly techniques which should be considered for each operation: manual, semiautomatic, robotic, and automatic. Similarly for other operations such as soldering, inspection, test, etc., comparisons may be drawn between the different types of equipment.

The application of FMS is not restricted to those parts of the manufacturing process which are completely automated and can be controlled remotely from a host computer. FMS is about getting the most out of each operation whether it is automatic, semi-automatic, manual or robotic. In many ways, the control of the human environment is the most difficult since the aim is to maximize the use of their particular skill and to reduce the scope for operator error whilst acknowledging their psychological requirements.

FMS in the human environment is best illustrated by the example given below.

Stores

The stores area is an essential part of manufacture and must therefore be controlled by the production control system. The first step taken in the stores area is to consider improvements which may be made to the operating conditions and procedures. The traditional aisles of racks patrolled by numerous operators may be rationalized by installing vertical carousel units which essentially bring the components to the operators. This has the advantage of saving floor space and making more efficient use of operator time.

The kitting sequence is now initiated by the scheduling function of the production control system so that material is supplied to the factory when it is needed and not before. The bill of materials supplied to the stores operator has three main parameters for each component entry:

- The source bin, i.e. the storage location of the component.
- The quantity.
- The destination bin, i.e. this is required if auto or semiauto component carousels are being loaded.

The operator's responsibility is therefore to pick components and load them into premarked destination bins according to a schedule presented to them on VDUs. This method of operation has been found to reduce picking time and also improve the accuracy of the prepared kits.

Conclusions

The nature of PCB manufacture in all but the highest volume factories has been shown to be an ideal environment for FMS.

In implementing this philosophy, emphasis must be placed on flexibility and on the 'manufacturing system' in its widest sense. Flexibility must be built into the actual assembly processes, which does not always mean that automation is the answer, and also into their support functions such as material supply and transport. The manufacturing systems must involve all stages of production from design through procurement, planning, productions and packing. The most important factors being control and integration with particular care being given to the capture, storage and dissemination of data.

No financial justification has been attempted here since it is unique to each situation. However reflection on the level of competition in a fast changing market, the influx of new technologies and the proportion of costs associated with WIP and inventory provide sufficient motivation for considering the implications of not using FMS for PCB assembly.

PCB ASSEMBLY AND TEST

As computers, printed circuit board testing equipment, industrial robots, robotic vision and inspection equipment grow and become more sophisticated, the concept of a computer-controlled PWB assembly and testing system will become increasingly feasible. More and more board manufacturers are looking to totally automate their component insertion, inspection and testing lines. This paper discusses concepts for incorporating robots and automatic loading for PWBs into

First published in 'Robots 8 Conference Proceedings', 1984.
Author: Ronald D. McCleary, Universal Automated Systems Inc.
© 1984 Society of Manufacturing Engineers.

ATE equipment into an integrated assembly system. Productivity gains through decreased in-process storage and increased quality will be the result of such an automated line.

Component insertion

Component insertion refers to insertion of electrical components into printed circuit boards. The components referred to in robotic component insertion are odd-shaped components which cannot be inserted by existing high-speed inserters. These consist of connectors, high-profile DIPs, SIPs, relays, transformers, potentiometers, etc. These

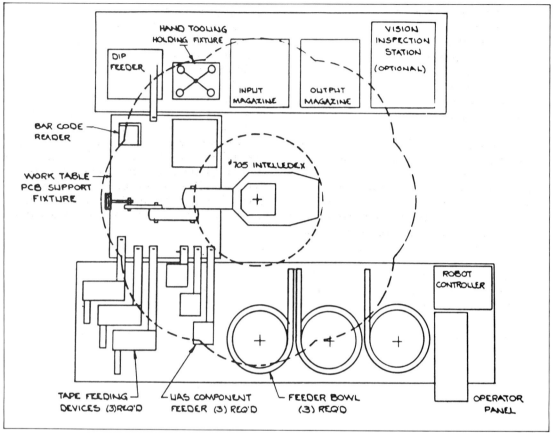

Single-robot workcenter

odd-shaped components are the topic of this section.

Typically, the PCB arrives in the robotic workcenter from the high-speed automatic inserters with all standard components inserted. There are several ways to approach the problem of odd-shaped component insertions, i.e. depending upon the batch sizes and total production. Two such methods are investigated. One, a single robotic system for low-speed high flexibility insertion, and the second, a multiple robot high production, large batch size system.

Single-robot workcenter. In situations where lot sizes are very small (one part to several hundred parts), the total number of components is low (15–20) and the production requirements are approximately one component every six seconds, a single robot may be the answer.

Since there is only one robot in the workcenter, it will require a large work envelope to accommodate all of the required parts feeders. The robot will require more dexterity (degrees of freedom) to handle all of the required tasks, and a sophisticated control system to handle the interface with feeders and other external equipment. As the lots change, the controller will receive program information requiring different program locations.

The system will include equipment to perform the following functions:

- Feed a range of component shapes and sizes.
- Preform the components for insertion.
- Insert the component.
- Cut the leads and clinch (if necessary).
- Inspect the board for presence of all components.

- Transfer the boards into and out of the workcenter.

Since the robot will be handling a large variety of component shapes and sizes, it will require several different grippers. A storage rack for the variety of grippers present in the workcenter is required. The robot must be capable of changing grippers in a timely fashion.

The robot in the single-station workcenter will be required to handle a full variety of the robotically inserted components. This means that feeders must be present for the variety of components required. For the system to operate as an effective flexible workcell, it must have all feeders required so all components can be run through the system. As a result, some of the feeders will be idle at times.

As lot sizes get larger (above 100 boards), it is feasible to have an operator intervene and make minor modifications to feeding equipment, potentially resulting in a reduced number of feeders being required and reducing the total cost of the system.

Single-robot workcenters are capable of inserting components at an approximate rate of one component every six seconds. This includes unloading feeders, straightening leads and inserting.

Since such a large variety of components and board types are being handled in one workcell, it is desirable to have a programmable clinching mechanism which can easily adapt to different board configurations.

In this workcenter, the robot would have the ability to load and unload the incoming and outgoing magazines should this prove economical. This consumes valuable robot cycle time and should be considered as an alternative only if the robot has available time.

Multiple-robot workcenter. More than one robot will be used in cases where: high-speed insertion is required, lot sizes are large, and a large variety of components are to be inserted.

A multiple-robot workcenter offers different challenges than a single-robot workcenter. Generally, each robot handles a smaller variety of components which can be split into

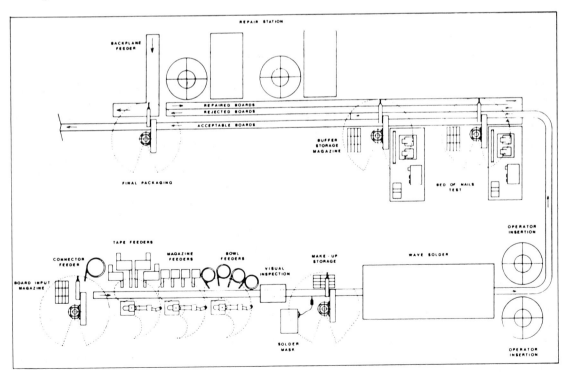

Multiple-robot workcenter

families, simplifying the robot gripper requirements.

Less sophisticated robots with less dexterity are required in the multiple-robot workcenter. These robots are generally quicker than the more sophisticated robots and insertion rates per robot can be as little as three to four seconds.

The overall product mix must be carefully evaluated so as to balance the cycles of all the robots across the entire product line. It is not a good system if all robots are not equally busy on all products.

The cut and clinch mechanisms will be less complicated due to the fact that each robot has less of a variety of products to insert.

Component feeders. Part feeders come in three basic types: vibratory bowl, magazine, and tape.

The average lot size and utilization of components will, to some extent, determine the type of feeder to be used. As an example, if only one style board requires a particular part and only several of the boards are made a day, it would not be economical to utilize a bowl or tape feeder. However, if every style of board has one of these components then it may prove economical.

The configuration of the component and the method of vendor delivery has a large impact on the type of feeding mechanism. Components with rigid, pre-cut, accurately formed leads may be bowl fed. A good example is connectors and sockets. Some components, such as transistors, may be fed, leads cut, and then formed.

Axial and radial components are most commonly fed on tape reels. The most flexible system would incorporate separate tape feeders for each component. This is the most expensive approach, and in applications where such flexibility is not required, a much more economical system would be to sequence the radial components on one tape and all axial components on a second. Since the forming equipment is incorporated in the feeder itself this approach restricts the flexibility of the system by requiring all components from each feeder be formed to the same spread distance.

Tape reel feeders do not exist today in a configuration that can feed a component, cut it, form it, and hand it to a robot. Existing feeders which cut and form the leads could be modified to present the component to the robot in an oriented fashion.

Some components such as DIPs, SIPs and a variety of connectors do not lend themselves to bowl or tape feeding. These components can usually be magazine fed, where an operator periodically loads the magazines manually. The feeder meters the part into a fixture that straightens the leads and holds them repeatably for robotic insertion.

It should be noted at this point that not all components can be automatically inserted due to a variety of problems, most generally poor tolerance or stability in the manufacturing of the components which does not allow the robot to properly locate and grip the part repeatably.

Feeding. The PCB itself may be presented to the robot in a number of ways. The least costly means is to have the robot itself remove the board from an input magazine and reload it into an output magazine. This method detracts from the time that the robot has for actual component insertion and should only be contemplated for prototype applications.

Boards are very commonly stored in standard board racks which are transported throughout the factory. Automatic magazine loaders and unloaders are available at which it will remove the board from the rack (and reload it) and present it to the workstation. It is then necessary to center and accurately

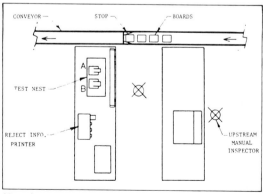

Manual system

locate the board utilizing accurately located tooling holes in the PCB.

Utilization of racks is desirable where small lot sizes are being run and the company wants to maintain lot integrity. However, in the case of very small or very large lots, utilizing racks means an increase in inprocess inventory which, is not desirable. In these cases it is much more desirable to feed the PCB directly to the workcenters for assembly or testing.

Component modifications to ease assembly. Until recently, there has been very little pressure on the suppliers of electrical components to manufacture in a way which will facilitate automatic assembly. With more manufacturers of electrical products pressuring the suppliers there has been some improvements in the design, manufacture, and packaging of the components.

Major characteristics which contribute to ease of assembly are:

- Cross-identification of polarity on components for easy alignment.
- Accurate case-to-lead tolerances in design and manufacturing.
- Preferably pre-cut and formed rigid leads.
- Chamfers on both male and female components.
- Packing in such a way as to maintain alignment of component and component leads.

Generally speaking, most odd-shaped components can be modified so as to ease the assembly application. This will most likely add to the unit cost of the component, but should reduce the cost of manufacturing enough to compensate.

At this time, it should also be suggested that all components undergo a full functional test by the suppliers on a 100% basis. If components going into an assembly are within tolerance, the assembly should pass its functional test with a minimum of rejects.

Clinching. Clinching refers to bending the leads of the components over after inserting the component into the board. Clinching is required if the board is to be handled after insertion. Clinching holds the components on the surface of the board during handling and through wave soldering.

Component leads may also need to be cut at the time of clinching if they are not pre-cut to length. The cutting equipment can be part of the same mechanism that does the clinching. This type of equipment is common to computer-assisted manual insertion stations.

With this type of clinching, the board must be moved by an *X-Y* positioner over the cut/clinching mechanism. This consistently places the point of insertion for the robot always at the same point in space, similar to the operation of the high-speed insertion machinery.

There are other mechanisms for clinching pre-cut components. One is known as the anvil method. In this method an anvil is preformed for the bottom of the board. As the component is inserted, the leads are formed in place (similar to the action of a stapler). This method requires that different anvils be used for different board configurations.

When the board assembly station is on the same assembly line as the soldering equipment, and no handling is required, it is common not to clinch. Since the board is not being handled, the components stay in place through wave soldering. In situations where the lot sizes are large, or the boards are similar in size, this is a preferable method.

Automatic bed-of-nails testing

Disadvantages of manual system. While the human operator is a very flexible form of automation, we tend to build inefficiencies in the way we work. The longer we work at a particular task, the more set in our ways we become. This results in an ever-increasing lack of proficiency on the job.

Some of the inefficiencies encountered in manual operations are not due to the operators themselves, but are due to the way in which the flow through the operators workcenter has been concepted. This relates to the way in which the material that the operator is working with is delivered to and taken from the workcell, as well as the way in which the machinery that the operator is working on is set up for operation.

The following is a list of items frequently found which contribute to the inefficiency of a manual workcell:

- Poor material handling on input and output of workcenter.
- Lack of in-process buffer storage.
- Typically, only single-station test fixtures are used.
- Operator tagging of rejected boards is inefficient both in time and accuracy.

Another item which should be mentioned at this point is the fact that the operators do not work at a continuous rate and as the day progresses the operator's efficiency drops off. Finally, the loss of production due to coffee breaks and lunch breaks is also a contributing factor.

Manual system improvements. There are some things which can be done in an existing system which is manned by human operators to increase productivity. Some examples of these items are:

- Improve input/output material handling – While presenting circuit boards directly to the operator on conveyors can be very cumbersome, it is not much less effective and still quite productive to have fully-automatic transporter systems which carry containers of untested circuit boards to the operator with a small buffer storage spur at the workstation. In the same way the finished boards, having been tested, may be removed from the workcenter whether good or bad via additional conveyorized transporter systems.

- *The use of database systems for repair information* – Some companies with bed-of-nails test systems, have gone to the use of computerized database storage systems, or paperless testers, which automatically pick up a serial number from the circuit board and then store the results of the test, either good or bad. If bad, the nature of the rejection is stored. This information in the computerized database system may then be recalled by the repair technician at the workstation. This removes the need for the operator to tape a computer printout to the

circuit board to identify the nature of the fault. This increases the efficiency of the operator and improves the reliability of the failed part information. As an added benefit, the manufacturer may develop an accurate histogram of the performance of the printed circuit board manufacturing facility through the use of this computerized database.

- *Use of dual-station test fixtures* – Another suggestion for increasing the productivity of manual and automatic testing of printed circuit boards is the use of two-station test nests. With a single-station test nest, the operator must wait while the tester checks the circuit board (approximately 30 seconds) before unloading and reloading another part to be tested. With a two-station test nest, the operator can improve efficiency by approximately 10% by being able to load a second test nest while the first is being tested. Once the test is complete, the test machine will automatically begin testing the board in the second nest and so on.

Flexible automation advantages. In the preceding section we discussed how the efficiency of the human operator may be significantly improved by altering the procedures currently in use in the typical workcenter. There are, however, still many areas in which the overall productivity of the workcenter can be greatly improved through the use of flexible automation, in this case robotics.

Consistency is one of the major overall benefits generally related to robotics. It is widely accepted that the introduction of an industrial robot to a process greatly improves the consistency of the process. This is due to the fact that once programmed, the industrial robot performs its tasks in a very repeatable manner. The robot is not only repeatable in physical positioning, but also in the fact that once the robot has been taught how to do a process correctly, it will continue to perform this process correctly until reprogrammed.

This consistency results in an increase in productivity. Since it is so consistent/accurate in the performance of its task, machine error can be ruled out. This will result in productivity increases due to not passing boards by error which should have been rejected, or rejecting boards which should have been passed, causing them to go through a repair area and eventually be re-tested.

The intelligent industrial robot allows the manufacturer to have as flexible a manufacturing system as he had when the human operator was present. This is a big advantage in companies where parts are not produced in lots of one million plus, but in lots ranging from one unit to thousands of units.

The robot, being a reprogrammable unit, allows for major installation alterations without obsolescence. Small, minor lot changes can automatically be compensated for by the robot. If one type of board is replaced by a similar board requiring different test characteristics, the robot can sense the change and inform the tester of the change so that the proper test can be performed.

Another major area often discussed is the increase in utilization of the expensive test equipment since the robot performs at a consistent rate all day and all night. The industrial robot, when coupled to a buffer storage system, does not stop for coffee breaks, lunch breaks or shift changes. In multi-shift operations, the robot is literally capable of running 24 hours a day giving the maximum possible utilization of the costly equipment.

While it is not always the case, the robot is usually capable of performing the same operation that the operator was doing at a faster rate allowing for a further increase in productivity.

The point at which the manufacturer realizes the greatest increase in productivity is when the technology of the flexible automation (robotics) is married to the productivity-improving items discussed in the previous section. The result of this marriage is a flexible manufacturing system (FMS) which can accommodate varieties of product in a very efficient manner.

Sample installation. While still employed by Unimation Inc. (Systems Division), the author installed a system utilizing a Unimate model 560 PUMA industrial robot. This robot has a

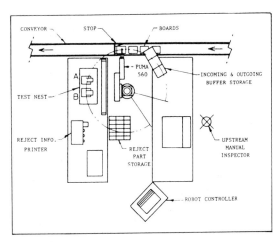

Automated system

weight capacity of approximately 5lb, repeatability of ±0.004in. and six fully coordinated degrees of motion.

The system as it existed prior to the introduction of the PUMA robot, implemented many of the items already mentioned.

In the manual operation, the operator has parts entering the workcell on a conveyor transporting individual parts. Good parts, after being successfully tested, were placed on another conveyor which took the parts to the next testing operation. Rejected parts, with the computer printout taped to them by the operator, were placed in racks which had to be manually emptied after approximately two hours. The workstation already utilized a two-station test fixture which allowed the operator to work more efficiently.

The disadvantages of the system with a manual operator are as follows. Parts entering the system on the input conveyor would become backed up when the operator was not operating at maximum efficiency, causing operations preceding the testing operation to be stopped since their output (the test stations input conveyor) was backed up.

Additionally, if the test station output conveyor becomes backed up due to a slow down further along the conveyor line, the testing operations in the workcenter had to be stopped since there was no place to put the acceptable parts after testing.

Also, it was necessary for the operator on

occasion to get up and replace full storage trays for the output of rejected parts, thus decreasing the operator's overall efficiency.

Taking into account these areas of inefficiency, as well as the fact that the operator would occasionally break for coffee, lunch, etc., the overall average cycle time for the workcenter was approximately 23 seconds per part while the actual testing timer per part was only approximately 20 seconds. With the dual-station test fixture, had the operator been working at 100% efficiency, the cycle time per part would have been approximately 20 seconds.

To flexibly automate this system using the Unimate PUMA 560 industrial robot, required very few modifications to the existing equipment. Those modifications that were required are as follows:

- An accurate and repeatable stop had to be installed on the incoming conveyor with a limit switch capable of detecting both part presence and orientation.
- An interface had to be developed for the robot input/output to communicate with the test unit.
- Accurate and repeatable buffer storage locations (existing board carriers) were utilized for storage of overflow from the input conveyor and for overflow going to the output conveyor.
- An accurate and repeatable system for rejected boards and the computer printout of rejected information had to be developed for the rejected parts.
- Photo switches and their interface had to be utilized to identify overload conditions on the conveyors as well as empty locations for part set down.
- A robot hand tooling gripper had to be developed and fabricated for: picking up the circuit boards and loading the tester, throwing a small switch on some circuit boards during test, and for picking up, tearing off and loading the computer printout reject information into the tote prior to loading the reject board into the tote.
- A VAL software program had to be developed, written and debugged for the PUMA robot to handle all operations.

In the final system configuration, utilizing the PUMA 560 robot, the robot waited at a home position and would constantly scan a variety of input signals monitoring the status of the workcell. A typical sequence of operation would be as follows:

- Robot receives signal from test station A that the test is complete.
- Test automatically begins immediately at test station B which was previously loaded.
- The robot checks to see if the board has passed or failed. If the board passes, the robot would move to station A (the tester) and remove the part.
- The robot checks to see if the output conveyor load station is empty. If 'yes' the robot places the part on the output conveyor station and retracts. In the event that the output conveyor load station is occupied, the robot waits one second then checks that station again. If this station is still occupied after the one second, the robot then proceeds to the output conveyor buffer storage and places the part into the conveyor buffer.
- If the robot receives a signal that the part in station A was a reject part, it then goes to the computer printout, tears it off and places it in the next available compartment in the reject part tote bin.
- The robot moves to test station A and removes the parts and places it into the same reject tote bin locations as the label, placing the part on top of the label.
- Once the robot has completed the above, it will check the input conveyor station to see if a part is present and properly oriented. If there is a part present and oriented at this location, the robot then picks up a part at the input conveyor and loads test station A.
- If the robot finds that no part is present or that a part is present but not properly oriented in the input conveyor, it will indicate this as a fault condition then proceed to the input conveyor buffer storage tray and remove a part from this tray and load it onto test nest A.

The robot then returns to its home position and scans through the signals, again waiting for a signal indicating that his services are required again. Should no signal be received, the robot will attempt to do some house cleaning chores.

As an example, if no signals demanding its operation are received, the robot will check to see if the output conveyor load station is now empty. If so, the robot will check to see if any parts are present in the output conveyor buffer storage tray. Again, if this condition is met, the robot will unload the next circuit board on the tray and load it onto the output conveyor load station. This will continue until either the output conveyor becomes backed up, the output conveyor buffer becomes empty or until the PUMA robot receives a signal from the automatic test set indicating that its services are required.

The services required of the robot will be of the following nature: either a test station has completed the test and the board has passed or failed, there is a circuit board in the test station which requires that an operation be performed on it as part of the test, (flipping a switch) or a board was immediately rejected by the testr which would indicate that it was not properly seated in the test nest, in which case the robot would pick it up and reload it.

As previously indicated, the system as automated by the use of the robot now contains the following elements which make up a good flexible machining workcenter:

- Parts are presented to the workcenter by conveyor.
- Buffer storage exists for incoming parts/ outgoing parts which have been rejected.

- A two-station test nest has been implemented as in the manual case, which allows the robot to maintain cycle times of 20 seconds per board, as no time was lost due to loading and unloading of circuit boards.

Two major improvements could be made to the system as it now exists:

- The robot still must place a computer printout with the boards that are rejected. There is still some room for error with this system. A better solution would be to implement the computer database system described earlier.
- It is still necessary for an operator to remove the full trays of rejected parts. A better solution would be a small section of conveyor which would transport rejects to the repair area. This may be hard to justify on a cost only basis as an addition. It would be less costly and more effective to implement this in the early planning stage.

Conclusion

Pressure exists on manufacturers of robots, vision systems, and control systems to introduce products which are more user-friendly, accurate, and flexible. The manufacturers of electronic components are being pressured by large PCB manufacturers to produce more readily assembled components. As these technologies converge, the fully-automated assembly of these components will become more feasible.

FACTORY CONTROL AND ROBOTIC SYSTEMS

Westinghouse Electric Corporation, a major producer of sensor systems for all military environments, has made a substantial change in the control and configuration of the factory for production of highly complex, low-

First published in *Robots 8 Conference Proceedings*, 1984, under the title 'New robotic systems change the electronic assembly factory'.
Authors: James A. Henderson and Robert N. Hosier, Westinghouse Electric Corp.
© 1984 Society of Manufacturing Engineers.

volume electronic assemblies. Using the control of high-speed computers, coupled with the advantages of sophisticated robotic systems, a new factory has been developed which reaches beyond the upgrade of existing facilities into the modular application of the technologies of today and tomorrow. This new facility, located in College Station, TX, already has been linked with a nearby university to provide continuing synergy between student development and new technology and innovation.

A major key to the development of the new facility has been the top-down systems architectural approach incorporated from the conceptual phase. This structural approach requires the maintenance of a total system perspective during generation of specifications, helping ensure that all controls and interfaces are included. The definition of the factory and lower level systems, the development of the required automation systems for highly flexible, low-volume production, and the results realized to this point are presented.

Factory control systems

In either a high-rate or small-batch automated manufacturing facility, in-process control is an integral part of the overall system design. In order to meet the quality requirement with minimum human intervention, all equipment within a workstation must be furnished with sensors to monitor performance and compensate for irregularities and uncertainties in the work environment. The sensory data must be processed and analyzed by each machine and workstation controller so that the flow and quality of the automated process can be maintained in the most efficient and effective manner. Therefore, the factory control architecture must have a high degree of sensory-interactive processing at the lowest level of control in order to make process control decisions and sophisticated strategic planning systems at the top level of control to make business decisions.

The complexity of this type of feedback-driven architecture increases as more workstations are added to the system. In order to manage the system and ensure its modularity and maintainability, Westinghouse has developed a hierarchical control architecture with appropriate sensory interaction and real-time computation at each level. The advantage of the hierarchical approach is that it allows the factory control system to be partitioned into managable modules, regardless of the complexity of the entire system.

The application of a hierarchical, real-time, sensory-driven control architecture to a manufacturing process is relatively new. It has been implemented in the Automated Manu-

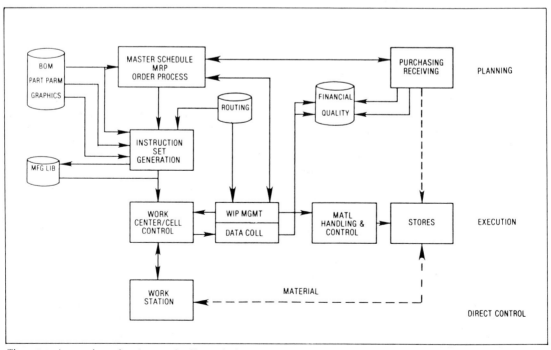

The computer-integrated manufacturing control architecture has been structured with three levels

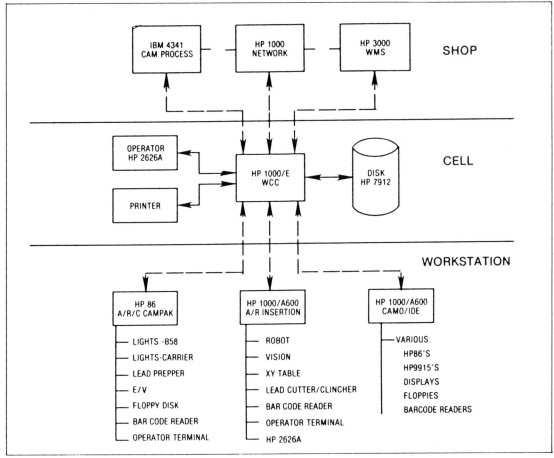

The assembly cell control architecture is hierarchically structured

facturing Research Facility (AMRF) of the National Bureau of Standards for a machining cell. In this paper, we will describe the factory control system Westinghouse has developed and is implementing at its Electronic Assembly Plant (EAP) in College Station, TX, for two major small-batch manufacturing operations. One of the operations is printed wiring board assembly and the other is electronic cable and harness assembly. Both operations are precision, automated assembly processes using state-of-the-art robotic and computer-aided manual workstations. Each workstation operates under micromini-computer control. Cell controllers deliver instructions to the workstations and continuously monitor the states of the machines.

Strategic planning activities occur at the highest level of control. This is where management systems are used to plan business activity and monitor, analyze, and control the performance of execution-level processes. For example, sophisticated planning algorithms which evaluate historical repair records and schedule work for each of the workcells are executed at the planning level. Engineering design data are entered and maintained at this level as is the inventory control information needed to ensure availability of parts and tools for forecasted business. Financial interfaces and procurement control functions also reside at the planning level.

The execution partition consists of cell controllers which control the factory floor based on real-time information required for short-term decision making. The cell controllers simultaneously direct the operation of multiple workstations which perform manufacturing, test, and material handling activities. Controlling execution of these activities includes comparing the actual and planned production schedules and dynamically adjusting production routing to maximize production throughput.

Direct control is the lowest level in the CIM hierarchical control architecture and is concerned with real-time, sensory-interactive systems that control workstation processes. For example, the Westinghouse axial/radial/can workstation utilizes force and visual feedback to control the robotic arm. This level of control demands that the workstation control computer monitor and coordinate the actions of intelligent devices (vision system and robot) and fixed devices (X-Y table controllers). In addition to providing commands and engineering data to the intelligent devices, the workstation controller also detects error conditions and recovers from them.

Printed wiring board assembly

Westinghouse is in the process of designing, building, and implementing an automated

printed wiring board assembly (PWA) system at EAP. This system will assemble both through-hole and planar boards. It has been designed to achieve low cost with high production flexibility and yield. To achieve these goals, certain manual processes such as component preparation, presentation, insertion, and placement have been automated. Other processes, such as mechanical assembly and inspection, have been improved. The workstations which compose the assembly cell will be controlled by execution-level software which interfaces directly to the factory planning-level engineering and business systems described in the previous section.

The PWA system has been designed to assemble through-hole and planar (flat pack) printed wiring boards (PWBs) with a lot size of one. This will be accomplished by a combination of robotic and manual workstations supervised by dynamic cell controllers. Engineering data needed by the cell and workstation controllers are retrieved from engineering databases which are developed and maintained at the planning level of control. These data include information such as part geometrical and electrical characteristics, bills of material, and preferred routings. Updated

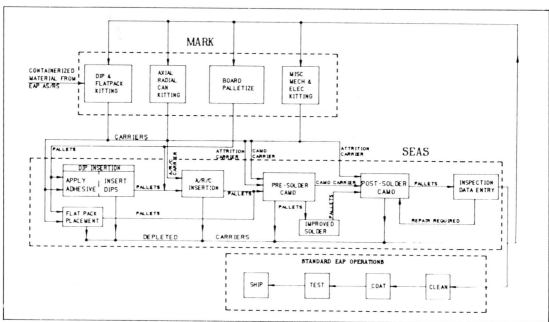

The PWA process flow integrates the MARK and SEAS cells into one system

production requirements, including priorities and due dates, are obtained from the work management system (WMS) on a daily basis. During normal operation, cell status is reported to WMS at the time a work order is completed by the cell. If an abnormal status is achieved, it is immediately reported. If necessary, cell status information can also be obtained by WMS on demand.

The process flow for the EAP PWA system can be divided into four major areas:

- Containerization – includes the processes of receiving, evaluation, preparation, allocation, and storage in part-specific containers.
- Preparation and presentation – includes final part preparation, electrical verification, and kitting for assembly.

- Main assembly – includes all the robotic and manual assembly operations, solder, and inspection.
- Final assembly – includes the clean, coat, and functional testing of the product before shipment to the customer.

For the purposes of this paper, only the preparation, presentation, and main assembly portion of the process will be described in detail. The PWA parts, boards, and components will be received at an off-site location where they will undergo a physical inspection and functional test. Parts which pass these tests will be allocated in the inventory database of the program which purchased them. Also, many of the components will have their leads trimmed, tinned, and formed. Next the parts will be stored in part-specific trays which

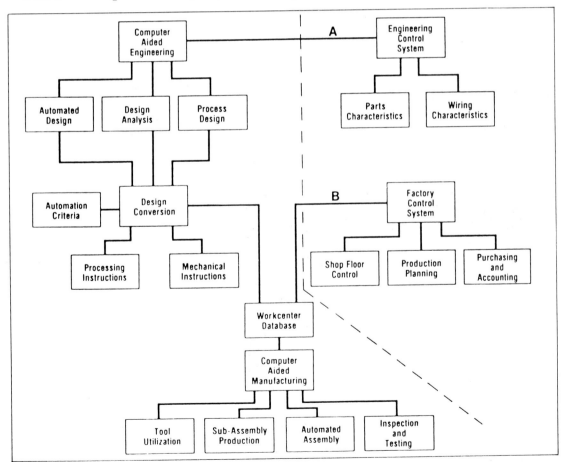

The REACH program requires software interfaces with design engineering (A) as well as production controls (B)

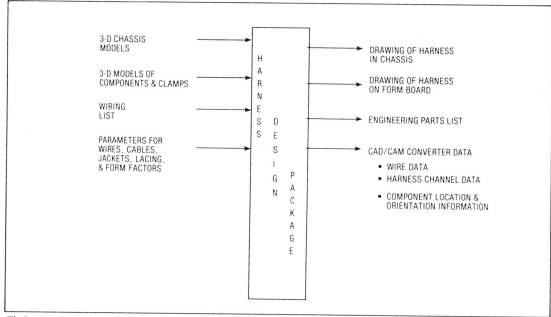

The Computervisionharness design algorithm establishes all of the data sets for required material, robot instructions, and test inspection

have identifying bar-code labels. Although parts from different programs will be co-located in the trays, their ownership will be traceable through the inventory database. After the trays have been filled, they will be placed in bins and shipped to EAP.

Preparation and presentation cell. The preparation and presentation functions are performed in the material accountability and robotic kitting (MARK) portion of the process. Upon arrival at EAP, the bins of trays will be stored in the factory automatic storage and retrieval system (AS/RS). Approximately seven days before a work order is to be started, the appropriate part bins will be automatically moved from the factory AS/RS to the MARK cell where they will be stored in one of three carousels. One carousel will contain dual-in-line packages (DIPs) and flat packs; a second, discrete components; and a third, miscellaneous mechanical and electrical components as well as the bare PWBs.

There will be approximately four robotic kitting workstations at each of the first two carousels. It will be the function of these kitting stations to retrieve the appropriate tray of parts from the carousel and then

remove, form, and test the part. After testing, the part will be placed in a carrier. This process will be repeated until all the parts needed for a single PWB are in the carrier. The carrier then becomes a kit for a particular board style and is automatically transported to the standard electronics assembly system (SEAS).

At the third carousel, where the miscellaneous electrical and mechanical components and PWBs will be stored, there will be computer-assisted manual workstations which also prepare board-specific kits. Each PWB will be fixtured in a pallet, one board per pallet, at one of these workstations. Each workstation will have a computer terminal from which the operator receives kitting or palletizing instructions. A graphics display of the PWB mounted in a pallet will be provided to the operator. The information presented to the operators is all tied to the engineering and process databases generated by the planning-level control systems. The prototype workstations which perform these tasks are currently undergoing final testing.

It is important to recognize that the pallet and the carrier, along with data, are the common elements which flow through the

PWA system. It also should be noted that each board will have several carriers associated with it at various points in the process. For example, a board which contains analog, digital, and mechanical parts will require at least one separate kit for each part type and may require an additional kit to be built to deliver attrition parts to the assembly workstations. The routing of the pallets and kits will be controlled by scheduling algorithms in the cell controllers. Bar-codes on the kits and pallets will be used to identify the material being transported, associate the kits and pallets with their work orders, and initiate access to engineering data files during setup at each workstation.

Assembly cell. The SEAS cell performs the main assembly portion of the process. It is composed of a combination of robotic and computer-aided manual workstations that assemble all the components on the PWBs. All DIPs, flat packs, and up to 90% of the axial, radial, and can components will be assembled at these workstations without operator intervention. The yield of these workstations will be at least 98%. If one of the robots is unable to assemble a component, it will place it in pockets on the pallet and log the failure and the location of the component in an exception file. These exception components will be assembled at a manual workstation later in the process.

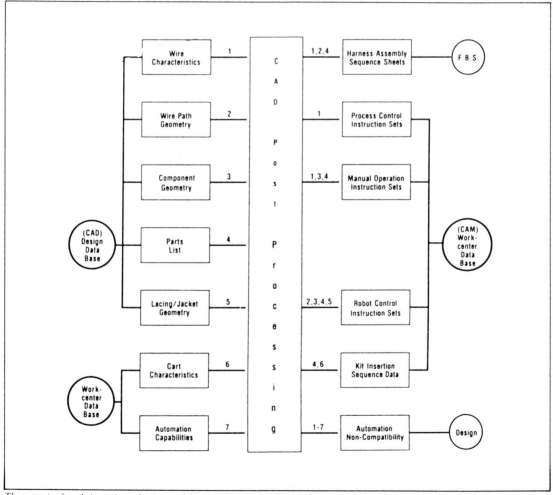

The conversion from design to the production capabilities requires many interfaces, with inputs (on left) combining in several ways to obtain outputs

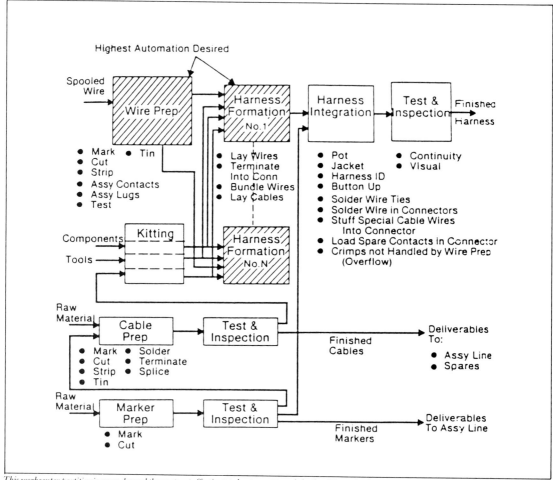

Highest Automation Desired

Spooled Wire

Wire Prep
- Mark ● Tin
- Cut
- Strip
- Assy Contacts
- Assy Lugs
- Test

Harness Formation No.1
- Lay Wires
- Terminate Into Conn
- Bundle Wires
- Lay Cables

Harness Integration
- Pot
- Jacket
- Harness ID
- Button Up
- Solder Wire Ties
- Solder Wire in Connectors
- Stuff Special Cable Wires Into Connector
- Load Spare Contacts in Connector
- Crimps not Handled by Wire Prep (Overflow)

Test & Inspection
- Continuity
- Visual

Finished Harness

Components
Tools

Kitting

Harness Formation No.N

Raw Material

Cable Prep
- Mark ● Solder
- Cut ● Terminate
- Strip ● Splice
- Tin

Test & Inspection

Finished Cables

Deliverables To:
- Assy Line
- Spares

Raw Material

Marker Prep
- Mark
- Cut

Test & Inspection

Finished Markers

Deliverables To Assy Line

This workcenter partitioning was deemed the most cost effective total system approach for feasible production automation

In order to achieve the high yield at the robotic workstations, extensive use is made of sensory feedback. Vision, tactile, and displacement sensors are used to align component leads with holes (for a minimum diametric lead-to-hole clearance of 0.005in.), detect successful component pickup, measure lead spacing, align flat pack leads and pads, and detect lead-to-board interference during insertion.

High production flexibility has been achieved at these workstations by designing them to economically produce lot sizes of one. This was accomplished by minimizing setup time to 60 seconds per board. This includes the time to automatically load and unload the PWB pallet and component kit, access en-

gineering data necessary to assemble the parts, and perform board-to-robot space coordinate transformations. This standard setup approach has been designed into all the robotic and manual workstations in the MARK and SEAS cells.

A computer-aided miscellaneous operations (CAMO) workstation has been built to assemble those components not assembled automatically. This latter category includes two types of parts: those which were not intended to be assembled automatically and attrition parts which failed robotic assembly. The attrition parts will arrive at the presolder CAMO workstation in either of two ways. As described previously, they first will arrive in the pallet pockets. The workstation controller directs the operator to pick the part from a specific pocket location, at a position indicated by a light-pointer, inspect it for cosmetic damage, electrically test it, and, if usable, assemble it on the PWB. If, on the other hand, the component is unusable, the operator will request an attrition part from the MARK cell. This part, along with any others needed for a particular PWB, will be loaded in an attrition kit and immediately delivered to the presolder CAMO workstation.

After the presolder CAMO operations are performed, the PWB will be transported in its pallet through the soldering workstation and on to the postsolder CAMO workstation where any remaining electrical and mechanical parts will be assembled. The completed PWA will then be transported to the inspection data entry (IDE) workstation where it will be inspected for mechanical defects such as incorrect component placement or unacceptable soldering. The IDE workstation is a computer-aided manual workstation designed to eliminate all inspection paperwork, thereby keeping the operator's attention focused on the inspection task. A color graphics display of the PWA is used to identify troublesome assembly areas and confirm the location of defects. Defect locations are measured by using a sonic digitizer. A nonconformance report will be issued for each detection, and the PWA will be returned to a CAMO workstation for repair. When the repair is completed, the PWA will be returned to IDE for final inspection and then routed through the final assembly processes of clean, coat, test, and ship.

During the preliminary system design, it became apparent that the most important criterion to be met in order for the system to be economically viable was high first-time-through quality. Each process and workstation design was evaluated to ensure that it could meet stringent quality requirements.

Electrical wiring harnesses

The project for robot-enabled assembly of cables and harnesses (REACH) is another example of the need for total top-down system engineering. The ability to produce a complex electrical harness with any degree of automation depends heavily on the capabilities of establishing the database of information required for each wire in the harness. Length, strip and crimp information, routing configuration, contact position in connectors, tying and jacketing – all these data and more must be established from design, incorporated in material ordering and handling requirements, provided in appropriate robot instruction sets, and made available to test and inspection. As a result, the REACH project baseline encompasses design and conversion from computer-aided design (CAD) to computer-aided manufacturing (CAM), as well as the development and factory installation of the production workcenter.

Computer-aided design. Computervision Corporation has been working jointly with Westinghouse to establish an interactive graphics design capability for harnesses which reduces the design and development efforts to 10% of that previously required.

This new program, with commands in engineering rather than drafting language, enables a designer to create a 3-D model of the structure, establish a harness within that model (checking for interferences), and remove the harness from structure and lay it out in a formboard plane (maintaining stiffness characteristics). A data output file is then created from which a bill of material, robot instruction sets, and assembly, test, and installation drawings can be obtained.

This development highlights an area which is often overlooked when factory automation is discussed. The database requirements for automated production cannot be economically met using traditional design practices. Substantial factory changes are now being seen within the product design area with the use of interactive graphics as a major design tool; interactive graphics enables product data to be established in a computer-generated, computer-usable file. Product change and documentation updates are also performed on this system.

CAD/CAM conversion. There is probably no production automation system which will be served by only one design capability, and thus be directly linked to the design programs. As a result, a conversion routine which can use the data from any design program and transform it into production usable information should be configured modularly. This will allow new design and/or production capabilities to be interfaced as they are developed, with minimal impact on existing systems.

The conversion routine for the electrical wiring harnesses is initially being configured to provide three interfaces. The interfaces from CAD-to-manual (existing hand-labor production) and CAD-to-CAM are to be used for new design harnesses. Data formats for each output depend on the input requirements of the subsequent production equipment. The third interface is being developed to prepare data for existing harnesses for production by the automated workcenter and will be a limited-use routine.

The manufacturing workcenter. Specifications for the automation of electrical wiring harnesses were generated for a total capability, design through test. As a result, the production workcenter not only addresses the automation areas, but also controls the work in the nonautomated workcells.

The selection of areas which were to receive the highest priority for automation was based on the breakdown of labor content and yield (successful assembly) information generated during the initial, detailed study phase of the program. The marking, stripping, and contact crimping of wires required more than 60% of direct hands-on labor and also was responsible for the greatest amount of test/inspection failures. The second leading area was lay and lace (harness formation). Prioritizing the bottoms-up development of work cells was based on the most viable, cost-effective systems configuration and the availability of equipment.

Within the total system scenario, individual workcell specifications were generated for the workcenter partitioning. The wire preparation workcell will automatically supply terminated (both ends) single conductor wires to up to five robotic harnesses formation cells, with two wire prep cells accounting for automation of 50% of present hands-on labor. The remaining 10–15% of wire prep labor will be performed in marker prep and cable prep (multiple conductor) by manual or semiautomated techniques.

Wire prep automation. The wire prep workcell will cut, strip, crimp or solder, and mark single conductor wires from AWG 16 through 24. Included in this workcell are several built-in product tests and equipment diagnostics, verifying the quality of both the processes

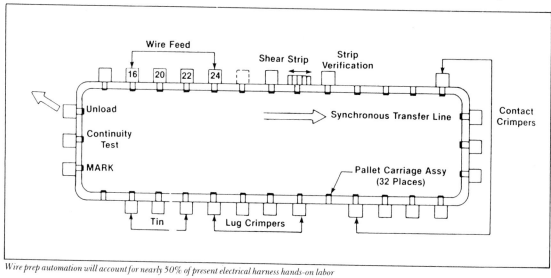

Wire prep automation will account for nearly 50% of present electrical harness hands-on labor

and final product. This workcell is 12ft wide by 40ft long and will produce terminated wires at the rate of one every 7 seconds.

Harness formation automation. Since present harness formation requirements have shown that more than 75% of our harnesses can be produced on planar boards of 16 × 48in. size, concentration at the MS&TC has been on very accurate robots with a smaller work envelope. Other Westinghouse divisions require larger work space; as a result, two further approaches to harness formation, extending the workable harness size to as large as 5 × 50ft, have been addressed. The details of the smaller robotics work cell are addressed here; extrapolation to the larger robots is obvious.

The formboard of plywood and nails has been replaced with a steel plate and magnetic tool approach, permitting the robot to automatically set up the formboards as required and eliminating the need for storage of any formboards (a space wasting problem).

The robot used is a Westinghouse 5000 long bed with dual-arm capability. At a total rate of one minute per wire (average), these robots will prepare the formboard, load connectors, route single wires, insert wire contacts into connectors (0.002in. accuracy required), route special cables, and bundle the harnesses. One of the layouts for this robot is shown, wherein the *X*-travel of the formboard would permit a viable 4 × 5ft harnesses configuration.

Acknowledgement

Portions of the Westinghouse efforts in production automation have been funded by the Air Force Systems Command's Electronic Systems Division, under contract F19628-81-C-0101 as part of the Division's GET PRICE program.

Authors' organizations and addresses

G. Bornecke
Siemens AG
Postfach 70 00 75
D-8000 Munchen 70
West Germany

C. Capillo
ECR Corporation
1525 McCandless
Milpitas, CA 95035-6841
USA

B. Corner
Harwin Engineers SA
Fitzherbert Road
Farlington
Portsmouth
Hants P06 1RT
UK

L. Cox
Manager, Instrument Services and Design
Tektronix Inc.
Howard Vollum Park
PO Box 500, M/S 19-390
Beaverton, OR 97077
USA

C. P. Cullen
Senior Systems Engineer
Automatix Inc.
1000 Tech Park Drive
Billerica, MA 01821
USA

J. G. Davy
Westinghouse Defense Electronics Center
MS 1637
Baltimore, MD 21203
USA

D. A. Elliott
Director of Advanced Technology
Electrovert Ltd
1305 Industrial Blvd
Laprairie, Quebec
Canada J5R 2E4

A. Filippone
Multiwire Division
Kollmorgen Corporation
250 Miller Place
Hicksville, NY 11801
USA

A. M. Freer
GEC Electrical Projects Ltd
Boughton Road
Rugby
Warwicks CV21 1BU
UK

M. Friedman
Integritek
Dylon Product Group
3670 Ruffin Road
San Diego, CA 92123
USA

H. Giltuz
Orbot Systems Ltd
Industrial Zone, PO Box 215
70650 Yavne
Israel

B. D. Hoffman
formerly of:
AT & T Technologies Inc.
Princeton, NJ 08540
USA

G. J. Horky
Manager, Dept 9G8
Packaging & Warehouse Automation
Entry Systems Division
IBM Corporation
1000 NW 51st Street
Boca Raton, FL 33431
USA

R. N. Hosier
Westinghouse Electric Corp
Defense Group
Manufacturing Systems and
Technology Center
9200 Berger Road
Columbia, MD 21045
USA

M. Kastner
Signetics Corp.
811 E. Arques Avenue
PO Box 3409, M/S 23
Sunnyvale, CA 94088-3409
USA

J. Keller
Joe Keller Associates Inc.
280 Torchwood Avenue
Plantation, FL 33324
USA

G. A. Koff
Mannesmann Demag
2660 28th St SE
Grand Rapids, MI 49508
USA

A. W. Koszykowski
AWK and Associates
Cromwell Court
New Road
St. Ives
Huntingdon
Cambs PE17 4BG
UK

K. M. Lin
AT & T Engineering Research Center
PO Box 900
Princeton, NJ 08540
USA

C-H. Mangin
CEERIS International Inc.
PO Box 939
Old Lyme, CT 06320
USA

Mike Marsh
Racal-Redac
Lyberty Way
Westford, Ma 01886
USA

R. D. McCleary
formerly of:
Universal Automation Systems
140 Pond View Drive
Meriden, CT 06450
USA

R. L. Myers
Omnimation Consultants
1049 - I Camino del Mar
Del Mar, CA 92014
USA

R. Osterhout
Universal Instruments Corp.
PO Box 825
Binghamton, NY 13902-0825
USA

E. J. Penn
V. P. Engineering
Tringon Industries Inc.
100 S. Milpitas Blvd
Milpitas, CA 95035-5459
USA

E. St. Peter
Xerox Corp.
555 South Aviation Blvd
M/S M1-35
El Segundo, CA 90245
USA

T. J. Petronis
Applied Robotics Inc.
301 Nott Street
Schenectady, NY 12305
USA

J. L. Turino
Logical Solutions Technology Inc.
310 W. Hamilton Avenue
Suite 101
Campbell, CA 95008
USA

C. M. Tygard
Everett/Charles Contact Products Inc.
700 East Harrison Avenue
Pomona, CA 91750
USA

L. Wallgren
Pace Inc.
9893 Brewers Court
Laurel, MD 20707
USA

E. Westerlaken
Cobar Europe BV
Tol 1
4791 BL Klundert
The Netherlands

R. Woodcock
Intelledex
536 NW 34th
Corvallis, OR 97333-1098
USA

Many of the papers contained in this book were written by the following staff of *Electronic Packaging & Production*:

N. Andreiev	J. Harris
R. Berson	R. Keeler
D. Brown	H. W. Markstein
S. Crum	R. Pound
T. Dixon	J. E. Sergent
G. L. Ginsberg	S. L. Spitz